Healthcare Technology Management

A SYSTEMATIC APPROACH

Healthcare Technology Management

A SYSTEMATIC APPROACH

Francis Hegarty • John Amoore • Paul Blackett
Justin McCarthy • Richard Scott

CRC Press
Taylor & Francis Group
Boca Raton London New York

CRC Press is an imprint of the
Taylor & Francis Group, an **informa** business

CRC Press
Taylor & Francis Group
6000 Broken Sound Parkway NW, Suite 300
Boca Raton, FL 33487-2742

First issued in paperback 2020

ISBN 13: 978-0-367-57396-6 (pbk)
ISBN 13: 978-1-4987-0354-3 (hbk)

Library of Congress Cataloging-in-Publication Data

Names: Hegarty, Francis, author. | Amoore, John N., author. | Blackett, Paul, author. | McCarthy, Justin (Clinical engineer), author. | Scott, Richard, 1962- author.
Title: Health Technology Management : A Systematic Approach / Francis Hegarty, John Amoore, Paul Blackett, Justin McCarthy, and Richard Scott.
Description: Boca Raton, FL : CRC Press, Taylor & Francis Group, [2016] | Includes bibliographical references and index.
Identifiers: LCCN 2016025952| ISBN 9781498703543 (hardback ; alk. paper) | ISBN 1498703542 (hardback ; alk. paper) | ISBN 9781498703550 (e-book) | ISBN 1498703550 (e-book)
Subjects: LCSH: Medical instruments and apparatus. | Medical technology. | Biomedical engineering.
Classification: LCC R855.3 .H44 2016 | DDC 610.28/4--dc23
LC record available at https://lccn.loc.gov/2016025952

Visit the Taylor & Francis Web site at
http://www.taylorandfrancis.com

and the CRC Press Web site at
http://www.crcpress.com

Contents

Preface

Healthcare is about people – those receiving care and their carers and the professionals providing the care. Healthcare is about resources – the financial and other resources required for care provision. Healthcare is about tools – the technology, the pharmaceuticals and the infrastructure through which care is supported and delivered. This book discusses managing the medical equipment, systems and devices that form much of healthcare technology and which is vital for the effective and efficient provision of care.

Medical equipment, from the simple to the complex, is integral to healthcare. The thermometer measures a vital sign of health – a humble device, yet indispensible, and whose accuracy depends on correct choice, correct use and correct maintenance. The spectrum of tools extends to sophisticated measurement systems, endoscopy and imaging technologies, ultrasound, x-ray and MRI scanners. Life support equipment, such as cardiac pacemakers, dialysis machines and ventilators, maintain and support the function of body organs. Treatment and therapy is provided by defibrillators and infusion devices. Surgery relies on increasingly sophisticated technology such as electrosurgery machines that simultaneously cut through tissue and coagulate tiny blood vessels to leave a clean cut. Surgical tools can be controlled remotely through the use of robots. Endoscopic surgical tools have revolutionized surgical procedures, with so called key-hole surgery extending the scope of day surgery. Technology can enable disabled people to lead more comfortable, fuller and more active lives.

The phrase 'medical equipment' covers this wide range of technologies that improve healthcare outcomes and effectiveness for all patients. We use the term 'healthcare technology' to include all of this equipment as well as the systems that bring individual items together to work effectively, the integration with the IT infrastructure and the less sophisticated but vital nonactive, usually disposable, devices such as syringes and giving sets.

Like any asset, medical technology must be carefully managed for it to be effective, a process we call healthcare technology management (HTM). This book describes and discusses HTM. It recognizes the context in which the healthcare technology is deployed in order to enhance healthcare for the benefit of patients and their carers. It recognizes that the management of healthcare technology must align with the strategic aims of the organization in which it is used, be that the personal family in a home environment, a community health centre or hospitals, small and large.

Therefore, HTM must be organized so as to support the operation of individual items of equipment at the point of care, where the performance of the equipment has a direct impact on the care and well-being of the patient. But the book goes further: it suggests that this management should not be passive, but active to add value to healthcare delivery, enhancing the benefits that the technology can deliver for patients, clinical staff and the healthcare organization. We define 'value' as the relationship between 'benefit' and 'cost'. The book addresses benefit (and who benefits) and cost, showing how value, in its widest sense, can be enhanced.

This book is also about the professionals who bridge the clinical and technical disciplines and deliver HTM. It is about the engineers, technologists and scientists who can add value to the medical equipment and its application, and it is about the management tools that support them in this activity. Historical and local organizational developments often dictate the form and structure of the operational systems that healthcare organizations put in place to manage their medical equipment. The names given to the groups managing medical equipment vary from institution to institution and from one part of the world to another. Terms include clinical engineering, medical engineering, medical physics, medical equipment management and medical equipment maintenance. Specialized groups managing particular types of equipment such as dialysis, radiology equipment or equipment for the disabled may be formed. In this book we use the term 'clinical engineer' to mean all or any of the technical, engineering and scientific staff who work in these areas. The general principles outlined in this book apply to all who manage healthcare technology, whatever they are called locally.

The book bases its approach to HTM on the international ISO 55000 asset management suite of standards. This defines asset management as an organization's coordinated activity that aims to realize value from assets. The Standard clearly links asset management to the strategic aims of the organization.

In the context of HTM, the strategic policies and operational procedures need to inform each other and provide a clear line of sight from the organization's management board right through to the individual piece of equipment being used at the bedside. This requires that those managing healthcare technology base their approach, their particular HTM system with its plans, decisions and activities, on the objectives of the healthcare organization and its raison d'être of caring for patients. The book suggests an HTM structure that includes a multidisciplinary overseeing group we have called the Medical Device Committee whose remit is to ensure that the healthcare technology is effectively managed and deployed to support the strategic aims of the organization. The name of this group and its structure and personnel composition will vary between organizations: for a small hospital or health centre, a single person may be allocated its responsibilities, whilst in a large academic hospital, it may comprise a team of professionals involved in the management and use of medical equipment.

The book describes the systematic and structured delivery of HTM, governed by a medical device policy, through HTM programmes each covering a range of equipment. Within each of the HTM programmes, the individual equipment types or group of similar types will be managed through equipment support plans detailing the practical support required.

The book defines and describes each of these terms and processes, recognizing that it is the principles underlying them that are important. It also stresses that the HTM and its processes must be holistically designed and implemented to add value, in the service of the organization, for patient care.

Healthcare continues to develop and evolve. Technological developments open up new opportunities, new methods of working. Individual people and populations seek new methods of accessing healthcare. Today the call across the world is for care in the community, be it a rural part of the developing world or an urban district in a large metropolitan area. HTM must be alert and open to these developments, basing its response on solid foundations of safe, effective and financially sound management in the service of patients and carers.

Engineering is not simply about technology. It encompasses 'people', 'finance' and 'machines' (in the very broadest sense of that last word). It develops and supports technologies, ensuring their safe and effective application, making the best use of the financial resources available, for the benefit of people. The management of medical technologies in healthcare is particularly about people, finance and machines, recognizing that the people include those receiving and those delivering healthcare. 'Human factors', 'ergonomics' and 'user friendly' are terms that recognize the importance of the interactions between technology and people. These interactions are particularly important in healthcare where technologies must be designed and applied to ensure inherent ease of use with built-in 'mistake proofing' to protect against inadvertent misuse.

This book is directed at all responsible for medical equipment and its management. Clearly this will include those who directly manage, care for and maintain the medical equipment, both those in training and those in practice. But the book is also directed at those with overall responsibility for this important collective asset, the executive and board members of healthcare organizations. It is applicable to those in other professions, medical, nursing, risk management, finance, general management and procurement who are involved in managing medical equipment and all aspects of healthcare technology throughout its life cycle. It is also applicable to industry, to those who develop, provide and support medical equipment. Regulatory authorities will find the book a useful guide in describing practical methods of safe and effective medical equipment management. And the influence of the book should help patients by delivering positive, effective and value-enhancing HTM.

Finally, the book will emphasize the importance of teamwork. Healthcare and indeed HTM covers a very broad range of disciplines. No one individual can understand the detailed intricacies of all the activities, all the components. Teams of professionals, each recognizing their strengths, their weaknesses and their reliance on others, work to provide effective HTM and healthcare. The systems approach, systems engineering, is advocated to help direct and manage the numerous elements for the benefit of healthcare to patients.

STRUCTURE OF THE BOOK

The book has been written collectively by the five authors, led by Francis Hegarty. We have all contributed to all parts of it with substantial internal reviewing, and we take collective responsibility for it all. In true clinical engineering fashion, we have worked as a team.

In the course of writing, we have asked various colleagues to peer-review some sections or chapters and have been greatly encouraged and assisted by the feedback they provided. We are very grateful for this input and acknowledge it later.

We have included a large number of case studies. These are 'stories to illustrate a point' and are linked to relevant discussions in the main text. They are all based on real situations but are not intended to be a complete description of actual events. Some poetic licence has been used to clarify or amplify the intended learning. You might describe them as 'parables'. Most have been drafted by one or other of the authors, but in some cases we have asked colleagues to contribute drafts which the authors have refined and for which we, the five co-authors, take full responsibility. We gratefully acknowledge those contributors to the case studies and their names are listed in 'Acknowledgements'.

In most case studies we have assessed the 'value' of the activities described, that is their relative benefits and costs. The 'value' is summarized in a diagram that shows benefits and costs increasing, decreasing or staying the same. Whilst recognizing the difficulty in assessing benefits and costs in healthcare, we believe that it is important that clinical engineers adopt an approach of questioning whether or not a proposed activity adds 'value' and for whom. This will require that clinical engineers develop skills in assessing benefits and costs.

We have also included self-directed learning points in both the main body of each chapter and in most case studies. These are designed to enhance the text by opening up points for discussion, emphasizing that the book is not simply a text to be passively read, but to be actively engaged in. We hope this will enhance the value of the book to students and trainees and to those who teach or mentor them. We also hope that these self-directed learning points will trigger thoughts and discussions by clinical engineering practitioners and by those responsible for managing healthcare technology.

To that end, we have set up a website at http://www.htmbook.com which may provide a space for further discussion and comment and will provide links to the authors through various social media platforms.

Finally, we have chosen to use UK English spelling except where words in the title of a reference differ, in which cases we have kept the original spelling. We have used the online Oxford English Dictionary as our source and have referred to the online guidance at http://www.oxforddictionaries.com/, particularly http://blog.oxforddictionaries.com/2011/03/ize-or-ise/.

Francis Hegarty
John Amoore
Paul Blackett
Justin McCarthy
Richard Scott

Acknowledgements

The authors are grateful for colleagues, clinical, technical and managerial, who have taught them, and continue to teach them, the practice of clinical engineering that adds value to healthcare technology for the benefits of patients, users and healthcare organizations. In doing so, the authors acknowledge the continuing learning and development from teamwork across the multidisciplinary clinical engineering profession. The book would not have been possible without this nurturing that each of the authors has received during their careers that has built on their more formal training.

During the development of this book, the authors sent drafts of chapters to colleagues for comment and advice. The authors are very grateful for the effort that these colleagues gave and that the authors used to improve the work, though the authors remain fully responsible for the content. These reviewers include Dr. Michael Appel (USA), Calum Campbell (UK), David Cook (UK), Peter Cook (UK), Duncan Ferguson (UK), Dr. Peter Jarritt (UK), Patrick Macaulay (UK), John Parker (UK), Dr. David H.T. Scott (UK) and Andrew Wong (UK).

We note in the Preface the use of case studies and the contributions which we gratefully acknowledge towards some of these from other colleagues as follows:

Nicola Aburto

Jim Blackie

Patricia Byrne

Andrew Cleaves

Samir Deiratany

Tim Foran

Jill Grey

Steve Keay

Conail McCrory

Barry McMahon

Geraldine McMahon

Andrew Norman

Colm Saidlear

Debbie Sell

Triona Sweeney

Patricia Brooks Young

We gratefully acknowledge the rehabilitation engineering front cover photograph provided for us by Paul Rogers and Nadeem Akram, which is used with their permission.

The authors record their grateful thanks for the support, encouragement and forbearance shown by their families and friends during the execution of this work.

Finally, the authors record their gratitude and thanks for the skilful support and encouragement of the Taylor & Francis Group, in particular from Francesca McGowan, Emily Wells and Rebecca Davies.

Author

Francis Hegarty is a founding member of the Medical Physics and Bioengineering Department in St James's Hospital, Dublin, Ireland. Over the course of a 30-year career, he has served within this department as a biomedical engineering technician, chief technologist and later principal physicist leading the clinical engineering group. In this time, he has managed teams providing equipment management services and he is familiar with the application of healthcare technology in a broad range of clinical departments. He was instrumental in establishing the department's management structure and is an advocate for multidisciplinary team working between clinicians and engineers. He led on the implementation of the department's healthcare technology management systems. Central to this was his development of an innovative medical equipment management database system. He has led a number of multidisciplinary hospital projects where medical equipment was integrated in clinical information systems.

He has contributed to hospital management committees and chaired the hospital's multidisciplinary medical device vigilance committee. During his career, St James's Hospital underwent significant redevelopment and as a result Francis gained extensive experience in the design, planning and commissioning of new hospital facilities. Francis has also acted as a consultant in this regard to a number of other hospital projects in Ireland and overseas. At the time of the publication of this book, he is the chief healthcare technology officer for Ireland's new national children's hospital project.

Francis studied electrical and electronic engineering at Dublin Institute of Technology and in 1994 completed an MSc in physical sciences in medicine through the Faculty of Health Sciences in Trinity College, Dublin. His research interests include clinical measurement, clinical informatics, medical optics, assisted living technologies, art in health and healthcare technology management. He has published and presented widely on these topics. In 1998, he led the establishment of a postgraduate diploma in clinical engineering at Trinity College, Dublin, and acted as the course coordinator until 2005. He lectures on a number of undergraduate and postgraduate courses and is a member of the Faculty of Health Sciences, Trinity College, Dublin, as well as a chartered engineer, a member of the Institute of Physics and Engineering in Medicine and an affiliate of the Royal College of Anaesthetists.

John Amoore retired in 2015 following a career spanning over 40 years, largely applying engineering to healthcare in academia, industry and healthcare. He gave nearly 30 years in service to the National Health Service in Scotland both in Lothian and in Ayrshire and Arran from where he retired as head of the Medical Physics Department where his responsibilities included budget responsibility for the procurement and maintenance of medical equipment. Academic training began with a BSc in electrical engineering which laid the foundation for his career and the recognition that engineering is about people and money as well as technology. Continuing professional development throughout his career followed, with part-time study in physiology (MSc) and biomedical engineering (PhD) gaining skills in healthcare technology, its deployment and management.

John's academic career included appointment as a senior lecturer in the Biomedical Engineering Department at the University of Cape Town, South Africa, developing lecture courses and supporting postgraduate students. He helped develop the new Biomedical Engineering Society of Southern Africa during the 1980s. Since moving to the United Kingdom, he has continued research projects and served as external examiner for PhD theses. His publications (over 60 journal publications, 5 chapters in books and over 100 conference presentations) span several areas: theoretical modelling to better understand the cardiovascular system, medical device safety and risk management, infusion device management, blood pressure measurement, characteristics of specific medical devices and healthcare technology management. He continues to serve as an external reviewer of manuscripts submitted to journals for publication. He has served on expert groups advising on blood pressure measurement technology.

Clinical engineering contributions include developing structured databases for medical equipment management, pioneering medical device training for clinicians, helping establish the first medical equipment library in his area in Scotland, leading the development of multidisciplinary groups to advise on the management of medical devices, developing service level agreements to provide clinical engineering services, developing medical device incident reporting initiatives and helping the successful move of a major academic hospital (~800 beds). He has experience of responsibility for all medical equipment for a regional health area, including budgetary responsibility for the procurement and maintenance of medical devices from clinical thermometers to major imaging equipment.

He recognized the importance of understanding the clinical situation in which medical equipment is deployed and the need for a team approach and ergonomic analysis in medical equipment selection. From the core clinical engineering aim of ensuring the safe application of medical technologies evolved an interest in understanding the causes and prevention of adverse events involving medical equipment. This, and in collaboration with clinicians, led to an interest in person-centred care.

During his working career, he was registered professionally as an engineer and clinical scientist and contributed to engineering, medical engineering/physics and medical professional societies, including the organization of professional conferences and seminars.

Paul Blackett started his career working for the inventor of the world's first blood glucose meter in a small business unit in Consett, County Durham. After spending several years in medical electronics manufacturing in South Tyneside, Paul joined the National Health Service (NHS) in 1985. Initially working at the Freeman Hospital in Newcastle upon Tyne as an electronics technician, Paul also spent some time at South Tyneside

District Hospital, before eventually moving southwards to take up a post at the Royal Preston Hospital as a senior medical engineer.

Paul is currently the medical engineering manager at Lancashire Teaching Hospitals NHS Foundation Trust which comprises two acute hospital sites at Chorley and Preston. Through workshops at these two hospitals, the medical engineering team delivers a comprehensive service to their own trust and, through service level contracts, to neighbouring organizations.

Paul is a member and former chair of the Clinical Engineering (North) National Performance Advisory Group and is an active member of the Institute of Physics and Engineering in Medicine (IPEM), working as an external moderator for the technologist training scheme and a former member of the Clinical Engineering Special Interest Group. He has presented at scientific meetings on the subject of risk, key performance indicators, standardization and cross-Atlantic differences in clinical engineering guidance. Paul has also been part of the Department of Health working party on Modernising Scientific Careers and a member of the Northwest Healthcare Science Workforce Board. Paul is currently an IPEM representative on the Register of Clinical Technologists Management Panel.

Paul graduated with BSc (Hons) after completing a part-time degree in computing and electronics. Keen on systems and standardization, Paul's current work interests lie in unique device identification and risk analysis.

Justin McCarthy has a BSc (Hons) in electrical and electronic engineering from the University of Manchester. He has had 40 years of National Health Service (NHS) experience in the field of clinical engineering. In his early career, he was involved in developing new medical devices and equipment. He worked as part of the multidisciplinary team in Cardiff that developed and brought into clinical use the first commercially available intravenous patient-controlled analgesia apparatus. This gained him a research MSc.

In later years, as the head of clinical engineering, he led the teams that provided the healthcare technology management and maintenance services to the Cardiff NHS Trust's medical equipment, introducing a formal ISO 9000 quality management system in 1991. He advised the Trust on risk, procurement and healthcare technology regulatory issues. He helped establish the clinical engineering MSc course at Cardiff University and continues as an honorary senior lecturer teaching on the undergraduate medical engineering course.

He has contributed to 14 peer-reviewed published papers, 7 journal articles and 5 textbooks as well as over 80 conference proceedings and presentations.

He retired from the NHS in 2009 but continues to present at meetings, publish and sit on professional working groups as well as to provide consultancy services to Trusts and small businesses. He acted as an expert for the NHS Confederation at a European Commission policy conference in 2010 and has recently been leading, on behalf of the Institute of Physics and Engineering in Medicine (IPEM), their input to the European Council, the European Parliament and the MHRA regarding the new Medical Devices Regulations. He continues to be involved in formal Standards at the United Kingdom and international level and is chair of the IEC SC62A subcommittee, responsible for IEC 60601-1, the primary safety standard for medical electrical equipment.

He is a chartered engineer and a fellow of the Institution of Engineering and Technology and of the IPEM.

Richard Scott is a consultant clinical scientist and chartered engineer, working as Head of Clinical Engineering at Sheffield Teaching Hospitals NHS Foundation Trust. Additionally, he is the professional lead for Clinical Engineering and Reconstructive Science at UK National School of Healthcare Science (part of Health Education England).

An engineering placement in the Division of Anaesthetics at the Clinical Research Centre, Harrow, completed as part of an electronics degree at North Staffordshire Polytechnic encouraged Richard to pursue a career in clinical engineering. Upon graduating in 1984, he joined the Wessex Regional Medical Physics Service, based at the Royal United Hospital, Bath, providing scientific support for clinical instrumentation and undertaking a range of research projects. He obtained an MSc in medical electronics and physics from St Bartholomew's Medical College, followed by a PhD at the University of Bath, investigating the frequency dependence of respiratory mechanics via the oscillatory airflow technique. Subsequently, he specialized in the management of medical devices and contributed to establishing the local medical equipment management service. In 1995, the opportunity arose to become head of the Medical Equipment Management Department at Sherwood Forest Hospitals NHS Foundation Trust in Nottinghamshire. In December 2016 he moved to Sheffield and continues to encourage healthcare scientists to work together to play a key role in adopting innovative practices and driving service transformation for patient benefit.

Richard has contributed to curriculum development for UK National Health Service (NHS) clinical engineering education programmes across the career framework and has been keen to develop engineering personnel with skills to ensure that healthcare technologies are effectively managed and care advanced for patient benefit. He led the Clinical Biomedical Engineering Higher Specialist Scientist Training scheme curriculum writing group on behalf of the Royal College of Surgeons, which has pioneered a new NHS consultant clinical engineer role.

Richard is active in the development of international electromedical safety standards, serving on a range of British Standards Institution committees and is immediate former president of the Hospital Physicists' Association. He is a Fellow of the Institute of Physics and Engineering in Medicine, served as their Vice President: Professional and has acted as a professional advisor to the UK Department of Health Modernising Scientific Careers team. Richard is one of the first UK clinical engineers to be admitted to the Academy for Healthcare Science's Higher Specialist Scientist Register.

CHAPTER 1

Clinical Engineering in the Healthcare System

CONTENTS

1.1 INTRODUCTION

Clinical engineering practice emerged to meet a very real need. In the late 1960s and 1970s, the increase in the number, functionality and range of electronic medical equipment used in healthcare brought with it a corresponding need for engineers and technicians to support this equipment which needed regular maintenance and repair. Over time the number and range of equipment continued to proliferate, becoming more complex, and the role of engineers, technologists and technicians based in hospitals evolved to include support for the application of medical equipment as well as its maintenance. Continuing developments in material science, electronics and instrumentation improved the reliability of the medical equipment; however, its complexity continued to increase. Safety was enhanced through the development and adoption of Standards which set out minimum operational and safety requirements for medical equipment which must be met before the item can be placed on the market.

However, the need for vigilance in the use of this equipment, which often makes direct physical connection with patients or delivers energy to patients, remained. Just as the introduction of x-rays into clinical practice resulted in a need to have physicists working in hospitals to develop safe ionizing radiation working practices, so the increase in electronic medical equipment led to the need to have engineers working in healthcare. Initially the priority was for engineers to develop and provide maintenance services both to repair faulty devices and, equally importantly, to ensure that the equipment remained safe and effective through regular maintenance checks and calibration. Whilst essential, this maintenance and repair support was increasingly seen as only meeting part of the requirements for safe and effective applications of medical equipment. By the end of the twentieth century, the role of those who practice clinical engineering had expanded well beyond these maintenance roles to include risk management, support for clinical governance and end user support, contributing to Standards development and contribution to research, development and innovation of new medical equipment, devices and systems. The growing medical equipment industry also needed design and production engineers, together with maintenance staff to repair faulty devices.

Clinical engineering is the name used to describe this specialist strand of engineering which is focused on excellence in the application of technology in the clinical environment. Clinical engineers are experts at solving the problem of complexity in today's healthcare industry, harnessing and adding value to the application of technology whilst assuring safe and effective healthcare delivery.

1.1.1 What Do We Mean by 'Medical Devices', 'Medical Equipment' and 'Healthcare Technology'?

These three terms have been given an internationally agreed meaning, and so for clarity, it is worth reviewing the World Health Organization definitions of the terms medical device, medical equipment and health technology (WHO 2011, p. 4). The term 'medical device' is used to describe all items or machines that are used to improve the health of an individual, excluding drugs. The World Health Organization brief definition for a medical device is,

> "An article, instrument, apparatus or machine that is used in the prevention, diagnosis or treatment of illness or disease, or for detecting, measuring, restoring, correcting or modifying the structure or function of the body for some health purpose. Typically, the purpose of a medical device is not achieved by pharmacological, immunological or metabolic means."

The term 'medical equipment' is used to describe active, powered medical devices and systems deployed to support the delivery of care. The World Health Organization definition for medical equipment is,

> "Medical devices requiring calibration, maintenance, repair, user training, and decommissioning – activities usually managed by clinical engineers. Medical equipment is used for the specific purposes of diagnosis and treatment of disease or rehabilitation following disease or injury; it can be used either alone or in combination with any accessory, consumable, or other piece of medical equipment. Medical equipment excludes implantable, disposable or single-use medical devices."

WHO (2011, p. 4)

The World Health Organization defines 'health technology' as

> "The application of organized knowledge and skills in the form of devices, medicines, vaccines, procedures and systems developed to solve a health problem and improve quality of life. It is used interchangeably with 'healthcare technology."

In this book we shall explore a systematic and structured approach to the active management of healthcare technology with the emphasis on devices, equipment, procedures and systems and show how, through this, clinical engineers can add value to the application of medical devices and equipment for patient care.

1.1.2 The Context in Which Clinical Engineering Now Operates

Where once complex medical equipment was confined to the acute care facilities in teaching hospitals, now it is present in all areas of hospitals, in the primary care setting and in the community and patients' homes. The increasing desire to shift care from hospital institutions into the community is and will continue to be supported by developments in mobile medical equipment. So the need for clinical engineers now extends beyond the walls of the hospital. Wherever the location of patient care, the core purpose of those practising clinical engineering remains the same: to help ensure the availability, at the point of need, of the appropriate technology that is safe, effective and understood by the users.

As medical equipment and its applications develop and expand, these technologies need to be actively managed, and their use supported. Active management includes technical maintenance, professional asset management including life cycle and financial management, expert scientific support for their use at the point of care and vigilance with regard to their safe use. Thus training of users was added to the repertoire of the clinical engineer, helping ensure that the equipment is applied by competent users who understand the equipment. There was also growing awareness that training alone would not suffice to ensure safe and effective equipment application. The equipment must be easy to use, and the role of the clinical engineer has extended to ensure that the importance of human factors was incorporated in industry at the design stage and, within healthcare organizations, when procuring medical equipment. Active management also includes feedback to manufacturers and regulatory agencies of problems detected during operational use, especially human factor aspects such as unanticipated human–device interface and human performance–based failure modes.

In this chapter we will discuss the role of clinical engineers today in the healthcare sector. In doing so we will discuss the context within which they work. The focus will remain on clinical engineering as practiced in hospitals, but the discussion acknowledges that there is a wider and growing role for clinical engineering in all sectors of healthcare. So in many instances where we use the term 'hospital', the points we are making apply equally to other parts of the healthcare systems as described in the following text. The discussion will be structured around two themes, namely equipment management and advancing and supporting care. We will also examine some of the other roles played by clinical engineers in supporting research and device regulation. We will see that clinical engineering today still meets a very real and expanding need in the healthcare sector. We identify the important role that clinical engineers play in healthcare delivery, helping organizations to meet the objectives of all their stake holders.

1.2 THE HEALTH SYSTEM

The health system is the collective term used to describe the people, institutions and resources that deliver health services to meet the health needs of the society it serves. The make-up and workings of health systems vary between jurisdictions. In some countries, health system planning is driven by government policy. In others the free market plays a bigger role. The health system is a broad term that encompasses health promotion, health education, disease prevention and the provision of care within the home, community and health institutions. The World Health Organization defines health systems as follows:

"A health system consists of all organizations, people and actions whose primary intent is to promote, restore or maintain health. This includes efforts to influence determinants of health as well as more direct health-improving activities."

WHO (2007, p. 2)

Health systems are typically viewed from the organizational perspective, but we must remember that healthcare is person focused, and as we describe the sectors of healthcare systems, we will also consider them from the perspective of the patient and the patient's journey.

Healthcare is the term used to describe the diagnosis, treatment and prevention of illness in society. Healthcare is usually described as consisting of three sectors, primary, secondary and tertiary care. However, considering that practices that promote public health are also included in the term, a fourth sector is necessary and this is healthcare in the community or home setting. In this book we are concerned with how clinical engineers support the delivery of healthcare through the application of technology, with common principles applicable to all sectors. Before discussing the role technology plays, it is useful to discuss the four sectors in more detail.

1.2.1 Home/Community Care

Ideally we would all like to live long healthy independent lives in our own homes, and for most of our lives, we do so. However, all have had the experience of occasionally getting sick with a brief illness like the flu and our family and friends caring for us during this time. Healthcare is provided informally by families and other social networks at home and in the community. Sometimes due to limitations associated with chronic illness or ageing, individuals need a higher level of care. Families may employ care assistants to help look after loved ones in their own homes. Independence can be maintained through the provision of expert care from professionals working in the community, providing home care, aids for independent living or specialist rehabilitation services. Where the illness or dependency is more challenging and beyond the ability of family and carers in the home to manage, individuals may be looked after within special care home facilities within the community. All of these types of activity are described by the term Home/Community Care.

1.2.2 Primary Care

When individuals have exhausted the limits of home healthcare, they consult with a community physician, family doctor or other licensed clinical professional such as a physiotherapist, nurse specialist or pharmacist. This part of the healthcare system is termed primary care. So if you get the flu and it does not clear up after a few days care at home, you may go to your family doctor or pharmacist for a consultation and advice. There is significant intersection between the home/community care and primary care sectors. The elderly and infirm, living at home managing one or more long-term conditions successfully, will no doubt also regularly consult with their primary care physician. These community-based physicians rely on medical devices to diagnose and provide first-line care and treatment. In order to have the best chance of living long healthy independent lives in our own homes, it is desirable where possible that chronic physical, mental and social health issues be managed in the community through a combination of home/community care and primary care. As the cost of healthcare increases, particularly the cost of hospital care, there is a recognition that society needs to invest in primary care to support individuals to continue to live independently in the community as long as possible. This has financial, health and social benefits for the individual and society. This is particularly so given the increase in the number of elderly people living in the more developed nations of the world.

1.2.3 Secondary Care

Secondary care is the term used to describe care provided by medical specialists and usually delivered from within a hospital setting. Typically a primary care practitioner will refer an individual to see a medical specialist. So if you see your family doctor after a week of having the flu, they might refer you to the local hospital for a chest x-ray to see if you have pneumonia. In doing so, you are entering the secondary care sector, and the x-ray will be taken by a radiographer who is a licensed health professional and reviewed by a radiologist who is a specialist medical practitioner. Referral to secondary care can also be to a licensed health professional such as an occupational therapist or dietician. However, secondary care may also be required in circumstances where there is no referral, for example, to provide specialist acute care. Examples include situations where as a result of an accident individuals need to be brought to the Emergency Department in their local hospital. This in turn might lead to a need for trauma surgery and intensive care medicine. Secondary care is usually associated with attending or admission to a hospital facility. Again the trend in many jurisdictions is to try and build primary care facilities that provide access to medical imaging and other specialist services such as physiotherapy and mental health services for the purpose of keeping people in the community and freeing up scarce and expensive hospital resources for those who need it most.

1.2.4 Tertiary Care

Tertiary Care refers to specialist consultative care. It usually involves admission to a hospital that can provide a high level of specialist care in the management of, say, cancer or provision of specialist surgery such as cardiac surgery or neurosurgery. For reasons of clinical effectiveness and economics, such specialist services are delivered in specialist hospitals or specialist units within large hospitals. Access to tertiary care is usually through referral from either primary or secondary care professionals.

Clinical and technological developments occurring within the context of evolving social, population and economic environments have and will continue to blur the distinctions between primary, secondary and tertiary care. Aspects of specialist tertiary care, for example, can be delivered by mobile portable units to remote areas. Nonetheless, the distinctions between these levels of care are useful when considering the spectrum of healthcare provision.

1.2.5 Healthcare as a Cycle

Figure 1.1 illustrates these four healthcare sectors which summarizes how societies structure the healthcare system optimally to meet the need for diagnosis, treatment and prevention of illness. The arrows indicate the referral pathways. These are bidirectional as the objective is always to improve the health of the individual no matter where they are in the healthcare system and in doing so to move the individual towards independent living in the community. So to continue the example introduced earlier, on attending the secondary care for the chest x-ray, the doctor might discover you have pneumonia and admit you for intravenous antibiotics in the secondary care hospital. A few days later when the condition improves, you would be discharged into the care of the primary care family doctor who would continue your management in the community until you were well. The bidirectional arrows ultimately start from and lead back to the individual, emphasizing that care is person centred.

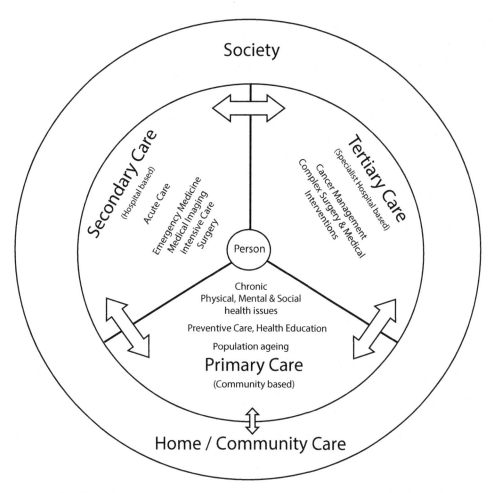

FIGURE 1.1   The Healthcare System.

The discussion in this book will centre on the role technology plays in supporting the delivery of care in secondary and tertiary hospitals where clinical engineers have traditionally played an important role for some time. However, recognizing the increasing need to support medical equipment in the community, we will also use the inclusive term healthcare system to describe hospitals, clinics, primary care centres, community physician surgeries, care homes and within the person's own home setting.

In an effort to optimize the healthcare delivery in society and control its associated cost, there has been a trend to develop primary and home/community care to manage chronic illness as close as possible to the individuals home. This has resulted in increasing use of healthcare technology including medical devices in the primary care setting, in community and increasingly in individual's own homes. Medical devices such as thermometers have been used in homes for some time. Over the years other devices such as glucometers which measure blood sugar levels have become common to help individuals with diabetes manage their condition at home. Blood pressure monitors are available for sale in most high street pharmacies.

So clearly there is a trend of increasing use by individuals of medical devices to manage their own health. Many technology companies are developing and bringing to market products which aim to help individuals monitor their lifestyles with a view to health promotion. Smartphones which can be connected to sensors provide a ready technology platform for software applications (apps) designed to help promote wellness. We are in an age where regulated medical devices are moving out of the hospital into the community and patient's homes, whilst at the same time consumer lifestyle products intended to promote health are becoming widespread. Whilst in the past the role of the clinical engineer was associated with the specialist medical devices and systems used in tertiary and secondary care, we acknowledge that the role of clinical engineer should now extend to all healthcare delivery settings. Whilst lifestyle products and some apps may not be regulated medical devices, their use in health promotion in the community may well be beneficial, and it is likely that clinical engineers will have a role in advancing care through the appropriate use of these devices. It is worth bearing in mind, however, that regulatory authorities such as the FDA in the United States and the Medicines and Healthcare products Regulatory Agency (MHRA) in the United Kingdom stipulate that some medical apps are regulated devices. As the focus of healthcare delivery shifts to maintaining wellness and care at home, it is likely to require tighter integration of personal health data between the individual, primary care and the hospital sectors. Clinical engineers are well placed to strategically advance the required data management framework, telemedicine interfaces and data analytical requirements necessary for the achievement of this new vision for healthcare delivery.

Although in this book we will discuss the use of regulated medical devices used in the hospital setting, we acknowledge that there is a growing need for such devices to be used in primary care and the home setting. In exploring this topic, we may use the term hospital to mean the healthcare organization. Where we do so, it is important to remember that the same practice and principles are applicable in other healthcare organizations and in other healthcare settings where medical devices are used, for example, senior care facilities, primary care practices and home care.

1.3 THE NEED FOR CLINICAL ENGINEERS IN THE HEALTHCARE SYSTEM

The development of healthcare technology over the millennia can be illustrated by considering the evolution of a simple cutting blade, through carefully designed surgical knives and by way of sophisticated electrosurgical tools, to precise robotic-controlled surgical arms enhancing but not replacing the clinical staff who provide the healthcare. Robotic systems can be integrated with information technology (IT) to allow operation at a distance and to allow software tools to assist the surgeon further by providing information derived from data collected in real time. As medicine and technology have advanced, medical devices and equipment have become more sophisticated, more complex and more interoperable, but their essential role of extending the ability of people to deliver healthcare remains. Today in hospitals, clinicians use complex technology in the delivery of care. This medical equipment may incorporate complex electronics, optics, lasers, a vast array of sensor technology and instrumentation, usually in a machine incorporating signal processing and software algorithms. Some deliver energy to the patient, others deliver powerful and potentially harmful therapeutic medication and other devices make direct electrical contact with the patient's body. Yet these same clinicians have not studied, as a central part of their training, the engineering or physics, upon which these clinical tools and technologies are based. The ubiquitous medical equipment brings with it not only benefits but also risks. Just as aircraft, bridges and other pieces of technology infrastructure need to be actively managed and used correctly, so to do medical equipment and systems, to ensure that the benefits they bestow far outweigh any risks associated with their use.

Clinical engineering is a specialist strand of engineering focused on excellence in the application of technology in the clinical environment. As a discipline it is concerned with the application of engineering tools and theory to all aspects of the diagnosis, care and cure of disease and life support in general, all of which are embraced in the term delivery of healthcare (Bauld 1991). By definition it is an interdisciplinary activity, and clinical engineers work closely with doctors, nurses, paramedics and anyone who uses medical devices to provide care. Clinical engineering delivered within health delivery organizations is provided by a range of individuals each with specialist expertise, qualifications and skill. This includes graduate engineers, biomedical and physical scientists, engineering and specialist technologists and technicians. The titles used to describe these talented individuals vary from country to country. In this book the term 'clinical engineer' will be used to describe all engineers, technicians and scientists who provide clinical engineering services regardless of their employment grade or level of professional development.

In managing the use of technology to support care, it is no longer appropriate to think about medical equipment in isolation. Medical equipment is used in a highly interoperable fashion with implantable, disposable or single-use medical devices and increasingly with clinical information systems. Clinical engineering has always evolved in response to changes in the development and availability of technology, developments in the life sciences and the interaction between these developments. This remains so today. Technology is increasingly about connectivity and information sharing and today, among the challenges facing the healthcare sector, are both the need for, and the difficulties arising from, the convergence of medical equipment and information and computer technology, both of which continue to proliferate. Today the term Healthcare Technology, defined in Section 1.1.1, is commonly used to describe the full range of equipment, devices and systems used to support the delivery of care.

The preceding paragraphs have recognized the strong links, the interoperability and convergence, between medical equipment and information and communication technologies (ICT). Traditionally the medical equipment has been managed by clinical engineering departments (CEDs), with the ICT managed by Information Technology (IT) or eHealth departments, though in some organizations these departments have or are merging. Whatever the structural arrangements within any healthcare organization, it is important that these disciplines recognize their common purposes and, where there are separate departments, that they coordinate their efforts to ensure the effective application of the Healthcare Technologies.

The development of medical equipment and healthcare technology over time is mirrored by the development of international Standards that guide their design, use and support.

When medical equipment started to be used more widely in hospitals, new risks emerged, such as the potential for electric shocks to patients and staff or injury due to device malfunction. This prompted engineers from industry, academia and healthcare to come together in the development of electromedical safety Standards developed under the auspices of international bodies such as the International Organization for Standardization (ISO) and the International Electrotechnical Commission (IEC); these will be described in greater detail in Chapter 3. A Standard is a document that provides requirements, specifications, guidelines or characteristics that can be used consistently to ensure that materials, products, processes and services are fit for their purpose. Standards help ensure that products and services are safe, reliable and of good quality. The key features of such standards are that design criteria are specified to minimize the probability of hazardous conditions arising. For example, the Standard dealing with electrical safety specifies the maximum allowable non-functional currents that may flow via earth leads, enclosures and patient applied parts, thus minimizing the risk of electric shock mentioned earlier. The standards additionally stipulate the requirement for safety under certain single fault conditions.

As medical equipment products became safer as a result of compliance with these Standards, the attention of governments and industry turned to what might be best described as management Standards. These documents looked at the equipment management life cycle and included guidance for those in hospitals who manage medical equipment on equipment selection, acceptance testing, user training, maintenance and disposal. The advent of such documents brought the need for active management of medical devices to the attention of organizations and action was required. The immediate question that arose was that organizations did not know what medical equipment holding they had, and exercises to create medical equipment inventories or asset registers began. Uniquely identifying all the medical equipment assets under the organizations control was essential to enable maintenance regimes to be put in place and records kept of equipment type and age to enable business planning activities. This led to the creation of services that focused on managing medical equipment assets; some grew out of existing hospital engineering or facilities management departments and others from scientific services such as departments of medical physics.

Having seen the evolution of technical Standards that have led to safer products and the development of management Standards that ensure robust organizational roles and responsibilities, we now see a new type of Standard emerging, namely the values-based Standard. Such Standards often draw on or use previous families of Standards, building on work done but having a clear focus on value, often looking from a customer perspective, incorporating responsibilities around environmental sustainability issues, being good corporate citizens and ethical trading. The ISO 55000 Asset Management suite of standards (ISO 2014) has 'value', as defined by the organization using the standard, as its centrepiece. In subclause 3.3.1 of ISO 55000, *asset management* is defined as "co-ordinated activity of an organization to realize value from assets".

Healthcare Technology Management (HTM), as practiced today by clinical engineers in healthcare, has much in parallel with the Standards journey – initially it was about using engineering principles and practice to ensure devices were safe and functional, then the roles were developed to ensure practice was effective and devices properly managed and now we see a need to achieve increased value from our healthcare technology assets. The name of the discipline has also varied over the years but a common understanding of 'Healthcare Technology Management' has emerged (AAMI 2011).

Figure 1.2 shows the healthcare system again but this time highlights that the care the patient received is delivered by many different specialists, usually working in concert as a multidisciplinary team. This team includes the clinical engineer whose primary role is to assist with the application of technology and thus extending the ability of other healthcare specialists to deliver healthcare. Clinical engineering emerged as a discipline in the secondary and tertiary hospitals where high technology medicine has developed. The grey-shaded area represents those parts of the healthcare system where tradition-ally there has been extensive use of technology by clinicians. Such clinical areas would

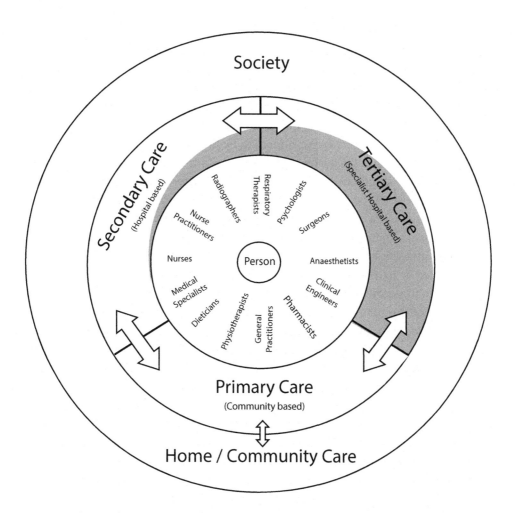

FIGURE 1.2 The Healthcare system and the people who populate it. The grey-shaded areas indi-cate where medical equipment first proliferated.

include surgical theatres, intensive care, cardiology, emergency medicine, diagnostic imaging, and radiotherapy to name but a few.

With the proliferation of medical equipment across all hospital areas and also into primary, community and home care situations, the clinical engineer is now in demand within all sectors of the healthcare system. It is now common for clinical engineers to provide services into primary care settings and indeed into the community and patient's homes. Healthcare delivery organizations are developing new way of delivering healthcare, using non-traditional means often predicated on technology. In particular they are also looking at how technology such as mobile phones and the Internet can bridge the gap between hospitals and patients in their homes, with a view to supporting elderly patients to live independently longer. Clinical engineers respond to these challenges not only through the development and implementation of home-based medical equipment systems or assisted living devices but also through the development and implementation of specific medical information technology (IT) systems.

Telemedicine is the term used to describe projects which use information technologies and telecommunication networks to provide healthcare at a distance. Connected Health is the term used to describe projects that use readily available consumer technologies such as mobile/cell phones, Internet and web-enabled medical equipment to deliver patient care and chronic disease management outside of the hospital to patients either in their own home or in a primary care facility. Where these technologies are deployed in the community and individual's homes, clinical engineers are required to implement and support these new and emerging ways of providing care. This requires them to be flexible, imaginative and prompters of change. Figure 1.3 illustrates that healthcare professionals are using technology to deliver care and that this is no longer confined to the hospitals. Clinical engineers collaborate with all of their colleagues across all the healthcare sectors to both support and advance care. As technology used at the point of care has moved from hospitals out into care homes, primary care facilities and patient's homes, there has been a requirement for the clinical engineers to follow. Developments in these information technologies and medical equipment connectivity will be increasingly relied on to meet the challenges facing healthcare, and clinical engineers need to be equipped with the knowledge and skills to support these developments.

At the centre of this diagram is the individual patient. It is of primary importance that the clinical engineer is at all times aware of the consequence that any of their actions can have on the patient, their care or their experience of receiving care. In doing so, the clinical engineer will be aware of his or her limitations and seek advice, as appropriate, from clinical colleagues. We have already seen that whilst the healthcare system is there to help the patient, it is a complex system made up of people, institutions and resources that deliver health services to meet the health needs of the society as well as individuals. Clinical engineers have to think about how their actions contribute to the care of patients; to the support of a wide range of healthcare professionals; to the support of the institutions that make up the healthcare system, including hospital management; and to society at large. As we explore the role of clinical engineers within the healthcare system, we will use this diagram as a reference point to help identify which of these people, entities or processes are affected by a particular action taken by a clinical engineer.

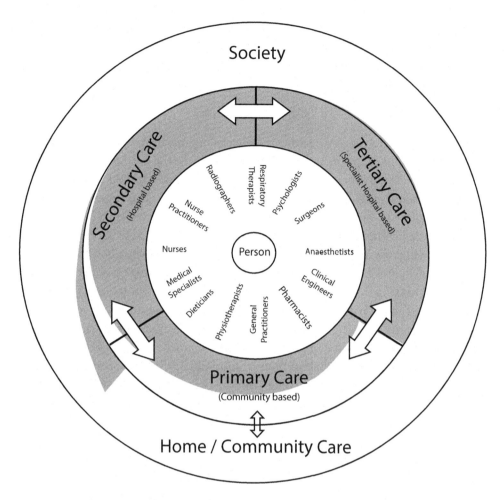

FIGURE 1.3 The Healthcare system and the people who populate it. The grey-shaded area indicates how technology has proliferated and is now used in all sectors.

1.4 VALUE IN HEALTHCARE

The delivery of healthcare is costly. How and by whom those costs are borne will depend on how the healthcare system is configured in a particular jurisdiction and on an individual's circumstances. It may be that the costs are borne directly by the individual, through a health insurance scheme, or by the government, with the latter ultimately funded through taxation. Regardless of the funding mechanism, the cost of healthcare must be found by society in some way. There is global concern that the cost of healthcare is too high and increasing, with calls to improve the effectiveness of healthcare using tools such as systems engineering, which we will discuss further in Chapter 2. For the moment we will concentrate on a more general discussion on benefits and costs.

When individuals need healthcare, they want to get timely access to the best possible care and have a successful outcome. For society, equality of access and timely effective healthcare should be an objective. The challenge for those with responsibility for healthcare management is how to improve care, including access and outcome, whilst controlling the cost so that the healthcare system is sustainable. Value is a concept that allows us to combine the two objectives of better and affordable care. Value in healthcare is simply the ratio of benefit to cost:

$$Value = Benefit : Cost$$

What constitutes value in any healthcare delivery organization will also depend on the needs and expectations of the organizations stakeholders. In the hospital context, the stakeholders would include the patients, the medical staff, the hospital administration and licensed health professionals through to regulatory bodies, funders and government. Consequently, in healthcare, value can vary depending on the concerns of different stakeholders.

Value can be difficult to define and measure. Benefit can be hard to express in monetary terms and the real costs associated with delivery of care can be equally difficult to assess. Nevertheless, this ratio which we call Value is useful and can be used in a qualitative way to inform decision-making that aims to improve the delivery of healthcare for its stakeholders.

If a new clinical procedure delivers the same clinical outcomes for lower cost, then value has increased. Clearly an initiative like this meets the objectives for all the stakeholders mentioned earlier and should be implemented. However, where a cost-reduction exercise is undertaken which decreases the number of individuals able to access the healthcare, the value is likely to decrease, in particular from the perspective of those not able to access care. From a purely financial perspective, the cost-reduction exercise may be seen as an improvement. However, looking at cost in isolation does not allow us to observe whether there is a resultant increase or decrease in Value. The concept of Value as the ratio of Benefit to Cost provides a tool for assessing change scenarios: Value may be increased by organizational changes that increase, or that decrease, costs, dependent on the effects of the changes on the benefits to the various stakeholders for whom the impact of the changes must be considered. So negotiating projects to increase value can be complex and depend on the perspective of the stakeholders. Value represents a holistic approach that requires ensuring that the views of all relevant stakeholders are carefully considered in arriving at proposed solutions.

In this book we will be looking at how healthcare technology is used and managed. We will see that clinical engineers can increase value by both controlling costs and realizing benefits associated with the use of healthcare technology. In doing so, they add value even though their presence increases the staff costs of healthcare. In exploring these topics through case studies, we will use the Value ratio as a qualitative metric to inform the discussion.

1.5 HOW CLINICAL ENGINEERING ADDS VALUE IN THE HEALTHCARE DELIVERY ORGANIZATION

The American College of Clinical Engineering (ACCE) proposed, in 1992, the following definition: "A clinical engineer is a professional who supports and advances patient care by applying engineering and managerial skills to healthcare technology" (ACCE 2015). The inclusion of managerial skills in this definition reflects the fact that within many large hospitals, the Clinical Engineering Department takes direct responsibility for actively managing the medical equipment over its life. This includes financial stewardship of these assets and support for their procurement and commissioning. The phrase 'supports and advances patient care' reflects the fact that the clinical engineering role extends beyond managing the equipment and includes application support, at the point of care. Healthcare technology management describes all of this activity, both the management of the medical devices and systems and the application support. Today this role extends beyond managing the medical devices and systems, to include the information technology (IT) networks into which they are being integrated and support for data analytics associated with clinical information systems. The term Healthcare Technology Management (HTM) describes the role more completely including as it does the supporting roles and the inclusion of medical IT support as well as the traditional medical equipment management activities (Wilson et al. 2014).

HTM can be considered as having two remits, two roles: equipment management and advancing and supporting care through the application of healthcare technology. This dual remit is important to recognize and appreciate: the remit is not simply to manage the technology, important though that is, as we will see in this book. The scope goes further to advance and support care through the technology and its application. The twin remits, 'Equipment Management' and 'Advancing Care', are discussed in more detail in Sections 1.5.1 and 1.5.2.

1.5.1 Equipment Management

The medical equipment in a hospital is essential infrastructure and represents a considerable financial investment. These assets need to be managed over time to ensure they can support the corporate and clinical goals of the organization. This includes planning for their upkeep and replacement over time, optimizing their utilization and ensuring they are maintained appropriately. Clinical engineers will often assume responsibility for medical equipment asset management. Through participation at corporate level, and as members of the department that coordinates and delivers the equipment management programmes, they ensure there is clear visibility of the issues associated with medical equipment asset management throughout the organization.

Clinical engineers are constantly updating themselves so they are aware of new medical equipment coming on the market. In doing so they are ready to act as independent advisers when planned replacement is warranted. They can identify opportunities where the organization can improve care through adopting new or better technologies. It is because they are familiar with the technical state of the art, the organizational objectives and current and evolving clinical practice that enables them to be effective in this regard.

Should the organization decide to procure new medical equipment, the clinical engineers play an important role in supporting the organization's Procurement Department to acquire equipment that is appropriate. This can include facilitating multidisciplinary teams to draw up equipment specifications and evaluate replies from companies (see Chapter 7). These multidisciplinary teams will include clinical staff (medical, nursing and paramedical as appropriate), clinical engineering staff, procurement staff and perhaps finance and management staff. Where large systems or groups of equipment are being purchased, it can be difficult for procurement professionals to ascertain the real costs of acquisition and ownership. To do so requires an appreciation of how medical equipment is used in clinical practice and maintained. Assessing the benefits offered by new equipment can also be difficult to determine. Nevertheless, with an in-depth understanding of the equipment and how it is used, clinical engineers can lead and assist the evaluation and selection process. In some healthcare organizations, the responsibility of clinical engineering departments extends to budgetary responsibility for medical equipment or to responsibility for leading and forming the multidisciplinary teams.

When equipment is purchased, the project of commissioning it and bringing it into service in the clinical environment can be complex. Again clinical engineers often play a lead role in this regard. They not only receive, check and document new medical equipment but also lead or assist in projects to bring the equipment into use in the clinical environment. This can include delivery to the clinical area, training, assistance with configuration, organizing the availability of associated consumables and planning for the ongoing maintenance. Where the project involves physical installations of large items such as critical care monitors, theatre equipment and large radiology equipment, clinical engineers may be called on to lead the installation team that may include Facilities Management, Infection Control, Health and Safety and the Fire Officer.

Over the course of its life, equipment will need to be maintained, and generally three types of maintenance activities need to be considered. Scheduled maintenance consists of all proactive activities whose purpose is to reduce the likelihood of failure of the equipment in service. Performance verification includes all proactive processes that assure equipment which appears to be working are in fact working optimally. Unscheduled maintenance covers all reactive actions which are initiated as a result of a reported real or suspected fault or failure of equipment or systems.

These maintenance activities can be delivered by in-house teams of clinical engineers or can be contracted to the equipment manufacturers or a combination of both. The Clinical Engineering Department will determine, on behalf of the organization, the most appropriate mix of service options for each type of equipment to meet clinical, corporate and financial objectives. They then design an appropriate equipment support plan (ESP) for each type of equipment or for groups of similar equipment and set it into action (Case Study CS1.1): ESPs will be discussed in more detail in Chapter 6.

Regardless of whether maintenance is delivered in-house or outsourced, the maintenance actions are documented and stored in the Clinical Engineering Department's Medical Equipment Management System (MEMS) database. This provides an archive of all maintenance activity which can be used for scheduling and planning. The database will also provide the basis for calculating key performance indicators (KPIs) that are used to manage the day-to-day service and provide management information for reporting on the effectiveness of equipment management activity. The process operates as a quality cycle which is reviewed regularly, usually annually. In this way the Clinical Engineering Department ensures the optimal mix of support options for the diverse range of equipment in its care.

This active management of maintenance leads to substantial economies. Wilson et al. report that in the United Kingdom, manufacturers typically charge 8%–10% of the purchase cost of the equipment for a comprehensive annual service contract (Wilson et al. 2014).

They go on to say that for a large teaching hospital, it is possible to reduce this to 6%–8% by reducing or varying the level of cover and augmenting it with in-house support. They suggest that by predominantly using an in-house support model, it is possible to reduce costs by a further 2%–3%, especially where evidence is used to match the level of support to the risk of failure and in doing so can simultaneously reduce equipment downtime. These percentages include the cost of the in-house clinical engineering function. So, by determining when and how to manage medical equipment using supplier or in-house maintenance, clinical engineers reduce the cost of ownership. This not only reduces cost which in itself adds value but also improves service quality, by decreasing equipment downtime and reducing risks associated with equipment failure or malfunction.

Equipment management takes place within a business model that has as its objective, increasing Value. The leadership in this regard is provided by clinical engineers but is achieved through working closely with the organization's management, finance and procurement departments. In doing so, they ensure that not only is the technical equipment management effective but also the equipment supports the strategic aims of the organization and that the maintenance arrangements are also an efficient use of financial resources.

1.5.2 Advancing and Supporting Care

Optimal and safe use of medical equipment requires more than for it to be properly acquired, commissioned and maintained. It requires the user to have an understanding of the technology, its characteristics and limitations and how it can be used to support healthcare. This knowledge and understanding will include an appreciation of the interplay between the patient, clinical and carer staff, the technology and the environment in which the equipment will be used. The clinical engineer will appreciate the importance of user-centred design, also known as human factors design, and its importance in ensuring safe and effective medical device use and hence be alert to instances of poor design, taking action as appropriate to inform manufacturers and regulatory agencies. In addition to the equipment management activity described previously, clinical engineers provide support that facilitates all hospital staff to integrate medical equipment and systems (better described as healthcare technology) into clinical practice effectively and safely. In this regard, the focus is on collaboration relating to the application of technology or using engineering skills to solve clinical, research or process problems rather than specific technical issues with equipment.

Clinical engineers have a role in facilitating the application of medical equipment, systems or novel methods at the point of care. For specific patients with particular clinical conditions, clinical engineers may be asked to advise on the appropriate use of equipment in the clinical management of that patient. The contribution can take the form of advice on the application of the equipment used in the care of the patient or the development of particular equipment solutions for the patient in question. Clinical engineers can also assist in analysis or interpretation of measurements, particularly where there is a question as to the validity of the measurement, suspected interference and other anomalies. Through these actions, they increase the benefit to the patient of the use of healthcare technology, adding value (Case Study CS1.2).

Where clinical engineers are based at the point of care, they are in a particularly strong position to contribute to quality improvement initiatives that focus on processes within clinical units. All clinical engineers can contribute in this way; with their systems analysis and measurement experience, they can be valuable members of quality improvement project teams. They can provide insights and identify ways of analyzing problems and developing solutions that may not be obvious to clinicians. Interdisciplinary by nature, clinical engineers can facilitate the coming together of different professional groups to reflect on and develop proposals for change. Often the change relates to the medical equipment and their role in supporting care, but increasingly the change can relate to existing processes for the purpose of improving clinical outcomes and cost control or improving the patient experience. One example involving redesigning a patient care process is illustrated in Case Study CS2.1. New ways of working supported by clinical IT systems are a case in point. Where such systems are being introduced at the point of care, clinical engineers can contribute to the systematic review of existing workflows and design of new ones enhanced and supported by the IT system. In doing so, they assist clinicians to develop new ways of working supported by the IT, rather than just computerizing existing paper-based systems. Their ability to do so is predicated on their engineering and systems science knowledge and also their familiarity with the clinical environment and work practices. This places clinical engineers in a privileged position, and often they can act as change agents and, through participation in process improvement initiatives, increase value for the organization's stakeholders.

Clinical engineers can also play a role in the ongoing support of clinical IT systems. They can assist in mining clinical databases which hold records of the care of patients. By mining this data set, doctors, nurses and engineers can collaborate to measure the effectiveness of care and, where evidence suggests, implement a quality improvement initiative. By operating the ongoing management of the clinical information system within such an evidence-based quality improvement cycle, safety, effectiveness and financial management can be improved.

One important role of the clinical engineer is to imaginatively foresee risks associated with the use of medical equipment. They should identify and highlight risk issues that need to be controlled and assist in identifying where other licensed health professionals should be involved in risk assessment, for example, the infection control team or facility engineers.

Adverse events associated with the use of medical equipment that can lead to patient harm do occur. We will explore later in this book in more detail the causes of adverse events involving medical equipment; see, for example, Chapter 2, Section 2.4.6. Suffice for the moment to note that problems can occur because of: device failure; operator error or omission; poor equipment design, in particular poor ergonomic design; and failures of supporting infrastructure. Two seminal publications, one from the U.S. Institute of Medicine *To Err Is Human* (Kohn et al. 2000) and the other from the NHS England Chief Medical Officer's report *An Organisation with a Memory* (Donaldson 2000), have prompted improvements in care processes to reduce the risk of harm. Whilst written to address patient safety in general, their messages are relevant to the safe use of medical equipment and the prevention of adverse events associated with medical equipment. The guidance from these documents can be used to inform the development and implementation of risk management practices that relate to the use of medical equipment. Both publications recognized human fallibility and consequently the need to build systems and operation procedures that incorporate checks and barriers that will prevent errors leading to incidents.

Clinical engineers support and promote safety in clinical practice through their participation in the corporate risk management processes. Where incidents occur involving medical equipment, clinical engineers are involved in investigating these events and performing the root cause analysis (Case Studies CS2.2 and CS7.7). They can also lead risk control projects, including leading on the reporting of adverse events to national regulatory authorities and thus helping ensure that lessons are more widely learnt. In doing so, they reduce the organization's exposure to litigation and increase the reliability and safety of medical equipment used in the care of patients.

Training in the optimal and safe use of medical equipment is required not only during commissioning but also over the full life of the asset. Training deals with the practical operational aspects of a particular item. Unless staff know the specifics of how to use a particular technology, they are unlikely to be able to use it effectively or safely. Training requirements will depend on the nature of the equipment and the clinical users. Clinical users may range from a small group of specialists (e.g. specialist ICU nurses using complex ventilators) to most nursing staff within the organization (using infusion devices, general purpose clinical thermometers). Training helps ensure patient safety and reduce adverse events. Without it, the risks of adverse events increase. Some clinical engineering departments, recognizing the importance of user training, employ staff to provide this training to medical and nursing staff, operating in cooperation with the medical and nursing leadership and training departments. Training ensures that the benefits offered by technology are realized in practice and so value increased.

Where new facilities are being built or existing ones upgraded, clinical engineers can play a pivotal role in developing the design brief and acting as facilitators of a conversation between architects and building engineers, and the clinical staff of the hospital. Even where the architect and builders have experience in building medical facilities, there is a need to critically review the intended use of all medically used rooms and the facility as a whole. This is best done by a multidisciplinary team comprising representatives of all groups who will use the space: doctors, nurses, licensed health professionals, general support staff, patients and their carers. Again in this context, it often falls to the clinical engineer to act as the synergist between these groups and the contractors.

Whilst the ongoing management of the services provided to clinical rooms (power, heating, medical gases, etc.) tends to be the responsibility of the organization's Facility Engineering team, it is usual for the clinical engineers to support this. The day-to-day activity of managing the equipment requires the clinical engineers to have an understanding of the specialized mechanical and electrical services to which the devices connect. Over time, as more and more equipment appears around the patient's bed, the clinical engineers need to assist the end users in optimizing the deployment of the technology both to improve care and to maintain a safe environment.

1.5.3 Healthcare Technology Management: Dual Remit

The activities described previously under both the supporting and advancing care and the equipment management headings are summarized in Figure 1.4. Together they make up an

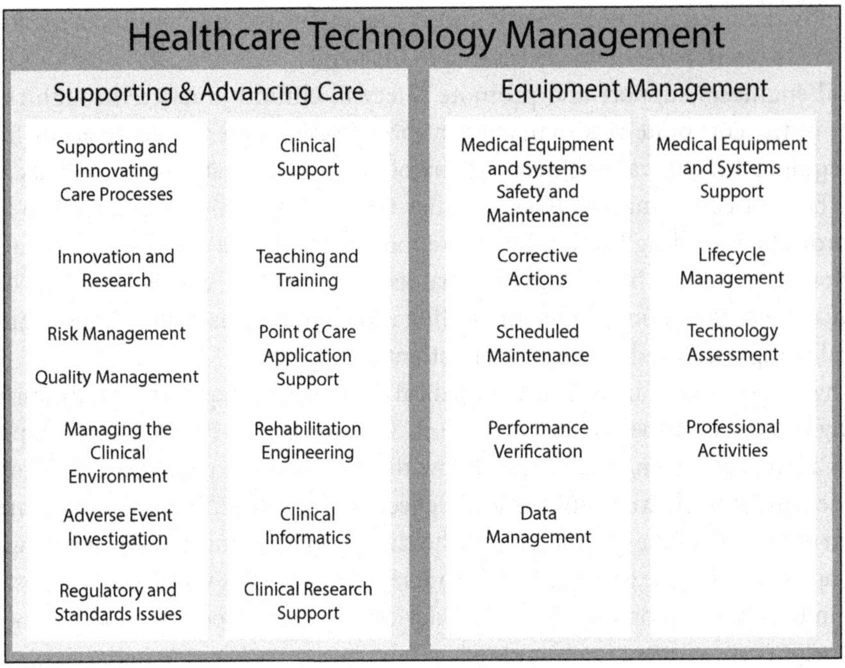

Healthcare Technology Management			
Supporting & Advancing Care		**Equipment Management**	
Supporting and Innovating Care Processes	Clinical Support	Medical Equipment and Systems Safety and Maintenance	Medical Equipment and Systems Support
Innovation and Research	Teaching and Training	Corrective Actions	Lifecycle Management
Risk Management	Point of Care Application Support	Scheduled Maintenance	Technology Assessment
Quality Management			
Managing the Clinical Environment	Rehabilitation Engineering	Performance Verification	Professional Activities
Adverse Event Investigation	Clinical Informatics	Data Management	
Regulatory and Standards Issues	Clinical Research Support		

FIGURE 1.4 The twin remits of Healthcare Technology Management: Supporting and Advancing Care and Equipment Management.

integrated healthcare technology management approach. In reality it is difficult to separate the activities out as described. In the delivery of healthcare technology management, these roles are tightly integrated and inform each other. In fact it is hard to imagine how excellence in one can be delivered without the other. So balancing the activities is essential. Here again the concept of value can be helpful in guiding the delivery of healthcare technology management.

1.6 CLINICAL ENGINEERING AND BIOMEDICAL ENGINEERING

We have discussed the role clinical engineering plays in hospitals and by association in primary care and in the community care settings. A description of the role of clinical engineers would not be complete without a discussion of how they contribute to the development of the wider field of biomedical engineering.

Biomedical engineering is a discipline that advances knowledge in engineering, biology and medicine and improves human health through cross-disciplinary activities that integrate the engineering sciences with the life sciences and with clinical practice. It includes:

- The acquisition of new knowledge and understanding of living systems through the innovative and substantive application of experimental and analytical techniques based on the engineering sciences.

- The development of new medical equipment, algorithms, processes and systems that advance biology and medicine and improve medical practice and healthcare delivery (Bronzino 2000).

The field of biomedical engineering combines engineering design and problem-solving skills with the life sciences to improve the delivery of healthcare. This includes research and development (R&D) activities into improved or new techniques for the diagnosis, monitoring and treatment of diseases. Working with partners in industry, healthcare and regulatory agencies, biomedical engineers develop, design, test and explore the use of medical equipment, recognizing the importance of human factors in engineering design. They develop, design and test prostheses and artificial organs. Biomedical engineers explore tissue, cellular and molecular bioengineering, medical genetics and advanced bio-manufacturing. Their work may also be theoretical as they apply mathematical and computational modelling tools to explore physiological systems and anatomical structures to better understand them.

Biomedical engineering is inherently multidisciplinary, involving collaborative research between the physical and life sciences and between the sub-disciplines that make up the biomedical engineering community. This research activity tends to be centred in universities and academic teaching hospital environments, but not exclusively so. Biomedical engineering extends to include the expert delivery of care through the application of technology at the point of care. In the hospital environment, biomedical engineering is practised where engineers and clinicians work together to optimize the delivery of healthcare through the application of technology. Figure 1.5 illustrates how biomedical engineering emerges from the physical and life sciences, encompasses both academia and healthcare institutions and shows a number of biomedical engineering sub-disciplines. The experience and knowledge gained by clinical engineers in supporting medical equipment and clinical

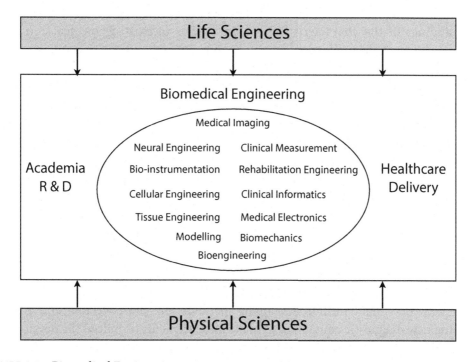

FIGURE 1.5   Biomedical Engineering.

practice within the hospital environment is valued by many who work in more academic areas of biomedical engineering research and device development. To develop new products and services, companies need to understand how they will be used in clinical practice. Researchers and product developers welcome the participation of clinical engineers and appreciate the unique perspective they can bring. Consequently, clinical engineers contribute to many stages of the process that develops, regulates and places medical equipment on the market. We will see that through participation in these ranges of activities the clinical engineer works beyond the hospital environment. At the same time, they acknowledge that their expertise in this regard comes from being based at the point of care working close to the clinician and patient.

1.6.1 Research and Development of New Medical Equipment

Research and development (R&D) is the term used to describe a group of activities that advance knowledge and deliver new technologies. In its broadest sense, the purpose of R&D is to discover and create new knowledge about scientific and technological topics. Furthermore, R&D is concerned with how to use existing and new knowledge to develop and deliver new medical equipment, processes and services. It requires reviewing the existing available knowledge on a subject, proposing a new theory and testing the validity of that theory. It has a strong academic focus and is usually centred in the universities. Teaching hospitals have close associations with the universities, and consequently research activity is a core activity within most teaching hospitals.

Clinical engineers, working within healthcare, will typically be involved in R&D activities with a direct impact on advancing the care delivered to patients that involve the application of technology. Their involvement may include developing improved understandings of the characteristics and limitations of existing medical equipment or the development or application of new equipment and systems or associated techniques used to diagnose and treat patients. It may include critical scrutiny of existing equipment to ensure that it delivers real benefits and that the claims of their developers and manufacturers are valid. Where the outcome of research involves the development of new equipment, systems or processes, again clinical engineers in hospitals are well placed to facilitate and actively contribute. Being hospital based, they can act to bridge the gap between clinical end users and university-based researchers, including helping to validate new equipment. Thus clinical engineers play an important role in supporting and initiating new research. This will be explored in more detail in Chapter 7 (e.g. see Case Study CS7.15). Their intimate knowledge of the technical, the clinical and the application of technology in the clinical arena may lead them to initiate R&D that develops improved technologies and better methods of applying existing technologies and of methods, including human–device interactions. Thus clinical engineers play an important role in supporting and initiating new research.

Their activities might include developing methods for data collection and analysis, developing new diagnostic equipment to make measurements or indeed the development of one-off devices or software to allow the research to progress. These endeavours will usually require collaborating with their clinical colleagues. Sometimes clinical engineers will work with clinicians to identify a problem that needs to be solved. They may identify an improvement arising out of clinical practice and articulate it in a way that serves to frame a research question. The resulting research and development can, and often does, happen within the hospital. Depending on the nature of the research, or the resources needed, these projects might require collaboration with biomedical engineers or others based in academia. Consequently, clinical engineers also work with university-based colleagues on R&D projects.

Clinical engineers can and do initiate original research themselves. This could be investigating new ways of measuring human function. It can also be evaluation of new or existing technologies. Given their role in the hospital, it is not surprising that often clinical engineers are involved with research and development projects relating to the development of new devices or software. Their close connection with clinical practice provides them with the opportunity to identify where medical equipment and systems used for diagnosis and treatment can be advanced. Their involvement with research and development projects can often focus on incrementing new features or functions with existing medical equipment and systems. Their input and insight in this regard is valuable, and often research and development departments within medical device companies are keen to avail themselves of the perspectives and contributions from hospital-based clinical engineers. This is one of the ways that clinical engineers can contribute to advancing care, by critically influencing the development of existing and new products.

1.6.2 Innovation

Innovation begins with recognizing a problem, an unmet clinical need or a process that could be improved and developing solutions. The solution may be the development of a new device or identifying a better way to undertake a task. Clinical engineers may be involved in innovating improved clinical methods and processes, for example, in the application of medical equipment, and we shall discuss in Chapter 2 the systems engineering approach that provides a tool for structured redevelopment of processes.

In this section, we will discuss the involvement of clinical engineers in the development of new medical equipment. We shall describe the steps that bring a new technology or device from the research space into the marketplace, a process in which innovation is aligned with concepts of 'technology transfer' and 'commercialization'. Whilst research and development in the university or teaching hospital space may discover and prove the effectiveness of a new idea, the refining of the discovery, matching it with an identified customer need and developing equipment, products and services that can deliver the solution to the market and meet regulatory requirements, is a different process. Often the idea or technology must transfer from one institution, say the university, to a commercial company for this to happen. The commercial company focus is on how to best exploit the idea, providing the start-up capital to refine the product and offering investors a return on future sales.

Figure 1.6 highlights the particular role clinical engineers play in advancing care through participation in research, development and innovation of new medical equipment. Based on the healthcare organization from where they draw their particular

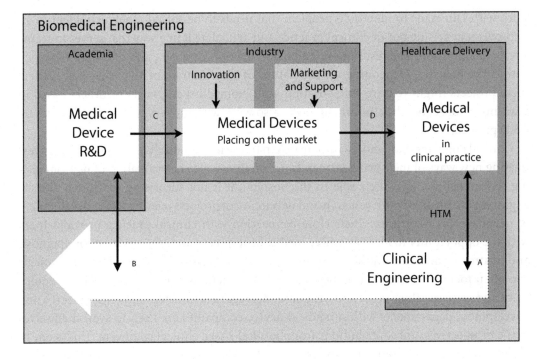

FIGURE 1.6   Clinical engineering extending beyond the health institution to support biomedical engineering research. Flow process stages A, B, C and D are explained in the text.

expertise, clinical engineers can extend their activities beyond the hospital to support research and development in academia. That participation can arise out of a clinician or clinical engineer initiating a research project to meet an identified need (Figure 1.6 process 'A'). It may also arise from a research initiative in the university which requires collaboration with clinical experts (Figure 1.6 process 'B'). This may be a development or the trial of prototype equipment, software or processes. Once the value and effectiveness of the new device, software or process is established, the invention moves from the research to the innovation space (Figure 1.6 process 'C'). Through innovation activity, the device is incrementally developed to be ready for placing on the market. Industry then brings the new equipment to the hospital. Clinical engineers are then involved once again during the medical device acquisition phase which includes commissioning, placing in the clinical environment and supporting the introduction of the new technology (Figure 1.6 process 'D').

So whilst the dominant contribution of clinical engineers is to provide healthcare technology management, their experience supports research and development and influences the development of new medical equipment and their successful implementation in clinical practice (Figure 1.6). This participation in research brings many benefits for healthcare in general and perhaps also for the organization where they work as it faces the challenge of keeping abreast of current and emerging technologies. Clinical engineers participating in research projects will be actively engaged with the wider biomedical engineering community, and the experience and knowledge gained through this activity will be valuable in the work they undertake within the healthcare organization. Direct participation in supporting company innovation is therefore another way in which clinical engineers contribute to advancing care through the application of technology.

During the innovation phase of device development, companies may need to consult with clinicians and other end users to better understand the clinical use and environment and processes within which the device will be used. This can extend to undertaking testing or trial of new and novel medical equipment in clinical practice, in a controlled and safe way. When biomedical engineers in innovation departments are trying to do this, their first point of contact is often the clinical engineers in the hospital. Clinical engineers are the experts in the application of devices in the clinical environment and can in their own right make valuable contributions to the innovation process. Very often the role of the clinical engineer in supporting this activity is to act as the bridge between the innovation engineers and the clinicians who will eventually use the equipment. Where this consultation extends to testing or trialling equipment, clinical engineers can provide the access to clinicians and the hospital environment to allow this to take place, within accepted hospital ethics, policy and safety guidelines. They can perform risk assessment and undertake essential safety testing of new equipment to ensure that the test or trial is conducted safely. This ability to give innovation departments access to both clinicians and the clinical environment, to broker the communications between industry and hospital experts and to give clinical and engineering feedback on the performance of the device, software or process being developed is valuable to industry.

Whilst supporting innovation, the clinical engineer must be careful to ensure that the priority remains the goal of improved patient care through a new product or process. Clinical engineers who themselves initiate new ideas must ensure that the resulting process is subject to critical scrutiny. Clinical engineers will need to be the gatekeeper ensuring that only innovations that meet the criteria of improving patient care are allowed to proceed to clinical trial. Regulatory authorities including the FDA in the United States and the competent authority in each European Union (EU) country will play their part in this process, but the clinical engineers may need to intervene early in the process, objectively raising concerns were warranted.

1.6.3 Clinical Engineers Contribute to Standards Development

The medical device market is highly regulated. Before medical equipment can be placed on the market, companies must ensure their design and manufacture meet essential minimum requirements which are set out in Standards. As part of the innovation process, companies must design and plan the manufacture and support of medical equipment to meet the requirements set out in the Standards. Clinical engineers participate in the writing of Standards as practitioner experts in medical equipment. In this capacity, they play an important role in ensuring that lessons learned or challenges associated with the application of medical equipment inform the documents that guide the development of the next generation of technology being brought to market. Like participation in research, the experience and knowledge gained through this activity informs clinical engineers as

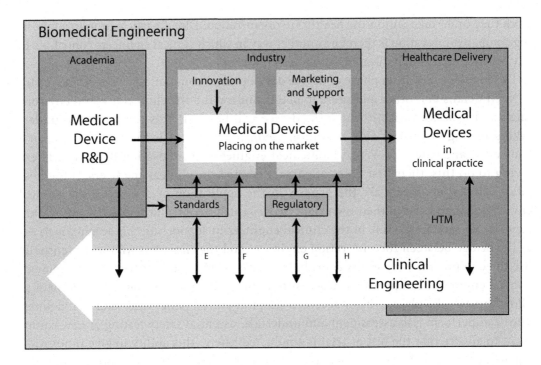

FIGURE 1.7 Clinical engineering extending beyond the health institution to support innovation and support post placing on the market. The process stages E, F, G and H follow process stages A, B, C and D in Figure 1.6.

to the latest developments in the field of biomedical engineering. It gives them insights in safety and risk issues associated with the use of medical equipment as well as keeping them up to date with new and emerging technologies. Whilst participation in Standards is voluntary, the benefits for hospitals that support their clinical engineers to participate are significant (Figure 1.7 process 'E').

1.6.4 Clinical Engineers Contribute to Post-Market Surveillance

When medical equipment is placed on the market, it is subject to ongoing regulation and post-market surveillance. Most jurisdictions will have in place a body which manages the regulatory environment. In the United States this is the FDA, the Therapeutic Goods Administration (TGA) in Australia, and in each EU member state there is a 'competent authority' that ensures compliance with the EU's Medical Devices Directive (see Chapter 3 for more details).

As part of this regulation, there is a requirement for companies to report incidents in which medical equipment may have caused injury. Where this occurs, the manufacturers must report the incident to the competent authority in that jurisdiction. Similarly the manufacturer must notify the competent authority of any field corrective actions mandated on its products already placed on the market. The goals of this ongoing surveillance and regulation are to detect and correct problems in a timely fashion. The regulatory authorities and clinical engineers work together to ensure that this regulation is effective within the health institution.

The regulatory agencies publish information of incidents which occur and final corrective actions being undertaken by companies. Suppliers of medical equipment issue field safety notices and regulatory authorities issue safety warnings. This information allows the clinical engineers to take appropriate actions to manage any associated risk. The actions could range from making the clinical end users aware of a possible risk and actions that they should take to mitigate the risk, ensuring field corrective actions are undertaken by the manufacturers, to quarantining and removing from the clinical environment medical equipment which has been identified as defective.

Clinical engineers, like all healthcare workers, and indeed the public are encouraged to report incidents involving medical equipment which give rise to harm, to the regulatory agencies. Clearly, within the health institution, clinical engineers should play a lead role in this regard. Both through their healthcare technology management role and also their role in risk assessment and control, they will become aware of incidents or indeed near misses and should report them as appropriate. The management of safety warnings and of incidents and near misses will be discussed in greater detail in Chapter 7, Sections 7.2.6 and 7.2.7.

This bidirectional activity with the regulatory agencies (Figure 1.7 process 'G') is another way in which clinical engineers advance the safety of medical equipment in practice.

1.6.5 Knowledge Sharing between Clinical Engineers and Equipment Manufacturers

There is benefit in manufacturers and healthcare delivery organizations communicating and sharing experiences with each other regarding the practical application of medical equipment. Whilst this is regulated in relation to incidents which result in harm, thankfully these incidents are rare. Where no harm occurs, sometimes termed near misses, it is very important to investigate the causes, to learn lessons and to share experiences to improve clinical processes and re-enforce positive actions. Near misses are free lessons that can improve the safety of care. This will be discussed in greater detail in Chapter 9.

Many clinical engineering departments have close working relationships with equipment suppliers, and this will be discussed and expanded on in the sections dealing with the delivery of healthcare technology management. Many well-established clinical engineering departments proactively partner with device manufacturers when it comes to the ongoing support of medical equipment, and this partnership might extend beyond technical issues. One of the tenets of a good quality management system is that the healthcare delivery organization should build mutually beneficial supplier relationships. Clinical engineers can build on existing arrangements for technical support to extend them to provide a more general support for the use of products. Sharing good practice is one example of this. Feedback from the manufacturer to the hospital as to how to optimally use their equipment is valuable, as is feedback to the manufacturers of how the device is performing in practice. These discussions help develop knowledge and understanding between the two groups who, although having different perspectives, share the common goal of advancing care through the application of technology.

Just as clinical engineers sometimes contribute to device development through partnering with researchers in universities, they can also contribute to the development of support mechanisms by partnering with manufacturers. For example, the Clinical Engineering Department might identify a need for an ongoing training programme to ensure that when new staff join the healthcare organization, they are trained in the use of medical equipment. Their insight into how the medical equipment is used within their particular institution might allow them to tailor the configuration of the equipment to the organization's requirements and then work with the manufacturer to develop a specific training programme. This is a mutually beneficial arrangement: the healthcare organization gets training input from the manufacturers' experts, framed and contextualized by the clinical engineers to ensure it is relevant to the organization; the manufacturer ensures that their product is used optimally, enhancing the products reputation, reducing the likelihood of adverse incidents and, through supporting ongoing training, enhancing the reputation of the company itself. By building these support relationships between manufacturers and the health institution (Figure 1.7 process 'H'), focused on promoting best practice, clinical engineers are both supporting and advancing care.

1.7 FOUNDATIONS OF A HTM SYSTEM WITHIN A HEALTHCARE DELIVERY ORGANIZATION

We have seen that healthcare technology brings both benefits and risks. Its cost of acquisition and ownership needs to be controlled and critically reviewed to ensure it is sustainable and adding value to the organization. Active asset management provides a framework for managing this. Traditionally medical equipment management has focused on technical issues and user support activities that reduce failures, misuse or other factors that affect safety or reliability. A comprehensive asset management system should seek to increase the value of the assets for the organization and its stakeholders. It encompasses all of the activities described in this chapter. It goes further than traditional technical maintenance and promotes other processes that enhance the use of assets so that they are optimally deployed to the benefit of the organization's stakeholders. Clinical engineers charged with delivering a comprehensive healthcare technology management programme can look to a number of sources to guide them. The ISO 55000 set of Asset Management Standards (ISO 2014) details the elements of a comprehensive asset management system that can be adopted by different industries. Other standards and guidance such as the U.S. AAMI/ANSI EQ56 'Recommended practice for a medical equipment management programme' (AAMI 2013) and the UK Medicines and Healthcare products Regulatory Agency's (MHRA's) bulletin 'Managing Medical Devices in hospital and community organisations' (MHRA 2015) provide information specific to the management of medical equipment. The ISO 9000 family of standards (ISO 2015) is designed to help organization implement quality management systems to ensure they meet the needs of their customers and other stake holders. All are useful and discussed in more detail in Chapter 3.

1.7.1 Clinical Engineering within the Corporate Structure

An effective Clinical Engineering Department should deliver the complete healthcare technology management role. It cannot do this on its own, but requires cooperation and support from the organization, both at senior corporate Board level and throughout the organization. The Clinical Engineering Department (CED) will need to explain and demonstrate how it can support top management in advancing the delivery of healthcare. The CED must show that it delivers a positive Return on Investment to the Board, adding Value to the healthcare organization. Working with senior management, the Head of Clinical Engineering will need to understand the Board's corporate plans and ambitions as well as the financial pressures and other current problem issues. Rather than simply reacting to problems, the Head of Clinical Engineering can proactively suggest solutions and should be encouraged to act in this way.

The hospital should consider where to place the Clinical Engineering Department within its corporate structure so that the staff can act effectively as independent internal experts. For the department to act effectively, it must be positioned so that it is independent of any of the clinical departments who may be competing for resources, but also in tune with and understanding their clinical needs and the organization's strategic objectives.

The position of the Clinical Engineering Department within the organizational structure should recognize the strategic roles it plays across the organization. It requires close working relationships with clinical leadership and clinical teams. It is thus often placed alongside other clinical support services such as Pharmacy, Laboratory Services or Radiology. However, placing the CED within one of these departments does not recognize clinical engineering's wider objectives. The 'repair' component of the work suggests a link with Estates or Facilities Management, but this does not recognize the direct clinical support role of the CED. Alternatively, the growth in clinical computing and increasing interaction between clinical computing and medical equipment has led to consideration of combining the Information Technology Department with the Clinical Engineering Department. The danger of any solution which merges the CED with another, such as Facilities, Engineering or the IT Department, which does not have a direct role in supporting clinicians, is that the core mission of the Clinical Engineering Department to advance and support patient care will get eroded over time. Such a merged department may function well initially. The concern is that over time the merged department may respond and develop to meet genuine organizational needs, particularly short-term resource-limitation pressures which can mask the strategic core focus on direct support to clinicians and patients.

However, there can be benefits of combining the Clinical Engineering Department with another hospital department that directly support healthcare technology with a clinical focus. Thus in some areas there are combined medical physics and clinical engineering departments. Whilst medical physics and clinical engineering both have different roles, they are both expressions of the need for hospitals to have engineers and scientists from the physical sciences supporting and advancing care through the application of engineering and physics principles. Between them medical physics and clinical engineering have an impact on nearly every patient pathway in the modern healthcare organization. Both clinical engineering and medical physics emerged from the intersection of the life and physical sciences. Merger with medical physics is likely to allow clinical engineering to flourish as being part of a bigger department allowing it to build scale and find efficiencies. Furthermore such a department can share scientific and engineering skills and management expertise. It is their common direct involvement in supporting care and close working relationships with clinicians that allows clinical engineering and medical physics to merge without fear of this core value being eroded over time (IPEM 2015a).

1.7.2 Clinical Engineering Leadership

Effective clinical engineering departments require strong leadership from a head of department who understands the importance of staying close to the clinical workflow, how to balance the advancing care and equipment management roles (the twin remit – Figure 1.4) and who is focused on building a department and culture which can support the organization. The optimal solution suggests an independent Clinical Engineering Department led by a qualified registered clinical engineer whose status within the organization is recognized as the organization's Chief Healthcare Technology Officer (CHTO) or equivalent title. To facilitate the department contributing meaningfully to the full gamut of activities described in this book, the head of clinical engineering must be part of the hospital's senior management team. Reporting to the Chief Executive Officer (CEO) or Board member with responsibility for healthcare technology, the head of clinical engineering will work with appropriate committees to ensure that the management of healthcare technology supports and advances the corporate strategy of the organization.

The exact governance structures may vary between organizations. The Board may also constitute a Medical Device Committee as described in Chapter 5 to ensure that cognizance is taken of the varying needs of the departments within the organization. This group will typically have a membership that includes leading medical and nursing staff with an interest in healthcare technology, as well as representatives from procurement, health and safety, nurse training, infection control and clinical engineering. Alternatively, or in addition, a reporting link may be established to a committee of the clinical heads of each of the medical departments (e.g. a clinical directors committee or forum).

The reporting structure of the Clinical Engineering Department should reflect the role. In some organizations, the Clinical Engineering Department may sit within a general 'service support' directorate, but the senior clinical engineers and certainly the head of department will require direct links with senior medical and nursing leadership. Healthcare technology finances, both capital and revenue, require similar links with the directorate of finance. It is thus expected that the senior clinical engineers will work with the Medical Director, the director of nursing, the finance director and the chief executive to guide the use of health technology within the organization. Regardless of the local management configuration, it is important to recognize that the role of clinical engineering extends to and affects the strategic management of the organization, and if the head of department is not appointed to a senior management level where he or she can contribute to the development of the healthcare technology management policy, then they may not be able to operate effectively to deliver a complete clinical engineering service.

1.7.3 Clinical Engineering Ethics

In daily practice, those involved in clinical engineering have to balance many concerns, financial, safety, clinical, technical, personal and institutional. Consequently, in clinical engineering as in other branches of engineering, the moral values guiding the work are myriad and can come into conflict. When this arises, it is not a sign that anything is wrong but rather is the consequence of the moral complexity inherent in healthcare. In thinking through such conflicts, clinical engineers need to keep to the fore the connection between excellence and ethics in healthcare technology management (Case Study CS1.3). At all times clinical engineers need to engage in and support responsible conduct.

Resources that might help clinical engineers when negotiating such conflicts are the codes of ethics developed by professional organizations. Codes of ethics such as those of the IEEE (IEEE 2015) in the United States and the Institute of Physics and Engineering in Medicine (IPEM) in the United Kingdom (IPEM 2015b) state the moral responsibilities of engineers as represented by a professional society. Many healthcare organizations have codes of conduct with which employees are required to comply and the clinical engineer should know, understand and comply with the relevant codes. Codes not only serve as a resource to help focus an engineer's responsibilities but also encourage the freedom to exercise professional skills to achieve goals. As such codes provide a positive stimulus and in a powerful way, they voice what it means to be a member of a profession committed to promoting the safety, health and welfare of the public (Martin 2005). The American College of Clinical Engineering has produced just such a code for clinical engineers (ACCE 2016). This code states that in the fulfilment of their duties, clinical engineers will:

- Hold paramount the safety, health and welfare of the public;

- Improve the efficacy and safety of healthcare through the application of technology;

- Support the efficacy and safety of healthcare through the acquisition and exchange of information and experience with other engineers and managers;

- Manage healthcare technology programmes effectively and resources responsibly;

- Accurately represent their level of responsibility, authority, experience, knowledge and education and perform services only in their area of competence;

- Maintain confidentiality of patient information as well as proprietary employer or client information, unless doing so would endanger public safety or violate any legal obligations;

- Not engage in any activities that are conflicts of interest or that provide the appearance of conflicts of interest and that can adversely affect their performance or impair their professional judgment;

- Conduct themselves honourably and legally in all their activities.

1.8 CLINICAL ENGINEERING ADVANCING PATIENT CARE

Clinical engineering should both support and advance patient care. Therefore, patient care is the ultimate reason that healthcare technology management programmes exist. To help ensure that the focus is kept on the patient (and carer) when managing complex medical equipment projects, the 'Keystone Model' was developed, the patient and carer being the keystone of the arch that keeps the whole process together (Brooks Young and Amoore 2012). Previously, the authors have used this arch as a metaphor for the delivery of healthcare technology management (Hegarty et al. 2014). The supporting and advancing care and the equipment management roles can be considered the pillars of the healthcare technology management programme, the arch which supports the delivery of patient care. It is vitally important to recognize and acknowledge always that these roles are complimentary and are tightly integrated in practice. The archway shown in Figure 1.8 represents the integration of these two roles, and it is only by fulfilling both roles that clinical engineering can completely support the delivery of patient care. The keystone at the apex of an arch is the final piece placed during construction and locks all the stones into position. This keystone, being the point of strength within the structure, allows the arch as a whole to bear weight. We suggest that the keystone of any healthcare technology management programme should be optimizing and improving the patient and carer experience through the application of technology for healthcare.

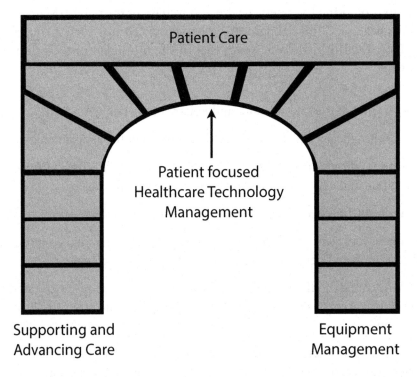

FIGURE 1.8 Patient-focused Healthcare Technology Management. (Reprinted from *Clinical Engineering*, Hegarty, F.J., Amoore, J., Scott, R., Blackett, P. and McCarthy, J.P., The role of clinical engineers in hospitals, pp. 93–103, Copyright 2014, with permission from Academic Press Elsevier.)

1.9 CONCLUSION

Engineers have supported healthcare technology in hospitals for over 50 years. As medical equipment became more sophisticated and widespread, the practice of clinical engineering has had to evolve. Those who deliver clinical engineering have had to learn, adapt and change their approach to keep pace with rapid developments in the technology and how healthcare is delivered. Initially clinical engineers focused on supporting the introduction of medical equipment into the hospital environment. This evolved into managing large fleets of equipment, including equipment grouped into complex systems that now include information technology. Today that challenge extends further to supporting medical equipment and systems that are placed in the community. The initial focus on technical maintenance and safety has expanded to include financial stewardship of medical device assets and technical and scientific support for clinical users of medical equipment. We have also seen that clinical engineers contribute to developments in the field of biomedical science through participation in research and innovation. So today clinical engineers are undertaking a wide range of activities, and the role differs from organization to organization in accordance with the local need. What unifies clinical engineering practice is its goal of optimizing the value of the medical device and equipment assets for the patient, organization and society at large. As discussed previously, value is increased when the ratio of benefits delivered from the use of medical equipment increases in relation to the cost associated with their use.

Whilst the authors acknowledge that clinical engineering practice varies between jurisdictions and organizations, in this book, we propose a generic approach, or model, as to how the myriad of activities that make up clinical engineering can be managed. This approach is described as a Healthcare Technology Management (HTM) system imagined in the context of a large hospital (Figure 1.9). It takes a life cycle view of asset management that will be familiar to anyone who develops and delivers clinical engineering services. The HTM system we propose consists of two interlocking management processes. The first focuses on the strategic management of medical equipment over the medium to long term, at least over a 5- to 10-year period. The second focuses on the operational delivery of support services for medical equipment management. Together they make up a complete HTM system that should deliver value for its stakeholders. Healthcare Technology Management has many stakeholders and each could have a different opinion as to what constitutes value in the context of medical device use. The diagram identifies four key groups of stakeholders and they are:

- Society, Government, Funders and Tax Payers;

- Patients, their Families and Carers;

- Healthcare Professionals;

- Healthcare System Managers.

The two HTM processes, each of which operates as a quality cycle, will require input and leadership from clinical engineers. The intersection of the two processes acknowledges that in Healthcare Technology Management it is impossible to completely separate the strategic and operational processes, and some clinical engineering actives such as replacement

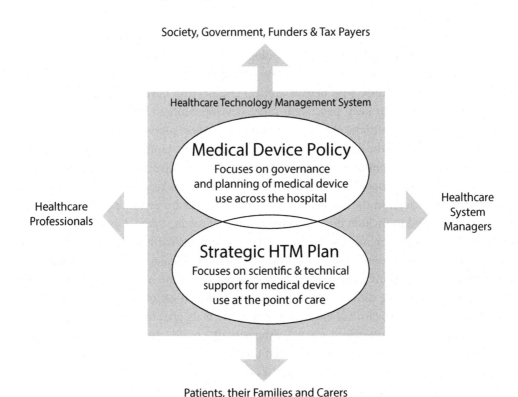

FIGURE 1.9   A model for describing a complete Healthcare Technology Management system to meet the needs of various stakeholders.

planning straddle both. This is not a failing of the model, rather a reflection of the complexity of the activity of Healthcare Technology Management which clinical engineers deliver.

This model of a Healthcare Technology Management system is consistent with contemporary guidance on good asset management practice. In describing these processes in later chapters, we have taken care to relate and align these activities to those described in the ISO 55000 series of Standards for Asset Management. This series of Standard outlines how best to manage assets and realize the value from them. The series sets out guidance on how to develop good asset management that maximizes value for money and satisfies the stakeholder's expectations.

REFERENCES

AAMI. 2011. Forum recommends unified name, vision for field. Final future forum 4.01.13. Arlington VA: Association for the Advancement of Medical Instrumentation. http://s3.amazonaws. com/rdcms-aami/files/production/public/FileDownloads/HTM/Final_Future%20_Forum. pdf (accessed 2016-04-01).

AAMI. 2013. AAMI/ANSI EQ56: Recommended practice for a medical equipment management program. Arlington VA: Association for the Advancement of Medical Instrumentation.

ACCE. 2015. Clinical engineer – ACCE definition, 1992. http://accenet.org/about/Pages/ ClinicalEngineer.aspx (accessed 2016-04-01).

ACCE. 2016. Code of ethics. Plymouth Meeting PA: American College of Clinical Engineers. http:// accenet.org/about/Documents/ACCE%20Code-of-ethics.pdf (accessed 2016-06-16).

Bauld T.J. 1991. The definition of a clinical engineer. *Journal of Clinical Engineering*, 16: 403–405.

Bronzino J.D. (Ed.). 2000. *The Biomedical Engineering Handbook*, 2nd edn., Vol. 1. Boca Raton, FL: CRC Press.

Brooks Young P. and J.N. Amoore. 2012. Subcutaneous infusions for pain and symptom control in palliative care: Introducing the Keystone model. *19th International Congress on Palliative Care*, Montreal, Quebec, Canada, October 2012.

Donaldson L. (Chair). 2000. *An Organisation with a Memory. Report of an Expert Group on Learning from Adverse Events in the NHS Chaired by the Chief Medical Officer*. London, UK: Her Majesty's Stationary Office.

Hegarty F.J., Amoore J., Scott R., Blackett P. and J.P. McCarthy. 2014. The role of clinical engineers in hospitals. In *Clinical Engineering*, eds. A. Taktak, P. Ganney, D. Long and P. White, pp. 93–103. Oxford, UK: Academic Press Elsevier.

IEEE. 2015. Code of ethics. New York: IEEE.org. http://www.ieee.org/about/corporate/governance/p7-8.html (accessed 2016-04-01).

IPEM. 2015a. Policy statement: Managing medical physics and clinical engineering services. York, UK: Institute of Physics and Engineering in Medicine. http://www.ipem.ac.uk/Portals/0/Documents/Publications/Policy%20Statements/Leading%20MPCE%20Services%20Policy%20Statement%20v2%20July%202015.pdf (accessed 2016-08-20).

IPEM. 2015b. Code of professional and ethical conduct, 6th issue. York, UK: Institute of Physics and Engineering in Medicine. http://www.ipem.ac.uk/AboutIPEM/JoinIPEM/Professional Conduct.aspx (accessed 2016-04-01).

ISO. 2014. ISO 55000, ISO 55001 and ISO 55002: Asset management suite of standards. Geneva, Switzerland: International Standards Organization.

ISO. 2015. ISO 9000, ISO 9001, ISO 9004 and ISO 19011: Quality management systems family of standards. Geneva, Switzerland: International Standards Organization.

Kohn L.T., Corrigan J.M. and M.S. Donaldson (Editors). 2000. To Err Is Human: Building a Safer Health System. Washington, DC: National Academies Press. http://www.nap.edu/openbook. php?record_id=9728&page=R1 (accessed 2016-04-01).

Martin W.M. and R. Schinzinger. 2005. *Ethics in Engineering*, 4th edn. New York: McGraw Hill.

MHRA. 2015. Managing medical devices. Guidance for healthcare and social services organisations, v1.1. London, UK: Medicines and Healthcare products Regulatory Agency. https://www.gov.uk/government/publications/managing-medical-devices (accessed 2016-04-01).

WHO. 2007. *Everybody's Business – Strengthening Health Systems to Improve Health Outcomes*. Geneva, Switzerland: World Health Organization. http://www.who.int/healthsystems/strategy/everybodys_business.pdf?ua=1 (accessed 2016-04-01).

WHO. 2011. *Development of Medical Device Policies*. Geneva, Switzerland: World Health Organization. http://apps.who.int/iris/bitstream/10665/44600/1/9789241501637_eng.pdf (accessed 2016-01-18).

Wilson K., Ison K. and S. Tabakov. 2014. *Medical Equipment Management*. Boca Raton, FL: CRC Press.

SELF-DIRECTED LEARNING

1. What do you understand by the term 'Healthcare Technology'? Can you illustrate your answer with five examples of technologies found in a hospital which would be included in this definition and five which would not?

2. Clinical engineering traditionally focused on the maintenance and support of medical equipment in the hospital setting. Can you describe three changes in that role that have arisen in response to how healthcare is delivered today?

3. Imagine you are the head of a clinical engineering service in a large teaching hospital. If the CEO asked you to justify why clinical engineers in your department were given protected time to contribute to the development of medical equipment Standards, how would you respond?

4. Can you identify three activities from your daily practice as a clinical engineer that add value for patients? For each, explain how your practice impacts on both side of the value ratio, the benefits delivered and the cost.

5. Figure 1.4 shows the twin remits of Healthcare Technology Management (Supporting and Advancing Care and Equipment Management):

   - Describe three examples of how the practice of one of the activities in the Supporting and Advancing Care pillar inform and add value to the activities in the Equipment Management pillar.

   - Describe two examples of how the practice of one of the activities in the Equipment Management pillar inform and add value to the activities in the Supporting and Advancing Care pillar.

6. Write brief notes that identify and describe the four sectors of the health system. In your opinion which is the most dependent upon medical equipment at this moment in time? How do you think this will change over the next 10 years?

7. The American College of Clinical Engineering defines a clinical engineer as 'a professional who supports and advances patient care by applying engineering and managerial skills to healthcare technology'. Explain the significance of the inclusion of managerial skills in this definition.

8. Write short notes on the contribution that hospital-based clinical engineers can make to Research and Innovation of medical devices. In your answer, identify the difference between Research and Innovation.

9. 'As a result of medical device Standards being developed and adopted, medical devices have become more reliable and safer; therefore, the need for clinical engineers in hospitals has decreased'. Critique and discuss this statement.

10. Choosing any medical device or medical equipment system you are familiar with, identify ways in which clinical engineers can add value to the management of that piece of equipment during the following phases of the equipment's life cycle:

    - Procurement.

    - Commissioning.

    - Performance verification during its use.

    - Investigation of an adverse event involving this device.

    - The disposal of the device at the end of its life cycle.

CASE STUDIES

CASE STUDY CS1.1: PLANNING A NEW EQUIPMENT SUPPORT PLAN FOR AN EXPANSION TO AN ENDOSCOPY DAY UNIT

Section Links: Chapter 1, Section 1.5.1

ABSTRACT

This study explores how a Clinical Engineering Department (CED) helped control maintenance costs as part of a planned expansion of an endoscopy day unit. By analyzing the existing equipment support plan and suggesting a different mix of in-house and external support for the expanded unit, the CED ensured the optimal mix was chosen which both controlled costs and improved the support provided.

Keywords: Endoscopy, Contract management, Calculating the cost of an equipment support plan

NARRATIVE

A hospital needed to build and equip a substantial endoscopy day unit to meet the increasing needs for gastrointestinal screening. The existing two-room endoscopy facility would be expanded to include eight extra rooms, 10 in total, with a pro rata increase in the number of endoscopes. The CEO was aware that the existing comprehensive support contract from the original equipment manufacturer (OEM) endoscopy supplier was costly, but she also understood and appreciated that repair of endoscopes was a specialist task. She feared the cost of maintenance might scale directly with the increase in the number of endoscopes proposed for the unit expansion. She asked the clinical engineer to look at how to best maintain the endoscopes so that they were available to support the busy unit but also look at ways of reducing the associated maintenance and cost of ownership.

The clinical engineer started by analyzing the existing equipment and support arrangements. First, he established capital cost of the existing endoscopy assets. The cost of the devices in each room was €205,000 or €410,000 in total (Table CS1.1A).

Next he looked at the cost of the annual OEM supplier service contract. This costs €30,000 per annum per room which is approximately 15% of the equipment capital cost.

The clinical staff in the unit were in favour of the existing support contract arrangements because any time an endoscope needed to go for repair, the supplier lent them an equivalent model. This was valued by all staff who were burdened with the challenge of keeping the services going to meet a constant demand.

The clinical engineer calculated that if the existing support arrangement were kept when the unit was increased to the 10 rooms, the cost of the annual contract would increase from €60,000 to €300,000 per annum (Table CS1.1B).

TABLE CS1.1A Cost of Devices in Each Room

Endoscopy Devices per Room	Number of Devices per Room	Cost per Device	Subtotal
Gastroscope	3	€25,000	€75,000
Colonoscope	3	€30,000	€90,000
Endoscopy imaging system	1	€40,000	€40,000
		Total:	**€205,000**

TABLE CS1.1B   Cost of Comprehensive Service Contact for 2 and 10 Rooms

Summary of support arrangements and cost for 2 and 10 endoscopy rooms based on a comprehensive service contract support model	
Cost of devices per room	€205,000
Capital cost of endoscopy devices for 2 rooms	€410,000
Cost of comprehensive service contract per annum for 2 rooms	€60,000
Service contract as a percentage of capital cost	14.63%
Capital cost of endoscopy devices for 10 rooms	€2,050,000
Cost of comprehensive service contract per annum for 10 rooms	€300,000
Service contract as a percentage of capital cost	14.63%

A more detailed review of the maintenance records for the endoscopy unit revealed that 40% of maintenance repairs were of a highly specialized nature that required OEM supplier expertise. A further 20% related to blockages in the small irrigation channels in the endoscopes and another 20% related to electronic faults or calibration issues with the associated light sources and video image processors and displays. The final 20% of reported faults were in fact user error or difficulty with operating the devices.

The solution proposed by the clinical engineer was to develop and implement a new equipment support plan for the equipment in the new unit, as follows. As part of the procurement negotiations, a new service was negotiated to cover only the major optical and mechanical repairs. The cost of this reduced contract was €135,000 or 6.25% of the capital cost of the equipment. Whilst the new contract controlled cost, the supplier would not commit to supplying a loan endoscope whilst one was out for a major repair. The clinical engineer recommended the hospital invest in extra endoscopes to be held as an internal loan stock to be released only when another device had to go off-site for repair. The reduced contract was to be complimented by an in-house maintenance programme established with appropriate resources. The clinical engineer proposed to employ an extra clinical engineering technician who would be dedicated to the new 10 bed unit. This clinical engineer would provide front line support to solve the other problems which are within the competence of the CED to deal with. A provision for spare parts was established as was another to cover the cost of occasional problems which occurred, not covered by the contract and beyond the competency of the in-house clinical engineering team.

TABLE CS1.1C   Cost of a Reduced Contract plus Clinical Engineering Department Support

Summary of support arrangements and cost for 10 endoscopy rooms based on a shared service model	
Number of rooms	10
Cost of devices per room	€216,000
Capital cost of endoscopy devices	€2,160,000
Cost of *major repair only* service contract per annum	€135,000
Service contract as a percentage of capital cost	6.25%
Clinical engineering technician salary	€50,000
Spares	€6,500
Extra call outs not covered by contract	€10,000
Training	€5,000
Total costs of *in-house* services	€71,500
Total cost of *comprehensive shared* service	€206,500
Shared service costs as a percentage of capital cost	9.56%

The cost of technical training for the staff was included in the proposal. The cost of the in-house component of the shared service model was €71,500 per annum.

Overall the shared service model reduces the annual cost of ownership from €300,000 if provided exclusively by the OEM to €206,500 if provided as a shared service, a reduction from 14.6% of capital value to 9.6% (Table CS1.1C).

ADDING VALUE

The solution proposed by the clinical engineer increases value by reducing the cost of the service provision. It also brings added benefits. The existence of a dedicated clinical engineer on site in the unit means that the 60% of faults that can be dealt with in-house are done promptly. Minor issues such as blown bulbs or user error can be resolved within minutes rather than waiting for the supplier representative to attend, and this in turn helped to maintain the unit's capacity to keep a busy service going and reduces staff stress.

$$\text{Benefits} : \text{Cost} \quad \therefore \quad \text{Value}$$

The CEO had the capital to invest in the expansion of the service, but was concerned that the associated operation expenditure was at risk of spiralling out of control, particularly as the hospital became responsible for providing the expanded service to the community. The solution proposed gave the CEO confidence that the assets were being actively managed, and the risk was being controlled. This gave her and the Board the reassurance needed to invest in the expanded service.

SYSTEMS APPROACH

In developing the solution, the clinical engineer did not look at the equipment in isolation. Rather he focused on how the equipment and existing support arrangements assisted the clinical staff in the unit to deliver the care programme to patients. By mapping this carefully, he identified that the OEM's practice of loaning an endoscope was important to the clinicians and central to keeping the unit going, and it was this aspect of the contract that was adding the most value for the clinicians. The clinical engineer rightly identified that by increasing the hospitals capital expenditure to have spare endoscopes in the asset base, the same tolerance of failures could be achieved, for a lower operational support cost.

PATIENT CENTRED

The solution proposed allowed the CEO to expand the facility which brought significant benefit for the community served by the hospital, as waiting times for procedures reduced significantly. The existence of the spare endoscopes and the clinical engineer on site in the unit meant that the delays to procedures associated with minor faults and user errors were eliminated. So patients were never left waiting for their planned procedure.

SUMMARY

The voice of the consultant gastroenterologist:

> "We were initially sceptical about the proposed new way of supporting the endoscopes as we had years of excellent support by [the company] who always lent us equipment when we needed it. Although, to be honest, we had no idea how expensive the service contract was until the business case was prepared for the CEO. Two years into the new arrangements we really see the benefit of the new arrangement. Daily, we are greatly reassured to have our own clinical engineer based in our unit. Always nearby and ready to sort issues as they arise, and we get the outpatient lists done on schedule, leaving more time to deal

with inpatient lists that in the past sometimes had to get deferred if we had 'scopes or stacks down. We also appreciate many other technical support they give us, helping with editing video for research presentations, and now helping us on a project to implement an image management system for the endoscopy unit next year."

SELF-DIRECTED LEARNING

In this case study, the clinical engineer calculated the cost of two alternative equipment support plans and expressed them as a percentage of the capital cost of the assets supported. This is a useful approach where you are planning future facilities and comparing possible support costs.

1. Are there any elements missing from the analysis?
2. If so, what difference might these make?
3. As an exercise, try and calculate the cost of the equipment support plan for the GI flexible endoscopy equipment in your own facility, and express it as a percentage of the capital cost of the equipment supported. (If you do not have an endoscopy unit in your facility, you can always pick another group of specialist devices.)

CASE STUDY CS1.2: THE CLINICAL ENGINEER AS AN APPLICATION SUPPORT EXPERT IN THE ICU

Section Links: Chapter 1, Section 1.5.2

ABSTRACT

When called to an apparent machine fault, the clinical engineer quickly identifies that the machine is working correctly. She uses her knowledge of healthcare technology, basic physics principle and life sciences to help identify that the apparent machine malfunction is in fact due to an interaction between the technology and a particular patient complication.

Keywords: Application support; Multidisciplinary; ICU ventilation; Clinical engineering ethics

NARRATIVE

In an ICU, an intensivist was ventilating a sedated patient using a pressure-controlled ventilation mode. The ventilator started behaving strangely and the graphical display of the patient's airway was atypical. The intensivist was preparing to change the ventilator as it was behaving strangely. However, he called the clinical engineer to get her input as to what could be causing the problem, thinking that it might be something the clinical engineer could quickly fix without having to go to the trouble and risk of changing the ventilator.

When the clinical engineer reviewed the ventilator, all looked to be in order. The clinical engineer discussed the issue at the bedside with the intensivist. Together they compared the expected normal flow, pressure and volume waveforms (Figure CS1.2A) to those seen on the ventilator screen (Figure CS1.2B).

They noted that the abnormal waveforms indicated that airway reached the prescribed pressure quicker than expected and the flow then decreased. This suggested to the clinical engineer that the available volume into which the gas was flowing was lower than that expected for the patient's airway and lungs. Shortly after reaching this peak, the pressure dropped suggesting that the gas in the volume suddenly equalized into another space. This pattern repeated a number of times during the inspiration phase. The clinical engineer confirmed that whilst the waveforms were unusual they did indicate that the control systems within the ventilator appeared to be

Normal flow, pressure and volume waveforms seen when ventilating an adult patient using a pressure controlled ventilation mode.

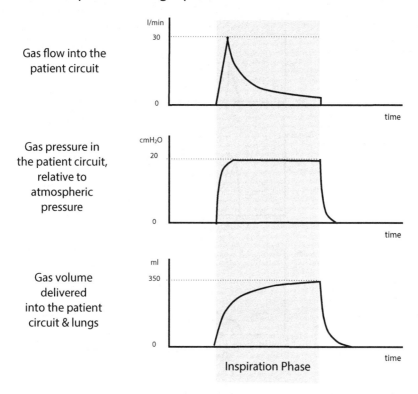

Gas flow into the patient circuit

Gas pressure in the patient circuit, relative to atmospheric pressure

Gas volume delivered into the patient circuit & lungs

Inspiration Phase

FIGURE CS1.2A Normal flow, pressure and volume waveforms seen when ventilating an adult patient using a pressure controlled ventilation mode.

working. She suggested that something might be blocking an area of upper airway or lung and when the pressure increased, the blockage temporarily opened, until the pressure falls at which time the blockage formed again. The intensivist agreed and suggested the blockage could be a sticky secretion in the upper bronchus of one of the patient's lungs. Rather than change the ventilator, the patient airways were suctioned and as expected there was plug of phlegm in the upper airway. Following the suction procedure, the ventilation waveforms were as expected.

Whilst the clinical engineer's role involved managing the equipment in ICU and theatres, she and her colleagues took a holistic view of HTM. Consequently, they not only managed the equipment support plans for the various equipment but were actively involved in teaching and training. This meant that they were familiar with how the equipment performed in the clinical setting. The clinical engineer in this scenario had a lot of experience supporting ventilation in practice and ran training courses for ICU nurses in the use of ventilation. Having an understanding of the physics principles that underpin pressure-controlled ventilation allowed her to hypothesize what might cause the device to behave as it did.

ADDING VALUE

When the apparent equipment failure occurred, the clinical engineer was able to quickly confirm that the equipment was working, however, clearly not as expected. Once this was established, it was the clinical engineer's ability to interpret the waveforms based on an understanding of

Abnormal flow, pressure and volume waveforms seen when ventilating an adult patient using a pressure controlled ventilation mode.

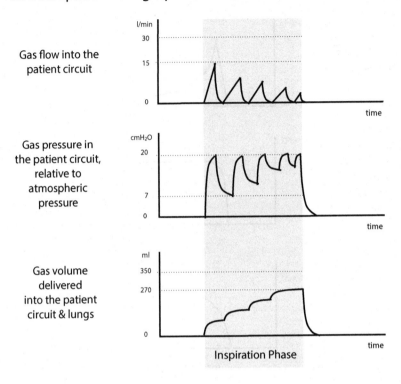

FIGURE CS1.2B   Abnormal flow, pressure and volume waveforms seen when ventilating an adult patient using a pressure controlled ventilation mode.

the physics principles and the anatomy of the airway that led to the suggestion that the problem might be an upper airway blockage. So her expertise and approach to problem-solving drawing on her knowledge of both the physical and life sciences allowed her to bring a different perspective to analyzing the abnormal situation.

Benefits : Cost ∴ Value

The benefit of this was to assist the clinicians in focusing in on the potential problem quickly, and without the delay and risks that would be associated with changing the ventilator. There was no increase in cost to the hospital as the clinical engineer was employed as part of a wider department providing equipment management. Rather, this sort of internal consultancy around physical principles of the operation of devices and the ability to critically review data is demonstrably an added benefit.

PATIENT CENTRED

The approach of the clinical engineer in this scenario was to act to ensure that the patient received appropriate ventilation. Had the engineer been technology focused, she might have walked away having confirmed the ventilator was working, feeling that her responsibility was discharged. The intensivist might then have wasted time swapping out the ventilator before being forced to consider other causes of the anomaly. However, the willingness of the clinical engineer to go further and facilitate an analysis that considered how the patient's condition might contribute to the anomaly helped resolve the situation quickly and to the patients benefit.

CULTURE AND ETHICS

Here the clinical engineer analyzed a problem and offered an opinion to the clinician as to the cause, which assisted in an early resolution, whilst respecting that the final responsibility for the patient and their care rested with the clinician. The culture supported a multidisciplinary approach to patient care, including the contribution from the clinical engineer. Such an approach to the working of the ICU is to be commended and is valuable but does not arise without strong leadership and willingness for professionals from different backgrounds to work across traditional boundaries. It also requires the different professionals to know where their skills and responsibilities lie and their limits.

SUMMARY

The clinical engineer's voice:

> "A few years ago I was asked to develop a training course for the nurses on how to use the ICU ventilators. Initially I taught the nurses how to set up and adjust the ventilators in response to the intensivist's instructions. Over the years we kept tweaking the course to make it more reflective of the actual problems that arose in the unit. This required my colleagues in clinical engineering and I to spend time on the ICU floor with the nurses seeing what the real challenges were for them. We also had to head back to the library and learn more about anatomy and physiology, and the nurses were great teachers. Out of all of this activity we learnt a lot about how the ventilators perform when connected to real patients. It was this knowledge and familiarity with the application of the ventilators that gives us the confidence to contribute to problem solving such as that described here. To be honest, the machines in ICU are very reliable now, but much more complicated for the nurses, so we see our role changing over time to be more about supporting the use of devices than just maintenance."

SELF-DIRECTED LEARNING

1. Using the clinical engineering codes of ethics referenced in the chapter text as a guide, write short notes on whether you consider the clinical engineer acted ethically in this case study.
2. Imagine this scenario happened differently. Instead of an intensivist being present, there was an ICU nurse in attendance. Also imagine that when the clinical engineer suggested to the ICU nurse that it could be an airway blockage, the nurse insisted it must be a machine fault and insisted the ventilator should be changed. What should the clinical engineer do in such circumstances and how do the code of ethics inform the decision-making in this situation?

CASE STUDY CS1.3: ETHICS AND THE CLINICAL ENGINEER

Section Links: Chapter 1, Section 1.7.3

ABSTRACT

Clinical engineers will encounter conflicts of interests and ethical dilemmas as they practice their profession. A scenario is described where a clinical engineer was called upon to advise on the procurement of novel medical equipment, requiring application of codes of professional conduct.

Clinical engineers must know the relevant codes and apply their professional knowledge and skills to enhance healthcare, treating patients, carers and colleagues with utmost respect.

Keywords: Ethics; Codes of practice; Clinical engineer

NARRATIVE

The two haemodynamic and cardiac electrophysiological cardiac catheterisation systems required replacement. The equipment underpinned the very busy case load of coronary artery investigations, with a growing list of electrophysiological investigations including those supporting cardiac ablation therapies. A weekly session was allocated to supporting the implanted cardiac defibrillator service.

These clinical services dictated the clinical and technical specifications. The catheterization systems must also be able to integrate with the x-ray facilities in the laboratory.

As the specifications were being finalized, pressure to add cardiac electrophysiological mapping grew. A respected senior cardiologist was interested in this emerging technology and gathered support from the hospital's Board, attracted by its novelty and the accolade the hospital would gain from it being the first to install it.

The Clinical Engineering Department (CED) head was told to add this to the specification, with expressed misgiving brushed aside; yes, there is only one supplier and they will fly you and a member of your staff to the hospital on another continent where the equipment was being developed. The CED was concerned that cardiology's clinical strategy did not include this application; literature research revealed that its clinical benefits were yet unproven, with clinical trials still ongoing. Regulatory authorities had approved use of the system. The hospital was keen to increase its Research and Development (R&D) activity, and the R&D lead had been encouraged to support the proposal. However, R&D had not been informed about the status of the early-stage clinical trials. Clinical engineering gathered unexpected support from the Head of Cardiology who had not been informed of the proposal and was concerned about the ability of the service to sustain this development given its current heavy and growing routine case load. Furthermore, the cardiology business manager had no funding to pay for the consumable and maintenance costs associated with the mapping system. A further concern was whether any of the other cardiologists would be willing to learn to operate it.

The suggested supplier learnt of clinical engineering's misgivings offering to fly two members to the hospital pioneering the system, with the added inducement of attending and presenting at a cardiology conference being held in the city. Ethical concerns about accepting the paid visit were met by a supplier who had obviously done some homework: "Your hospital Procurement department has approved your acceptance of the offer, subject to inclusion in the hospital's hospitality register".

The benefits of the proposal were clear: a first for the hospital nationally, research potential for the senior cardiologist with possible attraction of future cardiologists, and a 75% discount offer for the mapping component. Missing were clear proven benefits for patients. Costs were a problem: its operating costs were high, with no allocated funding; it would consume considerable cardiac catheterization laboratory time, conservatively estimated as one to two sessions per week – and it was not clear how this could be accommodated except by costly overtime.

The CED Head was told to present an analysis to the Board. Professional ethics requires clinical engineers to apply technology to improve the efficacy of healthcare; the mapping system came with promises of improved care, but the literature showed the benefits unproven. Clinical engineers must manage resources efficiently and effectively: the lack of approved consumable and maintenance costs, questions about the ability of cardiology to sustain the cardiac mapping service and lack of enthusiastic support from the Head of Cardiology argued against this as an effective allocation of resources. But crucially, there was no evidence that this development would enhance patient care; it was novel and experimental, but the proposal had not been rigorously tested through the hospital's R&D ethics system. The Head of CED presented the findings to the Board, recommending rejection of the mapping system; the Board, reluctantly, agreed.

Ethical considerations will frequently arise, often during the procurement process, from assessment of need to the evaluation and selection processes. These must be objective, faithfully and rigorously assessing the strengths and weaknesses of prospective equipment.

The procurement process is not the only area where clinical engineers face ethical considerations. Clinical engineers must not breach the confidentiality of patient information. They must never access healthcare records to discover information about anyone, including family or friends, except where required by their duties for the benefit of patient care. Seeing a friend attending hospital also constitutes obtaining confidential information which clinical engineers must not divulge. The friend may, for example, be attending a cardiac clinic, but might not want spouse or others to know.

Concerns of breaching impartiality arise when suppliers offer gifts to clinical engineers who must be aware of their organization's policy in regard to receiving gifts. It is not uncommon for gifts of calendars, diaries, pens, biscuits and chocolates to be given. Such gifts of a trivial nature may be acceptable within the organization's rules, but clinical engineers should guard against indirectly favouring the suppliers offering such gifts.

Maintenance records on medical equipment are legal documents that may be requisitioned during medical liability and other investigations. Clinical engineers must treat these records with respect and be diligent and honest when recording activities. Falsification of records may be a criminal offence, but is always an ethical offence. Records must be factual, avoiding subjective comments, particularly about members of staff, patients or equipment.

Clinical engineers should know the ethical policies of their healthcare organization and their professional societies. Their guiding principle should be to apply their clinical engineering knowledge and skills to promote healthcare and the well-being of all, including patients, carers and fellow professionals.

SUMMARY

Clinical engineers are bound by codes of ethics that can guide their professional conduct ensuring that it is directed to supporting medical equipment for the care of patients.

SELF-DIRECTED LEARNING

1. Describe any ethical dilemmas have you encountered in your professional lives.
2. Do you know the professional codes of conduct applicable to your healthcare organization?
3. A doctor wants to purchase novel medical equipment. How would you investigate the suitability of the equipment in your healthcare organization?
4. How would you respond if offered expenses to present your work at a conference by the supplier of the equipment that you will describe? Think about different scenarios: equipment already purchased and in use and equipment under consideration.

CHAPTER 2

Taking a Systems Engineering Approach

CONTENTS

2.1 INTRODUCTION

'Systems engineering' is a structured method of solving system problems or improving system performance. The systems can be physical, organizational, biological, social and economic, etc. Systems engineering views all systems and their properties holistically, each composed of identifiable constituent elements, sometimes referred to as component parts. It involves compartmentalizing complex systems, be they equipment, structures or processes, into manageable units or elements in a way that includes the relationships between the elements that together constitutes the whole. Once the system is analyzed, an improved system is synthesized by identifying changes to the elements or the system or how they interconnect. The purpose of proposing this new system is to improve the system as a whole. In improving the system, the focus remains on the quality of outputs of the system. The formal consensus definition of a system by the International Council on Systems Engineering (INCOSE) includes the three concepts of elements, their interrelationships and the objective and the output of the system.

> "A system is a construct or collection of different elements that together produce results not obtainable by the elements alone. The elements, or parts, can include people, hardware, software, facilities, policies, and documents; that is, all things required to produce systems-level results. The results include system level qualities, properties, characteristics, functions, behaviour and performance. The value added by the system as a whole, beyond that contributed independently by the parts, is primarily created by the relationship amongst the parts; that is, how they are interconnected."
>
> INCOSE (2016)

INCOSE goes on to define systems engineering as 'an engineering discipline whose responsibility is creating and executing an interdisciplinary process to ensure that the customer and stakeholder's needs are satisfied in a high quality, trustworthy, cost efficient and schedule compliant manner throughout a system's entire life cycle'. Some of the words and phrases in this definition are worth highlighting: 'interdisciplinary', particularly relevant for multidisciplinary healthcare processes; 'customer and stakeholder's needs', the healthcare system's focus on the needs of patient and carer; 'cost efficient', the need for healthcare to be affordable as well as effective; and 'throughout a system's entire life cycle', the high quality must be consistent, sustainable and lasting.

The medical equipment that engineers create and support are used by healthcare professionals and their patients and play a major part in the care experience. If a technology is technically operating correctly, but is not being used correctly, perhaps in such a way to lead to less than optimal outcomes or perhaps to be unsafe, should the engineer not make it their business to contribute to rectifying the situation? Engineering is more than simply about technology (Lawlor 2013): 'Essentially, it is important to understand that engineers don't just work with machines, designs or circuit boards, and engineering doesn't only require a good understanding of science and mathematics. Engineering needs to be understood in the context of its role in society, and your role as an engineer has to be understood in the context of your work within a company, and ultimately within society'.

This and the definition of engineering that follows prompts us to consider whether clinical engineers have a role in improving healthcare outside of the technical support they are associated with. 'Engineering is the discipline of using scientific and technical knowledge to imagine, design, create, make, operate, maintain and dismantle complex devices, machines, structures, systems and processes that support human endeavour' (Blockley 2012). The definition helpfully reminds us that we should apply our skills to 'systems and processes'. Given that healthcare organizations are 'structures, systems and processes that support human endeavour', we propose that clinical engineers can and should contribute to analyzing and improving the healthcare organizations and wider systems of which they are a part. That means that clinical engineers can contribute to supporting and advancing care through participation in organizational design and management beyond their traditional equipment management role. The definition also reminds us that the discipline of engineering has a purpose, an objective, namely 'to support human endeavour'.

Clinical engineers are part of the overall healthcare system. In this chapter we explore how clinical engineers, as one element within that healthcare system, can add value to the system as a whole. Their role in equipment management is well accepted but rarely considered are their contributions to the wider healthcare system. We suggest that beyond the traditional equipment management role, clinical engineers can add value by analyzing, developing and implementing holistic healthcare technology management (HTM) systems. Furthermore by contributing to interdisciplinary projects focused on improving other aspects of the healthcare system, clinical engineers can add value by extending themselves to the application of systems engineering beyond the equipment management role. We will explore this in greater detail in Chapter 7.

Healthcare systems are inherently complex combinations of people, resources, equipment, aspirations and emotions. The whole, the combination of these elements, should work in harmony for the common goal of improving the health of the people it serves. Separating and dividing such complex systems into their constituent elements can assist in their analysis and subsequent improvement. The complexities of healthcare systems can be unravelled, identifying the characteristics of each constituent element and their inter-actions. In doing so, the weakness and strengths of the elements and interactions can be revealed. The combination of the elements and their working together to achieve the common goal can be better understood, with shortcomings addressed.

The President's Council of Advisors on Science and Technology (PCAST) in the United States, in their report entitled 'Better health care and lower costs: Accelerating improvement through systems engineering', proposed that systems engineering principles are applied to improve healthcare systems (PCAST 2014). The report 'identifies a comprehensive set of actions for enhancing health care… through greater use of systems-engineering principles. Systems engineering, widely used in manufacturing and aviation, is an interdisciplinary approach to analyze, design, manage, and measure a complex system in order to improve its efficiency, reli-ability, productivity, quality, and safety. It has often produced dramatically positive results in the small number of healthcare organizations that have incorporated it into their processes'.

The PCAST 2014 report built on earlier joint work by two U.S. organizations, the National Academy of Engineering and the Institute of Medicine who, in 2005, published a report con-cluding that the U.S. healthcare industry had neglected engineering strategies and technol-ogies that have improved quality, productivity and performance in many other industries (IoM 2005). The result, of what the authors termed 'collective inattention', contributed to nearly 100,000 preventable deaths annually, with outdated procedures and an inefficient cost-wasteful system whose costs were rising at thrice the rate of inflation. It called for healthcare professionals and engineers to combine their efforts to find solutions.

This chapter recommends that clinical engineers adopt, where appropriate, a wider per-spective, applying their understanding of the systems engineering not only to a holistic equipment management methodology but also to support the wider improvement of the healthcare organization they serve. Systems engineering applied to healthcare organiza-tions enables greater understanding of the complexities of healthcare systems, of the over-all system's architecture as well as of the constituent components, the interrelationship between the components and how they work together, or not, to ensure that the strategic aims are achieved. It also provides a methodology for improving these systems and verify-ing that improvement has been achieved. Consequently, we propose that clinical engineers must not only focus on physical medical equipment and technical issues, important as they are, but expand their horizons to consider engineering solutions that aim to maximize value within healthcare, delivering optimal patient outcomes. The focus of this chapter therefore is to propose a 'philosophy of approach', making use of the systems view, which will inform many of the topics explored in the rest of this book.

The following are three general points to remember whilst looking at the details, the mechanics of systems engineering and its applications to healthcare:

- First, systems have objectives: for healthcare systems those are to enhance the global health of the population it serves and to provide high-quality care for individual patients.

- Second, systems must have performance measures. This implies that, as noted ear-lier, we include metrics and measurements in our systems approach. Chapter 1 has described the Value measure, the benefit delivered relative to the cost of achieving it. Introductory thoughts on measurements will be provided in this chapter, whilst Chapters 5 and 6 will expand on healthcare technology management (HTM) performance measures.

- Third, the clinical engineer, with their methodical working, instinctive tendency to measure and focus on objectives can apply this structured approach beyond the confines of HTM to the wider healthcare system.

2.2 INTRODUCTION TO SYSTEMS ENGINEERING

Let's start by restating the INCOSE definition of systems engineering as 'an engineering discipline whose responsibility is creating and executing an interdisciplinary process to ensure that the customer and stakeholder's needs are satisfied in a high quality, trustworthy, cost efficient and schedule compliant manner throughout a system's entire life cycle' (INCOSE 2016).

Viewing objects, organizations and operating processes from a systems engineering approach has many advantages. It offers a structured method of analyzing each in detail, of examining the interactions between their constituent elements, whilst maintaining focused on the system's overall aim, its purpose. The approach helps to ensure full understanding of each element's characteristics, strengths and weaknesses, and how the elements interact. It recognizes that, important as each element is, it is the working together of the elements, their collective interactions, that achieves the system's objective. The systems engineering approach, when applied to some systems, may reveal that optimizing each and every element that makes up the system may not improve the system as a whole. The goal should always be to optimize the system as a whole, even if that means that individual elements are less optimized.

So when taking a systems engineering approach at all times, we must remember the importance of the individual elements; the ultimate output of the system results from the relationships between the elements and from how the elements are interconnected, and the output is greater than that which can be obtained from the sum of the individual elements. That is, the whole is greater than the sum of its parts. Each element is important, the relationship between elements is important, and the system's output, its success, will depend on how the constructor, the conductor, brings all the elements together towards the desired objective. Consider, for example, the output from an orchestra with many different musical instruments (elements) in which all the musicians play at their own tempo, according to their individual desires. In contrast consider the marshalling together of all the individual elements, the individual talents and strengths, by a world-class conductor, subtly merging the diverse elements to create beautiful sounds. The elements by themselves may be exquisite, but without direction, without the harmony that comes from recognizing the strengths of mutually constructive working, the overall results can be a disaster. As the INCOSE definition puts it, the 'value... is primarily created by the relationship among the parts'.

We have seen that the elements of a system can include people as well as physical inanimate objects. To facilitate the understanding of the human interactions involved in these types of systems, Checkland and Scholes (1999) drew the distinction between hard and soft systems, the latter involving human interactions that are perhaps difficult to predict and not easy to define. In contrast, hard physical systems consist of elements that are more readily defined, with clearer boundaries and more predictable interactions between elements. In the context of soft systems, it is important to remember the contribution that human factors engineering can bring to solving problems.

As we have noted, systems engineering involves viewing a system as a whole, then dividing it into its constituent parts, analyzing each part separately, considering its relationships with the other parts. This process of analyzing systems so we can understand them is central to systems engineering. Each element has a function, with properties and characteristics that enable it to perform its function. The constituent elements interact with each other. In a simple system, the relationships between the constituent elements may be to pass information or signals from one to the next, typically flowing logically and linearly from an input to an output (Figure 2.1).

In a more complex system the relationships may not be as simple as this, with individual elements having multiple interactions with other elements. The systems analysis approach enables each constituent element and its boundaries and the interactions between elements to be subjected to dedicated analysis, whilst recognizing the role each plays in supporting the whole system to deliver its objective (Figure 2.2).

Each element might be further analyzed and reveal itself to be a system in its own right, a subsystem of the whole. So the systems engineering approach allows for nested analysis of complex systems.

Identifying the manageable component elements of the system and consequently the boundaries between them is a key step in the systems engineering approach. Go too deep

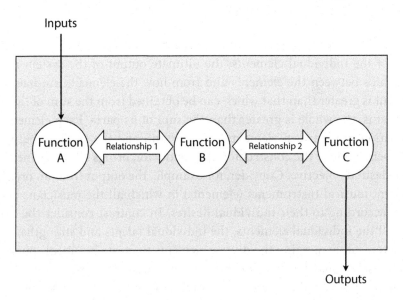

FIGURE 2.1 A simple system. A graphical analysis of the system showing its individual elements and their relationships to one another and the external environment.

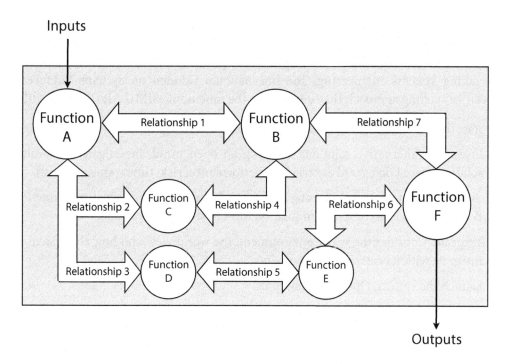

Inputs

Outputs

FIGURE 2.2   A complex system. A graphical representation of the system showing its many elements and the relationships between elements and the external environment.

and you can lose the sense of how the functional elements contribute to the whole. Go too shallow and you may not gain enough insights into the complexity of the system. Deciding what to include as part of each functional element may not always be obvious and will in general vary with the purposes for which the analysis is being undertaken.

Analyzing systems in this way is inherent in many aspects of engineering education and practice whether it is a circuit or software design, analysis of complex engineering systems or equipment fault finding. Each of these examples requires an understanding of how a circuit or block of code works and how each element interconnects with other parts of the circuit or software programme so that the system as a whole does what it should. This can help fully explain the system and how it works – and what prevents it from working, or working more effectively. Central to the systems engineering approach is its emphasis on identifying and clarifying the objective of the system, what the system is meant to do, what is its objective. Clinical engineers will instinctively adopt this type of systems analysis approach when fault finding and investigating problems with medical equipment. They should also adopt it to analyze and improve the methods and processes required to manage medical equipment. This will be explored in detail in Chapters 5 and 6.

Analyzing systems in this way is an important step but systems engineering involves much more. Systems engineering involves creating processes and putting them into action to ensure or improve a system. So how is this done, what are the steps? Intuitively we identify that there is a need to analyze the system in question, to identify its elements and their interconnections, and how the system as a whole delivers its outputs. Then there is a need to imagine and think creatively, generating ideas and ways for improving the system. Having identified potential new systems, these should be tested and critiqued and the best chosen for implementation. Various suggested methodologies have been proposed to assist those undertaking systems engineering. The International Council on Systems Engineering advocates a starting approach they describe by the mnemonic SIMILAR (INCOSE 2016):

S: State the problem. Identify what must be done, including the customer requirements.

I: Investigate alternative solutions. Keeping an open mind, investigate and evaluate solutions based on agreed criteria (e.g. performance, risk, time frame and cost).

M: Model the system. Use various techniques including tabletop discussions, functional flow diagrams and computer simulation tools to study the system.

I: Integrate. Consider the wider environment, the world view and how the system will integrate with its external environment.

L: Launch the system. Operate the system.

A: Assess performance.

R: Re-evaluate.

This approach is summarized in Figure 2.3 which we will return to as we discuss systems engineering in the next sections, focusing on defining the objective, identifying the elements and their interrelationships and then testing models of the system.

Figure 2.3 has echoes in the CATWOE process (Checkland and Scholes 1999) that evolved from consideration of soft systems involving people, though CATWOE can also be applied to hard systems. CATWOE is a mnemonic and the process begins by focusing on the system's beneficiary, the Client or Customer:

C: Customer – For healthcare systems, the ultimate beneficiary is the patient, but we can envisage system analysis of processes, such as developing healthcare technology management systems, where we will want the beneficiary and hence the direct purpose of the system to be an improved method of maintaining medical equipment.

A: Actors – These are the elements which could be nurses and clinicians or support staff or could be inanimate physical objects.

T: Transformation – What change, what purpose is the system designed to achieve. How does the system achieve the change, the transformation?

W: World view – What effect does the system have on the external environment and vice versa? How does the system perceive the external environment and how does the external environment perceive the system?

O: Owners – Who are the responsible owners and decision makers? Who can influence or resist change and how can their views be managed and harnessed?

E: Environmental constraints – What are the environmental constraints influencing the system?

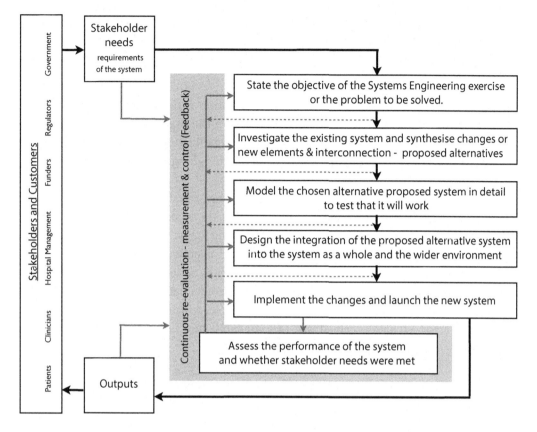

FIGURE 2.3   Using systems engineering to improve a healthcare process.

Patricia Trbovich (2014) emphasized the need for a systems approach to problem-solving when she wrote that problems do not occur in isolation, but in relationships to other processes. She noted, however, that all too often, the problems are studied in isolation, with the lack of a "systematic, holistic approach… leading to fragmented solutions that do not address the problem as intended and introduce new, unintended issues" (Trbovich 2014). She went on to note that this sometimes resulted in the cure being worse than the disease. To avoid this, she summarized five aspects required for a systems approach in healthcare:

- First, the approach must be holistic, recognizing interdependencies between different healthcare processes.

- Second, the approach will identify and clarify the interdependencies, showing how processes in one area can affect processes in other areas, possibly with intended and unintended consequences. To minimize the unintended consequences and strengthen the holistic approach, multidisciplinary cooperation is required.

- Third, leadership and 'Great Systems Thinkers' must be identified and developed – echoing the PCAST (2014) call.

- Fourth, identify areas where improvements can be effective, areas described as 'leverage points' where action taken can be most effective in resolving problems. This requires a proactive analysis of an organization and its weaknesses, rather than waiting for adverse incidents to highlight flaws.

- Fifth, and importantly, system working in isolation is not a systems approach; systems thinking must be integral and ingrained in the organization. It demands that there is no longer silo thinking, but a holistic, organization-wide approach that does not consider only the immediate future, but develops long-term sustainable solutions.

Recognizing the need for an improved understanding of systems engineering, Kopach-Konrad et al. (2007) outlined the methodology that the systems engineer can apply to improve healthcare. Noting that "…systems engineering focuses on the design, control, and orchestration of system activities to meet performance objectives…", the authors recognize that systems consist of entities (elements) performing functions whose interactions lead to a "global system behaviour". The entities include people, patients and clinical staff. The healthcare system is not static, but changes, for example, with the movement of patients.

Kopach-Konrad et al. described six fundamental steps that the systems engineer undertakes:

- Defining the system's purpose, its objective and its scope;

- Collecting data about the system;

- Modelling the system;

- Simulating the model to learn about the system;

- Using the simulation results to improve the system;

- Implementing and evaluating the improvement plans.

Various engineering tools can support the systems engineer in the system analysis and synthesis. The authors refer the reader to the INCOSE website (INCOSE 2016) for further information.

Based on the considerations mentioned earlier, including the SIMILAR approach described by INCOSE, the systems engineering methodology can be summarized in Figure 2.4.

Systems engineering does not view systems as set in concrete, but subject to continuing questioning and analysis. Techniques such as the 'Plan–Do–Check–Act' (PDCA) cycle, which we will discuss later (Figure 2.10), can be useful here. For example detailed analysis of a healthcare system (Figure 2.3) may lead to a realization that aspects of the needs of the stakeholders require to be better formulated or modified.

Appendix A of the PCAST (2014) report provides a useful summary of systems engineering with answers to frequently asked questions such as 'What is it?' and 'How are systems formed?'

Define the system	What is its purpose, its function?
	Specify the scope, clarifying the system boundaries.

Identify the system elements and interactions	Existing systems	Understand how it operates, strengths and weaknesses Collect data and information,identifying elements and interactions.	New systems	How will the system operate? What elements and interactions between elements are required?

Determine the system modelling tools	Determine tools for assessing the system: brain-storming; computational models

Use modelling tools to understand the existing system	How sensitive is the system to changes, internal or external? How robust is the system? What are the strengths? What are the weaknesses and how can they be turned into strengths?

Imagine and propose alternative solutions	What alternative ways can achieve the objectives? Develop criteria for evaluating alternative solutions and selecting the preferred solution.

Use modelling tools to understand the proposed change	How will the proposed change improve the system? What are the strengths of each? What are the weaknesses of each and how can they be turned into strengths?

Optimise the proposed solution	Iteratively propose changes to the system to optimize its performance

Select the preferred solution and implement	Agree the preferred solution Perform a test of the change to verify it works Once confirmed implement the change across the system

Measure the effect of the change	Measure the system performance to verify the improved system works

FIGURE 2.4   Systems engineering methodology in healthcare.

2.3 SYSTEMS ENGINEERING METHODOLOGY

As we discuss systems engineering and its methodology, we should remind ourselves of an important word from the INCOSE definition, repeated in the PCAST (2014) report: systems engineering is 'interdisciplinary'. It is holistic, the antithesis of the silo approach that limits itself to the needs and aspirations of any particular vested interest within the system. Clinical engineers, whose very title and existence brings together two distinct professional domains, will welcome this synergy.

Various systems engineering methodologies have been developed. In general the methods start by clarifying the problem, identifying the desired objective and recognizing who are the beneficiaries (c.f. the Chapter 1, Section 1.4 discussions on the Value of healthcare to its beneficiaries, its stakeholders). Having agreed the problem and the desired objective(s), the task moves on to identify the constituent parts and the elements of the system and then to clarify the interrelationships between the elements that realize its objective(s).

Describing the system with its constituent elements and their relationships lays the groundwork for improving and optimizing it. Whether we are analyzing an existing system to improve it or are developing a new system, we will want to improve the system. We will probably want to investigate alternative structures and interrelationships, and we will need to test these alternatives using modelling and simulation tools including the Plan–Do–Check–Act (PDCA) quality improvement methodology (to be discussed in more detail in Section 2.3.5).

The systems analysis must recognize that any system exists within an environment, with systems engineers stressing the need to remember these external boundaries, to keep open the 'world view'. (In theory, some systems are described as 'closed' systems where no interrelations external to the system need to be considered. In general, in healthcare we deal with 'open' systems which do exist within an external environment.)

2.3.1 Step 1: Defining the System's Objective

The systems approach starts by clarifying the desired objective of the system and recognizing who are the beneficiaries ('S' in SIMILAR, Figure 2.3). The beneficiaries of healthcare services include the patients and their carers, their families and friends and healthcare organizations and their funders. These are beneficiaries of HTM as are the clinicians who use the medical equipment.

Clarification of the objective requires a holistic interdisciplinary approach, recognizing the interdependencies of healthcare processes (Trbovich 2014) and the complexity of the healthcare system in which teamwork is essential. In a case study (Case Study CS2.1), we explore this teamwork by the different professional groups whose contributions were required to improve the delivery, by syringe pump, of medication for palliative care. We need to recognize the legitimate objectives of each of the professional groups (the elements in our systems engineering approach to analyzing the system), but also that their individual objectives must be subservient to the overall objective, namely good-quality person-centred palliative care.

Those involved in analyzing the system will need to get together and clarify and agree the overall objective – they will probably do this by forming a Project Team and we will discuss the role of clinical engineers in project teams in Chapter 7. Collective understanding and agreement of the objective is important to avoid unintended consequences (Trbovich 2014). Thus, for example a legitimate objective from a nursing perspective might be a solution that focuses on a safe consistent approach in which only 10 mL syringes are used with the syringe pumps to deliver the palliative care medication. Removing variability increases the confidence of the caregivers with the promise of enhanced quality of care. However, pharmacists will point out that some pharmaceutical combinations require to be dissolved in say 15 mL of saline. The 'unintended consequence' of focusing on the simplicity of only using 10 mL syringes is that patients may require to receive separate lots of medication from two or more syringe pumps, complicating their care and increasing clinical risk.

An example of silo rather than collective holistic equipment planning can arise when a department wishes to procure medical equipment that is reliant on healthcare IT, but where the medical equipment and IT procurement are kept separate. Consider, for example, an ophthalmology department wishing to procure new ophthalmic imaging equipment that will enable the optical images to be available in the doctors' consulting rooms. A silo approach is when clinical engineering focuses exclusively on the ophthalmic equipment, leaving the network interconnectivity to the IT department, compartmentalizing the responsibilities and procurement process. The result is likely to be unsatisfactory and may cost more. The interdisciplinary systems approach requires all to focus on the objective (good-quality images available in the consultant's office), with the common objective helping to remove departmental barriers and impediments that get in the way of the optimum solution.

The CATWOE process (Checkland and Scholes 1999) usefully starts with identifying the customers, the beneficiaries. Simplistically we may feel that the patients are the customers. However, when we examine some systems to improve them, the immediate beneficiaries for whom we are trying to improve the system may not be the patients. Consider for example a Clinical Engineering Department (CED) seeking to improve a hospital's medical equipment replacement planning so that the Department of Finance can plan its financial allocations. Finance requires robust accurate forecasts so that it can develop medium to long term financial plans. Thus for this HTM process, developing financial plans for rolling replacement, the clinical engineers will focus on the Department of Finance as the beneficiaries of the planning, though recognizing the world view with patients the reason for and ultimate beneficiaries of the medical equipment. We will explore some of the details of this in a case study that examines replacement planning (Case Study CS5.6).

As we explore the systems engineering approach, we will use a few diagrams to illustrate each step, applying this to an analysis of the HTM process. The first step is to define the objectives and stakeholders (Figure 2.5).

2.3.2 Step 2: Identifying the System's Constituent Elements

With the objective and the beneficiaries clarified, the systems approach continues by examining the details of the system: what constituent parts does the system require to deliver its output, to meet its objective? As we analyze the system, identifying and analyzing the

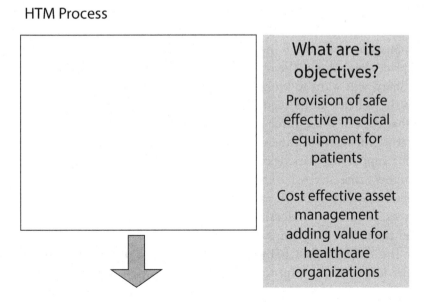

FIGURE 2.5   A systems engineering view of the Healthcare Technology Management process: Step 1 – identifying the objectives.

HTM Process

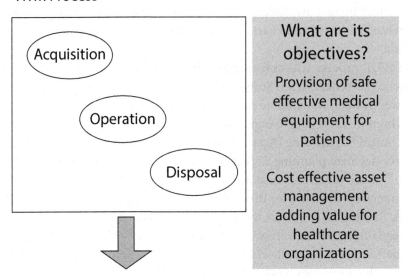

FIGURE 2.6   A systems engineering view of the HTM process: Step 2 – identifying the elements.

component parts, we might find that we want to modify the objective and review the ben-eficiaries. This is legitimate and in keeping with the cyclical 'Plan–Do–Check–Act' method that will be discussed in Section 2.3.5.

Identifying the elements, and how far to break them down, is a question often encoun-tered in systems analysis. The answer is often found by considering why and for whom the analysis is being undertaken. For example, Healthcare Technology Management (HTM) can be considered as a system with three constituent elements, that is the three overarching phases of the management of the technology: acquisition, operation and disposal (Figure 2.6). Let us consider that the analysis is being undertaken by the orga-nization's management Board. It may be sufficient for the Board to know that there are policies in place that achieve the objective of effectively managing the medical equip-ment. The Board will want to know the broad aspects of how the equipment is man-aged, that is the policies for its acquisition, operation and disposal. The Board will seek assurance that each of these three constituent elements (Figure 2.6) is well managed, that the relationships between them are well understood and that the system has clear objectives for supporting patient care. At this Board level, a systems analysis would view each of these three elements as functional entities, each with their own character-istics and interrelationships with each other, but the Board would not, under normal circumstances, require to understand the details of each of the three elements. We will consider the interrelationships in the next section.

However, the leader of the CED will want to dig much deeper into each of the three elements. The leader might want to review the Acquisition process to ensure that the medical equipment best suited to the hospital's clinical needs is acquired, analyzing in depth the acquisition's separate elements: identify need, specify, identify products, evaluate, select and procure (Figure 2.7). Similarly the CED leader in developing the

HTM Process

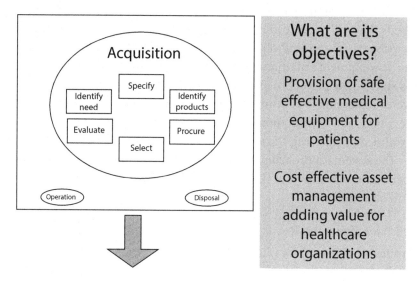

FIGURE 2.7   A systems engineering view of the HTM process: Step 2 – a more in-depth identification of the elements.

HTM Programmes that support the equipment (to be discussed in Chapter 6) will want to delve deeply into what constitutes the Operation phase. The objective might be to ensure that the clinical staff are provided with medical equipment that is comprehensively supported for patient care: the objective and beneficiaries of the systems approach. The systems approach will analyze what is required for comprehensively supporting the equipment, with elements such as commissioning and installation, staff training, planned and breakdown maintenance, supply of consumables and processes to manage adverse events identified. The analysis will delve into each of these elements and their processes, clarifying and analyzing them to ensure they are effective.

Earlier, in Section 2.3.1, we discussed who the 'customer' was when analyzing a medical equipment planning programme; the analysis was being carried out for the Department of Finance who needed the information for forward financial planning. But we did not discuss the constituents that are required for an equipment replacement planning system. Elements that are required will include an accurate inventory of the current medical equipment, funding provision estimates, knowledge of the expected lifespans of medical equipment, the strategic plans of the organization (e.g. entering the home healthcare sector) and the requirements of the individual clinical departments for whom the equipment will be procured. Each of the clinical departments will have their own aspirations and perhaps strategic plans (e.g. increasing endoscopy facilities – see Case Study CS1.1). The system elements may differ in nature, in this case inventory data, budgets, knowledge of lifespans, strategic plans and departmental requirements. The systems approach can handle them all, bringing them together to achieve the intended outcome.

Each element will need to be studied and analyzed on its own. For example, the replacement planning requires an accurate inventory. Does this exist and how is its accuracy maintained? Without the integrity of each element, the overall objective may be compromised. Thus a subgroup of those tackling the equipment replacement planning might focus on the methods that will provide and sustain an accurate inventory. In doing so, this subgroup must recognize the overall objective; if they identify that the inventory can also be the basis for the equipment's maintenance planning and history, they will need to state this and perhaps start a new project to examine this, so that this need does not distract from the objective of replacement planning. This example is given as it brings in the need to consider the 'world view'. The equipment replacement planning is not an end in itself, but is ultimately aimed at ensuring the availability of appropriate functioning equipment for patient care. The systems approach does not exclude the 'world view', but, recognizing the wider perspective, enables each of the processes involved to be examined and optimized for the overall objective.

2.3.3 Step 3: Identifying the Relationships between Constituent Elements

Having identified the Objectives (Figure 2.5) and the Elements (Figures 2.6 and 2.7), we now need to look at how the elements interrelate, in particular how the elements work together to achieve the objectives (Figure 2.8). The elements by themselves will not provide the desired output; it is how the elements work together to make the whole greater than the sum of its parts that is central to the systems approach.

For example the hospital's Board might seek reassurance that the processes required for operating the equipment will be considered during acquisition (Figure 2.6): 'Does your acquisition process take into account the cost of operating the equipment?' the Board might ask. 'Are you considering the disposal requirements when acquiring new large equipment?'

HTM Process

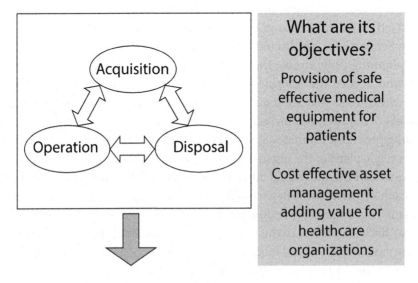

FIGURE 2.8   A systems engineering view of the HTM process: Step 3 – identifying the interrelationships between elements.

At this level, the details of compliance with environmental regulations would not necessarily be required, but the Board would be content that these issues have been addressed.

Similarly, interrelationships will be identified when exploring elements in more depth, for example the acquisition phase (Figure 2.7). The elements are: identify the clinical need; specify (which can in turn be divided into clinical and technical specification); identify products available on the market; evaluate (which can be divided into evaluation criteria, evaluation method and evaluation team); select (who will do this) and; procure. Each of these elements depends on the others. The specification will be driven by the clinical need; there are natural interplays between the evaluation and the specification and between the evaluation and the selection. Systems engineering demands that we take these relationships seriously and carefully ensure that the specification reflects the equipment that is actually needed. The evaluation and in particular its criteria will depend on the specification, but will also directly need to reflect the identification of need. To illustrate the latter relationship, consider the procurement of an arthroscopy system. The surgeons defining the requirements will want to include the handling of the camera head that attaches to the arthroscope. Unless this requirement is built into the evaluation criteria, there is a risk that a product with poor manual handling might be selected, with risks of stress injury to the surgeons' wrists. We briefly explore this in a case study (Case Study CS7.3) that discusses procurement of an arthroscopy system. The synthesis of the different elements will need to recognize the interdependencies, and so, whilst for purposes of explanation, we have divided these into separate steps, in practice the systems engineering approach will consider the three different steps in parallel.

Systems engineering also incorporates a 'world view', not blinkered by the objective of the system under investigation, but recognizing that the system exists within a real world. Thus for example our acquisition system will want to be influenced by the practical experience of its output, the procured equipment. Experience with the equipment in clinical use may reveal deficiencies that were not spotted during the evaluation, possibly because they were not included in the specification or evaluation criteria. This experience should be used to enhance the development of specifications and the evaluation process.

2.3.4 Step 4: Improving the System

Systems analysis is not an end in itself, but it is generally undertaken to improve a process. Systems analysis provides the methodology to enable a multidisciplinary team to critique the existing system and to suggest improvements. Suggestions might optimize specific individual elements or might suggest alternative relationships between elements or the need for different elements. Figure 2.9 describes the process of improving the system. As alternative ideas are suggested, they will need to be tested and evaluated and, if found to be beneficial, incorporated into the whole system. Their functionality within the system will need to be evaluated and measured, with the process perhaps suggesting improvements and redevelopments.

During this process of analysis, systems engineering provides the mechanism for keeping the overall system objective in mind.

Thus, for example, we can continue our discussion on the HTM system and the identification of its constituent elements and interrelationships as shown in Figure 2.8. The hospital's Board might be concerned as to whether new environmental waste regulations have been fully

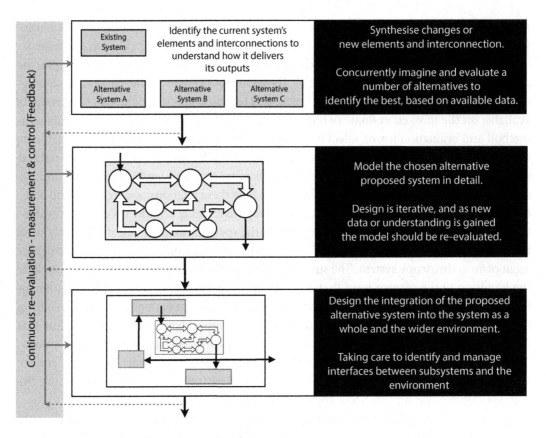

FIGURE 2.9   Systems analysis: improving the system.

incorporated into the HTM system leading to analysis of the Acquisition process and whether it is appropriately incorporating plans for end-of-life disposal. This will lead to analysis of these elements and their interrelationships, with perhaps suggestions for change. Suggestions could include paying a surcharge at the procurement stage to pay for disposal, making the environmental disposal a responsibility of the supplier at the time of procurement or deferring decisions as to how to dispose of the equipment till its end of life. These various suggestions will need to be evaluated, and we discuss the evaluation phase in the next section.

2.3.5 Step 5: Systems Improvement Methods

Proposed improvement suggestions need to be evaluated before implementation, with various techniques available. These will include desktop discussions for sharing and considering the views of the multidisciplinary team, calmly, constructively and objectively examining each view. More formal simulation tools are available including computer programmes that model systems and enable scenarios with different alternative sets of elements and interrelationships to be assessed. From these assessments, using evaluation criteria, a preferred solution will be selected. When synthesizing new products or processes, the team (remember the word 'interdisciplinary' in the definition of systems engineering) will want to assess different alternatives, keeping open minds and testing each alternative. A proposed solution will inevitably include a change to an element or a change to an interrelationship or both.

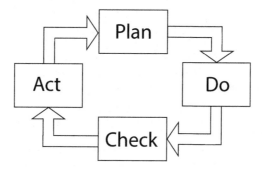

FIGURE 2.10   The 'Plan–Do–Check–Act' cycle.

Once the solution is selected, it needs to be tested. Some measure of the performance of the system elements and how they relate, as well as the output from the system as a whole should be made. Then the change introduced in a controlled way and its effect assessed by looking at how the metrics change. This stage is sometimes referred to as a 'test of change'. The selected solution needs to be implemented, a process in itself that will involve review with, where necessary and appropriate, improvements to the system. The 'Plan–Do–Check–Act' (PDCA) cycle of quality improvement methodology will generally be appropriate here (Figure 2.10).

Once the effectiveness of the solution is confirmed in the test phase, it can be implemented in a wider context using the same 'Plan–Do–Check–Act' approach to ensure that the desired outcome is achieved. Note the 'Plan–Do–Check–Act' cycle forms a loop, suggesting that it can be used in an iterative fashion to continue to improve the performance of the system. In complex systems, it is important to remember to keep an eye on the total system output metric when trying to improve the performance of a number of subsystems as changes made in one area can have an effect in another part of the system.

As a practical example, proposed changes to the Acquisition phase of the HTM system (Figure 2.8) were suggested at the end of Section 2.3.4 in response to new environmental regulations. The various ideas came out of discussion sessions and need to be evaluated. What are the cost implications of paying a surcharge that includes the cost of disposal at the time of procurement? Will suppliers accept the tender requirement of paying for equipment's final disposal – and will they charge for this to cover their risks? What are the disadvantages of awaiting 'end of life' to determine the method of disposal, paying any costs then? Each of these suggestions must be tested, typically guided by some measures such as cost and feasibility. What are the possibilities of selling or donating medical equipment no longer required? Different options for different groups of equipment might be decided. This evaluation phase will be used to determine the changes implemented for this aspect of the HTM system.

2.3.6 Measuring the Performance of Systems

Analysis and improvement of the system will require metrics against which we can assess and evaluate the improvements that the suggested changes provide. If you don't or can't measure the performance of a system, then you will not be able to assess it, nor can you assess the impact of any change. The metrics should be appropriate for the system that is analyzed and will thus vary with the system. The metrics should link to the objectives and beneficiaries of the system being examined, indicating whether the suggested improvements enhance the system fulfilling its objectives for its beneficiaries.

A global analysis of a HTM system (Figures 2.6 and 2.7) would look for assessment metrics that measure the provision of safe and effective medical equipment that caregivers have available at the point of care for supporting patient care. A simple metric for this may not be readily apparent. Records of unavailability of medical equipment at the point of care are not often kept, though, where the unavailability of equipment was associated with an adverse event or near miss, these may be accessible from incident reporting systems, providing trending information over successive years. (A near miss is an incident that would have occurred but for some intervention such as by caregivers.) The records kept by clinical engineers in the Medical Equipment Management System (MEMS) database might record service requests indicating the unavailability of equipment. Similarly annual trends of adverse event reports involving medical equipment may provide some insights into the performance of the HTM system. Survey questionnaires of clinicians may be a better way of obtaining information about the services provided by the HTM system. The questionnaires will need to be carefully constructed to minimize subjective assessments. Suggestions on questionnaire design are given in Creative Research Systems (2014) and StatPac (2014).

The HTM system could also be assessed against the cost-effectiveness of its services. Global costs of the HTM services as a percentage of the replacement cost of all the medical equipment are unlikely to yield informative metrics; the cost of maintaining medical equipment can vary with the type of equipment and the global average can mask differences in individual performances. Consequently it is more instructive to compare the cost of maintaining specific functional types, for example critical care ventilators or general purpose infusion pumps. Measuring the performance of HTM systems is crucial to assessing continuing improvement initiatives. Measurements will need to be made of the different HTM activities, covering the processes involved in the acquisition, operational support and disposal of the medical equipment. These measurements and the key performance indicator metrics that derive from them underpin benchmarking with other healthcare organizations which can be effective means of assessing HTM services. These will be discussed in more detail in Chapters 5 and 6.

Systems engineering, as we have seen with the PCAST (2014) report, can also be applied to enhance healthcare systems. Various measures for assessing healthcare have been proposed. For example, The Institute of Healthcare Improvement (IHI), who have advocated a Triple Aim of better population health, better personal health and affordability of healthcare, has developed measurement systems to assess progress in each of the three Triple Aims (Stiefel and Nolan 2012):

- Improving population health:

  o Health Outcomes: mortality, health and functional status, health life expectancy;

  o Disease Burden: incidence of disease types;

  o Behavioural factors: smoking, diet and eating habits and physical activity;

  o Physiological factors: blood pressure and obesity.

- Improving personal health and experiences of care:

  o Patient surveys and key indicators of care.

- Affordability of care:

  o Total cost per member of the population, per hospital department.

2.3.7 Discussion

The systems engineering approach has been discussed in Sections 2.3.1 through 2.3.5 using the example of the HTM system. We have shown how the approach starts with identification of the objective and the stakeholders and goes on to identify the system elements and their interrelationships. The analysis, the decomposition of the HTM system into its constituent elements, is carried out to better understand its underlying process with the purpose of improvement. We concluded the example by briefly looking at the need to assess whether the HTM system adequately complies with environmental waste regulations and how suggestions for change need to be evaluated.

In the course of this first part of Chapter 2, we have proposed various methodologies to assist those undertaking systems engineering analysis. There are many more techniques available and the reader should find an approach that works for them. But the reader should remember the key aspects of systems engineering: first, clarifying the objective, the purpose of the system and its customers; second, identifying the constituent elements that make up the system; and third, the interrelationships between the elements that generate the output.

2.4 SYSTEMS ENGINEERING AND HEALTHCARE TECHNOLOGY MANAGEMENT (HTM)

In this section we will explore how systems engineering tools can help the clinical engineer deliver healthcare technology management.

2.4.1 Systems Approach When Managing an Item of Medical Equipment

Medical equipment can in itself be analyzed as a system, a combination of different elements that are interconnected and interrelated to function as a device. For example, we can consider a vital signs monitor as the combination of several different elements – a system: the transducers and patient cables that acquire the patient signals; the signal acquisition units; the processing unit that processes the raw signals into meaningful information; the visual display unit that displays the processed vital signs to the operator; the alarm section that alerts the operator; and the control panel that enables the operator to control the monitor. The monitor has an objective: to faithfully measure, display and monitor the patient's vital signs, issuing audible and visual alerts if any are outside the desired set limits.

When equipment fails for a purely technical reason, we fault find it and rectify the problem so it can be returned to service. Fault diagnosis combines an understanding of how the equipment functions as a system with knowledge of testing methods that check the interconnections between the elements and the functioning of the elements themselves. The approach we apply to fault finding is a systems approach. We can take a number of different approaches but the intention of the exercise is to first locate the failed element (e.g. a circuit, or IC) or how it interconnects with other elements (cables, tracks on a board). The equipment can be broken down into fundamental functional blocks and interconnections as described earlier; by measuring the function of the blocks and how they interconnect instead of examining the device as a whole, it is possible to identify the faulty block or interconnection. Then this element is further examined to find the fault (Loveday 1994), or, as frequently carried out today, the faulty element replaced.

2.4.2 Systems Approach to the Sociotechnical Aspects of Equipment Management

The use of a vital signs monitor mentioned earlier may result in an erroneous measurement of blood pressure being made, even though the equipment itself is technically working exactly as expected. How equipment is used can influence its effectiveness and safety. So it is appropriate and prudent to consider medical equipment as part of a wider system, a sociotechnical system that consists of the equipment itself and the environment within which it is used, what it is being used for and by whom it is being used.

A perfectly calibrated and functioning sphygmomanometer, if used incorrectly, can result in an erroneous blood pressure measurement that in turn can dictate the care plan prescribed for the patient. Taking a systems approach to diagnose the 'fault' in this scenario requires the investigator to consider the knowledge and skill of the user as an element in the sociotechnical system that is the medical equipment in use. Experienced clinical engineers will be fully aware of the issue of user difficulty with operation resulting in report of equipment failure. This scenario is not uncommon across all medical equipment types, and given the increase in complexity and ubiquity of medical equipment is a real issue to be dealt with in assuring medical equipment benefits are delivered.

It may also be appropriate to consider the environment in which or the purpose for which equipment is used. A physician making a diagnosis of hypertension will take a very different view of accuracy requirements compared to the Emergency Department (ED) physician assessing a patient admitted following a road accident. The former will require errors of less than 3 mmHg, and the latter may be content with blood pressure measurements rounded off to the nearest 10 mmHg. In this case the clinical engineer, in developing the equipment support plan (ESP) (see Chapter 6), may decide that the most cost-effective solution will be to check for accuracies within 3 mmHg. However, in other situations, the circumstances of use of the equipment may influence decisions on equipment support plans.

So in developing a plan for how to proactively manage a piece of medical equipment, a systems approach challenges us to consider how and why the equipment is being used clinically, assessing not only the technical elements but also its socio-elements. Consider for example a vital signs monitor. A holistic plan will not only include the technical maintenance but also support of the device in its use, recognizing its users and the patients for whom the equipment should be giving benefits. Applying the CATWOE approach (Section 2.2), we might identify the following:

C: Customer – The patient;

A: Actors – The nurses and clinicians;

T: Transformation – Vital signs data added to and changing the patient's records;

W: World view – A measurement that will dictate the patient's care pathway;

O: Owners – The nurses and doctors;

E: Environmental constraints – Accuracy of the measurement can only be assured if:

- The equipment is technically maintained;
- The equipment is used by a competently trained user.

The CATWOE systems approach leads to the conclusion that training and verification of user competence is as important as the technical maintenance of the device. We will discuss in more detail the development of holistic equipment support plans in Chapter 6. However, the key learning from taking a systems approach is that we have to look at the equipment as one subsystem in a wider sociotechnical system.

2.4.3 Systems Approach to Managing a Fleet of Medical Equipment

Clinical engineers today are challenged to develop and implement a huge number of equipment support plans (ESPs), all of which must take the wider sociotechnical view. Often practices develop organically over time or are introduced in response to adverse events. Certainly there is no one approach that prevails; the resources and approach taken to implementing ESPs for different types of medical equipment used in different environments will vary.

So how can we use the systems approach to improve ESPs? Using the SIMILAR approach (Section 2.2), we might, for a particular equipment type, consider the following:

S: State the problem – Is its ESP appropriate and does it deliver a positive outcome for the patient?

I: Investigate alternative solutions – Keeping an open mind: look at how individual ESPs might be changed to improve outcomes, perhaps analyzing whether value (benefits delivered as ratio of the cost) could improve if aspects of the support were outsourced or if new aspects (user training) were added.

M: Model the system – Use metrics to critically evaluate any alternative solutions proposed, determining whether value has increased or decreased. Discuss in detail with colleagues how the new proposal might be implemented in practice.

I: Integrate – Consider the impact of the proposed improvements on the wider environment. For example will the suggested changes in the ESP for one type of equipment compromise those for other equipment (e.g. by reducing the time available)? Will optimizing one reduce effectiveness of the overall healthcare technology management programme, and if so how do you mitigate against this?

L: Launch the system – Once the proposed change and its implications have been positively assessed, implement it with metrics measuring its cost-effectiveness.

A: Assess performance – Review the metrics of the new ESPs and the wider programme as a whole to ensure improvements have been achieved.

R: Re-evaluate – Regularly recheck, perhaps using the PDCA cycle, assessing each individual ESP both in isolation and within the overall HTM Programme. Are they responsive to changes in the sociotechnical healthcare system they are designed to serve?

Essentially then, the systems approach is a method of analyzing and understanding a process by dividing it into manageable constituent elements. The details of each element can be analyzed separately whilst recognizing the relationships it has with other elements within the whole process.

2.4.4 Systems Approach to the Planned Replacement of Medical Equipment

Medical equipment has to be replaced periodically to ensure that the organization can meet its aims. Replacement may be prompted by the existing equipment being deemed obsolete, no longer supported by its supplier, the availability of improved technologies with better benefits or the clinical or financial risk associated with its continuing use considered too high. We will discuss these reasons in more detail in Chapter 5 and in Case Studies CS5.6 and CS5.8. The cost of all the replacement projects proposed nearly always exceeds available resources and so methods of prioritizing equipment replacement are required. Viewing the replacement programme as a quality improvement exercise and applying Trbovich's (2014) five-step systems approach, a hospital might consider the following:

- *A holistic approach*: Decisions should be made recognizing interdependencies between different clinical services. A multidisciplinary committee should be convened aiming to improve the medical equipment for the hospital as a whole.

- *Interdependencies*: The committee should understand the interdependencies between different clinical services and the resultant equipment needs. Obvious interdependencies will include the impact of diagnostic imaging functionality on surgical and critical care services. Less obvious interdependencies may include the need for adequate vital signs monitoring in general wards to take the pressure off critical care units. This should include possible intended and unintended consequences of equipment procurements such as staffing and maintenance impacts and changes in clinical care pathways. Equipment procurements should be designed to improve the clinical pathways; an example is given in Case Study CS2.1.

- *Leadership*: The competing demands from different clinical specialties will require a disciplined committee with strong informed leadership acting under an appropriate governance structure and reflecting the strategic aims of the organization.

- *Identify leverage points*: Proactive analysis of the organization looking for weaknesses in the ability of its medical equipment assets to support the organizational goals should be used to identify replacement projects that will deliver impact, increasing value and offering win–win solutions.

- *Organization-wide*: Allocating resources for medical equipment procurements should consider the needs of the healthcare organization as a whole, avoiding the clamour of vested interests and clarion calls for prioritization. Equally, decisions should avoid the apparently fair but ultimately flawed easy option of dividing the available resources amongst the competing bidders. The organization-wide view is the antithesis of the silo approach that can disrupt effective planning. Rather, decisions require sober assessments of the needs of the organization and how those needs can effectively be met by the acquisition of medical equipment. The assessment will need to consider the whole life cost of the equipment, a subject we will examine in Chapter 4.

A replacement programme that is based on these principles can effectively allocate the resources for the procurement of medical equipment for the benefit of both the individual clinical departments and the organization as a whole. The systems engineering methodology will take into account the requirements of the individual departments, viewing them as elements in the hospital organization system, elements whose effective functioning is vital to the overall functioning of the organization. But the needs of the individual departments must not take precedence over the needs of the organization as a whole. Rather, the interdependencies between the requirements of each department (element) can help share the available resources for the benefit of the organization.

2.4.5 Clinical Engineers Using Systems Engineering to Improve HTM Processes

We used the systems engineering analysis of HTM processes in Section 2.3 to guide the discussion on the systems engineering approach. Clinical engineers, when reviewing the operation of their Clinical Engineering Departments (CEDs), are recommended to use systems engineering methods. This can be applied to HTM as a whole (Figure 2.6) or to specific aspects of HTM (Figure 2.7).

The systems approach can help manage the procurement of individual items of equipment. It can help keep in focus the objective of the acquisition, particularly if we remember to keep the focus on the patient and the needs of the patient. This can help direct the specification, evaluation and selection of the equipment.

The systems approach can also assist ensuring that complicated arrangements are managed successfully for complex projects. In Case Study CS4.4, we will show how a complex operation is divided into discrete elements, recognizing their mutual interdependences, and with a strong leadership, the process can be managed effectively. The case study describes the disposal of old medical equipment, but the process can be applied to other HTM projects.

Similar structured system approaches can contribute significantly to continual improvement of most aspects of HTM. It provides the structured methodology for supporting quality management through the use of Standards such as ISO 55000 and ISO 9001 including managing the preventive and corrective actions programmes that are fundamental to an ISO 9001 quality management system. These quality management systems are discussed in more detail in Chapters 3, 5 and 6.

2.4.6 Clinical Engineers Using the Systems Approach to Investigate Adverse Events

The systems approach can also support the investigation and analysis of adverse events, ensuring that all aspects are considered. Clinical engineers are often asked to help these investigations, particularly those that involve medical equipment.

An adverse event is defined as an event that results in, or could result in, harm to patient, visitor or staff member. When an adverse event occurs, it is important to identify the causes (and there are often more than one); the process of identifying the causes is sometimes called root cause analysis, and many identification approaches have been proposed. Unfortunately, a typical reaction to an adverse event is to look for a person or persons to blame or, where medical equipment is involved, to simply to blame the equipment. However, in many cases, these blame-seeking approaches fail to really reveal the actual causes. Failure to fully understand the causes limits the ability to put in place measures to prevent recurrences.

Reason (2000) argued that adverse events are primarily the result of system failures, their prevention requiring an understanding of the nature of the system including its weaknesses. He advocated a systems analysis of the healthcare process where the adverse event occurred, using this to help identify the causes. Weaknesses in the process were described as latent flaws creating vulnerabilities, an environment 'waiting for an accident to happen'. Reason went on to describe how system flaws could be likened to the holes in layers of Swiss cheese. Multiple Swiss cheese layers provide protective barriers that prevent incidents, with the multiple layers compensating for flaw in other barriers. However, where multiple flaws align, a trigger event, an active failure, can break through the flawed protective barriers, the protective procedures, resulting in the adverse event. The importance of these latent conditions, these latent failures, was demonstrated nearly a decade earlier by Runciman et al. (1993) in their analysis of the first 2000 incidents recorded in the Australian Incident Monitoring Study. The investigators showed that whilst human error was involved in most (70%–80%) of the adverse events, its overall contribution to each adverse event was small. Most (90%) of the causes were found to be system failures.

Amoore and Ingram (2002, 2003) applied Reason's model to those adverse events involving medical devices. They showed how, by analyzing the discrete elements of the application of the medical device within the process of care, the wider causes can be identified and methods of prevention developed. This is shown diagrammatically in Figure 2.11 with latent background conditions combining with unsafe acts or omissions (triggers) and failures in barrier defences (holes in the Swiss cheese) to cause an adverse event. The consequences of the adverse event may be mitigated by compensating procedures and/or by actions by members of staff. The systems approach enables those analyzing the adverse event to focus in turn on each of these aspects. This will be explored in more detail in Case Study CS2.2.

An alternative systems approach for analyzing the causes of adverse events involving medical equipment is to consider the healthcare system in which the event occurred identifying the elements of the system and the interactions between the elements (Amoore 2014). In general four broad elements are involved (Table 2.1): the medical

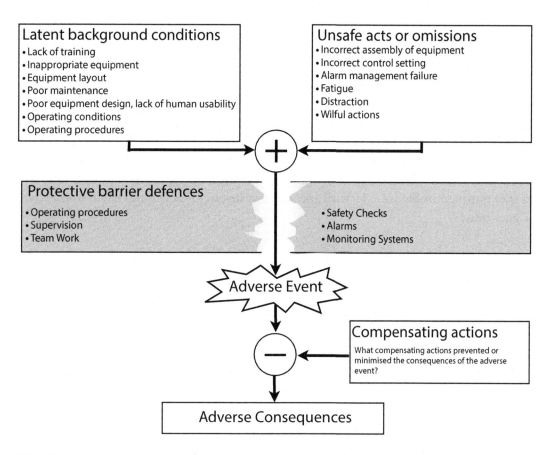

Latent background conditions
- Lack of training
- Inappropriate equipment
- Equipment layout
- Poor maintenance
- Poor equipment design, lack of human usability
- Operating conditions
- Operating procedures

Unsafe acts or omissions
- Incorrect assembly of equipment
- Incorrect control setting
- Alarm management failure
- Fatigue
- Distraction
- Wilful actions

Protective barrier defences
- Operating procedures
- Supervision
- Team Work

- Safety Checks
- Alarms
- Monitoring Systems

Adverse Event

Compensating actions
What compensating actions prevented or minimised the consequences of the adverse event?

Adverse Consequences

FIGURE 2.11   A systems approach to analyzing the causes of adverse events and development of methods of prevention.

TABLE 2.1   Systems Analysis of Contributory Factors to Adverse Events Involving Medical Devices: Identification of the Elements Involved

Medical Device	User/Care Provider	Patient	Environment of Care
• Device design (technical) • Device design (ergonomics) • Manufacture • Device failure • System failure • IT–medical device failure	• 'Use error' • Device set-up • Training • Following procedures • User maintenance • Distraction • Fatigue	• Patient's pathology or physiology interacting with the device • Patient tampering	• Procurement and commissioning failures • Device layout and mounting • Utilities • Lack of availability of device • Maintenance • Environmental interference
Note: A more detailed list of contributory factors can be found in Amoore (2014).			

device, the care provider (user or operator of the medical device), the patient being cared for by the medical device, and the environment in which the care is provided and it's supporting infrastructure.

Each of these four groups can be subdivided into more detailed subgroups as required and as information about the event allows as shown in Table 2.1. The approach involves considering, in turn, whether each of these elements could have contributed to the adverse event. This systems approach to the analysis of the causes of adverse events involving medical devices reveals that typically there are several causes for the incident that a simple unstructured approach may not identify. The approach is complementary to that outlined in Figure 2.11 with the analysis helping identify the latent and active trigger causes.

2.4.7 Health Technology Assessment

Systems engineering also supports health technology assessment (HTA), that is the systematic evaluation of health technologies, a multidisciplinary process that evaluates the social, economic, organizational and ethical issues of health technology (WHO 2011). HTA developed formally in the mid-1970s in response to the very high cost of the newly developed computer-assisted tomography (CT) systems, costing in those days in excess of $300,000. The U.S. Senate asked the then recently established Office of Technology Assessment to conduct a study of the requirements for justifying the implementation of costly new medical technologies and procedures, helping to formalize what is now an established discipline. (It is interesting to reflect that typical standard CT systems today cost similar dollar amounts, with the functionality of today's CT systems far superior in terms of image resolution and reduced radiation dose.)

Methods of HTA have been described, for example the *Health Technology Assessment Handbook* (Kristensen and Sigmund 2008). It notes that HTA incorporates four key elements: technology, patient, organization and economy. In addition to these, the HTA may often consider ethical aspects. The Handbook covers the methods of HTA that seeks to ensure that the best available evidence is used to base the conclusions. It recommends that any HTA begins by clarifying the topic, the problem to be analyzed, organizing the project including building the project team. It stresses that where possible available evidence should be used. The clarification of the topic leads to the formulation of the HTA question and the HTA project where evidence is gathered and analyzed, considering the four elements, technology, patient, organization and economic. The evidence is brought together and synthesized, leading to conclusions about the technology under investigation. The assessment will consider the efficacy of the technology (does it work and if so how effectively), its benefits for patients, its costs and its comparison with alternatives. The challenge is assessing the Value – and ensuring objectivity in those making the assessment. Health economics provides a powerful tool for carrying out these assessments, objectively comparing the benefits of a healthcare technology with its costs. A practical application of this is discussed in Case Study CS7.5.

2.5 INTRODUCTION TO HEALTHCARE AS A SYSTEM

Healthcare itself can be analyzed as a system. It is worth pausing to consider whether system approach thinking helped design healthcare or whether healthcare grew 'randomly' from the response to particular demands for specific care needs. Even in those jurisdictions where healthcare is more tightly organized and controlled, holistic thinking at national and local levels is not always given the prominence that it deserves. The National Health Service in the United Kingdom was a bold attempt, after the deprivations of World War II, to develop a coordinated healthcare service. The planners did have an objective, namely universal care, free at the point of need and paid for out of general taxation. And the development, whilst not without flaws, is often widely praised for a coordinated approach to healthcare. However, it is currently facing several pressures, the need for greater coordination between acute and community care and the mounting concern about the ability of care in the community to meet the health and social care needs of an ageing population.

When applying systems thinking to healthcare these realities, the emotions that run deep in those served by healthcare must be recognized. This is not to deny the validity of applying the systems approach, but to recognize the emotions as an important undercurrent pervading the system. Indeed the systems approach does not deny human aspects, as exemplified by the work of Checkland and Scholes (1999). Healthcare is not a simple mechanical arrangement of different processes, specialists and departments, but at its heart a deeply personal attempt to provide care for the individual in physical or psychological distress. The challenge in adopting systems engineering in healthcare as advocated in, amongst others, the PCAST (2014) report, is reconciling what appears on the surface to be a conflict between the personal healthcare priorities of each individual (patient-centred care) and the global priorities of populations, with the need to make healthcare affordable – the three corners of the Institute of Healthcare Improvement's Triple Aim (Stiefel and Nolan 2012).

2.5.1 Models of Healthcare as a System

When applying systems thinking, we must have in mind a conceptual model of the system including the environment in which it functions. We also need to include within these models measures of system performance. This will help gauge how effective interventions have been or where we next need to intervene. These measures can be qualitative as well as quantitative. In this section therefore, we build on the healthcare models outlined in Chapter 1, recognizing the patient focus stressed in that chapter. Clinical engineers will need to consider the implication of their stewardship of the healthcare technology on patient care and may need to assess and demonstrate the value of healthcare technologies to patient care.

Figure 2.12 presents a generalized model of a health system with patients entering and leaving. The system includes people, staff, infrastructure, medical technology, care plans and processes supported by financial and other resources. These are the constituent elements of our system. Healthcare is not delivered in a vacuum, but in organizational structures that exist within national socio-economic environments. Consider for example a

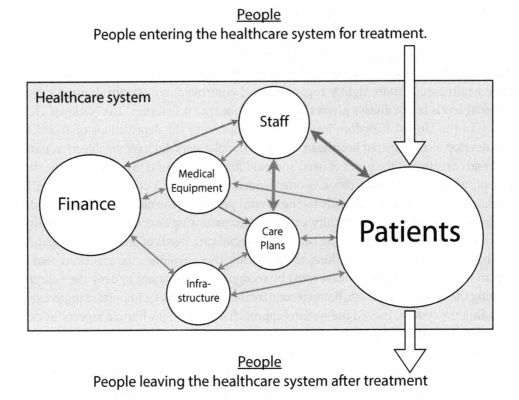

People
People entering the healthcare system for treatment.

Healthcare system

Staff

Medical
Equipment

Finance

Care
Plans

Patients

Infra-
structure

People
People leaving the healthcare system after treatment

FIGURE 2.12   The health system with people at its heart, those entering to receive treatment, those receiving treatment and those leaving after treatment. The patients are cared for by people, the staff, who are supported by the infrastructure, the medical technology, the Care Plans (the healthcare processes), in turn dependent on the financial resources available.

hospital, a healthcare system. From a functional perspective it receives patients, the sick and those requiring care. Its output is treated patients. If the patient flow in exceeds the flow of healed patients leaving, then the system is not coping. Analysis of the patient flow through the hospital can reveal parts of the process that are suboptimal. Attention can then be applied to improve those parts of the process. Patients might be delayed from being discharged from hospital because they are waiting for a routine test, say an ultra-sound or endoscopy test to be performed, before they can be discharged with confidence. Perhaps the care plan cannot be implemented because of a mismatch between demand and capacity that is between the workload and the resources, in this case the availability of equipment and staff to perform the test. If the processes can be optimized to increase the capability, the patient experience will improve in that they have an earlier safe discharge. This is a challenge that systems engineering is tasked with solving. Sometimes the solution requires additional funding; however, additional funding may not always solve the problem and should not be applied to cover over a deficient system. One of the Institute of Healthcare Improvement's Triple Aim is affordable care (Stiefel and Nolan 2012); solving problems by simply applying additional funding will not support that aim.

The issues are not always within the hospital system but at the boundaries between the hospital and the environment within which it operates. Simple analyses in some healthcare systems have shown, for example, that patients ready to be discharged home are remaining in hospitals because of insufficient capacity of care in the community or delays in arranging the appropriate package of pharmaceutical or technological aids to support the discharged patient during early phases of rehabilitation at home.

The system approach facilitates an understanding of the impact of the internal and external environment, whilst recognizing the patient focus, the patient at the centre, directly looked after by a Care Team operating within Organizational Structures subject to the external 'national' environment. This holistic approach has been described as a four-level model of healthcare, as shown in Figure 2.13 (Ferlie and Shortell 2001; IoM 2005). It presents an

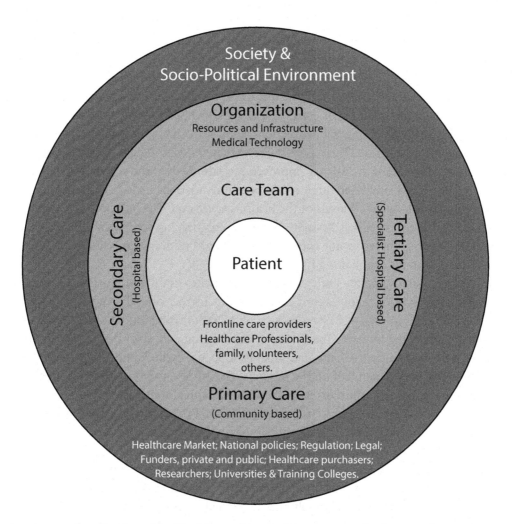

FIGURE 2.13 Four-level model of healthcare. The model recognizes the focus on the patient, supported by the whole healthcare system and front-line caregivers. (Adapted from IoM, *Building a Better Delivery System: A New Engineering/Health Care Partnership – Workshop Summary*, The National Academies Press, Washington, DC. With permission.)

idealized healthcare system with the patient at the heart, supported by the care team which will in general include healthcare professionals, family, friends and volunteers. The Care Team works within an Organization, either in a hospital building or in the community. In turn the organization operates within an environment composed of and constrained by interacting, conflicting and supporting social, economic, legislative and political factors. The health system is complex, facing competing and often conflicting demands, dealing with strong deep and heart-felt emotions and often strongly felt political and economic challenges.

The systems approach provides tools for considering not only the whole four-level model, but also the details of each level, identifying their characteristics, features, challenges and opportunities to improve the healthcare. Each level does not operate in isolation, but depends on its interactions with the other levels.

2.5.2 The Need to Improve Healthcare Systems

The importance of analyzing health systems using the systems engineering approach was more explicitly endorsed and called for in the 2014 report to the President of the United States by its President's Council of Advisors on Science and Technology (2014). The report recognizes that significant amounts of healthcare expenditure are unnecessary and wasteful, an unsustainable situation. It calls for a reassessment of the values of the healthcare process, with strategic changes in direction required to increase efficiency and ensure high-quality affordable care. The report suggests that systems engineering techniques can provide needed solutions. It calls for improved measurement of the system that should be judged on Value, not simply on quantity. Its recommendations include ensuring "systems engineering knowhow" at all levels of the health service and "the building of a healthcare workforce that is equipped with essential-systems engineering competencies". The report recognizes that healthcare is not confined to hospital buildings, providing examples of application of systems engineering in community care.

Healthcare systems operate under different approaches in different jurisdictions across the world. The PCAST team examined healthcare in the United States, including its payment model. However, the basic messages of the report transcend national boundaries and systems of paying for care. The first recommendation addresses the need to align the payment systems with desired outcomes, that is, to focus on the Value provided. The solution details will vary dependent on the financial structures of the healthcare systems, but the overall system goal is common across differing financial structures.

The IoM report (IoM 2012) which preceded the PCAST report highlighted how improved use of Information and Communication Technology (ICT) can support better care at lower cost, in particular by ensuring real-time access to patient information continuously and reliably captured at the point of patient care. It called for engaged, empowered patients, supported by a system anchored on patient needs that promoted inclusion of patients, families and carers. Health systems should be aligned to improve care for patients whilst minimizing costs. This requires a system operating with full transparency that monitors the safety, quality, processes, prices, costs, and outcomes of care. The achievement of these goals requires strong leadership instilling a culture of continuous learning as a core aim. It requires an underpinning of competences in systems analysis.

2.5.3 Quality Improvement Initiatives in Healthcare

Healthcare lags behind industry in implementing quality improvement initiatives. However, that is changing, and many hospitals now have continuous quality improvement programmes focused on innovating methods of care delivery with the aim of improving safety, controlling costs and making the service more accessible and effective for patients.

However, too often quality initiatives in healthcare are dictated by special interest groups that focus on one aspect of a healthcare operation and the professional accreditation needs of the associated discipline, without fully comprehending the interrelationships with others. This should not be construed as a criticism of professional accreditation schemes of which several good examples exist including that developed by the (UK) Royal College of Anaesthetists (RCOA 2015). The caution is however added here, as within individual healthcare organizations there is too often a narrow discipline-specific approach. The silo thinking often frustrates a systems approach to quality improvements. It can hinder the organization's attempts to solve problems of patient flow between departments, if individual departments concentrate on their own demands without recognizing the problems of the other departments. This 'silo' approach is the antithesis of the systems approach. The systems approach does identify and recognize the needs of each clinical discipline, each element, each 'silo', but views each as part of a greater whole; the boundaries of the silos are not barriers, but interfaces that recognize mutual interdependencies. The requirements of each discipline must be addressed, but in association with the other disciplines, recognizing mutual relationships that work together to support the common aims of the organization as a whole.

When considering the systems approach, it is helpful to consider the context of the system being studied. With Ferlie and Shortell's four-level model in mind (Figure 2.13), we can distinguish between what are sometimes described as macro and micro levels. The macro level equates to the outer socio-political layers, whilst the micro level involves direct inclusion of the patient and their carers who are at the centre of the system. Different skills, authorities and responsibilities are needed to affect change at these two levels. At the macro level, there may be a requirement to influence national policies and advice, whereas at the micro level direct changes to clinical service provision can be more immediately influenced. The clinical engineer must therefore consider the relevant extent of the system and how best to analyze and influence it.

2.6 APPLYING SYSTEMS ENGINEERING TO HEALTHCARE

Increasingly, as we have noted earlier, calls have been made for improvement to healthcare systems, advocating that systems engineering techniques can be used to achieve better healthcare. The systems engineering approach can be applied at all levels of healthcare systems (Figure 2.13), at the front line where the patient sees the care team and at the socio-political level. Particularly at the socio-political level, it is important to recognize that most healthcare organizations are primarily geared to 'broken-health' management. Hospitals, with their Emergency Departments and critical care facilities, are largely dedicated to dealing with the consequences of failed health, though it is often argued that maternity units have a different ethos. There are notable exceptions at national levels with campaigns for healthy living including healthy eating and exercise. Systems engineering, with its emphasis on the objective, can be very helpful here, to clarify what the aim is, particularly at this socio-political level where policy makers are caught by the increasing demands for managing broken health whilst recognizing the importance of disease prevention and health improvement. Whilst visionaries are exploring the ideal healthcare system that is personalized, integrated and yet distributed, flexible and responsive, it is recognized that there are barriers to the application of systems engineering at this very global level (Valdez et al. 2010). Barriers include inadequate current systems engineering knowledge, current healthcare policies that constrain change and lack of professionals with adequate knowledge of both healthcare and systems engineering.

Encouragingly, at many levels, the application of systems engineering has been shown to improve healthcare systems (Gabow and Mehler 2011; PCAST 2014), and hence, there is motivation to develop and pursue its application. In the rest of this section we will discuss systems engineering application at healthcare organization level and at clinical departmental level. We have recognized in Section 2.4 the use of systems engineering tools by clinical engineers in developing healthcare technology management. It is important as we discuss the application of systems engineering to healthcare organizations that clinical engineers have a role to play here: clinical engineers have the logical systematic thinking that can facilitate this application.

2.6.1 Systems Engineering at Healthcare Organization Level

Figure 2.12 shows diagrammatically the flow of patients through a healthcare system. This diagram is applicable whether that system is a hospital, a community healthcare organization or a regional system incorporating primary, community and hospital care. Patients will flow between these different care sites as their health changes. Hall et al. (2006) have discussed the flow of patients within this regional system, suggesting the regional planning objectives include minimizing cost, maximizing access and maximizing positive health outcomes. We have discussed in Chapter 1 how patients transfer between these care systems and within any of them patients will transfer between clinical departments, for example between a hospital's Emergency Department (ED) and its operating department. Transfers of patients between care sites can lead to delays, delays that are costly for the healthcare organization (the department ready to transfer the delayed patient incurs costs keeping the patient longer than necessary) and can lead to poorer outcomes for patients. There are various causes of the delays, lack of capacity in the receiving site, but also administrative delays including transferring patient records and processing medication prescriptions. Some delays are caused by lack of resources, for example transport (be it ambulance between care sites or trolleys within a care site). Patients also suffer delays because of lack of capacity to treat them; this occurs particularly commonly in EDs where peak demands for care can overload the care processes with the department.

We have painted a picture of a system with a goal (to affordably maximize health outcomes in a timely fashion) and with various components, each dependent and relating to the others. Thus it is no surprise that the call is for systems engineering to be applied to optimize the system. Hall et al. (2006) have described how the application of systems engineering can help develop solutions, planning the processes involved in the patient flows and measuring the success of changes, with performance measures based on the aims of the systems. Through patient flow maps and gathering data to quantify the flows and process steps, the authors show how systems engineering can generate recommendations for improving performance. They based their discussions on the Los Angeles County Department of Health Services, concluding that success requires an understanding of healthcare as a system composed of interactions between patients, clinicians and resources.

Systems engineering can also help plan development of healthcare facilities, be they community treatment centres or tertiary academic hospitals. When new facilities are being planned, systems engineering thinking can help ensure that its capacity and resources are sufficient to meet the clinical needs of the population. Systems engineering, considering patient flow patterns, can help suggest how the various component parts, ED, Radiology, Operating Theatres and Critical Care areas for example should be sited to best support the flow of patients. Clinical engineers can have a role here, being involved in the health-care planning process, contributing to the functional design as well as the actual building design and equipping. Clinical engineers can contribute to such projects analyzing data on how the existing and proposed developed hospital will function, turning the data into meaningful information that in turn supports quality decision-making.

2.6.2 Systems Engineering within Clinical Departments

Quality improvement initiatives can also be delivered within and between clinical units or wards. In fact it is at this level, where the patient interacts with the provider of healthcare, that quality, safety, reliability and efficiency are delivered and the patient experience of care is created. By working at this level, clinical engineers actively promote a culture of safety and quality improvement around the use of technology. These small front-line systems are called 'clinical microsystems' and are made up of small interdependent groups of people who work together regularly to provide care for specific groups of patients (Nelson et al. 2002; Batalden 2015). With their systems analysis and measurement experience, clinical engineers can be valuable members of clinical microsystems. They can provide insights and identify ways of analyzing problems and developing solutions that may not be obvious to clinicians based at the point of care. Their logical analytical minds can objectively view clinical processes from new perspectives suggesting different ways of providing clinical care. Teamwork is central to the systems engineering method and clinical engineers will generally cooperate with clinical colleagues in these system analyses. Through this, clinical engineers can be contributors to ongoing innovation of hospital management and clinical care programmes as we shall see in more detail as we consider their work in project teams in Chapter 7.

The investigations may also involve extracting data from clinical information systems (CISs), converting the data to useful information and presenting it in formats that shed light on the performance of clinical processes. The increasing amount of clinical data that is being digitally collected offers the opportunity of this data mining to measure and clarify the performance of clinical processes. We illustrate this in Case Study CS2.3 that shows how a clinical engineer was able to shed light on critical care functions in a way that supported improvements in patient care.

We illustrate in Case Study CS2.4 how clinical engineers, through systematically analyzing clinical processes, helped reduce the time delay been patients admitted to Emergency Departments with bone fractures and their subsequent surgery. This is but one example of systematic approaches that have been applied to improve the safety and quality of patient flows in hospital. The transfer of patients post-surgery from operating theatre to intensive care is often a stressful part of the care pathway. By systematically studying the process and applying expertise from other industries (aviation and motor car racing) to develop improved patient handover protocols, the safety and quality of the handover process has been improved (Catchpole et al. 2007). In applying the systems approach to improve the practices of care, the importance of understanding the human factors underlying the relationships between people and systems has been recognized (Catchpole and McCulloch 2010). To achieve safer care Catchpole and Wiegmann (2012) wrote: "Now, more than ever, we need good designs, a systems approach to improvement, and we need to measure the impact that this work is having on outcomes". As we have illustrated in Case Study CS2.3 clinical engineers can use their skills to develop meaningful ways of measuring clinical processes that can guide change.

2.6.3 Exploiting Developments in Technology to Transform Healthcare

Developments in technology have transformed much of daily life and have supported dramatic enhancements in healthcare. However, technology is being asked to do more, including supporting the patient-focused agenda, reducing costs whilst improving quality and supporting health and care in the community: a serious challenge. However, consider the transformation that mobile technology has effected in daily life; the early days of phone communication required a tethered telephonic device, initially with the requirement to call the operator who made the connection for you. Direct dialling and mobile phones changed all that putting the power into the hands of the user, with mobile devices having increasing versatilities and abilities. Healthcare looks forward to benefiting from the transforming power of technology.

This requires that those knowledgeable about technology and those knowledgeable about healthcare come together thinking innovatively. Systems engineering tools can powerfully assist this process. We have seen in Section 2.2 that these tools focus on the Customer and the Objective with a clear Statement of the problem to be addressed. Alternative solutions can be investigated and tested, with a multidisciplinary approach that provides fertile ground for original thinking and innovation. Clinical engineers, linking the clinical and the technical, have an important role to play here and should not shy away from becoming involved in transforming care processes. We have seen in Case Study CS2.1 how the initial involvement in replacing technologies rapidly widened into care process changes. Case Study CS2.5 perhaps demonstrates this more clearly with a clinical engineer using the SIMILAR toolkit (INCOSE 2016) (Section 2.2) to transform a healthcare service from hospital centred to community centred.

Case Study CS2.5 demonstrates the opportunities offered by new technologies that allow for practices which traditionally were hospital based to move into the community. In many cases, the clinical engineer might be the person who is most aware of the potential for new technology to improve care and patient experience. However, to unlock the value of these new technologies often requires multidisciplinary teamwork. It also requires a willingness to investigate openly and objectively changes to the way care is delivered, and this in turn can impact on individuals work practices. A systems approach will assist the thorough assessment of alternative process changes and guide the team to test and select the optimum solution.

Whilst we have introduced this topic of transforming healthcare processes exploiting technological developments, success requires the overall aim of the project to be kept front and centre; this will guide the project away from becoming purely technical, instead focusing on its person-centred core and thus recognizing it as a sociotechnical project (Case Study CS2.5).

2.6.4 Systems Approach and Data Flows

The aspiration for the fully digital hospital is that, as the volume of data increases and systems for transforming the data into meaningful information develop, the science behind this, clinical informatics, will help clinicians and hospital managers to improve services. Increasingly clinical engineers, with an inherent understanding and ability to apply a systems approach and with data analysis skills, are finding a role in clinical informatics groups. Clinical engineers can support senior healthcare management with quality improvement initiatives. This support can be provided for global healthcare projects and also for projects within individual (or groups of) clinical units (Case Study CS2.4).

Such endeavours benefit greatly from taking a systems approach, acknowledging that the clinical information system is a sociotechnical one and must be developed and implemented in the full knowledge of the importance that people play in it. When implemented well, these clinical information systems can improve the care to patients by supporting both the standardization of care and the measuring of its effectiveness and then the implementation of quality improvements in a controlled way (Case Studies CS2.3 and CS2.4).

2.6.5 Clinical Engineers' Contributions to Improving Healthcare Systems

Clinical engineers are one of the few groups in healthcare organizations who have ongoing multidisciplinary working relationships with colleagues across the organization and with the external healthcare industry. Within the hospital, clinical engineers work with clinicians, managers, Facilities, Procurement and Finance; external to the hospital, through their healthcare technology management role, they have relationships with the medical devices industry and regulatory sectors. Interdisciplinary by nature, they can facilitate the coming together of different professional groups to reflect on and develop proposals for change. Coupled with their innate ability to apply a systems approach, they are in a privileged position, and often they can act effectively as brokers between these different professional groups, encouraging change through identifying opportunities for quality improvement and fostering synergies between these different groups. Sometimes the action might be implementing a new technology, but they can also be agents of changes to existing processes for the purpose of improving outcome, reducing risk, controlling cost or improving the patient experience.

For their knowledge, skills and abilities to be applied to healthcare process redesign and improvement requires that senior clinical leaders and senior management respect and trust the abilities of clinical engineers to apply expertise and leadership outside their direct HTM roles. This requires openness from both clinical engineers, who must recognize their own limitations, and from the senior clinical and management leads. With this understanding and trust, the clinical engineer can apply the logic and structure of their systems approach to improve healthcare processes that perhaps do not directly involve medical technology. In general, this will only form part of the work of the clinical engineer, though those showing particular strengths in these areas may find themselves being increasingly tasked to support and develop healthcare process redesign projects.

2.7 CONCLUSION

The systems approach encourages organizations to change. Importantly, the objective is not change for the sake of change, but to improve the care provided to patients. Change requires understanding the current state and assessing it against the organization objectives. It requires the organization to be a learning organization: to learn more about itself, to understand it better and to learn how to improve. Perhaps more importantly and more subtly, it requires the organization to understand how to change and to embed the change and, having embedded it, to continue to analyze, change further and improve.

The call is for a positive cycle of improvement, with analysis and scientific advances feeding into evidence, leading to improvements in care processes. Smith et al. in Part II of their IoM report (IoM 2012) refer to this as a 'Learning Heath Care system'. The processes of care must be continuously analyzed, with lessons learnt to provide further evidence for further improving care, creating a virtuous circle of improvement. Success requires that this learning and improvement process takes place within a supportive culture, driven by strong leadership, with incentives to focus on improving care.

The emphasis on the need for a learning ethos is common to health services across the world. When the Chief Medical Officer for England reported on the need for learning from mistakes, the title chosen for the report 'An organization with a memory' (Department of Health 2000) is deliberately aspirational. It challenges us to consider whether we really do remember and learn from the past, stressing the need to learn from previous experience.

In Chapter 1 we identified that healthcare technology management can be described under two headings, supporting and advancing care and equipment management. Both these activities can benefit from a systems approach (Figure 1.4).

In Section 2.4, we have identified how a systems approach can influence healthcare technology management. The approach can be applied to the whole of the equipment management activity (Sections 2.3 and 2.4) and to individual HTM processes. So it is appropriate to take a systems approach to the holistic management of medical equipment over its useful life. This will be further explored in Chapters 4 through 6.

In Section 2.6 we have seen that many of the activities grouped under the supporting and advancing care heading can draw on a systems approach and benefit from it. These include equipment-related activity but extend beyond to include activity focused on improving the healthcare system itself. We identified the value of embedding a learning culture, a questioning culture within clinical engineering, a culture focused on aiming to improve the quality and effectiveness of care in ways that enhance value: that is the patient care benefit in relation to the costs of care.

We have reminded ourselves that engineering is about people, money and machines and that we can use our skills, our training and our competence to help ensure that the machinery of the health systems (tangible technical as well as systems and processes) work with the people involved (professional and lay carers and the patients) to ensure that the financial resources applied to health systems are not wasted, but enhance the value of care.

So we conclude that one of the base principles upon which a comprehensive healthcare technology management system is built should be a systems approach. This chapter introduced some of the knowledge, skills and techniques that will be required to equip the engineer working in healthcare to meet the challenges. Illustrative examples and references have been presented, the aim being to get clinical engineers started with a set of building blocks that will help in furthering their understanding of the system in which they work as well as providing practical techniques to assist in delivering solutions. From this starting point, clinical engineers should feel equipped to identify additional tools to add to their toolkit that add to their repertoire to enable them to analyze and solve the challenges they will face in their careers.

REFERENCES

Amoore J.N. 2014. A structured approach for investigating the causes of medical device adverse events. *Journal of Medical Engineering and Technology*, 2014: Article ID314138. http://www.hindawi.com/journals/jme/2014/314138/ (accessed 2016-03-31).

Amoore J.N. and P. Ingram. 2002. Learning from adverse incidents involving medical devices. *British Medical Journal*, 325(7358): 272–275.

Amoore J.N. and P. Ingram. 2003. Learning from adverse incidents involving medical devices. *Nursing Standard*, 17(29): 41–46.

Batalden P. 2015. Transforming microsystems in healthcare. The Dartmouth Institute, Lebanon, NH. https://clinicalmicrosystem.org (accessed 2016-03-31).

Blockley D. 2012. *Engineering – A Very Short Introduction*. Oxford, UK: Oxford University Press.

Catchpole K.R., De Leval M.R., McEwan A., Pigott N., Elliott M.J., McQuillan A., MacDonald C. and A. J. Goldman. 2007. Patient handover from surgery to intensive care: Using formula 1 pit-stop and aviation models to improve safety and quality. *Pediatric Anesthesia*, 17: 470–478.

Catchpole K.R. and P. McCulloch. 2010. Human factors in critical care: Towards standardized integrated human-centred systems of work. *Current Opinion in Critical Care*, 16: 618–622.

Catchpole K.R. and D. Wiegmann. 2012. Understanding safety and performance in the cardiac operating room: From 'sharp end' to 'blunt end'. *BMJ Quality & Safety*, 21: 807–809.

Checkland P. and J. Scholes. 1999. *Soft Systems Methodology in Action*. Chichester, UK: Wiley.

Creative Research Systems. 2014. Survey design. The Survey System. Petaluma, CA: Creative Research Systems. http://www.surveysystem.com/sdesign.htm (accessed 2016-03-31).

Department of Health. 2000. An organization with a memory: Report of an expert group on learning from adverse events in the NHS chaired by the Chief Medical Officer. London, UK: The Stationary Office.

Ferlie E.B. and S.M. Shortell. 2001. Improving the quality of health care in the United Kingdom and the United States: A framework for change. *Milbank Quarterly*, 79(2): 281–315. http://www.ncbi.nlm.nih.gov/pmc/articles/PMC2751188/ (accessed 2016-05-16).

Gabow P.A. and P.S. Mehler. 2011. A broad and structure approach to improving patient safety and quality: Lessons from Denver Health. *Health Affairs*, 30(4): 612–618.

Hall R., Belson D., Murali P. and M. Dessouky. 2006. Modelling patient flows through the healthcare system. In *Patient Flow: Reducing Delay in Healthcare Delivery*, ed. R. Hall, Chapter 1, pp. 1–44. New York: Springer.

INCOSE. 2016. What is systems engineering?. Washington, DC: International Council on Systems Engineering. http://www.incose.org/AboutSE/WhatIsSE (accessed 2016-03-31).

IoM. 2005. *Building a Better Delivery System: A New Engineering/Health Care Partnership – Workshop Summary*. Washington, DC: The National Academies Press.

IoM. 2012. *Best care at Lower Cost: The Path to Continuously Learning Health Care in America*. Washington, DC: The National Academies Press.

Kopach-Konrad R., Lawley M., Criswell M., Hasan I., Chakraborty S., Pekny J. and B.N. Doebbeling. 2007. Applying systems engineering principles in improving health care delivery. *Journal of General Internal Medicine*, 22(Suppl 3): 431–437.

Kristensen F.B. and H. Sigmund. 2008. *Health Technology Assessment Handbook*. Copenhagen, Denmark: Danish Centre for Health Technology Assessment, National Board of Health. http://sundhedsstyrelsen.dk/publ/Publ2008/MTV/Metode/HTA_Handbook_net_final.pdf (accessed 2016-03-31).

Lawlor R. (Ed.). 2013. *Engineering in Society*. London, UK: Royal Academy of Engineering. http://www.raeng.org.uk/publications/reports/engineering-in-society (accessed 2016-03-31).

Loveday G.C. 1994. *Electronic Fault Diagnosis*, 4th edn. Essex, UK: Addison Wesley Longman Limited.

Nelson E.C., Batalden P.B., Huber T.P., Mohr J.J, Godfrey M.M., Headrick L.A. and J.H. Wasson. 2002. Microsystems in health care: Part 1. Learning from high performing front line clinical units. *The Joint Commission Journal on Quality Improvement*, 28: 472–493.

PCAST. 2014. Better healthcare and lower costs: Accelerating improvements through systems engineering. President's Council of Advisors on Science and Technology. Washington, DC: Executive Office of the President of the United States of America.

RCOA. 2015. Anaesthesia Clinical Services Accreditation (ACSA). London, UK: The Royal College of Anaesthetists. https://www.rcoa.ac.uk/acsa (accessed 2016-03-31).

Reason J. 2000. Human error: Models and management. *British Medical Journal*, 320: 768–770.

Runciman W.B., Webb R.K., Lee R. and R. Holland. 1993. System failure: An analysis of 2000 incident reports. *Anaesthesia and Intensive Care*, 21: 684–695.

StatPac. 2014. *Questionnaire Design – Considerations*. Pepin, WI: Statpac, Inc. https://statpac.com/surveys/questionnaire-design.htm (accessed 2016-03-31).

Stiefel M. and K. Nolan. 2012. A Guide to measuring the triple aim: Population health, experience of care and per capita cost. IHI Innovation Series white Paper. Cambridge, MA: Institute of Healthcare Improvement. https://www.ihi.org/engage/initiatives/TripleAim/Pages/MeasuresResults.aspx (accessed 2016-03-31).

Trbovich P. 2014. Five ways to incorporate systems thinking into healthcare organizations. *Horizons*, Fall: 31–36.

Valdez R.S., Ramley E. and P.F. Brennan. May 2010. Industrial and systems engineering and health care: Critical areas of research – Final report. (Prepared by Professional and Scientific Associates under Contract No. 290-09-00027U.) AHRQ Publication No. 10-0079. Rockville, MD: Agency for Healthcare Research and Quality.

WHO. 2011. Health technology assessment of medical devices. WHO Medical device technical series. Geneva, Switzerland: World Health Organization. http://whqlibdoc.who.int/publications/2011/9789241501361_eng.pdf. (accessed 2016-03-31).

SELF-DIRECTED LEARNING

1. Regarding an ECG machine (recorder) as a technical system, can you state what the objective of this system is? Draw a block diagram of it identifying the main subsystems or elements of which it is comprised. Consider a subsystem as functional blocks such as power supply and printer. Describe the function of the power supply and analogue front end instrumentation amplifier, identifying any interconnections these elements make with other elements.

2. Regarding the same ECG machine (recorder) in use as a sociotechnical system, can you extend the block diagram (question 1) to include both the hard and soft elements? Can you identify three possible causes of the system failing to meet its objective as a result of user error?

3. Clinical engineers in hospitals regularly have to order spare parts from an equipment supplier. Can you draw a process flow diagram for the complete ordering process from the clinical engineer making the initial request, to the part being delivered and the supplier being paid? In your diagram identify the following people: the clinical engineer who needs the part, procurement officer for the hospital, the parts dispatcher for the supplier, the delivery person, the accountants and the parts supplier.

4. Considering a volumetric infusion pump used in an intensive care setting as a system, can you state what the objective of the system is? In Section 2.4.2 we used the CATWOE acronym to help identify the components of a holistic equipment support plan for a vital signs monitor. Can you apply the same approach to identifying the components of a holistic equipment support plan for a volumetric infusion pump used in an intensive care unit?

5. What does the International Council on Systems Engineering mnemonic SIMILAR stand for?

6. Write short notes showing you understand each step of the SIMILAR approach.

7. What do you understand by the phrase 'a learning organization' as used when applied to a hospital? In your opinion is there a difference between 'a learning organization' and a 'quality organization'?

8. In this chapter we propose that clinical engineers have a role in improving healthcare organizations by applying a systems approach to the review of processes and systems

beyond those associated with medical equipment management. What unique attributes do clinical engineers have that make them effective in this regard? If a clinical engineer worked at healthcare process improvement exclusively for 5 years, and was not involved in any medical equipment management activity, would you still describe them as a clinical engineer? Explain your answer.

9. Can you identify a process in your daily work that has evolved organically over time and can be described as a system? If asked to analyze and suggest an improvement to this process using a systems approach, how would you go about it?

10. What do you understand the term 'adverse event' to mean in the context of the use of medical equipment? Write brief notes on the advantages of taking a systems approach to the investigation of adverse events.

11. In your opinion, what are the top three barriers preventing clinical engineers from contributing to process improvement initiatives aimed at improving clinical care processes?

CASE STUDIES

CASE STUDY CS2.1: REDESIGN OF PALLIATIVE CARE SERVICES INVOLVING MEDICATION DELIVERY BY SYRINGE PUMP

Section Links: Chapter 2, Section 2.3

ABSTRACT

Palliative care, for patients with life-limiting illness, is delivered across all healthcare sectors, primary care and acute hospitals, often requiring subcutaneous medication delivery by syringe pump. To improve the experience of medication delivery across all sectors, a clinical engineer and palliative care specialist convened a multidisciplinary team to redesign the process.

Keywords: Care process; Quality; Syringe pump; Medication delivery; Palliative care; Team

NARRATIVE

Palliative care is multidisciplinary improving the quality of life of patients and their families facing problems associated with life-threatening illnesses. Supporting living well and dying well, it includes relief from pain and other symptoms (diarrhoea and vomiting). When oral and other administration routes are not possible, medications are delivered subcutaneously, often by a syringe pump delivering low-volume diluted medications over defined period (e.g. 20 mL over 24 h). A narrow-bore tube (infusion line) connects the syringe to a flexible cannula inserted under the skin.

An audit of over a thousand syringe-driver medication infusions across a region (population 600,000 covering acute hospitals, community treatment centres, hospices, care homes and patients' homes) revealed many not delivered to time. Inconsistent policies and procedures hindered practice and patient transfer between care sectors, with different equipment in use: syringe pumps, infusion lines, cannula and protocols. The reasons for the frequent failures to deliver the prescribed medication were poorly understood, vaguely ascribed to 'equipment' and 'user'.

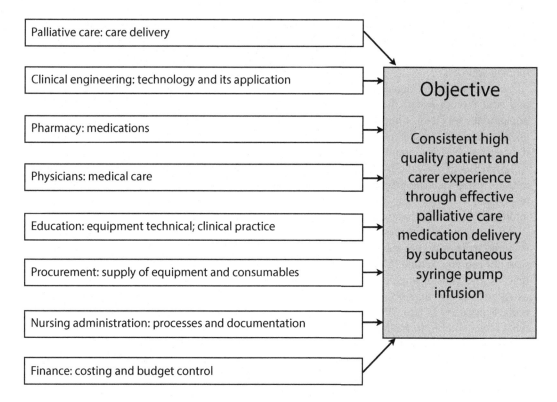

FIGURE CS2.1A   Multidisciplinary team, all signing up to the Objective.

Palliative care and clinical engineering set about finding a solution to improve patient care, but soon realized that achieving it required contributions from professionals with various skill sets, and by carers in the different sectors. A multidisciplinary project team was assembled, listening to views, first clarifying and confirming the objective (Figure CS2.1A).

High-quality care requires coordinated care comprising several elements: patient and carer support systems, consistent care processes, common equipment, appropriate medication lists, documentation, technical support, logistics and procurement, finance (start-up and operational) and nursing and physicians. All elements and their interrelationships had to be carefully considered.

Equipment problems included no standardization, with the existing syringe pumps not complying with current standards. Replacement required funding (links to Finance and Procurement). The pumps would be used in all sectors, including patients' homes where there would be no immediate professional support. Consequently, device selection would require nursing and technical staff to judge clinical, ergonomic, technical and aesthetic merits, guided by procurement and financial considerations.

Pharmacy reviewed current palliative care pharmaceuticals and their compatibilities, developing guidance lists of medications that could be safely combined for delivery over a 24 h period. Pharmacy supported Procurement and the technical team in selecting multidrug compatible infusion lines and cannula.

Palliative care reviewed the care bundles (care processes, protocols and documentation), taking account of the audit evidence. Bundles included charts and documentation to support practice and movement of patients between care settings, ensuring seamless transitions.

Clinical engineers, in discussion with nursing staff, developed equipment support plans (Chapter 6) consistent across care settings. The project leads, Procurement and Finance investigated start-up and operational resource and financial requirements, identifying and agreeing funding sources.

Core education curricula and training plans (start-up and continuing) had to be developed and education provided. Training for both clinicians and technical staff covered not just the equipment's mechanics but safe clinical use. A patient–carer teaching pack was developed enabling patients and carers, particularly in the home setting, to self-manage their own device if they so wished, importantly supporting independence and autonomy.

Specialists in the various elements took the lead in developing optimum solutions for their areas of expertise. The project leads did not require in-depth knowledge of each element, a responsibility delegated to each element lead. However, the project leads did require to be informed of problems in any element and of boundary issues between elements.

An example of cohesively working through common problems was determining the required syringe pump accuracy, typically expressed clinically as what constitutes an infusion not running to time. This is of interest to palliative care specialists and clinicians, pharmacists and clinical engineers. The technical accuracy of the pump is typically expressed as an error of less than 5%. But over a 24 h period, that equates to an infusion finishing 1 h and 24 min early or late. What inaccuracy is pharmaceutically acceptable to the patient and what is acceptable to palliative care nurses, perhaps visiting home-patients once per day? The three professional groups, working through the implications affecting each other and across the disciplinary boundaries, helped define a pharmaceutical and clinically acceptable working definition.

The myriad of clinical, pharmaceutical, technical, document control, procurement, financial and senior management issues that had to be addressed can obscure the objective. A structured systems approach, with a clear objective whilst enabling each element to be thoroughly examined, is required. This led to the Keystone model, with its keystone, which is the patient and carer, the focus and the arbiter for determining solutions to issues (Brooks Young and Amoore 2012).

ADDING VALUE

Patients and carers benefit from a care process designed to support their needs and from a system operating seamlessly across care settings. Professional carers benefit from a structured process supporting safe practice in busy environments. The healthcare organization benefits from improved care quality. The project and the shared education developed mutual understanding and appreciation between nurses and clinical engineers, leading to improved cooperation in resolving post-implementation problems.

Increased disposable battery costs were compensated for by reduced infusion lines costs and reduced maintenance costs arising from the improved technical support. The net effect is an increase in benefits with no increase in cost.

Benefits : Cost ∴ Value

SUMMARY

Consistent services for subcutaneous palliative care medication delivery across all care settings were developed. The structured methodology of the systems approach kept patient and carer needs as the focus, whilst enabling detailed consideration of each of the complex elements and their interrelationship involved.

SELF-DIRECTED LEARNING

1. How can you as a clinical engineer work with clinical and other colleagues to improve clinical care processes using the systems approach? What skills would you require and how would you develop those skills?

2. Think of a clinical care process that needs improving; it could be patient flow between theatre and critical care or the flow of patients through the Emergency Department. What different team members will be required to analyze the system and to test improvement models? How can you ensure that the objective is clearly defined and that all the necessary elements are identified? Discuss which disciplines would be required to clarify different elements in the clinical care process.

REFERENCE

Brooks Young P. and J. Amoore. 2012. Subcutaneous infusions for pain and symptom control in palliative care: Introducing the Keystone model. *19th International Congress on Palliative Care*, Montreal, Quebec, Canada, October 2012.

CASE STUDY CS2.2: CLINICAL ENGINEERS USING A SYSTEMS APPROACH TO INVESTIGATE ADVERSE EVENT CAUSES

Section Links: Chapter 2, Section 2.4.6; Chapter 1, Section 1.5.2

ABSTRACT

Investigators of adverse events involving medical devices typically suggest that the causes are use error or device fault. However, a fuller understanding is achieved if the associated process of care is analyzed holistically from a systems perspective. This will examine any latent background factors, the trigger event and key elements, namely the medical device, the user, the environment of care and the patient.

Keywords: Adverse events; Systems approach; Safer healthcare; Learning; Open culture

NARRATIVE

During routine surgery, a patient suffered abdominal burns from an electrosurgery device. Investigation revealed that the burn was caused when a finger-switched electrosurgical pencil (an electrosurgical tool), placed on the patient's abdomen, was inadvertently energized by the surgeon intending to use forceps. The surgeon, intending to use the forceps, pressed the foot-switch assuming it was connected to the forceps; lack of electrosurgical current to the forceps caused the surgeon to increase the electrosurgery power. However, the footswitch was connected to the pencil, energizing it and burning the patient's abdomen where it had been placed. Figure CS2.2A shows the connections of the monopolar surgical tools and footswitch to the electrosurgical device.

Investigation of the causes begins by reviewing the care environment and its elements: patient connected to an electrosurgery device controlled by the surgeon, operating theatre and its supporting infrastructure. Two surgical tools were attached to the electrosurgery device: the surgeon wished to use the forceps, leaving the unprotected pencil on the patient's abdomen.

Trigger event. The immediate cause was the surgeon inadvertently energizing, at increasing power, the pencil left on the patient's abdomen rather than in its safety plastic holder.

Front Panel
View

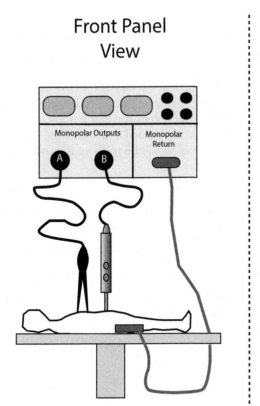

Rear Panel
View
(Hidden from surgeon's view)

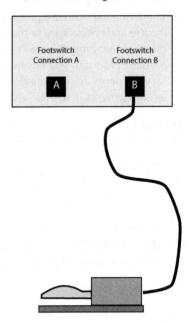

FIGURE CS2.2A   Connections to electrosurgery device.

Latent causes.
1. Care environment: The electrosurgery device had been returned from routine service, with the hospital's Equipment Support Plan (ESP) (Chapter 6) stipulating checking the footswitch operation in both sockets 'A' and 'B'; it was returned to theatre with the footswitch connected to the socket last checked (socket B).
2. Closer investigation of the manufacturer's operating manual revealed instructions that the device should not be stored with the footswitch connected. The hospital's electrosurgery device ESP stipulated checking of the footswitch functionality, but not that the footswitch should be disconnected after the service.
3. The theatre orderly placed the electrosurgery device on its theatre pendant for the next operation, without noticing that the footswitch was connected to the wrong socket.
4. The surgeon first used the pencil tool, controlling it with its own *on–off* switch, afterwards placing it temporarily on the patient's abdomen whilst intending to use the forceps to seal some bleeding blood vessels. The theatre table had plastic holders for surgical tools not in use, but theatre time pressures led the surgeon to place the pencil on the patient's abdomen.
5. The electrosurgery machine's design precluded the surgeon from seeing to which rear socket the footswitch was connected.
6. There was no routine theatre checklist for the connections to the electrosurgery device.

The patient's burn was associated with five elements: (1) the pencil tool, when not in use, had been placed on the patient's abdomen and not in its safety holder whilst the surgeon attempted to use the forceps; (2) the medical device did not clearly enable the surgeon to see which tool the footswitch controlled; (3) the theatre orderly had not ensured that the footswitch was connected to the socket corresponding to the forceps; (4) the repair technician had returned the electrosurgery device to theatre with the footswitch connected; and (5) protocols: the theatre checklist did not include checks of the connections, nor did the technician's ESP stipulate removal of the footswitch after service.

James Reason's 'Swiss Cheese Model' (Reason 2000) can usefully summarize the combination of failures which led to the patient being burnt. The identification of the failures can help develop lessons for preventing recurrences: (1) surgeons should never place tools not in use on a patient, but in the protective plastic holder; (2) protocols in theatres and in Clinical Engineering Departments (ESPs) should ensure that electrosurgery devices are never stored or transported with the foot pedal connected; and (3) electrosurgery device design should indicate to the user (i.e. on the front panel) which tool(s) will be energized by a foot pedal.

ADDING VALUE

Healthcare is made safer by fully identifying the causes of adverse events, followed by learning and implementing the lessons learnt to minimize the risk of recurrence and/or mitigate the consequences. Patient, family and friends gain an understanding of what went wrong, with reassurance that lessons have been learnt and implemented; medical and nursing staff gain an understanding the causes and how to prevent future recurrence; the healthcare organization learns how to make its healthcare safer; and manufacturers learn how to design safer products.

Full details of the incident should be reported to national reporting agencies, sharing lessons. The national reporting agency may then wish to issue a safety alert warning users in other healthcare organizations.

In this example, there are benefits of fully understanding the causes of the adverse event to all the stakeholders. There is a time cost to the clinical engineer's involvement in the incident investigation, but this is outweighed by the benefits realized.

Benefits : Cost ∴ Value

SUMMARY

The patient was burnt by an electrosurgery tool that was not in use but left on his body. The electrosurgery device was returned from service with the footswitch connected to the 'incorrect' socket outlet on the back panel hidden from the operator's view. The device was used with two tools connected to the two sockets on its front panel. In consequence, when the footswitch was operated, it energized the wrong tool.

SELF-DIRECTED LEARNING

1. What systems approach tools or guidance can support adverse event investigation?
2. How can 'the Swiss Cheese Model' help understand the causes and identify barriers that can help prevent recurrences?
3. Apply the systems approach to investigate an adverse event looking for protective barriers or procedures that can minimize the risk of recurrence. This adverse event could be one you have encountered, heard about or read about.

REFERENCE

Reason J. 2000. Human error: models and management. *British Medical Journal*, 320: 768–770.

CASE STUDY CS2.3: USING PHYSIOLOGICAL MEASUREMENT DATA STORED IN CLINICAL INFORMATION SYSTEMS TO OPTIMIZE CARE DELIVERY FOR POST CARDIAC SURGERY PATIENTS PASSING THROUGH A CARDIAC INTENSIVE CARE UNIT

Section Links: Chapter 2, Section 2.6.2; Chapter 7, Section 7.3.4

ABSTRACT

Patient care is improved by reducing the time required to wean critical care patients off mechanical ventilation. Analysis of care processes in a Cardiac Intensive Care Unit suggested improvements whose impacts were assessed by measuring the length of stay in the unit.

Keywords: Run chart; Weaning strategy; Mechanical ventilation; Data mining; Length of stay

NARRATIVE

Cardiac surgery intensive care units (CICU) aim to stabilize patients and wean off mechanical ventilation as quickly as possible so that patients can be transferred back to general cardiac surgery units. This is good for the patients and ensures optimal use of the expensive CICU facility. The CICU multidisciplinary team, which included a clinical engineer, met to review the performance of the unit aiming to reducing the length of stay of patients post cardiac surgery.

The clinical engineer, who managed the CICU's clinical information system (CIS), suggested that the CIS could be used to measure the actual length of stay. As all patients in the CICU are monitored constantly, their length of stay can be calculated by comparing the timestamp of the first and last heart rates recorded for each episode in the CICU. The average length of stay of all patients who went through the CICU in any 1 week was calculated to reduce the influence of individual patient variations on the performance measure.

Progress was reviewed by plotting the weekly average length of stay on a run chart (Figure CS2.3A). A run chart, which is simply a line graph of these data plotted against time, is useful in process improvement because it helps identify trends or patterns, revealing important information about how the process as a whole is performing. Being time defined, it also allows for changes in the system's performance to be identified and associated with process changes introduced at a particular time. Easily understood and good for sharing information, run charts not only show the effectiveness of change, but hint at what improvements might be made.

Baseline data were collected for 15 weeks prior to initiating any improvements. The run chart for time period (A) reveals that not only is the target 12 h average length of stay exceeded but there is a wide week-to-week variation in the average length of stay.

The large week-to-week variations in length of stay revealed in the baseline measurements (time period A, Figure CS2.3A) suggested to the multidisciplinary team the need to check for possible variations in practice. This review showed different medication prescription practices leading to the first improvement plan of standardizing practice across all the doctors in the unit. The impact was to reduce the week-to-week variation in the length of stay over the next 15 weeks (time period B). However, it did not lower the average length of stay. Nevertheless the intervention improved the system as a whole.

Encouraged by this early success, the multidisciplinary team reconsidered the practices of care, leading to the recognition that weaning off mechanical ventilation, a crucial stage in critical care processes, was not being actively managed. So, at week 30, the second improvement intervention was made, the introduction of an early ventilation weaning policy where nurses were encouraged and facilitated in implementing an early weaning strategy.

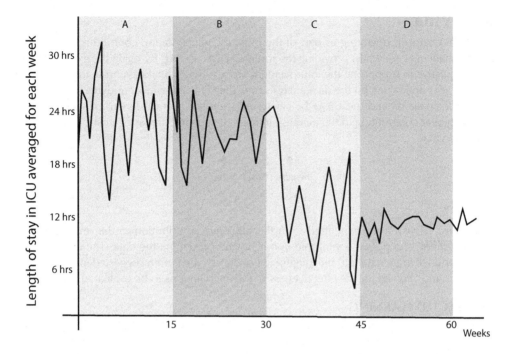

FIGURE CS2.3A   A run chart used to explore the effectiveness of changes in the performance of an Intensive Care Unit in meeting its length of stay targets. The time axis is divided into four periods (A to D) each associated with a particular phase of a quality improvement project.

Early weaning strategies aim to get patients off their ventilator support as soon as possible, without compromising their care. This led to a dramatic decrease in the average length of stay from about 20 h to about 12 h (time period C, Figure CS2.3A). However, the week-to-week variations persisted.

The new strategy of supporting early weaning had produced some positive outcomes, but nurses at the multidisciplinary review meetings suggested the need for better training in how to safely wean patients. This led, at week 45, to the third quality improvement change, namely an intensive period of training for all nurses in how to safely apply the early weaning strategy. In the following 15 weeks, the variation was reduced and the unit was meeting its target of having an average length of stay of 12 h (Figure CS2.3A, time period D).

The run chart is a simple yet powerful tool for demonstrating effects of process change in a system. The multidisciplinary team was able to use it to clarify the system performance, with its displays of performance variability suggesting the need to standardize various practices.

SYSTEMS APPROACH

This exercise followed a classic CATWOE systems approach. The Customers were the patients, whose care process is enhanced by reducing the time being mechanical ventilated; hence, the measured objective of reducing the length of stay. The Actors were the multidisciplinary team caring for the patients. Reviews led to suggested changes or Transformations, each of which was measured and displayed on the run chart. The Owners were the CICU managers and leaders, keen to improve performance, taking an overall World view of the CICU as a tool for providing patient care that would enable each patient to safely return home after high-quality safe care. The process is subject to External constraints, the demands on critical care beds and resources.

ADDING VALUE

The clinical engineer, operating as part of the multidisciplinary team, contributed to supporting and enhancing care whilst providing the routine HTM support. By going the extra mile and showing initiative in supporting the improvement process through taking part in the multidisciplinary reviews and taking on the data analysis role, the clinical engineer added value. Benefits increased for patients and the CICU by safely reducing the average length of stay, with the clinical engineer contributing to the increase in benefits without increasing costs, leading to an increase in Value.

Benefits : Cost ∴ Value

SUMMARY

The multidisciplinary team benefitted from the objective contribution to the reviews by the clinical engineer who played a vital role in mining and presenting the data. The clinical engineer converted the raw data into meaningful information, which both suggested improvements to clinicians and allowed for the effectiveness of the interventions to be evaluated.

SELF-DIRECTED LEARNING

1. Discuss this case study by describing in detail each of the CATWOE elements. Can the CATWOE process facilitate quality improvement projects?
2. How can you as a clinical engineer support and enhance care by mining data and presenting it as meaningful information? This could be data from measurements of healthcare technology processes or healthcare processes.
3. In this example, the CICU viewed itself as a system and optimized its elements to produce a better output and good care with reduced length of stay. However, it is also possible to view the CICU as an element or subsystem in the wider hospital system. Comment on how the CICU staff can design the integration of the new CICU care process with other elements of the wider hospital system?

CASE STUDY CS2.4: REDESIGNING A CLINICAL CARE PATHWAY

Section Links: Chapter 2, Section 2.6.2; Chapter 7, Section 7.3.1

ABSTRACT

A multidisciplinary team were convened to run a quality improvement project whose aim was to improve the care pathway for patients presenting in the Emergency Department with a hip fracture.

Keywords: Process flow; Mapping diagram; System change; Emergency Department

NARRATIVE

Hip fractures (proximal femur fractures) are one of the most common fractures in older people and the most common cause of injury-related mortality. Hip fractures commonly result in significant functional decline (Tinetti and Williams 1997). The National Strategy to Prevent Falls and Fractures in Ireland's Ageing Population reported that less than 50% of people who survive a hip fracture regain their pre-fracture level of function, less than 50% return directly home and over 20% are admitted to long-term care (Department of Health and Children 2008).

The total in-patient cost for all fractures in the over 65-year age group is €58 million; hip fractures represent two-thirds of this cost. Factors that contribute to variation in length of stay are generally organizational rather than clinical: that is delayed surgery, lack of integrated care and early rehabilitation, availability of downstream beds and community rehabilitation (Moran et al. 2005; Sund and Liski 2005; Sund et al. 2011).

The evidence base for improved hip fracture care is growing, and in general prompt, effective, multidisciplinary management improves outcomes and reduces overall costs (Kumar 2012). There have been substantial developments in the care of hip fractures in many acute hospital settings though collaborative practice between the orthopaedic service, orthogeriatricians (Vidan et al. 2005) and anaesthesia (Swanson et al. 1998; Thwaites et al. 2005; Khan et al. 2013).

The clinical lead of the Emergency Department (ED) initiated a quality improvement project to examine and improve the care pathway for patients presenting with hip fractures. The focus of the project was to have 100% of medically fit, emergency patients presenting with a hip fracture admitted to theatre within 48 h of ED registration. The starting point for analysis was the ED viewed as a clinical microsystem that connected to other microsystems of an end-to-end pathway.

A multidisciplinary team from within the ED department was convened to conduct the project. The team consisted of doctors, nurses, department administration and a clinical engineer. The aim of the project was to reduce the time interval from first presentation at the ED to having the necessary surgery. A study had been conducted which measured the presentation to surgery times, so there was a quantitative measure of the current clinical microsystem process so the effect of changes could be assessed by comparing the times before and after the change.

The ED was viewed as a system with the staff and their function described as elements of the system, and the processes and information flows viewed as the interconnections between these elements. Whilst the clinical and administrative staff could name and describe the activity that comprised the patient journey, the clinical engineer was able to visualize it as a process flow diagram that identified the interdependencies between different people and the times at which different actions or decisions were made.

The clinical engineer was able to use the process flow map to highlight a delay in how the system functioned. Front-line staff in the ED could quickly identify patients with suspected hip fractures who might need orthopaedic surgery, yet could not initiate the surgical booking process until the fracture was confirmed by imaging and the patient had been seen by a member of the orthopaedic surgical team. Only then could the operation be booked on the theatre schedule. This part of the process took several hours, and sometimes this delay meant that patients needed to be scheduled for surgery the following day, and this in turn resulted in significant delays.

The multidisciplinary team proposed a change to how the system worked. They identified that if the patient could be booked for surgery early, before imaging and orthopaedic consultation, the process as a whole would be faster. Once imaging confirmed the hip fracture, technology could be used to speed up the orthopaedic consultation, in particular using teleconferencing and remote viewing of the images using the PACS system. If these changes could be introduced, then once the fracture and suitability for surgery were confirmed, the process could proceed without delay to the pre-booked theatre, significantly reducing the overall time to surgery.

To implement this proposed improvement required change to the care plan and how the triage and ED doctors and nurses acted. It also required a change outside the ED clinical microsystem, in the wider hospital system, specifically, how the orthopaedic team worked and managed the scheduling of the orthopaedic theatres.

To implement the changes within the ED, the care plan was amended and the purpose and operation of the changes were explained to all involved. To implement the changes outside the department required the ED clinical lead to work within an interdisciplinary model where patients' needs were prioritized as a new end-to-end clinical pathway was implemented. The changes were introduced and measured using the concept of multiple 'Plan–Do–Study–Act' (PDSA) tests of change (Gruettner et al. 2012). This was assessed by measuring the time from ED presentation to time of surgery. The shorter the time, the better the outcome.

The data following the improvement processes revealed a substantial improvement. In the 2 months after the implementation, the performance went from 50% of patients getting to theatre within 48 h to 80%. This has now increased to in excess of 90% with the delays medically justified in those who breached the targets.

The voice of the ED lead physician:

> "The engagement of clinical engineers in quality improvement projects is central to the successes to date. The process-mapping tool was an invaluable tool to drive improvement. This tool formed the basis of structured process meetings for each step of the Hip Fracture Journey at which meaningful solutions to optimise the flow of patients were identified. It allowed us to implement changes using both parallel and sequential changes developed and implemented in different parts of the pathway."

ADDING VALUE

The clinical engineer added value by taking initiatives that reduced the delays between patients attending the Emergency Department with fractures and their surgery. The changes in processes did not increase the costs which are expected to have decreased through lower complications caused by delays in treatment – though these reduced costs are difficult to quantify.

Benefits : Cost ∴ Value

SUMMARY

This project focused on improving outcome for patients. The multidisciplinary team drew on their collective knowledge and experience to improve the care pathway. Clinical insight and experience informed the process throughout and the project was clinically led. The clinical engineer's contribution was not related to healthcare technology. It was their ability to analyze the existing functioning state of the ED as a clinical microsystem and to represent that graphically, which allowed them to contribute significantly to the multidisciplinary team. As a result of their participation in the team, the benefits for patients were increased, increasing value.

SELF-DIRECTED LEARNING QUESTION

1. This project was delivered using the methodology for a quality improvement project as described in Chapter 7, Section 7.3.1. It also followed a systems engineering approach (Section 2.3). Can you analyze this case study and identify which of the system engineering steps from Figure 2.3 were followed and which were not?

REFERENCES

Department of Health and Children. 2008. Strategy to Prevent Falls and Fractures in Ireland's Ageing Population. Report of the National Steering Group on the Prevention of Falls in Older People and the Prevention and Management of Osteoporosis throughout life. Dublin, OH: Health Service Executive, National Council on Ageing and Older People and Department of Health & Children. https://www.hse.ie/eng/services/publications/olderpeople/Executive_Summary_-_Strategy_to_Prevent_Falls_and_Fractures_in_Ireland%e2%80%99s_Ageing_Population.pdf (accessed 2016-03-31).

Gruettner J., Henzler T., Sueselbeck T., Fink C., Borggrefe M. and T. Walter. 2012. Clinical assessment of chest pain and guidelines for imaging. *European Journal of Radiology*, 81(12): 3663–3668.

Khan S.K., Weusten A., Bonczek S., Tate A. and A. Port. 2013. The Best Practice Tariff helps improve management of neck of femur fractures: A completed audit loop. *British Journal of Hospital Medicine*, 74(11): 644–647.

Kumar G. 2012. Protocol-guided hip fracture management reduces length of hospital stay. *British Journal of Hosp Medicine*, 73(11): 645–648.

Moran C.G., Wenn R.T., Sikand M. and A.M. Taylor. 2005. Early mortality after hip fracture: Is delay before surgery important? *Journal of Bone & Joint Surgery*, 87(3): 483–489.

Sund R., Juntunen M., Luthje P., Huusko T. and U. Hakkinen. 2011. Monitoring the performance of hip fracture treatment in Finland. *Annals of Medicine*, 43(Suppl 1):S39–S46.

Sund R. and A. Liski. 2005. Quality effects of operative delay on mortality in hip fracture treatment. *Quality and Safety in Health Care*, 14(5): 371–377.

Swanson C.E., Day G.A., Yelland C.E. et al. 1998. The management of elderly patients with femoral fractures: A randomised controlled trial of early intervention versus standard care. *Medical Journal of Australia*, 169(10): 515–518.

Thwaites J.H., Mann F., Gilchrist N., Frampton C., Rothwell A. and R. Sainsbury. 2005. Shared care between geriatricians and orthopaedic surgeons as a model of care for older patients with hip fractures. *New Zealand Medical Journal*, 118(1214): U1438.

Tinetti M.E. and C.S. Williams. 1997. Falls, injuries due to falls, and the risk of admission to a nursing home. *New England Journal of Medicine*, 337(18): 1279–1284.

Vidan M., Serra J.A., Moreno C., Riquelme G. and J. Ortiz. 2005. Efficacy of a comprehensive geriatric intervention in older patients hospitalized for hip fracture: A randomized, controlled trial. *Journal of the American Geriatrics Society*. 53(9): 1476–1482.

CASE STUDY CS2.5: CHANGING THE WAY CARE IS DELIVERED THROUGH INTRODUCING NEW TECHNOLOGY

Section Links: Chapter 2, Section 2.6.3; Chapter 7, Section 7.2.3

ABSTRACT

Patient requests for care in the community led a hospital to include in its strategic aims moving care, where appropriate, into the community; this would also help manage the increasing pressure on hospital services. As part of a planned replacement programme exercise, a clinical engineer identified a new technology that could transform the delivery of 24 h ambulatory ECG investigations, moving services from hospital clinics to the community.

Keywords: Transforming care; Ambulatory ECG investigations; Cardiology Outpatient's Clinic; General Practitioner's Clinic; Telemedicine

NARRATIVE

A hospital Cardiology department was planning to replace its ambulatory ECG monitors (commonly referred to as Holter monitors). Holter monitors are small battery-operated ECG recording devices that, with minimum disruption to patients, continuously record ECGs whilst the patients carry on with their normal daily lives, typically for 24 h periods. Patients may be prescribed these investigations to detect episodic cardiac arrhythmias that may occur during the stresses of life or the quiet of sleep. A patient's ECGs are recorded on the Holter device which is then brought back to the hospital's Cardiology department where the ECG records are analyzed by specialists who then send the results to the patient's primary care physician or general practitioner (GP) for discussion with the patient.

The process that was in place involved patients attending the GP practice and hospital twice: (1) patient visits GP; (2) GP refers patient to hospital Cardiology department;

(3) patient visits Cardiology department where the Holter recorder with ECG electrodes is attached; (4) the next day, the patient again visits the Cardiology department to return the Holter recorder; (5) whilst patient returns home, the Holter recording is analyzed in the Cardiology department; (6) the Cardiology department sends the test results to the patient's GP; (7) the GP asks the patient to attend the GP office to discuss the results; and (8) the patient attends the GP practice.

The traditional SIMILAR systems engineering methodology (INCOSE 2016; Section 2.2) begins with clarifying and Stating (S) the problem, followed by Investigating (I) alternatives. In this example, however, the clinical engineer first identified, through an understanding of technological developments, the opportunity for an alternative care approach that could overcome the lengthy complicated process of Holter ECG investigations. Whilst reviewing possible replacement equipment, the clinical engineer identified that new Holter technology could speed up the patient flow process, reducing the patient travel burden.

Investigation (I) confirmed that these new Holter devices do not need to be physically connected to the reader station in the hospital; the recorded ECG can be downloaded in the GP clinic and transmitted over the Internet to the hospital's Cardiology department. This promised simplification of the patient's journey, with the patient not required to attend the hospital, only the GP clinic. The patient has the Holter monitor attached at the GP clinic, returning the following day to return the Holter monitor with its recorded ECGs. The GP then organizes the Internet transmission of the ECGs to the Cardiology department where they are analyzed and the results returned to the GP.

The clinical engineer discussed this with a Consultant Cardiologist who convened a meeting with the GPs, hospital physicians and the clinical engineer. After discussing and clarifying the problem, those present analyzed the advantages and disadvantages of the proposed service change, comparing it with the existing service ('Modelling' in the SIMILAR method). The impact of the change on patient care was discussed, assessing how this proposal would Integrate (I) with the wider healthcare service. The proposed change was approved and the clinical engineer was asked to lead a project team to procure the new technology and Launch (L) the new process.

Once the change process is launched, systems engineering demands that its effects be Assessed (A) and, where necessary, changes made and the system Re-evaluated (R). The assessment, supported by a patient survey, showed that the new service reduced the burden for patients attending the hospital's Cardiology department twice, with the associated anxiety and time out from work. The service was speeded up, with the delay between the GP prescribing the investigation and the hospital visit eliminated; the Holter recorder can be attached immediately at this clinic visit. Whilst attaching the Holter recorder increased the GP workload, this was partly offset by the reduced need for arranging patients' hospital appointments.

From the hospital's perspective, it reduced its need to hold Holter clinics, reducing the associated administration and freeing hospital staff time. The Assessment revealed that some minor changes were needed in the Cardiology department to schedule the reporting of Holter data received over the Intranet. These changes were made and the revised process positively Re-evaluated.

The clinical engineer's contribution was in identifying the potential of the newer technology and leading on the implementation project. This case study illustrates one way in which clinical engineers can contribute to transforming care and how they can facilitate telehealth solutions.

The success of the new process required more than changing the technology; it required changing roles and responsibilities. The change process is best considered as a sociotechnical system consisting of people, processes and technology (medical and IT) that together support the delivery of healthcare across different healthcare sectors.

ADDING VALUE

The new process simplified and speeded up the 24 h ambulatory Holter investigations benefiting patients and their carers. The hospital benefitted by the reduction in Holter clinics and a more effective process of acquiring the ECG data. Hospital efficiency improved.

The process did involve some additional costs, chiefly the higher cost of the more complex Holter system and the need for more Holter recording devices to equip the GP clinics than if Holter monitors were all attached at the hospital. The clinics also needed extra hardware and software and training in fitting the Holter recorders and downloading the data; the training was incorporated into the procurement tender. The net effect benefited patients and the healthcare organization, thus increasing value.

Benefits : Cost ∴ Value

SUMMARY

The process from prescription to receipt of results of 24 h Holter recordings was simplified and speeded up through the imaginative use of new technology. The change benefitted patients and healthcare organizations, with care moved from hospital clinics to GP practices. Developments in medical technology have the opportunity to transform aspects of healthcare, but realizing the opportunities requires initiative and a willingness to change established patterns of care.

SELF-DIRECTED LEARNING

1. Can you draw system diagrams of the transformation of the service identifying the patient at home, the GP, the Cardiologist in the hospital and the Cardiac Technician in the hospital who does the initial analysis? Show how these people interact and also show the flow of information. Do the system diagrams help in the planning of the change process?
2. Discuss in more detail the scenario using the systems engineering SIMILAR approach, detailing each of the steps. Refer to the INCOSE website (see references) and Section 2.2.
3. Discuss the process changes from the perspective of the patient, the GP practice and the hospital's Cardiology department. Comment on the extent to which the application of the new technology improved efficiencies.

REFERENCE

INCOSE. 2016. What is systems engineering? San Diego, CA: International Council on Systems Engineering. http://www.incose.org/AboutSE/WhatIsSE (accessed 2016-03-31).

CHAPTER 3

Key Standards, Regulations and Guidelines

CONTENTS

3.1 INTRODUCTION

Standards, regulations and guidelines affect most Healthcare Technology Management (HTM) activities. In this chapter we will look at what exactly these are, the differences between them, how each is developed and what impact they may have on the work of clinical engineers. We will discuss the key ones in each of these categories and provide you with further references and examples.

All the Standards that we will mention in the text in this chapter are listed with their full titles in 'Standards Cited' section, immediately following the formal references section.

3.2 STANDARDS

3.2.1 What Is a Standard?

In this chapter, we use the term Standards (with a capital S to distinguish them from standards of behaviour, dress, etc.) to mean those formal documents drawn up by national or international Standards bodies such as the American National Standards Institute (ANSI), the British Standards Institute (BSI), the International Organization for Standardization (ISO), the International Electrotechnical Commission (IEC) and Standards Australia. We have included relevant website addresses in 'Web Links' section towards the end of this chapter.

We take for granted that many things in life are standardized, we buy electrical appliances and expect that they work when we plug them in, we purchase fuel for vehicles and expect that it is correct for the engine and we expect shoes or clothes of a certain size to fit – an area very poorly standardized. Have you ever thought about how this is achieved? A Standard is a document that sets out best practice, the state-of-the-art or minimum performance criteria for devices and systems. An example would be the standardization of car tyre sizes. A coding system is used that has been accepted by all manufacturers of tyres that shows width, profile, construction, diameter, load, and speed rating. This allows tyres to be purchased anywhere whilst ensuring compatibility with the car manufacturers' requirements. The relevant Standard facilitates the manufacturer in designing the product, knowing that, by adhering to the Standard, the manufacturer can be sure that the product will be compatible with what the market and consumers require.

The same is true of a medical device. If you need to buy an electronic blood pressure monitor, you would prefer to buy one, which not only works and but also safe, but which has been designed, built and assessed to meet essential performance and safety criteria set out by experts. If you buy one, that is, labelled as meeting certain Standards, the company selling you the device is letting you know that these essential requirements have been met. This is good for you as the customer. It is also good for the company as conformity with the Standard is an indicator of the quality and safety of the product. Where devices are sold worldwide, compliance with agreed international Standards ensures that the device is safe and functional wherever used and that it will be compatible with infrastructure such as the power supply in different jurisdictions. So Standards are good for everyone, providing assurances of quality, safety and functionality.

3.2.2 Who Writes National Standards?

The need for standards developed in individual countries as industrialization spread in the nineteenth and twentieth centuries. When the locomotive was first invented, there were different gauges set by competing inventors and entrepreneurs for the track on which they ran. Obviously, at some point an agreement would have to be made on a standard gauge or passengers would have needed to change train at the boundary of each company's rail track. Similarly, the need for standard thread sizes for nuts and bolts was recognized as a requirement for industrial development. Pressure for standardization can come from within an industry or be consumer driven, and many are entered into voluntarily for mutual benefit. However, for more formal standardization or perhaps where safety is of concern, bodies have been set-up to oversee the setting and agreement of national standards. The agreement and adoption of a standard for a domestic electric mains plug is such an example, and in this case adherence to the Standard is clearly in the national interest (Mullins 2006).

Where national bodies have been created to oversee standards, these are usually set-up as independent associations or institutions, recognized by national governments with representatives of industry and the public contributing as necessary. Funding comes from membership subscriptions, the sale of Standards and grants from government.

3.2.3 Who Writes International Standards?

With globalization and a global marketplace for technology, the drive for international standards is both inevitable and desirable. Working in a similar way to national standards

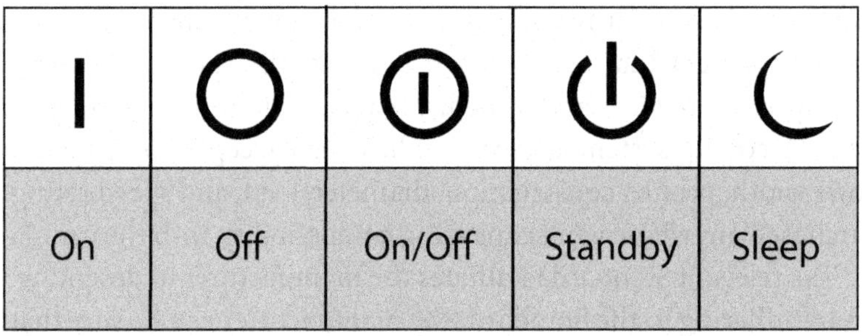

I	**O**	**⊙**	**⏻**	**☾**
On	Off	On/Off	Standby	Sleep

FIGURE 3.1 Standard symbols now used globally on many types of equipment.

bodies (NSB), international standards organizations bring together experts nominated by NSB from many countries to discuss and agree on a wide range of issues from car safety and the design of medical equipment to telecommunications and the international standard book number system.

The two major international standards bodies, the ISO and the IEC, have complimentary responsibilities: IEC deals with standardization in any area, that is, electrical or electronic in nature. ISO deals with all other areas. In practice many Standards result from formal joint working groups and either carry a joint ISO/IEC number and the logo of both organizations, or are designated as an ISO or IEC Standard depending on which organization led the joint working group.

By creating and adhering to global standardization, the exchange of technology and information becomes easier, safer and more economical with benefits for consumers and industry. Consider the symbols we now see on appliances. In the past the words 'on' and 'off' were commonly used but in a global context, symbols are now used as shown in the top row of Figure 3.1, and these are recognizable across language barriers. The symbols have been in use since 1973 and have been readily adopted and understood, appearing even on fashion accessories.

Standards cover detailed specifications of particular devices (as for the car tyre dimensions or functional characteristics of blood pressure measurement devices). They also cover management processes, and we illustrate in Case Study CS3.1 how those who manage assets realized the need for a Standard to guide their management of assets and how their recognition of this need developed the appropriate Standard (ISO 55000).

3.2.4 Types of Standard and How They Are Applicable to HTM and Medical Devices

In general, Standards can be categorized into four different types:

1. Basic Standards;

2. Group Standards;

3. Product Standards;

4. Process Standards (which may be basic or group Standards).

The ISO 16142-1 Standard, which provides guidance in the use of standards relevant to medical equipment and its management, explains these four types in the context of medical devices.

Basic Standards cover broad issues and have applicability across multiple industrial sectors: examples would include IEC 61140 addressing protection from electrical shock – a technical standard, or ISO 55000 for asset management systems – a process standard.

Group Standards cover the essential principles of a distinct group of equipment. Examples include the basic safety and essential performance requirements generally applicable to medical electrical equipment, IEC 60601-1, and the basic safety requirements for household electrical equipment, IEC 60335-1, both of which are technical standards. IEC 62353, which covers the in-service electrical safety testing of medical electrical equipment, is a process Standard in the HTM field.

More detailed Standards, specific to a particular type of product within a group, for example, all infusion pumps (IEC 60601-2-24), are called product Standards. Most product Standards deal with safety of design and construction but may also include functional requirements where these are deemed to be essential to the overall safe performance of the equipment. A process Standard applicable to a group may also be applicable at the product level and this is the case for IEC 62353.

Figure 3.2, adapted from ISO 16142-1, shows this relationship.

In their technical report (TR) paper, Vincent and Blandford (2014) describe the large number of product and process Standards that must be applied to the design and development of an infusion pump. These include the technical Standard specific to the safety of infusion devices (IEC 60601-2-24). But this particular Standard requires the support of other Standards to ensure safe and effective infusion devices. Importantly, there is the human usability Standard (IEC 62366-1) that has been developed in response to the recognition that human factors and errors associated with poor ergonomic design contribute

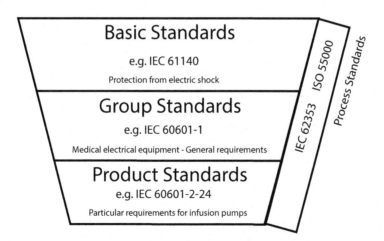

FIGURE 3.2   Hierarchy of Standards. (Adapted with permission from ISO 16142-1, Medical devices – Recognized essential principles of safety and performance of medical devices – Part 1: General essential principles and additional specific essential principles for all non-IVD medical devices and guidance on the selection of standards.)

significantly to adverse events, not least for infusion devices. The design, manufacture, and distribution of infusion devices should be governed by risk management processes (ISO 14971) and good manufacturing practice (ISO 13485). The design of the symbols on controls (ISO 15223) and of alarm systems (IEC 60601-1-8) are supported by Standards. Increasingly, infusion devices are used in community and home settings and in this context IEC 60601-1-11 applies.

Hegarty et al. (2014) describe the process of assessing a hospital's medical information technology (IT) network risk management practice by using the IEC 80001-1 process Standard. The authors describe the need for the Standard and stress that conformance with it requires the hospital to take ownership of the risk management of its medical IT network. The authors discuss assessing conformance of the network system against the Standard, making recommendations where non-conformity was found. As an aside, but pertinent to the general theme of this book, it is worth drawing attention to the remarks made by Hegarty et al. (2014) characterizing the resulting clinical information system as a "sociotechnical system consisting of people, processes and technology that together deliver a care process that is standardized, measurable and operates within a quality cycle".

In summary, Standards at all three levels, basic, group and product, as well as process Standards, have been developed that support the design and manufacture of safe and effective healthcare technology and the management of it in use.

It is important that you understand how the group and product Standards in the IEC 60601 series work together and with the other Standards mentioned, providing inputs to and taking outputs from each other. Dealing with these Standards as separate, unrelated documents undermines their real value in assuring safety in an efficient and effective way (M. W. Schmidt, 2015, personal communication).

Table 3.1 presents a few examples of Standards that have particular relevance to HTM activities with some that are more directly relevant to the design and manufacture of

TABLE 3.1   Examples of Standards Relevant to HTM

Standard Number	Abbreviated Title	Process Standard?
Basic Standards		
ISO 55000	Asset management systems	✓
ISO 9000	Quality management systems	✓
Group Standards		
IEC 60601-1	Medical electrical equipment (MEE) – Basic safety	
IEC 62353	Electrical safety testing of MEE in-service	✓
ISO/TS 19218-1 and -2	Coding structure for reporting adverse events	
IEC 62366-1	Application of usability engineering to medical devices	✓
IEC 80001-1	Application of risk management to IT networks incorporating medical devices	✓
Product Standards		
IEC 60601-2-x	Particular requirements for safety of specific types of MEE	
ISO 7176-1	Stability of wheelchairs	

medical devices and others dealing with quality and other aspects of management. Further details of three of these key Standards are presented in Section 3.5.

It is worth remembering that Standards are developed as voluntary documents and are not in themselves legal documents. They put forward best practice and state of the art as developed by consensus in the Standards committees. We describe the process of generating international standards in the Appendix 3A. It may be that legislators choose to refer to Standards or even make adherence to a Standard a legal requirement as part of a regulation. This process is described in the next section.

In discussing Standards, we have given as examples some of those that are relevant to the design and manufacture of medical devices and some that are relevant to the delivery of HTM services. There is unfortunately no substitute for reading Standards, regulations and guidance documents and being clear about their scope and intent which may be more restrictive than their title implies. It is vital that clinical engineers build a knowledge base to advise their organizations on matters of compliance and professional best practice. To stand one's ground demanding resources to achieve compliance and organizational effectiveness requires that the ground one is on is firm!

3.3 WHAT IS A REGULATION?

Regulations are legal documents that set out requirements for items or processes that are based on national law. In many legal jurisdictions, it is very common for a formal law – often called an Act – to be written and passed through the legislative process in such a way that allows designated authorities to write regulations to deal with the detailed implementation of the Act. The Act will specify the method by which these regulations are to be drafted and consulted upon. When finalized, these regulations will be approved by the legislature without the need for the complex procedure required to amend the enabling Act. Revisions to regulations can also be made in the same less complex way.

In many jurisdictions there are regulations that affect the HTM process. Mostly, these are general regulations that must be applied in all workplaces covering, for example, aspects of health and safety such as working on electrical equipment or the control of chemicals. One role of the clinical engineer is to understand these general regulations and how best to apply them in practice in the HTM context. Guidelines (see Section 3.4) often provide more context-specific advice.

3.3.1 Process from Act to Regulation

Regulations cover a vast range of subjects and in our context cover broad engineering safety issues; issues regarding safe systems of work, and issues regarding fair and equitable trade. In many jurisdictions an Act (a law) gives a regulatory agency of the government the authority to regulate medical devices. That agency will then have the power to make detailed regulations. Case Study CS3.2 outlines how regulations pertaining to medical devices are dealt with in the European Union (EU), the United States and in Australia.

In the United States, the historical background to the granting to the Food and Drug Administration (FDA) the power to regulate medical devices is interesting and sheds light on the relationship between legislation and regulation. The need in the United States to regulate medical devices became increasingly apparent during the early decades of the twentieth century. The legislature in the United States passed the Pure Food and Drugs Act of 1906, giving authority to protect the public against threats from harmful substances and deceptive practices. Medical devices were not covered, largely because the medical devices at the time were regarded as being comparatively simple. The existence of fraudulent medical devices was recognized, but not to the extent that regulation was warranted. However, the increasing prevalence of medical devices and the threats to health posed by fraudulent devices prompted the FDA within about 10 years to report that the 1906 Act "has serious limitations… which render it difficult to control… fraudulent mechanical devices used for therapeutic purposes" (Rados 2006).

The growth in the medical use of radiological products further strengthened the call for regulation, leading to amendments to the legislation beginning in the late 1930s that gradually extended the authority of the FDA to regulate medical devices. Initially, the FDA was given limited authority, excluding any requirement for pre-market testing of medical devices; its authority was initially limited to policing devices in use. The post–World War II developments of medical equipment with increasing functionality, such as life support equipment, and the reported failures of cardiac pacemakers in the early 1970s, prompted the U.S. Congress to pass the 1976 Medical Device Amendments into law. The FDA was given the authority to classify medical devices according to their risks and to exercise appropriate enforcement over each class, with devices in the high-risk class requiring pre-market approval. Further legislation followed, including the 1990 Safe Medical Devices Act that requires healthcare facilities using medical devices to report to the FDA incidents where it was suggested that a medical device caused or contributed to a patient's death, serious illness or serious injury. It also required manufactures to conduct post-market surveillance and empowered the FDA to order device recalls.

The amendments to the legislation governing the FDA to cover medical devices illustrates how the legislature provides authority to an agency to regulate certain products, with amendments to legislation being required to extend the authority of the agency. The authority of the regulatory agency is stipulated by the legislature, typically in response to technological and clinical developments.

An example of this process is the UK Health and Safety at Work etc. Act of 1974 (HSWA 1974). This Act sets out the broad health and safety at work principles and legal requirements and established a government agency called the Health and Safety Executive with powers to make regulations which have to be approved by the UK Parliament. One such regulation made under this act is the Electricity at Work Regulations of 1989. These do not specify in detail particular safety measures to be put in place but require those responsible to asses and mitigate risks, and thus, these regulations have a direct impact on how electrical work is carried out in HTM departments and on the safety facilities that must be provided.

3.3.2 Relationship between Standards and Regulations

Regulations often cite compliance with Standards as a means of meeting the regulatory obligation. Thus, within the EU the Medical Devices Directive (MDD) (European Council 2007a) is enacted into law in each EU member state by regulations made within that state's legal system. The legal requirement in the MDD is that a medical device must meet the essential requirements referred to in its Article 3 and set out in Annex 1 to the directive. The MDD goes on, in Article 5.1, to cite Standards: "Member States shall presume compliance with the essential requirements referred to in Article 3 in respect of devices which are in conformity with the relevant national standards adopted pursuant to the harmonized standards". European Standards are known as European Norms (ENs); see Section 3A.3. A 'harmonized standard' is an EN Standard that has been given greater official status by being listed in the Official Journal of the EU. EU Regulation 1025/2012 (European Council 2012a) requires, in Article 3, clause 6, that any harmonized EN Standard must be published as a national Standard (though translated into the national language) and any conflicting national Standard must be withdrawn.

Thus, the equivalent national Standard, identical to the EN Standard, can be used in part or in whole to meet the medical devices regulations.

It should be noted that at the time of writing, negotiations are taking place within the EU to replace the Medical Devices Directive with a Medical Devices Regulation (MDR). The Europe-wide legal impact of an EU regulation is somewhat stricter than a Directive, but in this case, the fundamental provisions are much the same, though exact articles and clauses cannot be quoted here because only drafts are available. It is expected that the new MDR will come into effect in the first half of 2017 with a 3-year transition period.

There are also examples where meeting a Standard is explicitly called for in a regulation. For example, in Part P of the Building Regulations for England and Wales, there is an explicit requirement for electrical installation work to meet BS 7671 *Requirements for electrical Installations*. In this case, meeting the Standard is a legal requirement.

3.4 WHAT IS A GUIDELINE?

Guidelines are published documents (PD) that usually go into the more practical application of the regulations themselves or are linked in some way to a regulation. They are less binding in a legal sense than regulations but are likely to be more context specific. The weight that a guideline carries will depend on its source. Guidelines that come from government agencies are often referred to as Codes of Practice. These are usually linked to specific regulations and therefore carry significant weight. Practitioners don't have to adopt in detail the methods and procedures described in the guidelines, but if they don't, in the event of problems, they will need to show that their way was at least as safe and effective in meeting the legal regulation. Even stronger in the United Kingdom are Approved Codes of Practice.

In Case Study CS3.3, we illustrate how both a professional body (the Association for the Advancement of Medical Instrumentation [AAMI] in the United States) and a government agency (the Medicines and Healthcare Products Regulatory Agency [MHRA] in the United Kingdom) separately produced guidelines for the management of medical devices. Both draw upon the expertise of clinical engineers and other relevant bodies, and with consideration given to national context and international regulations, recommended practice is established and documented.

3.4.1 What Is the Relationship between Guidelines and Standards?

Some Standards are developed specifically as guidance documents and are then designated by ISO or IEC as TRs. When such Standards are adopted and published in the United Kingdom, they are given the designation PD; whilst in the United States, they are termed Technical Information Reports (TIRs). An example is IEC/TR 80001-2-4:2012 *Application of risk management for IT-networks incorporating medical devices. Application guidance. General implementation guidance for healthcare delivery organizations.* This is published in the United Kingdom by BSI as PD IEC/TR 80001-2-4:2012. In the United States it is published as ANSI/AAMI/IEC TIR 80001-2-4:2012. It is usual for international Standards to be adopted and published by NSBs but the exact designation may well be slightly different in different parts of the world. However, the origin and title will always be clear as in the earlier example.

Guidelines from national or international expert groups may influence Standards. A good example is the influence on the Standard for and application of blood pressure measuring devices, ISO 81060. Physicians, concerned by perceived accuracy limitations of the emerging automatic non-invasive sphygmomanometers, worked with professional groups including AAMI in the United States and the British and European Hypertension Societies to produce consensus guidelines for the non-invasive measurement of blood pressure. These influenced the development of the relevant Standard including the methods and validation protocols for assessing the clinical accuracy of these devices.

3.4.2 What Is the Relationship between Guidelines and Regulations?

Codes of Practice that come from government agencies are usually linked to specific regulations and carry significant weight.

Guidelines that are not linked to specific regulations also may originate from government or quasi-government agencies. These are often the result of consultative exercises by the agency with professional bodies but have the weight and authority of the agency behind them. An example is the publication in the United Kingdom by the MHRA of a bulletin Managing Medical Devices described in Case Study CS3.3 (MHRA 2015). Sometimes regulations are issued by government agencies without consultation, and this can lead to difficulties if what is being suggested is impractical or expensive to implement.

Most professional bodies issue guidelines on matters relevant to the work of their members. Indeed, this is one of the expected roles of a professional body. Members with expertise in a particular field contribute to the drafting of these guidelines which often address areas of activity covered by regulations and, as such, provided they are kept up to date, can be regarded as 'best practice'. Basing departmental or personal practice on such guidelines would provide a strong defensive argument in the event of some untoward incident.

Finally, there are guidelines written within departments. These may cover topics where no other guidelines exist, or they may put externally available guidelines into a local context. They may describe a practical methodology for meeting regulations, or they may set out a consistent way of carrying out a particular task – often referred to as standard operating procedures (SOP). It is always a good idea to include reference to the source regulation or guideline in any internal documents. For example, a Clinical Engineering Department (CED) may want to ensure a standardized approach to carrying out preventative maintenance by all members of its team. This may be achieved through the development of an SOP as described in Case Study CS3.4.

3.5 KEY STANDARDS THAT INFORM THE PRACTICE OF HEALTHCARE TECHNOLOGY MANAGEMENT

3.5.1 Introduction

There are many Standards that inform the practice of HTM, with clinical engineers using them to guide the planning and execution of HTM. However, there are three suites of Standards that impact significantly.

The ISO 55000 Asset Management suite is a key tool that will enhance structured and professional HTM, contributing to healthcare organizations by enhancing the value of their medical equipment assets. Enhancing value, the ratio of benefit to cost, is an important objective of HTM as we have discussed in Chapter 1.

The ISO 9000 suite provides a methodology for quality management in an organization, whether it is a production or a service delivery organization. It is readily applicable to HTM.

Both ISO 55001 and ISO 9001 are complimentary. Their authors have taken care to ensure that they are compatible, and so implementing one does not impede the implementation of the other. ISO 55000 focuses specifically on asset management whilst ISO 9000 can be implemented to cover the detailed processes and procedures to ensure good quality asset management and also other activities within an organization such as manufacturing or marketing.

Third, the IEC 60601 suite of Standards is specific to healthcare and is a combination of a group Standard (IEC 60601-1) and a whole series of product Standards (IEC60601-2-xx) dealing with all aspects of the design, construction and safety of medical electrical equipment.

Case Study CS3.5 illustrates a scenario in which all three Standards mentioned earlier are made use of in the purchase of an item of medical equipment.

3.5.2 The ISO 55000 Suite of Standards

This suite of Standards is intended to be used by any industry sector and sets out the principles and concepts to be considered by those with responsibility for optimizing asset management within their organization. It is not addressed specifically to the healthcare sector. The Standard takes a holistic view and considers how best to realize the value of assets for an organization and its stakeholders. So it fits well with the challenge of managing medical devices as assets for the benefit of patients and healthcare organizations. We consider that it is the key standard for HTM activities.

The ISO 55000 series consists of:

ISO 55000 *Asset management – Overview, principles and terminology;*

ISO 55001 *Asset management – Management systems – Requirements;*

ISO 55002 *Asset management – Management systems – Guidelines for the application of ISO 55001.*

Clinical engineers need to understand this Standard as it will become adopted widely and is probably the most relevant current Standard in HTM. The ISO 55000 series provides a universally applicable, general purpose, best practice specification and methodology for managing an organization's assets which is readily applicable to the management of healthcare technology assets.

The origins of the ISO 55000 Asset management suite, first published in 2014, have been described in Case Study CS3.1. ISO 55000 sets out principles and terminology that support ISO 55001; the thrust of which is to provide requirements for an asset management system within an organization. Although assets have traditionally been seen as physical in nature, ISO 55001 also includes other types of assets such as software, intellectual assets, brands and agreements; indeed anything, other than people, which has value to the organization.

ISO 55001 requires the establishment of an overall asset management system in which assets and the value they bring to the organization are considered in association with identified stakeholders, leadership, evaluation and evidence of continuous improvement. The standard itself is a 'high-level' approach to planning, managing, monitoring and improving a portfolio of assets. The clinical engineer should be aware of the documentation and processes proposed in order to understand the wider picture of how the healthcare organization's objectives translate into asset and department objectives, plans and procedures.

The Standard suggests that an asset management system (AMS) can focus on all the organization's assets or on a sub-set defined in the scope of the system. The paradigm proposed in this book is that a formal AMS should cover all medical devices, or at least, all reusable medical devices – medical equipment – used throughout the healthcare organization.

The top management of the healthcare organization, having the wider view of projected patient activity, new initiatives and changing demographics, are responsible for the setting of overall clinical and resource plans and objectives. The development and implementation of the healthcare organization's strategic objectives are outside the scope of ISO55001 which focuses on an asset management system. It is important to note that whilst the ISO55000 Standard does not dictate the organization's strategy, it lays down methods that help to ensure that the assets and the system for managing them support the organization's strategy. Further, the management of the CED will have a part to play in contributing to the organization's objectives through technical input and perspective at a variety of levels.

The strategic objectives developed by the healthcare organization will set the scene, the parameters, within which the organization will develop its asset management policy (AMP) and to which tools such as ISO55000 can help ensure that the medical equipment and its management provide enhanced value.

Responsibility for translating and agreeing the organization's strategic plans and objectives into an AMP including asset management objectives (AMOs) and a strategic asset management plan (SAMP) will most likely be through a multidisciplinary group concerned with medical equipment and its management. In some organizations this is called a Medical Device Committee (MDC). The function and composition of this group will be described in more detail in Chapter 5, Section 5.2 where we have used equivalent terminology more directly relevant to HTM. At this point, it is important to stress that this group, which we are calling the MDC, must be mandated at Board level to oversee an asset management system and be responsible for a Strategic HTM Plan. This needs to be aligned with the organization's corporate objectives from which it derives its AMP (ISO 55001, 5.2) – which we have called the Medical Device Policy (MD Policy). The Strategic HTM Plan will be instrumental in directing and evaluating the performance of the clinical engineering asset management system. Typically, the head of the CED will be a member of this group, along with other technical department heads and clinical user representatives. Figure 3.3 shows the role and activity of the Medical Device Committee which we develop further in Chapter 5.

The MD Policy is used by the MDC to set a Strategic HTM Plan (equivalent to the Asset Management Plan of ISO 55000). This then leads to the creation of HTM Programmes run by various technical service departments (e.g. the Clinical Engineering Department) and on to appropriate equipment support plans (ESPs) for individual assets or groups of assets which we explore in more detail in Chapter 6. All these plans must be aimed at meeting the organization's AMOs. A feedback channel must exist in the Medical Device Committee to report to the healthcare organization at Board level on performance and evaluation of and improvements to the system.

Of course, other technical services can adopt the standard, and it may well be that the Medical Device Committee decides to encourage or require all to do so. If this happens, the boundary of the ISO 55001 system will be wider than just the CED.

Figure 3.4 shows how the Medical Device Committee and technical service departments could interact with the ISO 55001 asset management system.

The MD Policy is a Board-level document, which sets out (amongst other things) the principles of how the MDC will apply asset management in order to achieve organizational objectives. It can contain:

- References to following any local or national guidelines or standards;
- Criteria for how decisions are to be made that affect assets;
- Guiding principles of a commitment to patient safety, confidentiality and level of service;
- How resources such as staff and finance are to be employed;
- Descriptions of key performance data used to monitor the effectiveness of the system;
- Acknowledgement of and commitment to stakeholders such as clinical users and patients;
- Reference to roles and responsibilities.

Based on the MD Policy, the MDC develops a Strategic HTM Plan. It will authorize the arrangements for the management of different types of assets, for example, ward-based equipment,

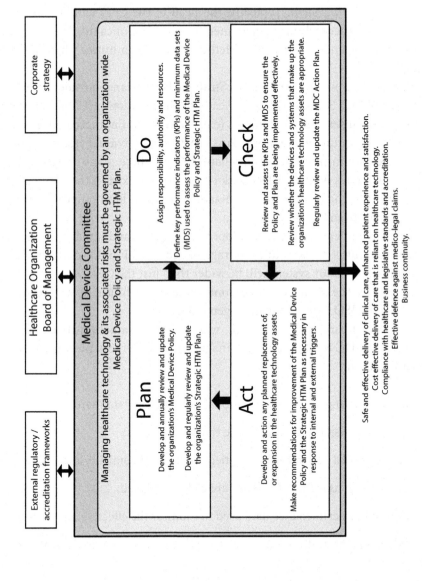

FIGURE 3.3   The Medical Device Committee.

Healthcare Organization

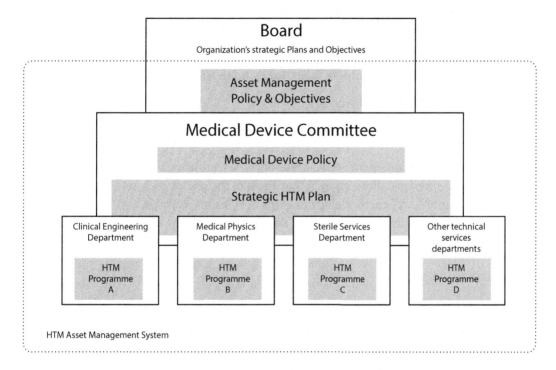

FIGURE 3.4   The Medical Device Committee in an HTM Asset Management System.

operating theatre equipment, dialysis equipment and radiology equipment. It will also establish objectives for those departments charged with delivering the services. The plan must be:

- Responsive to changing organizational policy;

- Inclusive of stakeholders' requirements and expectations. These could be different for different clinical areas or partnering departments;

- Aware of external influences, for example, changes in government policy;

- Clear about how decisions are made in regard to assets.

The Strategic HTM Plan must also set out:

- A statement of scope defining clearly what is in and out of scope;

- The required outcomes for the asset management system; (This is not an objective as referred in the following but an outcome that has been arrived at after consideration of external and internal issues.)

- The ability of changing circumstance or unanticipated events to influence the asset management system;

- The other organizations that are involved and how relationships are managed, for example, outside contractors;

- Which internal parts of the organization are involved in the system;

- Liabilities;

- Interactions with any other standards, for example, ISO 9001;

- A defined asset portfolio/inventory;

- The identity of internal and external stakeholders and their needs and expectations.

The objectives must be consistent with and support the healthcare organization's objectives and extract and interpret those aspects that involve the use of its assets. The objectives should follow the SMART methodology of being specific, measurable, achievable, realistic and time-bound. Objectives are interpreted and devolved into the individual scientific and technical support departments; objectives for the CED could include:

- Customer satisfaction.

- Uptime of a particular asset, group of assets or the entire inventory.

- Completion of a set percentage of scheduled maintenance within a timeframe.

- Monitoring and reporting on the age profile of assets.

- Identification of those assets that are unreliable or financially inefficient.

- Delivery of the management system within a financial limit.

- Attainment of a level of competence amongst internal staff.

- Benchmarking against other organizations.

- Level of service provided.

- Database uptime.

Having an MD Policy and a Strategic HTM Plan in place, these in turn inform the requirements for the remaining parts of the asset management system.

HTM Programmes need to be developed to articulate how the AMOs are to be achieved. This could be through the authorization of several departments and would include plans for aspects such as how:

- Preventive and corrective work is to be carried out and adherence to any standards.

- Finance is to be managed.

- Resources are sourced, allocated and sustained.

- Assets are to be assessed.

- Replacements are identified.

- Documentation/data are recorded.

- Activities are monitored.

These plans could also include reference to specific groups of assets. Some CEDs classify their assets into low-, medium- and high-risk types. Plans could be developed for each of these groups, providing information on how each group is to be managed.

Furthermore, these plans should individually include:

- A clear link to the relevant asset management objective(s).

- A role responsible for their upkeep.

- An indication of the intended audience, perhaps with sections designed for different groups.

- Any special environmental or interdependencies with other assets.

- Identification of risks and how this is managed.

For practicality of smaller organizations, the MD Policy, Strategic HTM plan and HTM Programme may all be produced and maintained as one single document. However, it is advisable to have clear sections for each.

3.5.3 The ISO 9000 Suite of Standards

The second Standard discussed is the ISO 9000 suite of quality management Standards. When this book went to press, the most up-to-date version was dated 2015; clinical engineers should always refer to the current version. ISO 9001 sets out the steps necessary to adopt a quality management system (QMS). Like ISO 55001 it is written to be applicable to a broad range of organizations and is designed to help them ensure they meet the needs and expectations of both their customers and stakeholders. The core of the Standard is a quality process and a set of systems and principles. One interesting feature of this Standard is that it is not a once-off recommendation to conduct activity in a certain way. The Standard recognizes that the development of quality should be an ongoing process. Built into it is the concept of regular review and continual improvement of process.

The ISO 9000 suite of Standards consists of:

- ISO 9000 *Quality Management Systems. Fundamentals and Vocabulary*;

- ISO 9001 *Quality management systems. Requirements*;

- ISO 9004 *Managing for the sustained success of an organization. A quality management approach.*

Many CEDs have developed a formal and externally audited QMS which has improved and continues to improve the quality and management of the services they deliver. If they are involved in the development of medical devices, they will have also used ISO 13485 *Medical devices – Quality management systems – Requirements for regulatory purposes.*

CEDs that operate within formal quality management systems typically adhere to agreed procedures covering the operational functions of their department. Procedures would include technical and administrative procedures for maintaining the medical devices themselves and would normally also include all those aspects found in the equipment lifecycle, therefore linking to ISO 55000. We have called these the equipment support plans (ESPs), describing them in more detail in Chapter 6. Demonstrating a commitment to quality in this way gives the healthcare organization's leadership confidence in the work carried out by the CED.

The advantage of using a system such as ISO9001 is that the principle of audit is built into the system. Internal audits are conducted by internal staff, with the audit designed to ensure that processes are being followed correctly and any inconsistencies are dealt with swiftly. For the formal, registered recognition of quality management systems, ISO 9001 also requires external audits which are carried out by an organization, that is, itself accredited. These take place to ensure that everything is going on as it should and that the accredited organization (e.g. the CED) is consistent in its delivery of services, listens to its customers to ensure it is meeting their needs and those of other stakeholders, learns and seeks to improve quality on an ongoing basis.

ISO 9001 includes seven management principles on which the standard is based:

1. Customer focus;

2. Leadership;

3. Engagement of people;

4. Process approach;

5. Improvement;

6. Evidence-based decision-making;

7. Relationship management.

These headings provide a useful framework upon which to base a discussion of the management of the CED.

3.5.3.1 Customer Focus

In all their activities clinical engineers must always remember that care of the patient is the ultimate goal, and so the patients, their carers and their families are the clinical engineers' primary 'customers'. The treatment, diagnosis and management of healthcare conditions are all about the patient and their families and friends. So, when making decisions or planning activities, it is both important and useful to consider their impact on patients and families. In Chapter 1 we pointed out that the complexity of applying clinical engineering in practice can lead to situations where conflicts arise due to competing pressures to meet different needs of the organization. When conflicts arise, considering the impact on the patient of different courses of action can be useful in identifying a solution which reflects the core values of the organization. Patients are the end beneficiaries of all that clinical engineers do.

Healthcare is about people and clinical engineering is no different. Clinical engineering places a considerable focus on managing technology, but, that is only half the story. Clinical engineers also directly support doctors, nurses, allied health professionals and managers in their work. These professionals can be regarded as the clinical engineers' second group of customers. Visibility and availability of the clinical engineer within the organization, especially in those areas with complex medical technology, is important to provide assurance and build rapport. The CED should be managed in a way that supports this approach, encourages the development of multidisciplinary activity and values discourse with clinical colleagues. Participation in teaching, training and research activities builds relationships with clinicians in a way that simply maintaining devices cannot. Being visible also allows the clinical engineer to give support and offer advice to clinical and managerial colleagues who may otherwise not seek it, nor be aware of the valuable resource that clinical engineering can provide.

Hospitals are places where people experience emotional highs and lows and simple general helpfulness can go a long way. Assisting lost visitors and being the face of the organization leaves a lasting impression with people. Pride in the working environment, both in the workshop and in the larger organization, is important. Reporting damage and missing signage, faulty lights, etc. are part of any employee's duties and contribute to a good work place.

3.5.3.2 Leadership
The head of the CED should provide leadership and act to enlist the help of others in the department, supporting these individuals to work together to accomplish the department's goals. He or she should establish a clear and accountable management structure that supports this. Through providing an effective management structure and defining and articulating a clear vision of the department's goals, the leader creates the framework and culture which invites others to work together to meet the common goals. Leadership in clinical engineering is not confined to the head of the department. Most organizations of scale will have teams within the department to deliver the ongoing processes that make up the HTM Programme. These teams also need leadership. An effective team requires an effective team leader. Teams should be led by qualified, competent staff with support and direction from the head of the department.

3.5.3.3 Engagement of People
Although clinical engineering services may be organized through departments and teams, the service is delivered by individuals, each of whom plays a valuable role. The diversity of equipment and clinical environments means that most clinical engineers are actively involved in decision-making in the day-to-day delivery of the HTM Programme. Within the framework of the management system, there is great scope and value in encouraging individual clinical engineers to act in a self-directed capacity. This not only motivates individuals but fosters a culture where their personal attributes can contribute to the work of

the department. Allowing people to use their creativity and critical faculties inspires innovation, enhancing the quality of the department's work. This self-directed activity needs to be balanced by consistency and critical review of procedures that, out of necessity, do not follow established protocols. Including all members of the department in decision-making is important as it motivates staff and fosters a sense of shared purpose. It also values the particular perspective of those who 'work at the coal face'.

3.5.3.4 Process Approach

The process approach promoted by the ISO 9001 Standard includes the plan–do–check–act (PDCA) cycle and risk-base thinking. We have described the application of these concepts to HTM in more detail in Chapter 6.

The particulars of the approach might vary between CEDs, but each should have a structured and documented process that manages activities and resources, clearly identifying the goals to be achieved. It is only by having a defined process which can be measured and controlled that the desired outcomes of the programme can be achieved. ESPs, discussed in Chapter 6, are a major contributor to this process approach. These clearly set out the processes by which the services for all medical equipment will be delivered, measured and controlled.

Many activities undertaken by the CED are projects rather than processes and better managed as such; these are explored in more detail in Chapter 7. These require active project management. Often, these projects are episodic but are placed and undertaken within a larger ongoing process. An equipment acquisition project, for example, is part of the wider Strategic HTM plan process. A risk investigation is a project which is undertaken within the MD Policy process. Understanding how projects relate to processes is important to understanding how the service is optimally delivered.

All this occurs within the HTM system, and we describe the systems approach to a clinical engineering's work in Chapter 2.

3.5.3.5 Improvement

In this book we emphasize the need to develop processes which are subject to regular review. The MD Policy, Strategic HTM plan and the HTM Programmes, as we have outlined them, have periodic review and continual improvement philosophies at the centre of their design. In Chapter 5 we discuss how the Strategic HTM Plan is reviewed and revised annually. In effect with each revision, a number of changes to the Strategic HTM Plan are identified and these changes form a quality improvement plan for those charged with implementing it. Figure 3.3 earlier illustrates this process of continual improvement.

In Chapter 6, we see that each ESP will also be reviewed and performance evaluated. Arising from this review, each plan will be updated and again these changes are in effect a quality improvement plan for those charged with delivering that particular ESP. The head of department and team leaders should ensure that these process reviews, and quality improvement plans are developed and implemented. By continually improving these processes, the overall performance of the CED and the organization can be improved.

3.5.3.6 Evidence-Based Decision-Making

When designing and reviewing both the individual ESPs and the HTM Programme as a whole, it is important that, when possible, decisions are based on empirical data. This provides the basis for robust and defendable risk-based decisions. Key performance indicators are important to measure the technical effectiveness of the HTM Programme, whilst financial indices support financial analysis. These indices should be used to guide reviews and quality improvement decision-making. The use of qualitative measures such as scoring systems and customer feedback is also valid, as is the experience and intuition of committed individuals acting to guide the development of the programme. The quality of decision-making is improved by combining analysis of performance measures of the programme with the experience and opinion of the members of the CED.

3.5.3.7 Relationship Management

The ISO 9001 QMS recognizes and is concerned with the ongoing relationships that the CED has with its customers and suppliers.

In many places in this book, we talk about the need to work proactively and collaboratively with both clinical and managerial colleagues in the healthcare system. Many of these activities are in the nature of projects rather than pre-planned equipment support activities. The carrying out of these activities must be guided by principles that support and encourage good working relationships between the CED and its clinical and managerial colleagues. In order to achieve the greatest value from the activities, the processes and outcomes must be agreed and documented. It is helpful to have standard processes in place for recording these interactions.

Good working relationships are also required between the CED and external suppliers of medical equipment and services. In Chapter 5, Section 5.9, we look at indices that allow for comparison of in-house and external support services. As part of that discussion, we identify that often the optimal solution is a shared service between an in-house team and an external service provider. In Chapter 1 we identified the continuum between the healthcare technology industry and the clinical engineering function in the hospital. In reality both the industry- and hospital-based clinical engineers are working together to harness technology for the benefit of healthcare. In the short term there may be opportunity and benefits for either the industry to make a significant profit, or the clinical engineers to make a considerable savings. However, it is important to keep long-term consideration in mind when making a key business decision. The equipment suppliers and the hospital's clinical engineers who are involved in the procurement of devices and support services are in a symbiotic relationship, and the ability to create sustainable value depends on the relationship between the two. Through open dialogue and balancing short- and long-term considerations, it should be possible to increase flexibility and in turn optimize costs and the use of resources.

The head of department and team leaders should pay careful attention to the management of relationships with customers and with suppliers and endeavour to build and sustain mutually beneficial relationships.

3.5.4 The IEC 60601 Suite of Standards

The third suite of Standards we discuss is the IEC 60601 suite. Unlike the ISO 55000 and ISO 9000 Standards, which are basic process Standards as described in Section 3.2.4 and Figure 3.2, the IEC 60601 suite specifically addresses the healthcare sector and the industry that supplies it. The Standards relate mainly to the engineering and technical specification for medical electrical equipment. They set out minimum general requirements for safety and essential safety performance of medical electrical equipment and also requirements for specific types of devices. They also set out the processes manufacturers must undertake as part of developing and placing medical electrical equipment on the market. Clinical engineers use these Standards when specifying devices for purchase and also when designing technical performance verification procedures.

The Part 1 Standard in the series, *IEC 60601-1 Medical electrical equipment – Part 1: General requirements for basic safety and essential performance* is the fundamental document, often referred to as 'the general standard' and is applicable to all medical electrical equipment. This is supported by a series of Part 1 Collateral Standards which are numbered IEC 60601-1-x. These Collateral Standards either deal in much more detail with an aspect of general applicability to medical electrical equipment (e.g. *IEC 60601-1-2 Collateral standard. Electromagnetic compatibility. Requirements and tests*), or deal with additional requirements that are applicable to a broad sub-set of medical electrical equipment (e.g. *IEC 60601-1-11 Collateral standard. Requirements for medical electrical equipment and medical electrical systems used in the home healthcare environment*).

The 60601 series also includes a set of Part 2 Particular Standards which modify the Part 1 general standard requirements and the collaterals, as appropriate for specific functional types of medical electrical equipment. These are mostly numbered IEC 60601-2-x, for example, *IEC 60601-2-4 Particular requirements for the basic safety and essential performance of cardiac defibrillators* and IEC 60601-2-25 for electrocardiographs. When a particular Standard based on IEC 60601-1 has been developed by a joint working group with ISO, the particular Standard is given the designation 80601-2-x. Two examples would be *IEC 80601-2-30 Particular requirements for the basic safety and essential performance of automated non-invasive sphygmomanometer* (produced under IEC lead) and *ISO/IEC 80601-2-13 Particular requirements for the basic safety and essential performance of an anaesthetic workstation* (produced under ISO lead).

It is, however, important to appreciate certain key features regarding the series:

- The 60601 Standards specify type tests and are predominantly aimed at designers and device manufacturers. Some of the tests are potentially damaging so not every device manufactured is tested to the Standard; rather one example device is tested, possibly to destruction.

- Unlike many other Standards, IEC 60601 specifies that, under certain single fault conditions, the product continues to be safe and, for those aspects deemed to be 'essential performance', the product continues to operate at a safe minimum level.

- Each part of the IEC 60601 Standard specifies very precise testing conditions. Therefore, acceptance testing prior to putting equipment into use, or in-service testing of equipment, is not testing **to** the IEC 60601 Standard but may be based on the 60601 test arrangements. The IEC 62353 Standard, *Recurrent tests and tests after repair of medical electrical equipment* makes this clear.

When using this suite of Standards, designers and manufacturers must identify possible hazards associated with the medical electrical equipment under consideration and assess the risks arising from those hazards. In doing this they must be guided by the ISO 14971 Standard *Medical devices – Application of risk management to medical devices*.

Many hazards associated with electrical equipment are dealt with directly in the IEC 60601-1 Standard including issues such as aspects of electric shock, leakage currents, strength of insulation and earthing. The risks arising from these hazards are addressed by specifying relevant, measurable limits. If the device meets those limits when tested in the specified way, then the manufacturer and the eventual user are assured that the relevant risks have been mitigated to an acceptably low level.

However, the Standard cannot cover all the possible hazards and associated risks across all the different types of medical electrical equipment. If the manufacturer knows of or foresees a hazard not explicitly addressed by the general Part 1 Standard or by a Particular Standard relevant to that type of equipment, then they must deal with the consequential risk, using the methodology and guidance in ISO 14971, and document the facts and risk mitigation details in a risk management file.

The clinical engineer also needs to have an understanding of this IEC 60601 suite of standards and be able to navigate, interpret and apply them when necessary. There are a number of situations in which this knowledge will be useful. These include:

- Establishing equipment specifications as part of purchasing projects.

- Designing medical electrical equipment safety testing regimes.

- Debating with manufacturers and suppliers regarding safety, performance or quality issues.

- Constructing in-house or modifying equipment and debating with end users who wish to do so.

- Combining individual medical electrical and/or other electrical equipment together to create a medical electrical system (Case Study CS7.21).

A range of depth of knowledge within a CED is required. For example, whilst every clinical engineer would not need to have a detailed knowledge of or be measuring creepage and clearance distances between internal components on commercially manufactured equipment, all should understand the requirements for and the meaning of labelling, symbols and indicators, etc. that should be present on medical equipment. Those carrying out construction or modification should have greater knowledge and understanding.

3.5.5 Other Standards Which Impact Significantly on HTM Work

Several of these Standards have been discussed earlier, but are included here for completeness.

ISO 14971 *Medical devices – Application of risk management to medical devices*;

IEC/TR 80002-1 *Guidance on the application of ISO 14971*;

IEC 62353 *Recurrent tests and tests after repair of medical electrical equipment*;

IEC 62366-1 *Medical devices – Application of usability engineering to medical devices*;

IEC 80001-1 *Application of risk management for IT networks incorporating medical devices – Part 1: Roles, responsibilities and activities*;

ISO 14155 *Clinical investigation of medical devices for human subjects. Good clinical practice*;

IEC 62304 *Medical device software – Software life cycle processes*.

3.6 KEY REGULATIONS IN HTM

3.6.1 Introduction

We have described in Section 3.3 the status of regulations as legal requirements that are based on national or sometimes trans-national law (as in the EU), and that in many legal jurisdictions, it is very common for a formal law – often called an Act – to be written and passed through the legislative process in such a way that allows designated authorities to write regulations to deal with the detailed implementation of the Act. Using this process, regulations can be created, reviewed, adapted and revised without the need for being put through the more complex processes of changes to primary legislation. Examples have been given in Section 3.3.1 and in the Case Study CS3.2.

3.6.2 Medical Devices Regulations

By their very nature, regulations and their legal underpinning differ from nation to nation. However, in our field of medical devices and equipment, there is a great deal of voluntary international co-operation and co-ordination through the International Medical Device Regulators Forum (IMDRF) [see web links]. IMDRF "is a voluntary group of medical device regulators from around the world who have come together to build on the strong foundational work of the Global Harmonization Task Force on Medical Devices (GHTF), and to accelerate international medical device regulatory harmonization and convergence".

Members of the IMDRF are:

- European Commission;
- U.S. FDA;
- Japanese Pharmaceuticals and Medical Devices Agency;
- China FDA;
- Health Canada;

- Australian Therapeutic Goods Administration;

- Brazilian National Health Surveillance Agency;

- Russian Ministry of Health.

Within each jurisdiction there are key sets of regulations that relate to the safe construction, marketing and use of medical devices. These may be based on the work of the IMDRF or national organizations, and clinical engineers must access and be familiar with them.

Within the EU and the European Free Trade Area, these regulations are common across all the nations involved. The key one for medical devices is the Medical Devices Directive (MDD) (European Council 2007a), but there is also the Active Implantable Medical Devices Directive (European Council 2007b) and the In Vitro Medical Devices Directive (European Council 2012b). As we have discussed earlier, each member state enacts these directives into regulations in their own jurisdiction. For a web link to all three directives see the EU Medical Devices Directive web link.

In the United States, medical devices are regulated by the U.S. FDA. For an introduction to their regulatory regime, see the FDA web link.

In Canada, Health Canada is the government agency that has the responsibility for regulating medical devices. The legal underpinning of the regulations is the Food and Drugs Act 1985 (as amended to 2012) and the regulations are the Medical Devices Regulations (SOR/98-282) (as amended to 2011) [see the Health Canada web link].

In Australia, the Therapeutic Goods Administration is the government agency responsible. The Therapeutic Goods Act 1989 and the Therapeutic Goods (Medical Devices) Regulations 2002 are the applicable legislation (see the Australian Therapeutic Goods Administration [TGA] web link).

It is important in the field of HTM that clinical engineers have a knowledge and understanding of the regulations that impact on the safety, design and marketing of medical devices within their own national jurisdiction. These regulations will impact on their strategic input to the purchase of new devices and equipment and on their input to the design of new, prototype or research devices or accessories and the repair or modification of existing devices and accessories.

It is also important that clinical engineers keep up to date with the changing landscape of regulations in their field. For example, as has been mentioned, at the time of writing (2016), the European Medical Devices Directive is under revision and will be replaced by Medical Devices Regulation, probably in the first half of 2017.

3.6.3 Health and Safety Regulations

Health and safety at work legislation has become a key feature in most well-developed jurisdictions, and the legislation is usually put into place through the use of regulations. Many of these regulations will have an impact on the work of a clinical engineer. A common feature, key to safe and effective HTM, is the requirement to identify hazards and hazardous situations, carry out risk assessments and put into place risk mitigation methods so that residual risks are acceptable.

In the EU the Framework Directive on Safety and Health at Work (Directive 89/391 EEC) was adopted in 1989, together with a series of related directives covering workplaces (89/654 EEC), work equipments (89/655 EEC), personal protective equipments (89/656 EEC), manual handling of loads (90/269 EEC) and display screen equipments (90/270 EEC) [see the EU Occupational Safety and Health web link].

As described in Section 3.3.2, being EU directives, they have to be put into national law in each member state of the EU in accordance with national systems of legislation. In the United Kingdom these six directives were given legal form by the publication of six equivalent UK regulations:

Management of Health and Safety at Work Regulations 1999;

Workplace (Health, Safety and Welfare) Regulations 1992;

Provision and Use of Work Equipment Regulations 1998;

Personal Protective Equipment at Work Regulations 1992;

Manual Handling Operations Regulations 1992;

Health and Safety (Display Screen Equipment) Regulations 1992.

Other examples of UK health and safety regulations (again derived from EU Directives) that are relevant to HTM work are:

Electricity at Work Regulations 1989;

Control of Substances Hazardous to Health Regulations 2002;

Lifting Operations and Lifting Equipment Regulations 1998;

Health and Safety (Sharp Instruments in Healthcare) Regulations 2013 (implementing Directive 2010/32/EU – prevention from sharp injuries in the hospital and healthcare sector).

These regulations will have equivalents in other jurisdictions. For example, in the United States the Occupational Safety and Health Administration (OSHA) administers the Occupational Safety and Health (OSH) Act of 1970. Under the Act OSHA can draft regulations, directives and plans covering particular areas of activity and risk. There are direct U.S. regulatory equivalents to the areas of regulation noted earlier in Europe.

Because of the federal structure of the United States, state governments also have the right to draft occupational safety and health plans. These state plans must be either identical or have standards, enforcement policies and procedures that are at least as effective as those of federal plans.

3.6.4 Other Health and Safety Regulations and the Responsibility of Clinical Engineers

Health and safety regulations are wide ranging, covering diverse areas of work activity. It is important that clinical engineers understand the legal arrangements in their own jurisdiction under which such regulations are made. They need to have a good working knowledge of those that have a direct, obvious and regular impact on their work but also be familiar with the whole range of regulations in order to be alert to regulations with which they are not so familiar that may be applicable in new HTM situations or activities that may arise.

3.7 KEY GUIDELINES IN HTM

3.7.1 Introduction

Guidelines are publications that are less binding in a legal sense than regulations. The weight that a guideline carries will depend on its source. Formal guidance from government agencies, directly linked to a regulation, must be taken very seriously, but to be put into the HTM context may need additional guidance from a professional body or need a local standard operating procedure.

As we have seen in Section 3.2 earlier, some guidance documents are issued by Standards organizations, national or international. These are often linked to more formal Standards, providing additional guidance and interpretation. The IEC 80001 series is a good example, where the base Standard, IEC 80001-1, is supported by a series of IEC TR 80001-2-x application guidance documents.

3.7.2 Relevant Professional Bodies

Professional associations of clinical engineering practitioners have been formed in many countries, with some of these associations having multi-national memberships. The professional bodies help ensure the standards of conduct of clinical engineers in their areas. They develop and provide training programmes, meetings and conferences to support the continuing professional development of their members and, through their members, enhance the profession as a whole. The professional bodies may develop ethical guidelines and may have criteria that must be met for membership at differing grades. Professional bodies may mandate expert members to advise or lobby governments and regulators.

In the United States the American College of Clinical Engineering (ACCE) is the professional body most relevant to the HTM field [see the ACCE web link]. There is a Publications and References tab on their home page which links to a range of reference material, guidelines and white papers.

In the United Kingdom the most relevant professional body is the Institute of Physics and Engineering in Medicine (IPEM) [see the IPEM web link]. A list of guideline documents, the Report Series, which cover a wide range of subjects in medical physics and clinical engineering can be downloaded or purchased through the publications/IPEM Report Series tab from the home page.

In Canada there is the Canadian Medical and Biological Engineering Society (CMBES) which is "… Canada's principal society for engineering in medicine and biology" [see the CMBES web link].

In Australia/New Zealand the equivalent of IPEM is the Australasian College of Physical Scientists & Engineers in Medicine (ACPSEM) [see the ACPSEM web link]. This "… has a mission to advance services and professional standards in medical physics and biomedical engineering for the benefit and protection of the community". It is fair to say that this organization appears to be more oriented towards medical physics than clinical engineering.

Engineers Australia's National Panel on Clinical Engineering is more relevant to the theme of this book and produces some guideline documents [see the Engineers Australia web link].

3.7.3 Not-for-Profit Organizations

Two U.S. 'not-for-profit' organizations that are of relevance are the Association for the Advancement of Medical Instrumentation (AAMI) and the Emergency Care Research Institute, commonly referred to as ECRI [see the respective web links].

AAMI is a multidisciplinary membership organization covering "… all those who place patient safety and quality healthcare as the highest priority". It is one of the organizations mandated by the ANSI to manage certain Standards, and various AAMI Standards committees serve as the U.S. mirror committee for the relevant ISO or IEC committees. AAMI also holds the secretariat of the IEC SC62A Standards committee responsible for the 60601-1 Standard and the SC62D Standards committee responsible for a large number of 60601-2-xx Particular Standards, both highly relevant to our field. AAMI publishes the U.S. version of the IEC 60601-1 Standard. AAMI also publishes a range of guideline documents, one of which is highly relevant and is discussed in the following [see the AAMI web link].

ECRI Institute (see the ECRI web link) is a totally independent and not-for-profit organization which since the late 1960s "has been dedicated to bringing the discipline of applied scientific research" to improve patient care (ECRI 2016). It prides itself on combining "The Discipline of Science" with "The Integrity of Independence" (ECRI 2016).

ECRI Institute is an institutional membership organization which provides a wide range of publications and consultancy services within its stated field. It operates globally from its headquarters in Philadephia, USA, with an Asia Pacific office, a Middle East office and a European office in the United Kingdom, supplying products and services tailored to Europe and the United Kingdom [see the ECRI Europe web link].

3.7.4 Sources

Clearly, it is not possible to list all the guidelines relevant to HTM and in any event such a list would soon get out of date. It is expected that the information given earlier and in the web links will point you in the right direction to search out relevant guidelines to support the regulations that are appropriate for your jurisdiction.

However, two guideline documents warrant specific mention because they cover the broad area of HTM, and we will be analysing both in comparison with the ISO 55000 suite of Standards discussed earlier. We make a comparison between them in respect of one particular equipment management issue in Case Study CS 3.3.

In the United Kingdom the MHRA has over the years issued well-written and practical guideline documents with a general theme of managing medical devices. The most recent is: *Managing Medical Devices - Guidance for healthcare and social services organizations – April 2015* (MHRA 2015).

In the United States AAMI has issued very similar guidance in the form of the ANSI/ AAMI EQ56 Standard. *Recommended practice for a medical equipment management program.* The most up-to-date version was issued in 2013 (AAMI 2013).

3.8 CONCLUSION

It is important that clinical engineers have a clear understanding of the differences between formal Standards, regulations linked to law and guidelines and an understanding of the relationships between them.

With all three types of documents, it is important to establish what the scope and intent of a particular document is. Some Standards are engineering and technical in their nature, others more process oriented and some are, in effect, internationally agreed guidelines. Some are written with healthcare in mind whilst others have general applicability across any industry or business sector.

Regulations are linked to legal requirements and therefore differ greatly in detail from jurisdiction to jurisdiction. However, broadly similar themes are addressed in all the developed and developing nations, and regulations around health and safety and medical devices are not far apart.

Guidelines also vary from one jurisdiction to another but as we have seen with *Managing medical devices 2015* from the MHRA and *EQ56* from AAMI, common themes emerge. A piece of advice that will stand the aspiring clinical engineer in good stead is, 'Don't re-invent the wheel'. Most legislative regulations have accompanying guides to aid in the interpretation that will assist in achieving compliance, and the enforcing regulators often make further guidance documents available or may answer questions directly. It is well worth spending time and effort establishing whether a guideline exists to help compliance with a regulation prior to implementing an unnecessary initiative.

Finally, learn from colleagues, both local and those in other organizations and other countries, and make use of Professional Body publications; usually, you will be able to 'tread in the steps of others'.

3A   APPENDIX: WHAT IS THE PROCESS FOR GENERATING STANDARDS?

3A.1 Introduction

The reality in this global, interconnected world is that most Standards are now drawn up at the international level through ISO and IEC. IEC, the older of the two international bodies that was founded in 1906, deals with all standardization matters in electrical, electronic and related fields. ISO, founded in its present form in 1947, deals with standardization in all other fields including (non-electrical) engineering standards, commercial standards and management system standardization.

There is an increasing number of joint working groups between the two organizations who produce 'joint logo' or dual-numbered Standards when the subject crosses both electrical and non-electrical technologies. A good example is the Standard for anaesthetic work stations ISO/IEC 80601-2-13. This standard was produced by a joint working group of ISO/TC121/SC1 *Breathing attachments and anaesthetic machines* and IEC/TC62/SC62D

Electromedical equipment. The 80601-2-xx number indicates that this Standard is structured around the IEC 60601-1 medical electrical equipment general standard but was the responsibility of a joint IEC/ISO working group.

Co-operation between ISO and IEC has grown over recent years, helped by them both having their headquarters in the same street in Geneva. Both organizations now work to a joint directive; Part 1 covering procedures for the technical work and Part 2 covering rules for the structure and drafting of international Standards (ISO/IEC 2016a; ISO/IEC 2016b).

The national standards body (NSB) of individual countries can choose to join an IEC or ISO technical committee as either a participating or an observer member. Both categories of membership allow an NSB to send experts to meetings of a technical committee or subcommittee or a working group, but only participating NSBs can formally vote on the draft documents produced by these committees. Most NSBs have chosen to arrange their own committee structure to mirror the ISO and IEC structure. So, for example, in the United Kingdom, the BSI committee CH/62 *Electrical equipment in medical practice* mirrors the IEC committee TC62 with the same title.

3A.2 Proposal for a New International Standard

The Standards-making process starts with a proposal. This may come from an NSB, or it may come from an existing international standards committee. The proposal may arise from an existing national standard that is deemed to require updating and the NSB judges that the work might have international interest and relevance. Or as happened with ISO 55000, it might arise from preliminary work done and implemented at a national level as guidance but which is thought to have international application. Or a proposal might come from within an existing international committee or working group who feel that some further level of standardization or technical guidance would be helpful in their mandated area of work. A proposal is very often accompanied by a draft but must at least include a clear statement of the scope of the proposed Standard.

Whichever route, a new work item proposal (NP) will be circulated by the relevant international standards body, ISO or IEC, to all members of NSB. They will vote approval or otherwise of the proposal and if approving, whether to participate and if so, will nominate experts from their country.

If approved, the work will be allocated to a relevant technical committee and within that to a working group. The nominated experts will meet under the chairmanship of an agreed convenor and the work will commence. Working groups meet face to face from time to time but increasingly much work is done by electronic communication.

The first stage is for the international group to draft and agree on a Committee Draft (CD) of the proposed Standard. At this stage a draft accompanying the proposal is very influential because the international group is then not starting from a minimalist document containing only a scope statement. The CD is then circulated electronically to the participating NSB who will have allocated the work to their relevant mirror committee. The national mirror committees review the draft, either in face-to-face meetings or electronically, and draw up a set of comments which are sent back to the international committee in a standard format. The international group meets to discuss all these comments.

Comments may be editorial, in principle agreeing with a technical aspect of the draft, but suggesting, for example, better ways of wording or clearer diagrams. Or the comments may be technical in nature. The detailed technical discussion takes place in the international meetings, and the objective is to reach consensus. It is rare that votes are taken at such meetings. To be influential, nominated experts from NSBs must be present at the meetings.

There may be more than one iteration of the CD stage, but usually after the meeting of and discussion amongst the nominated national experts, the first CD is refined into a second draft that the convenor and technical committee secretary deems has a sufficient level of agreement that it can be put to a formal vote of the NSB. So a Committee Draft for Voting (CDV) is circulated to the NSB. Their national mirror committees meet to consider and can again put in both editorial and technical comments, but at this stage must decide whether in principle they are in general agreement with the draft standard and vote on it.

If the vote of the national committees is in favour of the CDV, within the margins for voting set out in the rules, the draft proceeds to the next stage. This involves the working group meeting to consider all the comments they have had, agree on a resolution to each and agree a Final Draft International Standard (FDIS). If the vote on the CDV has been negative, then the working group meets to consider the comments and produce a second CDV (CDV2).

Once an FDIS is prepared by the working group, this is circulated to the national committees for a final vote. If a national committee still has serious technical objections to the draft, they have to vote negative, but knowing that at this stage they are likely to be outvoted, hence the importance of participation in the working group. A positive vote can be accompanied only by editorial comments.

Once the FDIS is approved, the convenor of the working group and the secretary of the parent committee will consider and resolve any editorial comments together with editorial or formatting input from the professional editors at the central office of ISO or IEC.

The final version is then published.

The process for making International (IEC or ISO) Standards is illustrated in a simplified form in Figure 3.A.1. The process is iterative and can be pretty slow. Typically, the time taken from a NP to the publication of a Standard can be 5 years.

Various abbreviations are used:

NSB	National Standards body
NP	New work item proposal
CD	Committee draft
CDV	Committee draft for voting
FDIS	Final draft international Standard

It is worth noting that:

- A NP may include a proposed first draft.

- In the United Kingdom, at the CDV stage, a 'draft for public comment' is also circulated in the public domain.

- IEC terms have been used; ISO terms differ slightly but the process is essentially the same.

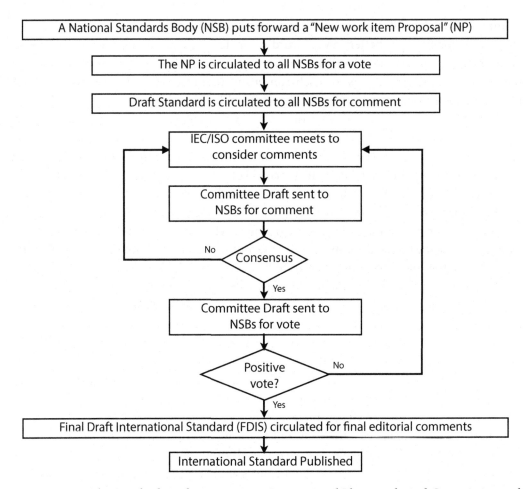

FIGURE 3.A.1   The Standards-making process in International Electrotechnical Commission and International Organization for Standardization.

People nominated by an NSB to an ISO or IEC committee or working group attend meetings as experts in their own right, not as delegates or representatives of companies (see ISO/IEC 2016a, 1.12.1). Clearly, they will be arguing for the points that have been discussed and agreed in their own mirror national committee meetings, and these are reflected in the formal comments sent in from that NSB. The objective at a meeting is to achieve consensus as to the content of a draft Standard. Consensus is defined in 2.5.6 of the ISO/IEC Directives, Part 1 (ISO/IEC 2016a).

The experience of two of the authors who are involved in IEC Standards work in the medical electrical equipment field is that most of the experts are from the relevant industry and the companies involved. The number of users, for example, doctors, involved is small. There are a small number of clinical engineers and medical physicists involved, and they are sometimes drawn in as chairs of committees because they are, and are seen to be, independent of any manufacturer. The contribution of clinical engineers and medical physicists is important because they understand the technology, the safety implications and the clinical use of the equipment under discussion.

3A.3 National and Regional Standards

We have described earlier the role of NSBs in contributing to international Standards. Experts from NSBs contribute to international Standards, and national committees scrutinize and comment on drafts because the NSB intends to adopt the international Standard as a national one. Most of the major industrialized nations and developing nations actively contribute to the development of international Standards and then adopt them as national Standards. Countries active in our field include the United Kingdom, the United States, Germany, France, Italy, the Netherlands, Sweden, Norway, Denmark, Ireland, Canada, Japan, China, South Korea, Brazil and Australia.

Additionally, there are a number of regional groupings in which the NSBs of two or more countries work together to produce standards suitable for that grouping.

The largest regional groupings are the two European Standards organizations, the European Committee for Standardization (CEN) which mirrors ISO and the European Committee for Electrotechnical Standardization (CENELEC) which mirrors the IEC. These two organizations have membership wider than the EU and are not formally part of the EU but have a mandate from the EU to produce standards suitable for use in connection with EU directives and regulations (see Section 3.3.2 earlier). NSBs which are members of CEN and CENELEC undertake to adopt their Standards, known as EN, as their own national Standards and to withdraw any conflicting national Standards.

However, a key organizational feature is that each European NSB can contribute individually to IEC and ISO Standards and so collectively ensure that they are suitable for use in Europe. A parallel voting system between the European and the International Standards bodies is in place for each stage of the development of an International Standard. This results in the majority of EN standards being identical in content to the equivalent International Standard with the same number. Over 80% of CENELEC Standards are the same as IEC Standards. As we have seen when discussing regulations, Standards play an important role in regulations such as the European MDD – soon to be revised and issued as the MDR.

CEN and CENELEC may develop stand-alone EN Standards, but by formal agreement, they will not start such a process without first consulting ISO or IEC to see whether the international organization is willing to put the proposed work into its work programme in a timely manner.

Another regional group of note is the agreement that exists between Standards Australia and Standards New Zealand to develop joint standards that are of mutual benefit to both countries. These may be 'home-grown' standards or common adoptions of international standards. Standards Australia is also active in the Pacific Area Standards Congress.

REFERENCES

AAMI. 2013. ANSI/AAMI EQ 56 Recommended practice for a medical equipment management program. Arlington, VA: Association for the Advancement of Medical Instrumentation.

ECRI 2016. About ECRI Institute. https://www.ecri.org/about/Pages/default.aspx (accessed 2016-09-20).

European Council. 2007a. Council Directive 93/42/EEC (as amended) concerning Medical Devices. http://eur-lex.europa.eu/legal-content/EN/TXT/HTML/?uri=CELEX:01993L0042-20071011&rid=1 (accessed 2016-03-31).

European Council. 2007b. Council Directive 90/385/EEC (as amended) relating to active implantable medical devices. http://eur-lex.europa.eu/legal-content/EN/TXT/HTML/?uri=CELEX:01990L0385-20071011&rid=10 (accessed 2016-03-31).

European Council. 2012a. Regulation (EU) 1025/2012 on European standardisation. http://eur-lex.europa.eu/LexUriServ/LexUriServ.do?uri=OJ:L:2012:316:0012:0033:EN:PDF (accessed 2016-03-31).

European Council. 2012b. Directive 98/79/EC of the European Parliament and of the Council (as amended). http://eur-lex.europa.eu/legal-content/EN/TXT/HTML/?uri=CELEX:01998L0079-20120111&rid=6 (accessed 2016-03-31).

Hegarty F.J., MacMahon S.T., Byrne P. and F. McCaffery. 2014. Assessing a hospital's medical IT network risk management practice with 80001-1. *Biomedical Instrumentation and Technology* 48(1): 64–71.

HSWA. 1974. The Health and Safety at Work etc. Act. http://www.hse.gov.uk/legislation/hswa.htm.

ISO/IEC. 2016a. ISO/IEC Directives, Part 1 – Procedures for the technical work. Edition 12.1. http://www.iec.ch/members_experts/refdocs/ (accessed 2016-05-03).

ISO/IEC. 2016b. ISO/IEC Directives, Part 2 – Principles and rules for the structure and drafting of ISO and IEC documents. Edition 7. http://www.iec.ch/members_experts/refdocs/ (accessed 2016-05-03).

MHRA. 2015. Managing medical devices – Guidance for healthcare and social services organisations. Ed. 1.1. https://www.gov.uk/government/publications/managing-medical-devices (accessed 2016-03-31).

Mullins M. 2006. The origin of the BS 1363 plug and socket outlet system. *IET Wiring Matters* 18: 6–9.

Rados C. 2006. Medical Device and Radiological Health Regulations Come of Age. *FDA Consumer Magazine; The Centennial Edition*. January–February 2006. http://www.fda.gov/aboutfda/whatwedo/history/productregulation/medicaldeviceandradiologicalhealthregulations-comeofage/default.htm (accessed 2016-05-16).

Vincent C. and A. Blandford. 2014. Infusion pump standards guide. Technical report IP002. Chimed. http://www.chi-med.ac.uk/research/bibdetail.php?PPnum=IP002 (accessed 2016-03-31).

STANDARDS CITED

Note 1: Undated references to Standards are given. It is important to always be aware of the most up to date version. These can be found by searching the IEC or ISO website under the respective store tabs.

Note 2: Both the IEC and ISO are based in Geneva, Switzerland.

IEC Standards

IEC 60335-1. Household and similar electrical appliances – Safety – Part 1: General requirements.

IEC 60601-1. Medical electrical equipment – Part 1: General requirements for basic safety and essential performance.

IEC 60601-1-2. Medical electrical equipment. Collateral Standard: Electromagnetic disturbances – Requirements and tests.

IEC 60601-1-8. Medical electrical equipment. Collateral Standard: General requirements, tests and guidance for alarm systems in medical electrical equipment and medical electrical systems.

IEC 60601-1-11. Medical electrical equipment. Collateral standard. Requirements for medical electrical equipment and medical electrical systems used in the home healthcare environment.

IEC 60601-2-4. Medical electrical equipment. Particular requirements for the basic safety and essential performance of cardiac defibrillators.

IEC60601-2-24. Medical electrical equipment. Particular requirements for the basic safety and essential performance of infusion pumps and controllers.

IEC 60601-2-25. Medical electrical equipment. Particular requirements for the basic safety and essential performance of electrocardiographs.

IEC 61140. Protection against electric shock – Common aspects for installation and equipment.

IEC 62304. Medical device software – Software life cycle processes.

IEC 62353. Medical electrical equipment – Recurrent test and test after repair of medical electrical equipment.

IEC 62366-1. Medical devices – Application of usability engineering to medical devices.

IEC 80001-1. Application of risk management for IT networks incorporating medical devices – Part 1: Roles, responsibilities and activities.

IEC/TR 80001-2-4. Application of risk management for IT networks incorporating medical devices – Part 2-4: Application guidance – General implementation guidance for healthcare delivery organizations.

IEC/TR 80002-1. Medical device software – Part 1: Guidance on the application of ISO 14971 to medical device software.

IEC 80601-2-30. Medical electrical equipment... Particular requirements for the basic safety and essential performance of automated non-invasive sphygmomanometer.

ISO Standards

ISO 7176-1. Wheelchairs – Part 1: Determination of static stability.

ISO 9000. Quality Management Systems. Fundamentals and Vocabulary.

ISO 9001. Quality management systems. Requirements.

ISO 9004. Managing for the sustained success of an organization. A quality management approach.

ISO 13485. Medical devices – Quality management systems – Requirements for regulatory purposes.

ISO 14155. Clinical investigation of medical devices for human subjects. Good clinical practice.

ISO 14971. Medical devices – Application of risk management to medical devices.

ISO 15223-1:2012. Medical devices – Symbols to be used with medical device labels, labelling and information to be supplied – Part 1: General requirements.

ISO/DIS 16142-1. Medical devices – Recognized essential principles of safety and performance of medical devices – Part 1: General essential principles and additional specific essential principles for all non-IVD medical devices and guidance on the selection of standards.

ISO/TS 19218-1. Medical devices – Hierarchical coding structure for adverse events – Part 1: Event-type codes.

ISO/TS 19218-2. Medical devices – Hierarchical coding structure for adverse events – Part 2: Evaluation codes.

ISO 55000. Asset management – Overview, principles and terminology.

ISO 55001. Asset management – Management systems – Requirements.

ISO 55002. Asset management – Management systems – Guidelines for the application of ISO 55001.

ISO/IEC 80601-2-13. Particular requirements for the basic safety and essential performance of an anaesthetic workstation.

ISO 81060. Non-invasive sphygmomanometers.

WEB LINKS

American College of Clinical Engineers (ACCE)
    http://www.accenet.org/

Association for the Advancement of Medical Instrumentation (AAMI)
    http://www.aami.org/

Australasian College of Physical Scientists & Engineers in Medicine (ACPSEM)
    http://www.acpsem.org.au/

Australian Therapeutic Goods Administration
   http://www.tga.gov.au/industry/devices-sgp.htm#.U-IMdaNeJEI
Canadian Medical and Biological Engineering Society (CMBES)
   http://www.cmbes.ca/
ECRI Institute
   http://ecri.org
ECRI Europe
   http://www.ecri.org.uk/index.html
Engineers Australia
   http://www.engineersaustralia.org.au/clinical-engineering
EU Medical Devices Directives
   http://eur-lex.europa.eu/search.html?instInvStatus=ALL&text=medical%20
   devices&qid=1407232874060&DTS_DOM=EU_LAW&textScope=ti-te&type=
   advanced&lang=en&SUBDOM_INIT=CONSLEG&DTS_SUBDOM=CONSLEG&page=1
EU Occupational Safety and Health web link
   https://osha.europa.eu/en/legislation/directives/the-osh-framework-directive/the-osh-frame-
   work-directive-introduction
Health Canada
   http://laws-lois.justice.gc.ca/eng/regulations/SOR-98-282/index.html
Institute of Physics and Engineering in Medicine (IPEM)
   www.ipem.ac.uk
International Electrotechnical Commission (IEC)
   http://www.iec.ch/
International Medical Device Regulators Forum (IMDRF)
   http://www.imdrf.org/
International Organization for Standardization (ISO)
   http://www.iso.org/iso/home.html
UK Machinery Regulations:
   www.hse.gov.uk/work-equipment-machinery/new-machinery.htm#what-you-should-know
UK Health and Safety at Work etc Act 1974
   www.hse.gov.uk/legislation/hswa.htm
U.S. Food and Drugs Administration (FDA)
   http://www.fda.gov/MedicalDevices/DeviceRegulationandGuidance/overview/

SELF-DIRECTED LEARNING

1. Show how Standards support the quality, safety and effectiveness of medical equipment and its clinical use.

   - Reference your answer with specific Standards, showing how in particular the Standard supports medical equipment safety and quality.

   - *Tip*: You may get ideas and inspiration from reference Vincent C., and A. Blandford. 2014.

2. Prepare a 20 minute presentation on a particular Standard for colleagues in your CED.

3. Show how reference to Standards can support the development of a purchase specification for a medical device and show how the Standard can support the evaluation and selection of the preferred make and model.

4. Access and write notes on the medical device regulations that apply in your jurisdiction.

CASE STUDIES

CASE STUDY CS3.1: DEVELOPMENT OF A STANDARD

Section Links: Chapter 3, Section 3.2.3

ABSTRACT

A formal Standard starts with a new work item proposal (NP) and develops through a series of iterations to reach a consensus. A UK *publicly available specification*, a pre-standard, on asset management, numbered PAS 55, was developed. It was used in the United Kingdom for several years, following which it was proposed and accepted as a first draft of a new ISO standard. It developed into ISO 55000.

Keywords: Standards development; PAS 55; ISO 55000

NARRATIVE

In the field of HTM we manage many items of medical equipment which fall under the commonly used term 'assets'. Assets have value to an organization, both financial and operational, and require management to ensure their contribution to strategic objectives and adherence to financial governance and, importantly, to ensure their safe and effective functionality and availability at the point of need. The processes involved in the identification, maintenance and lifecycle management of medical equipment assets are similar to the processes required within industry for managing other types of assets.

Within UK industry the need for a standard on asset management was recognized by the United Kingdom's Institute of Asset Management. The Institute worked with the UK national Standards organization the BSI, to develop and publish in 2004 a Publicly Available Specification, PAS 55 *Asset Management*. BSI uses this designation for a document that has been developed, often in collaboration with another institution, and put into the public domain for use and further comment, but is still in the development stage as a fully worked up Standard that has achieved consensus. The intention of PAS 55 was to create a document that provided a level of best practice in asset management that was equally applicable to all industries, public and private. After revision in 2008 it was proposed to the ISO that PAS 55 form the basis of an international standard. This proposal was accepted, with discussion and drafting taking place at the international level resulting in the publication, in 2014, of the ISO 55000 series of asset management Standards. These agreed international standards superseded the national PAS 55. The details of the Standards development process are described in the appendix of this chapter.

The development of this international standard from its beginning, the recognition of a need for a standard, through the development of a national Publicly Available Specification (PAS 55) to its international development and adoption as ISO 55000 is a good example of a range of industries working together to create a common set of Standards that can be used by all in the management of their physical assets. For the management of healthcare technology, the adoption of this international Standard is a valid aim.

ADDING VALUE

The adoption of the ISO 55000 Standard for HTM will add value to a healthcare organization by reducing cost through a systematic approach to asset management and adding benefit by ensuring that healthcare technology assets are purchased appropriately, used correctly and maintained and supported optimally.

Benefits : Cost ∴ Value

SYSTEMS APPROACH

ISO 55000 provides a framework for a systematic approach to HTM

SELF-DIRECTED LEARNING

1. Investigate the origin and development of the IEC 60601-1 Standard.

CASE STUDY CS3.2: MEDICAL DEVICES REGULATIONS IN THE UNITED STATES, AUSTRALIA AND EUROPE

Section Links: Chapter 3, Section 3.3.1

ABSTRACT

The International Medical Device Regulators Forum (IMDRF) brings together medical device regulators from seven nations plus the EU and the European Free Trade Association (28 plus 4 nations).

This Case Study looks at the regulatory systems in the United States, Australia and Europe.

Keywords: Medical devices; regulation

NARRATIVE

THE UNITED STATES

In the United States, the Federal Food Drug & Cosmetic Act (FD&C Act) gives the FDA the legal authority to regulate both medical devices and electronic radiation-emitting products, granting it regulatory and enforcement powers.

Most of the regulations that the FDA develops, publishes and implements in order to carry out its responsibilities for medical devices under this act are contained in Title 21 of the Code of Federal Regulations Parts 800-1299. For more detail see www.fda.gov/MedicalDevices/DeviceRegulationandGuidance/Overview/ucm134499.htm

AUSTRALIA

In Australia, the Therapeutic Goods Act 1989 is the relevant law, and the regulatory agency is the Therapeutic Goods Administration (TGA), part of the Australian Government Department of Health. The TGA regulates medical devices with reference to the Australian Therapeutic Goods (Medical Devices) Regulations 2002. A web search for *Australian medical device regulations* will turn up further detail.

EUROPE

The situation in Europe is more complicated because of the nature of the EU and its links with four other European nations outside of the Union who apply EU directives and regulations.

European directives require each member state to put the content of a directive into local law therefore allowing for some national flexibility and interpretation. European regulations apply throughout the EU as promulgated. A European regulation becomes EU wide law, only being translated into member state languages. Therefore the application of a regulation is more consistent across the whole of the EU.

In 2007 the European Council issued Directive 2007/47/EC which revised the original Medical Devices Directive 93/42/EEC (European Council, 2007). Therefore, the United Kingdom had to revise its UK regulations implementing the original directive and other member states made similar adjustment to their law.

In the United Kingdom, the European Communities Act 1972 and the Consumer Protection Act 1987 give the Secretary of State for Health, the Minister of Health, power to make regulations regarding medical devices. Therefore the revised MDD was put into UK law by the UK Secretary of State for Health issuing *The Medical Devices (Amendment) Regulations 2008* under his executive powers using what in the United Kingdom is called a statutory instrument, in this case S.I. 2008 No. 2936.

In the United Kingdom, the principal enforcement authority for these regulations (called the 'competent authority' in the Directive) is the MHRA.

Change is on the way because at the time of writing (2016), negotiations within the EU are at an advanced stage to change the rules on medical devices from a directive to a regulation. This has implications particularly for in-house development of medical devices to be used exclusively within a healthcare organization. The new regulation seems likely to come into force in the first half of 2017, though there will be a transition period of 3 years. Changes in the UK's relationship with the EU may lead to changes in regulation in the UK.

ADDING VALUE

The benefit to a healthcare organization of medical devices regulations is that devices being considered for purchase are assured to meet a basic level of safety and effectiveness. However, there are many other aspects to be taken into consideration prior to purchase as described in Chapter 5. The cost of devices is increased by regulation but there are benefits so, overall, value is probably neutral.

Benefits : Cost ∴ Value

SUMMARY

There is a move towards commonality of regulations regarding medical devices, but the legal systems differ from jurisdiction to jurisdiction so the implementation of regulations is complicated and varied. Global harmonization and therefore mutual recognition of regulations is a long-term goal.

SELF-DIRECTED LEARNING

1. Investigate and write notes on the medical devices regulations in your own jurisdiction.
2. If you are in an EU nation or use the EU regulations in your own country, compare and contrast the requirements of the MDD and the new MDR (when available) in respect of the in-house manufacture and use of a medical device.

CASE STUDY CS3.3: PUBLISHED GUIDELINE DOCUMENTS FOR HTM

Section Links: Chapter 3, Sections 3.4 and 3.4.2

ABSTRACT

Two guideline documents are considered, one from a professional organization and the other from a government agency. Although broadly comparable, a key difference in their approach is highlighted.

Keywords: Guidelines; medical equipment maintenance; manufacturer's instructions

NARRATIVE

Two guideline documents relating directly to HTM have been published, one by a professional organization (the United States's AAMI) and the other by an arms-length government body (the UK's MHRA). Both aim to promote best practice in the management of healthcare technologies. The guideline documents produced by each of these organizations have been compared, and though broadly comparable, a key difference is highlighted.

AAMI has produced *ANSI/AAMI EQ56: 2013 Recommended practice for a medical equipment management program* (AAMI 2013). This document seeks to reduce certain risks from medical equipment by specifying criteria that such a programme should meet. It sets out a framework of formal procedures which are recommended to be in place to ensure an effective medical equipment management programme.

Through dealing with subjects such as inspection and repair, inventory, leadership and resources, EQ56 clearly sets out the minimum standards required with the rationale behind each criterion explained. In this way the document provides a useful tool in setting up and developing a CED with the assurances that many professionals have agreed that these are acceptable and defendable as best practice.

The equivalent document in the United Kingdom is *Managing Medical Devices – Guidance for healthcare and social services organizations* from the MHRA (MHRA 2015). This considers the broader, whole lifecycle aspect of managing medical devices and suggests structures within the healthcare organization that should be in place to ensure effective HTM. It deals with the same topics related to the practical issues of managing and maintaining medical equipment and provides similar detail.

A key difference is in the treatment of the issue of maintaining medical devices/equipment 'in accordance with the manufacturers' instructions'. EQ56 has no explicit requirement to do so. In section 7, 'Inspection and Planned Maintenance Program', the recommendation is to develop and implement procedures for testing and inspection and for inspection intervals and "… requires that the inspections follow the procedures established by the organization…". The emphasis is on the organization establishing and documenting procedures and then following them. This is a good QMS approach.

By comparison, the MHRA document in section 8, 'Maintenance and Repair', states that, "… The frequency and type of planned preventive maintenance should be specified, in line with the manufacturer's instructions and taking account of the expected usage and the environment in which it is to be used…". In the key points summary table at the end of this section, two points stand out:

> "All medical devices and items of medical equipment are to be maintained and serviced in line with the manufacturer's service manual and advice from external agencies e.g. Medical Device Alerts."

> "Maintenance procedures are in line with manufacturer's maintenance instructions and timescales."

The original draft of the MHRA document used a variety of terms: *in accordance with, in line with, taking account of, based on*. In the final document the phrase '… in line with manufacturer's instructions…' is used consistently throughout.

It is worth noting that the second edition of the IEC 62353 Standard *Medical electrical equipment – recurrent tests and test after repair of medical electrical equipment* has a Note in section 4.1 which reads:

> "NOTE: A responsible organization having appropriate expertise can also take responsibility for modifying manufacturer's proposals based on local conditions of use and risk assessment."

SYSTEMS APPROACH

A systems approach to the issue of maintaining in accordance with manufacturers' instructions would suggest that the manufacturer's instructions are the key reference source when designing maintenance procedures, but a slavish following of them does not add value. There may be circumstances, based on experience, when doing more than the manufacturer's instructions is beneficial. Equally, experience and risk assessment may indicate that safety and reliability can be maintained with less use of scarce resources.

CULTURE AND ETHICS

Ethically and professionally, complete disregard for the manufacturer's instructions cannot be justified. Any deviation from them must be based on evidence, risk assessed, documented and periodically reviewed.

SUMMARY

The two guidance documents considered take a similar approach to the management and maintenance of medical equipment. However, the absence of any reference to the manufacturer's instructions in EQ56 is in contrast to the insistence on following them in the MHRA document. The right professional approach is to use the manufacturer's instructions as the strong basis for procedures developed in-house, based on experience and risk assessment.

SELF-DIRECTED LEARNING

1. Consider a situation in which your procedures deviate from the manufacturer's instructions. Based on the earlier documents and discussion, are your procedures adequate, appropriate and professional?
2. If not, what changes are needed?

CASE STUDY CS3.4: THE DEVELOPMENT OF A WORK INSTRUCTION FOR CARRYING OUT PREVENTIVE MAINTENANCE

Section Links: Chapter 3, Section 3.4.2; Chapter 6, Section 6.2

ABSTRACT

Standard Operating Procedures (SOPs) are key to consistent implementation of repeated tasks. They should be drafted adhering to a standard format, peer reviewed and updated, based on evidence and experience, as necessary. SOPs are also known as work instructions.

Keywords: Standard operating procedure; SOP; work instruction

NARRATIVE

The repeatable nature of planned maintenance lends itself to the creation of a SOP document for this work. Carrying around hardcopy service manuals or trying to find the relevant chapter in an electronic copy can be time-consuming and is not practical when wanting to tailor procedures locally. The SOP will clearly define the processes that should be carried out. Typically, several SOPs will be written, each for a specific procedure. SOPs may also include instructions for issues that are not directly maintenance related such as arrangements for providing replacement equipment following a breakdown so as to ensure continuity of clinical service.

The SOP for local use should be clear and straightforward to use, following a logical process and an agreed template and, for planned maintenance procedures, with unambiguous pass/fail criteria. The format chosen to standardize the work procedures comprises the following sections.

Title: Obvious but necessary to include the equipment for which the procedure was written. The title should also include the device type, manufacturer and model and be followed by an image to aid recognition.

References: The source data of the service manual and where relevant, training material is recorded. A traceable issue number and print date should be on the document, and a note on each page stating that the procedure is only valid on the day of printing. This is to make sure there are no out-of-date physical copies lying around that might be used; this is a document control issue in a quality management system.

Before starting work: Details should be included of any particular hazards associated with the device. These may be inherent (high-voltage outputs) or environmental (infection issues). Advice should be given on any required personal protective equipment, such as gloves and safety eyewear. Test equipment required should be listed, advising a check of its calibration due date.

Physical inspection: A thorough inspection of the outer case, cables and accessories should be carried out looking for damage and listening for rattles.

Test procedure: This section details the actual technical test required. It will include the applicable pass/fail criteria and what test measurements should be recorded. An SOP can refer to another SOP, for example, the test procedure may stipulate electrical safety testing of plugged-in medical electrical equipment with the electrical safety testing procedure covered in a general SOP dealing with that topic.

Completion of work: This should include procedures to ensure that the equipment is returned in a condition in which it can be safely put back into service. Typically, this may include returning settings to zero and switches to off positions. It may also stipulate that appropriate labels are affixed as per local policy.

Other issues: Location of spare equipment; out of hours arrangements following a breakdown.

SOPs should be peer reviewed in draft before being agreed and issued.

ADDING VALUE

The benefits of having and using SOPs are considerable. They ensure a standard approach to often repeated tasks and provide consistency over time and between staff. Used within a QMS, their review and updating in the light of experience contributes to continual improvement. There is an initial cost in time of drafting SOPs, but this should be set against the potential cost of failures arising from non-consistent processes. They add value to HTM.

Benefits : Cost ∴ Value

SUMMARY

SOP or work instructions provide a consistent and repeatable approach to standard tasks. They should be written to a consistent template, peer reviewed in draft and reviewed and updated regularly. They are essential for working in a QMS and add value to HTM.

SELF-DIRECTED LEARNING

1. Write a draft SOP for a new item of medical electrical equipment.

CASE STUDY CS3.5: STANDARDS AND THE PURCHASE OF A NEW BLOOD PRESSURE MONITOR

Section Links: Chapter 3, Section 3.5.1

ABSTRACT

The use of Standards to inform purchasing decisions leads to better outcomes.

Keywords: New equipment purchase, process Standards, product Standards, usability

NARRATIVE

A new blood pressure monitor is required by a ward to enable it to carry out regular observations of patients' vital signs. What part do the standards play in this process?

ISO 55001 is applied behind the scenes in many ways, setting the context within the bigger picture of asset management. The need for this blood pressure monitor may have been identified in a business case and will be purchased with regard to standardization policies previously identified or a need may identified as part of a scheduled replacement programme. Consideration will have been given to the support of the asset; who will maintain it and who will provide training and assessment of competence.

Overlapping in many ways is the ISO 9001 Standard. As a quality management systems standard, its value is in ensuring that the quality of the HTM system is maintained and that there is a culture of continual quality improvement. How does this affect the purchase of our monitor? Customer focus and supplier relations are particularly relevant in this process so it would seem appropriate to discuss with the clinical users and patients how they find different models of equipment as part of evaluations of the equipment. Supplier relations are equally important, and both these aspects are taken into consideration by ISO 9001 and ISO 55001. ISO 9001 procedures in place will also mandate that if this equipment is of a new type, then an ESP with its associated SOPs to deal with maintenance and calibration will have to be written.

It could be argued that IEC 60601 is not relevant here, but the technical characteristics of the medical equipment purchased are important throughout. In this example, the product Standard IEC 80601-2-30 will also be relevant (NOTE: See Section 3.5.4 for an explanation of the '80601' designation). Without the safety and effectiveness of the equipment then comfort, usability, support and cost fade into the background. The manufacturer will have used the IEC 62366-1 Usability Engineering Standard called up in IEC 60601-1, in the development of the equipment. Thus, IEC 60601 is very much centre stage.

At the intersection of these standards is our patient, benefiting from safe, accurate, reliable, well-maintained equipment, managed in a quality system and a strategic asset management system.

ADDING VALUE

Using Standards in the purchasing process adds value. Standards do cost to purchase but they provide important information on medical equipment. The benefits have been articulated earlier.

Benefits : Cost ∴ Value

SYSTEMS APPROACH

As described, the systems approach is evident in the bringing together of multiple factors to arrive at the best decision.

SUMMARY

In this example, the systematic use of process standards, for example, ISO 55001, ISO 9001 and IEC 62366-1, product standards, for example, IEC 60601-1 and IEC 80601-2-30 leads to better purchasing decisions, adding value to the organization and benefit to the patient.

SELF-DIRECTED LEARNING

1. Take another medical device, for example, a powered wheelchair, and investigate which Standards would be relevant to consider when you were contributing to a purchase decision.

CHAPTER 4

Life Cycle Management of Medical Equipment

CONTENTS

4.1 INTRODUCTION

In Chapter 1 we recognized that medical equipment is vital for healthcare and the importance of ensuring the availability of the appropriate equipment at the point of clinical care. 'Appropriate' implies not only that the equipment is safe and operational but also that it provides the functionality required for the associated clinical care. We also introduced the concept of Healthcare Technology Management (HTM), the strategic management and ongoing support that helps ensure the availability, at the point of care, of the appropriate safe and effective equipment. In this chapter we explore three general aspects of HTM:

- The life cycle management, in particular, understanding the total cost of ownership, which includes the costs associated with acquisition, operational use and disposal;

- Financial funding models;

- The essential principles of managing medical equipment reviewing the slogan 'Buy it Right, Use it Right, Keep it Right, Dispose of it Right' (Abraham 2000, McCarthy 2015).

Having introduced the principles that underpin a healthcare technology management system, we go on to describe the attributes of an ideal system. We propose a model that takes both a strategic and an operational view of healthcare technology managment. The solution proposed provides a clear 'line of sight' view of the responsibility and accountability from the hospital Board through the executive, the Clinical Engineering Department (CED) and end users, and finally to the person receiving care. The system is holistic in that it is concerned with supporting clinical outcomes and corporate goals as well as achieving excellence in the management of the equipment itself. The system proposed in Section 4.4 allows for the principles set out in the first three sections of this chapter to be implemented. Whilst the solution may challenge some clinical engineers to extend their role further from the traditional maintenance one into the roles of supporting and advancing care, it is based on current thinking around how an organization manages assets to deliver better value. It is closely aligned with the ISO 55000 asset management standard. We have defined key elements of a holistic HTM system and named these appropriately for medical equipment management, whilst ensuring they map onto the key elements suggested in ISO 55000.

4.2 MANAGING MEDICAL EQUIPMENT OVER ITS LIFE CYCLE

All equipment has a finite lifespan, dictated by its continuing ability to offer the required clinical functionality and its technical robustness (e.g. safety, reliability and availability of support). A lifespan implies the need to manage the equipment's life cycle, in particular the financial resources required. We show in this book how the clinical engineer, through the application of engineering principles and practice, can support the equipment's life cycle management in such a way as to optimize the value of the equipment.

4.2.1 Life Cycle Management

The lifespan of a healthcare organization is usually longer than the working life of its medical equipment. So after initially equipping the facility, the organization will need to plan for the periodic replacement of the medical equipment. Over a period of say 20 years, all the medical equipment procured to initially commission the facility may end up being replaced, with some items replaced several times. The active management of the medical equipment assets over their lifespans is referred to as life cycle management. It includes all the activities needed to support the equipment throughout its life.

Acquisition:

- Assessing the requirement;
- Specification;
- Evaluation and selection;
- Procurement;
- Installation and deployment together with associated project tasks.

Operational life support:

- Training;
- Supply of consumables and utilities;

- Technical and scientific support;
- Maintenance;
- Asset management;
- Governance.

Disposal:

- Removal from service;
- Possible sale or donation;
- Decommissioning;
- Disposal as waste.

The ultimate goal of life cycle management is to optimize the value of the medical equipment assets for the organization's stakeholders.

There are costs associated with the medical equipment over its life cycle, often referred to as the Cost of Ownership, Total Cost of Ownership or Whole Life Cost:

Cost of Ownership = Acquisition Costs + Operational Costs + Disposal Costs

This formula shows that costs are associated with each of the three phases of medical equipment's life cycle. Acquisition costs include not only the actual cost of the equipment itself but also the costs associated with the procurement (including tendering and evaluation costs), installation (including, for some equipment, specialized build work) and commissioning. Operational costs include training, consumables, maintenance and utilities. Disposal costs include costs of compliance with environmental guidelines and legislation. Management costs will be incurred in each of these phases, as we will see later.

There are many acquisition and life cycle management models, each with advantages, limitations and financial implications. The clinical engineer should have an understanding of these to assist the organization to select the most appropriate to meet its specific goals.

4.2.2 Life Cycle Medical Equipment Costs

When identifying the costs associated with ownership of medical equipment, it is best to start with a simple description of the acquisition and operational costs. The cost of purchase is usually a one-off payment to the supplier. To purchase the equipment, the hospital needs access to a source of capital funding. In the private sector, the capital can be obtained from the organization's capital reserve or borrowed from the financial markets and paid back over time with interest. In government-funded organizations, funding is often a capital grant from the government. The government grant funding must also be sourced, typically either from exchequer reserves (from taxation) or borrowed from financial markets. This capital expenditure is shown in Figure 4.1 prior to the equipment going into operational use. For simplicity we ignore inflation in this discussion.

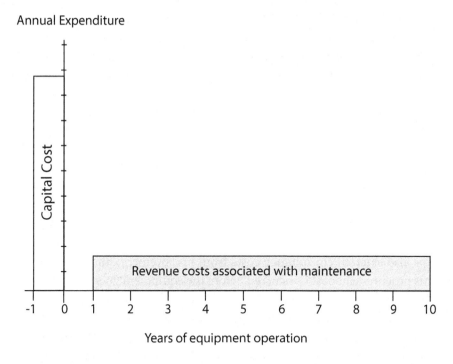

FIGURE 4.1   Life cycle costing showing initial capital and annual revenue expenditure.

In subsequent years the equipment will need to be maintained with an associated annual recurring maintenance cost (Figure 4.1). This ongoing funding is sourced from the organization's revenue resources. The maintenance gap in year one reflects the fact that it is usual for any required unscheduled maintenance to be covered under a warranty. There are operational costs in this first year and possible scheduled maintenance costs, and suppliers may offer various warranty terms, but for simplicity we shall assume a 1-year warranty with no maintenance costs.

An organization benefits from the acquisition of new equipment, increasing the resources available for its stakeholders and its ability to provide healthcare. The acquisition thus increases the organization's assets and therefore its capital worth. Hence, the expenditure involved is termed a Capital Cost. Capital expenditure is a one-off expense which provides benefits, through the application of the equipment, for many years, for its so-called financial lifespan. For example, the procurement of a CT scanner enables the organization to offer CT scanning, gaining utilization (CT imaging of patients) with 'income' from this asset. Capital expenditure increases the number or the capability of medical equipment assets. In accounting terms, items purchased using capital are accounted for in the organization's financial capital asset register. The capital values of assets on this register are often depreciated over time; the financial lifespan typically stipulated as less than the anticipated usable life of the asset. Medical equipment financial lifespans are often between 5 and 15 years, dependent on type of equipment, though for convenience a typical 10-year asset life is often assumed. As the capital value of medical equipment depreciates, its capital asset value, often called its net book value, will reduce, becoming zero at the end of its financial lifespan. Depreciation methods vary between healthcare organizations and the clinical engineer should consult with their Finance department to understand the local depreciation rules and any implications for replacing medical equipment prior to the end of its financial lifespan.

The maintenance expenditure does not provide extra benefits. Rather, maintenance ensures that the benefits expected from the equipment continue to be realized during the equipment's lifespan, rather than changing the nature of or increasing those benefits. Consequently, the associated costs are funded from revenue which, as they do not add capital benefits, are not recorded in the financial capital asset register, but rather are recorded as operational revenue costs.

This simplistic model is useful in that it allows us to identify that both capital expenditure and operational (revenue) expenditure need to be included in the cost of ownership of medical equipment. However, we need to develop a more complex picture of the costs of ownership to compare and assess different acquisition and life cycle models.

The acquisition of medical equipment is complex. Assessing the exact requirements, specifying and tendering all incur costs. These costs will include staff time, tendering costs and perhaps also external consultancy costs from procurement, financial or legal professionals. This procurement cost is incurred before the equipment goes into operation and is shown as block P in Figure 4.2.

The support and use of medical equipment is also more complex. For simplicity in Figure 4.1, we restricted operational costs to maintenance expenditure, without consideration as to who provides the maintenance. Maintenance might be provided by a combination of external and in-house maintenance. The latter is associated with staff costs (salaries and employment costs), infrastructure costs (buildings, utilities, communications and test equipment)

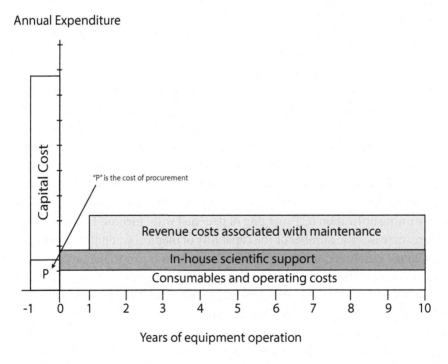

FIGURE 4.2 Life cycle costs showing acquisition costs and running costs. P represents the costs associated with procurement.

and other revenue costs (spare parts, unplanned breakdown repairs). In addition medical equipment may require in-house scientific and technical support to assist clinicians to use the equipment and derive the benefits from it. Furthermore, consumables, purchased out of revenue funding, are required for some equipment (e.g. infusion, dialysis and ventilation equipment). Figure 4.2 shows these various ongoing operational revenue costs.

The total cost of ownership of the medical equipment is the sum of all these costs over the lifetime of the equipment. Determining the actual costs can be difficult. Often, the infrastructure costs of in-house maintenance and scientific support are not reported as part of the medical equipment support costs. Consumable costs may not be easy to determine, particularly where consumables are not device specific. Utility costs may be difficult to measure and may be significant for some equipment (e.g. large imaging equipment with high electricity costs and requiring cooling and air-conditioning costs or the costs of water purification systems for renal dialysis equipment). Nevertheless, for clinical engineers who are actively trying to optimize the value delivered from medical equipment, it is important to try and identify as completely as possible the total cost of ownership. Furthermore, understanding these costs enhances the knowledge base of the clinical engineer, so as to undertake more accurate value for money comparisons when it is time for replacement planning.

4.2.3 Reviewing Maintenance Costs

Life cycle management of medical equipment includes reviewing the operational costs. Figure 4.3, for example, shows maintenance costs rising increasingly from year 6 to year 10,

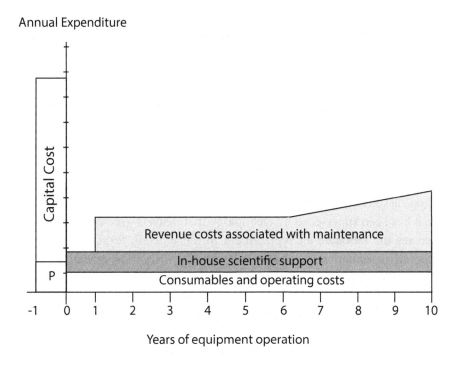

FIGURE 4.3 Life cycle costs showing increasing maintenance costs. P represents the procurement programme costs including tender and evaluation costs.

suggesting ageing of the equipment, increasing wear and tear or the supplier increasing support costs for older equipment. In comparison, the other operational costs remain the same. Decisions on replacement may need to be taken.

As noted earlier, maintenance may be provided by external contract and/or by in-house staff. Deciding the optimal balance of in-house and external contract maintenance is an important task that clinical engineers will be asked to address. Decisions will be made on the value provided (ratio of benefits to costs). Clinical engineers will need to consider different models of external maintenance contract (e.g. comprehensive, breakdown or scheduled maintenance only) to optimize value. Partnership working with manufacturers allows in-house clinical engineers to provide first-line support backed up by the manufacturer for more complex repairs or supply of costly parts. In-house maintenance builds internal knowledge and experience which is valuable to the organization, adding value to the advancing and supporting care roles of the clinical engineer (see Chapter 7), and to their roles as leaders in the hospital's medical equipment management processes (see Chapters 5 and 6).

4.2.4 Maintenance Cost Data Can Support Replacement Decision-Making

The maintenance cost is often expressed as a percentage of the equipment's cost. The cost can be literally the cost when it was acquired or more often the current cost of replacing the item referred to as the 'Replacement Asset Cost' (RAC). We will discuss the use of the ratio between maintenance costs and the RAC in more detail in Chapter 6 (see Section 6.2.4.1 and following). For the present purposes, we note in general terms the relationship and how it can be used to guide decision-making. For example, if medical equipment fails with a repair cost of 40%–50% or more of its RAC, it may be prudent to replace rather than repair, particularly where the existing equipment has a history of rising maintenance costs (Figure 4.3). Replacement rather than repair may be further supported by evidence of improved functionality of newer equipment.

4.2.5 Replacement Planning

The useful life of medical equipment varies between equipment types and with the context within which it is used, typically as noted previously, from 5 to 15 years. Equipment may reach the end of its life due to technical factors (wear and tear, ageing or lack of support) or by clinical obsolescence when its functionality is no longer clinically adequate or required or because it has been superseded by technology that delivers greater benefits. Case Study CS5.8 discusses the rationale for replacement in more detail. Professional organizations such as those of anaesthetics, endoscopy and radiology may issue guidelines on the clinically acceptable lifespan of the medical equipment used in their speciality. For the organization to continue to meet its goals, it will need to reinvest in equipment, replacing it regularly.

Figure 4.4 shows how the life cycle costs for a single item or group of items might look when replaced at the start of year 10. Not only does this require capital funding for the equipment but also funding of the necessary procurement programme represented by block P.

Healthcare organizations may spread out the replacement of the different types of equipment over time to smooth out the required annual financial costs. Developing trust and

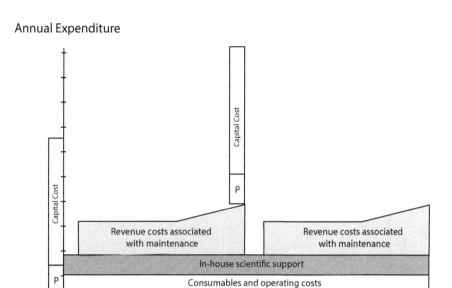

Annual Expenditure

Years of equipment operation

FIGURE 4.4   Life cycle costs associated with the initial provision and replacement of equipment at the start of year 10. P represents the procurement programme costs including tender and evaluation costs.

understanding between Finance and Clinical Engineering can lead to long-term funding projections that can help ensure the availability of funds for future replacements and can lead to plans that smooth out the financial demands associated with replacements over several years (see Case Study CS5.6). The planning should be based on the total cost of ownership including maintenance costs to ensure that the overall revenue costs of the equipment are also predictable and smooth, avoiding large peaks and troughs in maintenance costs. This is expressed diagrammatically in Figure 4.5 with capital funding required annually, sitting on top of the operational costs.

For an organization to be assured that its medical equipment continues to be optimized, a medical equipment reinvestment strategy is required. The strategy must align with the clinical needs of the healthcare organization, focused on ensuring the availability at the point of clinical care of the appropriate safe and effective equipment. The development of the detailed replacement plan will need to consider the current state of the equipment, what equipment is needed for the clinical services, equipment's life cycle costs and financial resources (capital and revenue). The Medical Device Committee (MDC) (Chapter 5) can support formulating the plans which will also require detailed discussions with Finance and with clinical leadership.

Clinical engineers, with their detailed knowledge of the equipment, its applications and its capital and operational costs, are essential to this overall asset management planning.

Their knowledge of new and emerging technologies with added benefits (clinical, safety, ergonomic or financial) can also guide the detailed replacement planning. They can take a

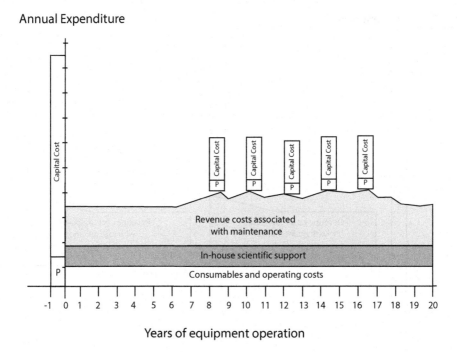

FIGURE 4.5   Life cycle costs showing equipment replacement staggered over a number of years. P represents the procurement programme costs including tender and evaluation costs.

leadership role in identifying the equipment for consideration to be replaced and in initiating the dialogue with both Finance and clinical leadership to ensure that the funding is available and that the procurement plans support the clinical developments of the healthcare organization.

4.2.6 Lost Opportunity Cost of Medical Equipment Awaiting Repair

A cost of ownership element that is not often considered is the 'lost opportunity' cost associated with medical equipment awaiting repair. This becomes tangible when it impacts on the ability of an organization to continue to deliver services as a consequence of the equipment being unavailable for use. Examples include major equipment of which the organization has only one or only a few; failure of an MRI scanner prevents the delivery of that clinical service, with the organization forced to hire scanning time elsewhere and/or delay clinical services – with costs, including staff costs, incurred when catching up once the equipment has been repaired. Other examples include endoscopy rooms, operating theatres and bed-spaces in critical care areas.

4.2.7 Medical Equipment Management System for Life Cycle Management

It is well recognized that healthcare organizations have an obligation to have a register of all medical equipment, including each of the many low-cost electronic thermometers or of the single high-cost MRI scanner. Managing equipment over its life cycles requires accurate asset information, purchase details, location, nominal lifespans or replacement dates, associated financial information and maintenance histories.

The recording of information about medical equipment requires a database system that can hold the data and enable it to be analyzed, typically called the Medical Equipment Management System (MEMS). At its core the database consists of both asset and maintenance information which together describe the life of any particular item of medical equipment. It provides the core information required for replacement planning, for revealing the cost–value of the medical equipment and for overall replacement costs.

The inclusion in the MEMS of standard maintenance operating procedures, or links to these, can enhance the ability of the recording system to support the planning and management of maintenance activities. Similarly, the inclusion of repair histories can support evaluating the equipment and calculating costs of ownership. The expansion of the database to include a log of all support actions associated with each asset, when combined with a flexible report generator, yields a powerful Medical Equipment Management System. This type of comprehensive MEMS provides the basis for in-depth analysis of the HTM system and the medical equipment. It can, for example, be used to review the use and failure patterns of different groups of devices, generating information that can tailor the equipment support plans for optimum value.

4.3 APPROACHES TO FINANCING THE LIFE CYCLE OF MEDICAL EQUIPMENT

In the following sections, we will compare and contrast different approaches to funding medical equipment. There is no single right choice, rather a range of approaches which should be considered. Indeed, a healthcare organization may simultaneously employ several approaches for different types of equipment. We identify the costs at a high level for the purposes of illustration; we make no attempt to provide actual cost comparisons. These vary widely depending upon the technology and the jurisdiction. So the representations of expenditure patterns presented here are intended to illustrate the principles only.

Regardless of the funding mechanism chosen, we feel it important to identify the resources that are required to provide holistic HTM Programmes, recalling the discussions on the twin remit in Chapter 1 (Section 1.5.3). Therefore, we identify the need for in-house clinical engineering resources for two activities. The first is the in-house maintenance role which we describe as the traditional equipment management role including external supplier contract management and delivery of in-house maintenance activities. The second is the need for scientific and technical support outside the traditional equipment management role, activities that both support and advance care.

4.3.1 Traditional Capital-Funded Acquisition with Revenue-Funded Support

The capital-funded acquisition and revenue-funded support model directly procures the medical equipment for 'cash'. Each procurement project is a competitive process in which suppliers bid against each other to win the business, abiding by agreed tender procedures. This competitive process encourages suppliers to offer reduced costs (directly reduced upfront and/or maintenance costs and/or increased warranty durations) and added benefits (continuing training options, software upgrade options, additional functionality). The process offers flexibility as each procurement allows the organization to purchase from a different supplier, responding to technological developments.

This process also provides flexibility in how equipment is managed. Should the in-house or external supports fail to deliver, there is an annual opportunity to move the revenue resources around to deliver a better solution, changing from in-house to external support and vice versa. This in itself is a competitive process which drives quality and efficiency. The flexibility supports running the HTM Programme as a quality cycle with annual review and change based on past performance.

The model's weakness is its dependence on a guaranteed funding stream. It suffers during periods of capital funding shortage, with many organizations unable to assure funding availability over the 5- to 10-year timescales required for medical equipment replacement planning. Whilst equipment's life cycles have some capacity to be stretched to weather brief funding droughts, this capacity is limited and the clinical care reliant on medical equipment can suffer if aged and unreliable equipment is not replaced. This funding model is illustrated by the diagrams in Figure 4.2.

The warranty period and what is covered by any warranty agreement deserves attention. The warranty period is often assumed to be 1 year, but extended warranty durations can be negotiated prior to procurement. On the other hand, it may be beneficial to negotiate reduced annual service contract charges in lieu of a warranty period; this may have financial advantages for the healthcare organization by reducing and clarifying future revenue costs, particularly where the annual service contract costs over the nominal life of the equipment can be fixed at the time of procurement.

4.3.2 Funding Equipment through Consumable Purchases

Where medical equipment use is associated with a consumable, there is the possibility of funding more of the life cycle cost out of revenue. By consumable we mean an accessory or item that is consumed during the process of using the equipment. For example, most volumetric infusion pumps require a dedicated single-use administration set that has to be changed between patients and in accordance with the organization's control of infection policy. Automated laboratory analyzers require the purchase of reagents or testing kits used once as part of the normal analysis. Often, the cumulative consumable cost over a few years can exceed the capital acquisition cost. In these circumstances the supplier may well offer to provide the equipment to the organization without any initial upfront acquisition cost, provided the organization contracts to procure a specific volume of consumables over a given period. In such a contract, often referred to as a consumable funded or reagent contract for laboratory equipment, the equipment's capital cost is rolled into the consumable cost and spread over a number of years (Figure 4.6). This approach can be used to fund both high-cost (laboratory diagnostic analyzers) and lower-cost items (electronic thermometers, feeding and infusion pumps, nebulizers). It shifts the cost of acquisition from capital to revenue, with Figure 4.6 showing no initial capital cost 'C', whilst recognizing that the contract negotiations do entail an often overlooked procurement programme cost 'P' that may include substantial legal advice costs.

Prior to agreeing any contract of this nature, it is important to clarify who has legal ownership of the equipment and who is responsible for maintenance and other operational support. Often, these contracts incur an extra cost for the consumables. Does this

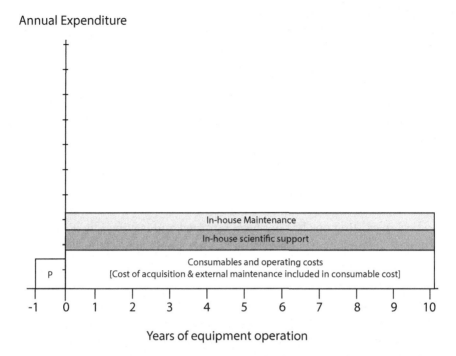

FIGURE 4.6   Life cycle costs showing cost of acquisition and supplier maintenance rolled into the consumable cost. P represents the procurement programme costs including tender and evaluation costs.

extra consumable cost change over time? Once the extra cost paid for the consumables has exceeded the normal capital cost of the equipment, is the excess removed? It is important to obtain clear procurement advice and seek information from other organizations that have experience of similar contracts with the supplier. Recalling the Total Cost of Ownership formula from Section 4.2.1, it is important that all life cycle costs are investigated when considering this procurement method as operational costs can be easily overlooked when a healthcare organization is offered 'free' medical equipment.

Ongoing costs should be carefully monitored with regular contract reviews, typically annually, particularly if the use or cost of consumables increases over time. Whilst these types of contracts can be relatively easy to set up, they can result in excessive costs if activity increases and/or if the consumable costs do not reduce once the cost of the equipment element has been paid off.

Maintenance costs may or may not be rolled into the consumable cost. Whichever the case, it is important for the hospital to have oversight, ensuring that any 'silent service contract' is carried out to its satisfaction. There will remain a need for some in-house control, maintenance and scientific support to respond to front-line requests for help. The approach carries the risk that at the end of the fixed duration contract, it is allowed to be tacitly renewed without diligently assessing the perhaps new operational needs and different supplier options that may offer improved benefits. In general, the market should be re-explored prior to any renewal.

4.3.3 Renting Medical Equipment

Medical equipment can be rented. This approach is usually adopted where there is occasional need for specialist equipment when it does not make sense for the hospital to own the equipment which might otherwise lie idle for considerable periods of time. Examples include specialist beds and surgical equipment. Renting can also be used to manage fleets of devices, where the number required varies over time. For example, some hospitals rent specialist mattresses whose required quantity varies in response to demand. Equipment may also be rented during major hospital rebuilds or replacements. It is not uncommon for mobile CT and MRI scanners to be rented for periods of several weeks during replacement of the organization's own fixed scanners. Often, the rental includes the provision of expert application support from the supplier.

Renting arrangements need to be carefully coordinated so that additional equipment is not hired when an existing one has not been returned and is lying idle. Responsibility for returning rented equipment once the contract is over needs to be carefully considered and allocated. This also involves an often hidden cost. The Clinical Engineering Department can provide this coordination.

Renting involves revenue rather than capital funding of the equipment (Figure 4.7) and many of the comments applicable to the consumable procurement approach are applicable to rental agreements. It may be difficult to evaluate the total cost of ownership, particularly if rental quantities vary over time. Rental agreements should be subject to annual reviews, testing for value for money and appropriateness of continuing with the agreement.

The supplier retains legal ownership of the equipment which is rented on the basis that it is serviced and ready for use. Rental agreements should ensure that the Clinical Engineering

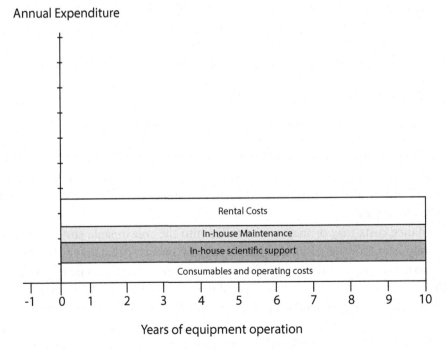

FIGURE 4.7   Life cycle costs associated with renting equipment.

Department is involved in any new deliveries, substitutions and removals of rented items so that an up-to-date record of items is held. Some maintenance costs are included in the rental cost; however, some in-house maintenance work, including safety checks and formal acceptance procedures, may still be required (Figure 4.7). The healthcare organization may also be required to provide utilities (power, water, medical gases) and may be responsible for waste removal (including disposal of radioactive substances). The healthcare organization will also want to exercise oversight of the contract, including ensuring that the rental company's service and cleaning processes comply with hospital requirements.

4.3.4 Leasing Medical Equipment

In this option the equipment is funded by a leasing company in return for an annual fee over an agreed contract term. In this model the capital cost is also moved to revenue. However, the burden of contract management for the healthcare organization increases, with associated increased contract management costs. Leasing requires negotiating with a finance company for the funding stream in addition to the normal competitive bid processes associated with the evaluation and selection of the equipment, similar to that required for capital purchases.

In this model the legal ownership of the equipment remains with the finance company for the period of the lease. Usually, lease deals are shorter than the life of the assets. When the lease is complete, the healthcare organization may have the option to acquire the asset at its residual value (shown in Figure 4.8 as block C in year 6). Alternatively, the lease deal may roll up this deferred acquisition–residual cost into the annual financing, so that at the

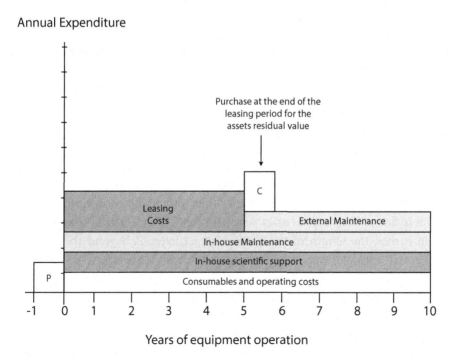

FIGURE 4.8 Life cycle costs associated with leasing equipment. P represents the procurement programme costs including tender and evaluation costs.

end of the lease period the asset transfers into the ownership of the healthcare organization. Where supplier maintenance is provided as part of the lease, the organization must make provision to fund it out of revenue when the term of the lease is complete. Most leases include clauses requiring the healthcare organization leasing the equipment to maintain and keep it in 'good condition'; the healthcare organization may have to pay to repair any damage to the equipment. Clarity should be obtained prior to signing lease contracts about the responsibility for software and other product upgrades during the course of the lease.

4.3.5 Managed Equipment Service

The Managed Equipment Service (MES) is another life cycle management approach. It involves the healthcare organization contracting for the delivery of a service (provided by equipment) rather than contracting for the supply of equipment. With an MES, the healthcare organization partners with a company who act as a broker to supply the required equipment. It should be emphasized that the contract is for the supply of a service not for the capital acquisition of the equipment.

For example, a hospital might contract with the broker to provide all the equipment required to perform clinical services for a given care pathway over a 15 to 20 year period. Usually, the hospital specifies the output required from the service provision rather than the equipment itself, for example, the annual provision of 5000 CT scans. The broker agrees to source, supply and maintain the equipment to support the clinical activity over an agreed period, typically extending to more than the typical life expectancy of that type of equipment. Thus, inherent in the approach is the agreement that the broker will manage the periodic reinvestment required to deliver the clinical output specified.

The MES approach benefits the healthcare organization in that it secures provision of the clinical services offered by the equipment for the 15–20 years of the typical contract. It eases the organization's financial planning with a known defined cost profile, removing annual sporadic variable burdens of sourcing capital funding, with the associated procurement and management costs of perhaps complex installation for initial and refresh equipping projects over extended periods. The broker assumes the risk of providing future equipment, including the financing of this risk in its annual charge. In some jurisdictions, the MES approach may offer taxation benefits to a healthcare organization as it is sometimes deemed to be the provision of a service rather than the provision of capital equipment. As the organization contracts for the provision of a service rather than procuring equipment, the associated costs are not included in its capital asset costs and capital depreciation charges.

As noted earlier, the length of the MES contract extends beyond the lifespan of the equipment and it is usual to specify an equipment refresh at specific dates within the contract period. The contract may stipulate replacement to equivalent specification or the healthcare organization may describe its requirements in terms of functionality bands. The band indicates the equipment's functionality in relationship to the current state of the art. For example, band 1 might indicate up-to-date equipment with full current functionality appropriate to a leading teaching hospital, whilst band 2 might provide general functionality appropriate to a local general hospital. So an imaging MES contract might specify the provision, for a large healthcare organization, of one Band 1 and two Band 2 CT scanners,

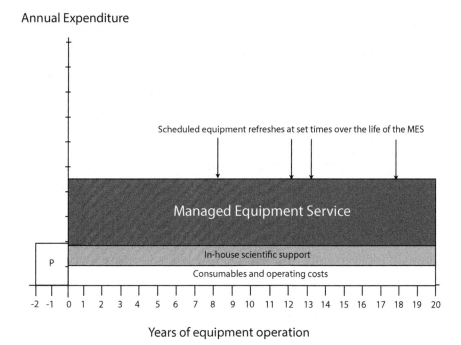

FIGURE 4.9   Life cycle costs associated with a Managed Equipment Service. There will also be associated continuing contract management costs over the life of the MES. P represents the procurement programme costs including tender and evaluation costs.

to be refreshed every seven years over the lifespan of the contract. Band 1 scanners can provide detailed neurological and brain scans, whilst Band 2 scanners provide more general routine CT scans. As with any contract, legal clarity on this type of specification should be sought. The make and model are not predetermined, with most MES contracts designed to allow for equipment to be sourced from different suppliers. The cost of financing, procurement, installation, maintenance and cost associated with managing the risk over the term of the MES is rolled into one contract.

Besides the annual MES contract cost, the healthcare organization is also likely to face high initial contract negotiation costs, including legal fees associated with the broker negotiations. It will also be subject to consumable costs and will require continuing in-house contract management fees and some in-house costs associated with medical equipment management and scientific support and perhaps even some in-house maintenance (Figure 4.9), the latter subject to compliance with any restrictions imposed by the nature of the MES contract and its restrictions consequent to it providing a service rather than providing equipment.

The MES is a contract for the supply of a service. The development and wording of the service specification requires different skill sets from the specification of equipment for competitive bidding for a defined item of medical equipment. The MES contract specification must clearly articulate the services required. It should come as no surprise that this requires careful detailed discussion with the clinical staff providing the service and with the healthcare organization's business managers. Clinical engineers can provide useful advice on the technology of the equipment, typical problems encountered and methods of managing and resolving them. Clinical engineers should be involved in ensuring that in the MES contract key performance indicators (KPIs) for equipment support are established that will meet the objectives set out in the Medical Device Policy (MD Policy) and the Strategic HTM Plan. The contract will need to include regular methods of assessing service delivery including assessment criteria, review and audit, and penalties for non-delivery of the full specified service.

MES contracts vary in scope. They can be designed to supply all the medical equipment required in a healthcare organization or just particular groups (e.g. imaging equipment or endoscopy equipment). The decision as to whether to enter into an MES contract is complex and requires detailed negotiations and clear understanding of the implications. The associated financial and legal considerations vary between jurisdictions and even in a single jurisdiction may vary over time as accountancy and taxation regulations change. Consequently, decisions to enter MES contracts should not be based solely on prevailing tax–benefit considerations. Detailed discussions on the merits or otherwise of MES contracts and the optimum length of these contracts are beyond the scope of this book, but a number of key principles are generally applicable.

MES contracts involve the transfer of some risks from the healthcare organization to the MES supplier. The supplier takes on the risk that equipment may fail prematurely, that equipment costs will rise faster than anticipated and that equipment fails to deliver the specified functionality. This latter point is important as the MES is contracted for the delivery of a specified functionality. The supplier takes on the technical risks of the equipment. However, the healthcare organization retains legal responsibility for the care of its patients. It is not always clear who takes legal responsibility for adverse events associated with the use of medical equipment under an MES, the supplier of the service, the healthcare organization or both.

A well-constructed robust MES contract offers clear advantages in terms of financial planning, avoiding peaks and troughs in financial demands on the organization which is also freed from the burden of ownership and replacing ageing equipment. Clinical staff may be reassured that they can concentrate on clinical care, freed from concerns about having to argue for regular replacement of their equipment.

However, there are also risks. First, it is very likely that the MES provider will incorporate certain assumptions about inflation which will be reflected in indexation costs. Second, it is inevitable that over such a lengthy term the organizational needs will change, and variations to the contract will be required. Therefore, a well-constructed MES must include reassurance that it has flexibility to change as clinical priorities change and to continually demonstrate value for money as changes or variations to contract occur. Both the inflation indexing and the inevitability of change can result in significant changes to what was initially perceived to be a fixed cost. The healthcare organization needs to be aware of the costs of monitoring the contract.

The Value concept is a helpful tool for judging MES contracts. When you compare the total life cycle costs of an MES with that of a traditional capital acquisition and revenue-funded support approach, there are some obvious differences (Figure 4.10). The total life cycle cost of each is the area under its curve. The MES might well cost more; however, it

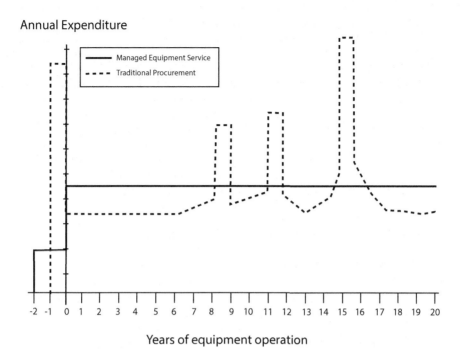

FIGURE 4.10 Comparison of life cycle cost profile of a traditional procurement and a Managed Equipment Service.

does not exhibit the peaks in funding associated with the capital injections of the capital-funded approach. Patients and clinicians may benefit from easier and more-timely access to new technologies. The agreed partnership between the hospital and the broker (often a medical equipment company) over a long period can bring benefits in expertise transfer between the organizations and be an engine for research and innovation. However, this can be to the detriment of the healthcare organization in terms of vendor independence and loss of internal expertise.

MES contract negotiations must also carefully consider the details of ongoing operational support for the medical equipment, including that for scientific support, maintenance, user support and training and healthcare technology management. There is a risk with MES projects that the clinical engineering led support programmes get absorbed or managed through the MES broker. Whilst this might seem prudent financially, it fails to recognize that healthcare technology management is as much about supporting the application of the equipment as its maintenance. So where MES brokers make claims such as 'hospitals no longer need to worry about maintenance', there is the risk that this leads to healthcare organizations losing their expertise in independently evaluating technology and managing the risks. This is a classic example of focusing only on the cost control side of the value equation. The risk of losing the added value of an in-house support team also extends to the rental, leasing and consumable-funded models. However, since these are rarely chosen as a total approach for a hospital, this risk only arises for particular groups of equipment.

4.4 EXTRACTING OPTIMAL BENEFIT FROM MEDICAL EQUIPMENT OVER ITS LIFE CYCLE

Medical equipment delivers benefits for patients by providing better methods of diagnosis and treatment. But how can we be sure that the benefits it is capable of delivering are realized and continue to be optimized over the full equipment's life cycle? Common sense suggests that selecting the appropriate equipment for the task, using it appropriately and maintaining and managing it over its full life, will ensure realization of the expected benefits. The slogan 'Buy it Right, Use it Right and Keep it Right' goes a long way to summarizing this approach (Abraham 2000). The addition to this of one more phrase 'Dispose of it Right' is necessary (McCarthy 2015) to include extracting any end-of-life residual value from the equipment and/or ensuring compliance with ecological and sustainability principles and relevant legislation when physically disposing of it.

4.4.1 Asset Management: Equipping and Procurement – 'Buy It Right'

4.4.1.1 Will the Technology Add Value, Will It Benefit the Organization?

All equipment should be justified. If the equipment does not deliver a benefit, then its acquisition and support consumes resources that might be better applied elsewhere. The required benefit should be identified, showing clearly how that benefit will be realized. The 'evaluation of clinical need stage' of the process should also explore whether any care processes will need to be modified to deliver the benefit (Case Study CS2.1). This identification of need should extend to an examination of how the equipment will be applied over its useful life.

The potential life cycle benefits offered by equipment need to be critiqued and tested. As equipment becomes more complex and specialized, and considering that it is used within a healthcare environment which itself is complex, the rationale for its need can be complicated and difficult to clarify. Health technology assessment is an activity which examines the value delivered by medical equipment (WHO 2011; see also Chapter 2, Section 2.4.7). By comparing the benefits delivered in relation to the costs, the value able to be delivered is assessed. This activity is usually done at national levels, particularly for new emerging technologies. However, even where national technology assessments judge equipment to deliver value, each organization needs to conduct internal technology assessments, using the evidence from national and international assessments, to clarify the benefits and added value applicable to its own organization.

Judging the Value requires a careful assessment, optimizing both sides of the Value ratio (Benefits in relation to Costs – see Chapter 1, Section 1.4). Cost must include the total financial cost of ownership described earlier in this chapter but may need to include non-financial costs as well. In practice, the resources are finite for both acquisition and operational funding and these will put finite limits on procurement. However, cost alone should not be the sole decision arbiter. Rather a mechanism that assesses the value delivered to the organization should be used, with decisions made taking into account both life cycle benefits and life cycle costs.

Procurement of wonderful technology for a particular hospital is wasted if the hospital does not provide the associated healthcare or if the clinical teams competent to use the technology are not available. It is also wasted if the maintenance support required is not provided.

4.4.1.2 Selecting the Right Product

Once it has been demonstrated that procurement of particular equipment will add Value, effort can be concentrated on selecting the right product, not simply the cheapest. The selection process will be discussed in more detail in Chapter 5. The associated Case Study CS5.11 shows how a weighted selection criteria system can help choose the Most Economically Advantageous Tender (MEAT) that is the best suited to the clinical needs.

4.4.1.3 Installing, Deploying and Commissioning the Equipment

When new equipment is acquired, its introduction needs active management, a process termed installation or deployment and commissioning. These activities should have been considered and planned for prior to the procurement of the new equipment, with outline consideration given at the time of clarification of need.

Installation can mean physical installation of large equipment such as an imaging scanner or placement and integration into a work environment such as bringing a new anaesthetic machine into an existing operating theatre. In either case, the logistics of equipment movement and connection to supplies and Information and Computer Technology (ICT) systems needs careful management. Linking medical equipment to ICT systems can be complex and careful checking and verification of the integrity of the networking is required, including assuring that there will be no conflict with existing systems on the network and no data security issues. Teamwork with the organization's Information Technology (IT) or eHealth department will be important as will be cooperation with the Facilities Department.

Once placed and verified, there are a number of actions that may be required before the equipment can be brought into action. The equipment may well prompt a change in the way care is delivered and so change management projects may be required (Case Study CS2.1). Often, the consumables required to operate the new equipment differ from those associated with the replaced equipment and so the logistics or material management departments need to change the stock control processes.

Many equipment types now offer a degree of configuration to make them appropriate for local operating conditions. The configuration options need to be reviewed, with selections optimized to enhance the benefits of the device in the particular workflow. Configuration selection requires close collaboration between the clinical engineering staff and the staff who will operate the equipment. Wherever possible, configurations should be standardized, so that as either equipment or staff move between different locations, equipment that appears to be the same behaves as expected. The configuration of medical equipment is explored in Case Study CS4.1.

Finally, there needs to be clear accountability as to who is responsible for different aspects of the equipment's support. In Chapter 6, we will discuss how this is documented in an equipment support plan. Suffice to highlight here that this needs to include consideration of the responsibilities of users (who may be patients), formal maintenance teams and those charged with compliance with regulatory bodies.

4.4.2 Asset Management: Operation and User Support – 'Use It Right'

To derive the full benefits expected from medical equipment requires that it be used correctly. The equipment should be deployed for use where it is most beneficial, with the users trained to operate it correctly. The installation and commissioning process lays the foundation for effective use, with both physical installation and option configurations ensuring that the equipment is optimally ready for the local use.

4.4.2.1 End-User Training

Medical equipment is often complex and its use may not be intuitive. (The selection process should have considered ease of use and manufacturers should address human usability during their equipment design.) Consequently, training is required to ensure the correct use of equipment, with assurance that end users (who may be patients) are competent in its use. Training is part of the commissioning process and also the ongoing operational support for the equipment (Case Study CS4.2). The training programme may need to incorporate technical training on the equipment, in addition to clinical training on how the equipment supports clinical care (see the dual training discussed in Case Study CS2.1).

Even with technically simple equipment, if inappropriate equipment is purchased (due, for example, to a lack of a multidisciplinary approach to procurement) and inadequate training is provided, then detrimental consequences can occur (Case Study CS4.3).

Additional training will be required over the equipment's life cycle both to assure continuing competency of those trained and to train new users. Training may need to include technical training on the operation of the device and training on its clinical application in relation to care processes and procedures in the organization.

Those charged with medical equipment maintenance often get requests for service support where no fault is found. The equipment appeared to the user to be not functioning. There are several reasons for this. On the face of it, the reason could be that the staff do not know how to operate the equipment – with appropriate training solving the problem. But the equipment might also be poorly designed, with a less-than-intuitive user interface. The importance of ergonomics and human factors in the design of medical equipment is now recognized and is the subject of medical device regulations and Standards. Comprehensive training is not a substitute for poor human factors design.

Clinical engineers should address this by offering training or refresh training for the users, particularly to those who requested the technical support. But equally importantly, clinical engineers should be aware of the human usability aspects of medical equipment, listening to clinical users' concern about poor usability. This should be used to improve the specification and selection processes. Supporting the end user to effectively use equipment is an important role of all medical equipment support.

Some equipment is so complex or its clinical application so specialized that it is operated only by dedicated professionals specifically trained in that clinical application and the associated medical equipment. Cardiac Perfusionists, Sonographers, Respiratory Technicians, Dialysis Nurses and Radiographers are examples of these professionals.

Clinical engineers may also directly support the application of some equipment; examples include the specialist renal dialysis technicians and those in some critical care units and operating theatres helping to set up and apply specialized equipment. They may prepare equipment for use, assist with implementing it in clinical practice or assist with analyzing and interpreting the data.

4.4.2.2 Responding to Changing Clinical Applications

Healthcare is not static, but evolves and changes. The applications for which particular medical equipment was procured may change; or equipment procured for one area may need to be redeployed in another area. Consequently, organizations need processes that respond to these changes to ensure continuing realization of optimal benefit from the equipment.

Clinical protocols may require changes to the configuration of the equipment, for example, revised alarm settings. Whilst it is a seemingly trivial example, it highlights the need for active configuration management. Ensuring appropriate alarm settings on medical equipment is not trivial; national and international clinical advisory agencies highlight alarm management as a high priority to ensure safe medical equipment application (the Joint Commission 2013, ECRI Institute 2014) and we will discuss further in Chapter 7 and Case Study CS7.19. Configuration features not required at time of purchase may be activated as the use of the equipment develops within the organization, or as the needs of the organization change. Configuration management extends to so-called 'smart' features on infusion devices designed to prevent infusion adverse events; their drug libraries with associated hard and soft safety limits may need to be updated with pharmaceutical developments.

Equipment may be redeployed to different clinical areas during its lifespan. For example, a patient monitor purchased for Critical Care may no longer be considered appropriate since more advanced functionalities are now considered the minimum for these clinical environments. Meanwhile, patients in general wards presenting with greater morbidity require continuous monitoring, not available with their simple spot check vital signs measurement devices. Thus, as part of the replacement programme, it might be decided to redeploy the existing critical care monitors to the less acute area where its function is appropriate to the clinical need. Note, however, that this has an impact upon support costs which will increase overall when replaced equipment is not withdrawn from use.

4.4.2.3 Auditing the Application of Equipment

We have discussed examples of the applications of equipment changing within the same area and with equipment procured for a particular purpose in one clinical unit being redeployed for a similar or different purpose in a different unit. Responding to these changes helps optimize the benefits derived from these assets. We have noted the importance of managing medical equipment configurations as part of this process.

Benefits realization requires a continuing underlying audit and management of the use of medical equipment. Optimizing the deployment of the equipment needs active management. A key tool for managing the equipment is the Medical Equipment Management System which is discussed in Section 4.2.7.

4.4.2.4 Responding to Adverse Events Involving Medical Equipment

Seminal work towards the end of the twentieth century recognized that errors do occur in healthcare, often caused by latent system failures, and that understanding the causes can help prevent recurrences with safer healthcare attainable (Kohn et al. 2000; Department of Health 2000). These reports triggered an increased emphasis on reporting, not hiding, adverse events, with the reporting leading to investigations aimed at understanding the causes. The philosopher Professor James Reason emphasized that the incidents often resulted from latent systems flaws (an accident waiting to happen) with errors or omissions combining with the trigger events to cause the incident (Reason 2000). Alphonse Chapanis (1980) coined the phrase 'error-provocative' when discussing system flaws that may, in some cases of design, be literally inviting people to commit errors. Healthcare is not without risks, but so are other industries, in particular the airline industry where the emphasis of actively learning from accidents by putting into effect risk management measures has significantly improved safety. These and related thoughts led to efforts to improve incident reporting and analysis, with the emphasis of understanding the causes to prevent recurrence, not on finding someone to blame. Concepts of no-blame reporting have matured, with organizations such as Patient Safety Organizations set up to understand the causes and learn preventative measures (https://www.pso.ahrq.gov).

Incidents associated with medical equipment account for 2%–4% of all adverse events (Amoore 2014). Initial responses of blaming the 'user' or the 'equipment' are giving way to a more open investigation, with clearer understanding of the real causes. Reporting is encouraged with direct reporting to national agencies available in several countries; examples include the following:

- *United States*: The Manufacturer and User Facility Device Experience database managed by the FDA (https://open.fda.gov/data/maude).

- *England*: The Medicines and Healthcare products Regulatory Agency reporting system (https://gov.uk/report-problem-medicine-medical-device). Information on patient safety incidents, covering all incidents, not just those involving medical equipment, is accessible at http://www.nrls.npsa.nhs.uk.

- *Australia*: The Incident Reporting and Investigation Scheme managed by the Therapeutics Goods Administration of the Department of Health (http://www.tga.gov.au/medical-device-incident-reporting-investigation-scheme-iris).

Consequently, most healthcare organizations have approved procedures for responding to adverse events. It is important that clinical engineers understand the local procedures and that the Clinical Engineering Department is proactive in helping to respond to and investigate the causes of adverse events. The objectives of the procedures are to understand the causes and to help develop methods, including ways of working, which minimize the risk of recurrences. Clinical engineers in general should not report to the national bodies mentioned previously without initially reporting to the incident reporting system in their own organization.

Clinical engineers, through their understanding of the technology and its risks and the clinical application of the technology, are well placed to make the application of medical equipment safer. Inherently, their work is aimed at safe deployment of medical equipment for patient benefit. When adverse events do occur, clinical engineers can help by promptly coordinating the safeguarding of evidence. This includes the medical equipment itself, noting the position of and (where safe to do so) not changing the settings of any of its controls. Accessories and consumables and their packaging should be preserved. A digital camera can be useful for recording equipment and its settings. It is also important to download event logs from the equipment. Equipment should not be handed over to manufacturers or suppliers without careful consideration and authorization.

Clinical engineers will liaise with internal risk management and other relevant departments as well as the department where the incident occurred. They may be responsible for reporting to national reporting systems and liaising with regulatory authorities and third-party agencies such as the Patient Safety Organizations mentioned earlier. By supporting the incident investigation, they can use their expertise to identify the causes of the incidents and the latent background factors that created the environment in which the incident occurred (Case Study CS7.7). From this understanding, they can support risk management and clinical leads in developing procedures to help prevent recurrences and thus support safer healthcare. They can also help develop standard operating procedures for managing incidents, including preserving all the evidence which might help the investigations. Where police or prosecutor investigation may take place, the medical equipment and the consumables should be quarantined until permission is given by the relevant authorities for the equipment to be released for further use.

4.4.3 Asset Management: Maintenance – 'Keep It Right'

Traditionally, maintenance is associated with technical activity to prevent failures or respond to failures by corrective actions. Medical equipment, which often makes an intimate connection with the human body, brings with it a particular need to assure that its use does not cause harm and that it is safe in use. Consequently, most hospitals have programmes in place which are focused on preventing failures, identifying hazards associated with equipment before they arise and processes for responding to failures, incidents or user difficulty. However, the effectiveness of these programmes varies widely.

4.4.3.1 Elements of Maintenance Activities and Tailored Equipment Support Plans

Medical equipment maintenance is an umbrella term for a number of activities, including performance verification, scheduled support and unscheduled support. We will discuss these in more detail in Chapter 6, Section 6.2.2, as we develop these as part of the equipment support plans for particular items of equipment, but here we briefly summarize them.

Performance verification describes activities undertaken to assure that equipment which appears to be working is in fact doing so within specification. This would include verification of calibration and safety testing.

Scheduled support describes activities which are known to be necessary to reduce the probability of failure of equipment in service. These include planned replacement of parts that age and carrying out routine calibration adjustments, particularly where calibrations are known to drift with time or usage.

Unscheduled support refers to the activities responding to occasional breakdown failures, apparent failures or user difficulties. These events, which by their nature occur randomly, require a timely response and so any support service must include provision for these unscheduled events.

Whilst some medical equipment requires a comprehensive support plan that incorporates all three of these supports, not all equipment does. It is important that the myriad of equipment types in a modern hospital are grouped and the support programme for each type reviewed to ensure it is appropriate. It is not always necessary or prudent to put in elaborate performance verification or scheduled support programmes. They may not be warranted and take resources away from other activity which adds more value such as proactive end-user training. The technical support of medical equipment should be part of a quality cycle with equipment records providing evidence of how different groups of equipment are used and their failure frequencies and mechanisms analyzed to develop tailored equipment support plans that optimizes the equipment support and controls costs. We will discuss these equipment support plans in greater detail in Chapter 6.

4.4.3.2 Medical Equipment Software Management and Upgrades

Much medical equipment is now a combination of software and hardware, with year-on-year equipment developments often affected by software changes. Manufacturers will produce a robust mechanical device designed to provide reliable service over many years with functionality improvements and developments often software driven. Software revisions will be issued in response to operational glitches and also to provide improved performance without adding functionality, whilst some software revisions will address safety issues that have materialized after the equipment was first marketed. These software revisions will typically be made available free of charge, particularly during the early years after the launch of a product.

Manufacturers will also enhance their equipment by providing additional functionality not available at earlier product releases. Examples are enhanced image processing in endoscopic camera systems and new models of intravenous anaesthetic delivery in anaesthetic syringe pumps. Whilst the outward appearance of two devices may give the impression of two identical devices, one with the enhanced software may in practice operate as a completely different device.

These medical device software developments require careful management. First, the organization must know the software status of all its devices; this is conveniently managed through the Medical Equipment Management System database. Second, the routine audit of the medical equipment must manage the software revisions in fleets of equipment, ensuring compatibility between devices in the same area. This is particularly important where equipment is procured over a period of years. Third, the existence of the software developments provides an opportunity for refreshing medical equipment, enhancing its functionality without the expense of complete replacements. It is also worth pointing out that hardware upgrades, particularly of computer components integrated into medical equipment, are also offered.

These upgrade options can address problems associated with keeping pace with clinical functionality requirements, prolonging the life of fleets of equipment through the lower cost investment in software enhancements rather than complete replacements. This can add value to existing assets. Careful consideration and management is required when extending the effective life and enhancing the benefits from existing medical equipment through software and hardware upgrades. This will include planning the training and operational changes required to safely apply the updated equipment; software upgrades may require a commissioning programme similar to the procurement of new equipment. Software upgrades may change the diagnostic data generated by the equipment which may impact on serial observations of a patient's condition and data collected for clinical trials. Records of the impact of the software upgrade on the diagnostic data should be kept, perhaps including the results from normative test data sets.

4.4.4 Asset Management: Decommissioning – 'Dispose of It Right'

When equipment is replaced, the old equipment should be removed from service and disposed of in a managed way. This will require appropriate recording of the disposal in the Medical Equipment Management System. This process, at the opposite end of an asset's life to the commissioning process, can be challenging and does need resources. Decommissioning includes disposal that is consistent with the organization's environmental policy and with the national recycling and sustainability regulations. Case Study CS4.4 shows a team approach to disposal of a large asset in a difficult location.

Some medical equipment contains patient sensitive data and the disposal process should ensure that patient sensitive data is removed from medical equipment prior to disposal. Clinical engineers will increasingly have to be careful in their management of the equipment to ensure patient data confidentiality with the increasing move to electronic patient records and medical equipment storing patient data. The healthcare organization's Information Governance (IG), IT or eHealth department should be consulted for advice and appropriate standard operating procedures put in place. As the data storage system of the medical equipment (e.g. hard disk) may contain both program data and patient data, it is typically not possible to destroy the data storage system without damaging the medical equipment. The clinical engineer can support a safe solution by ensuring close liaison between the equipment supplier and the IT department. Similar considerations apply to ensuring patient data are kept confidential in other HTM processes, including the evaluation of medical equipment as explored in Case Study CS4.5.

4.4.4.1 Residual Value

Medical equipment no longer required in one organization may still have residual value for other institutions. Looking at the decommissioning process from a life cycle value perspective, we should endeavour to recover any residual value. Where equipment still has residual value, disposal may take the form of trade-in, sale or donation. Equipment can be traded in against new equipment, reducing the cost of the subsequent replacement. If sold, there is a monetary return to the hospital; but before doing so, the organization should seek local advice about any legal obligations it may have as a 'supplier' of medical equipment. It may be prudent to minimize these obligations by selling through a third-party distributor. We discuss some aspects of the sale of old medical equipment in Case Study CS4.4.

Clinical engineers may sometimes wish to donate medical equipment no longer required in their organization. Advantages to the organization can be positive public relations, with it gaining reputation by being associated with the gifting. Control needs to be exercised when donating equipment; the World Health Organization provides guidance on donating medical equipment (http://www.who.int/medical_devices/management_use/manage_donations/en) designed to ensure that donations are managed in such a way as to ensure that the recipient of the donation gains benefit from, and is capable of gaining benefit from, the donated equipment.

4.4.4.2 Disposal Compliant with Environmental Regulations

Where equipment is truly at end of life with no realizable residual value, it will be scrapped. This should be managed in accordance with local disposal regulations. Disposal may require breaking up the equipment to extract certain materials or components, with specialist companies offering to extract valuable commodities and dispose of the equipment in an environmentally sensitive and compliant way.

4.5 HTM PRINCIPLES

Having reviewed the requirement for comprehensive asset management for healthcare technology, we now must consider how to do this in practice for healthcare organizations. In a small organization, it is easy to imagine how all aspects of asset management could be managed and delivered by one or two people who have visibility of all the equipment and knowledge of how it is used. In a large teaching hospital with thousands of medical equipment assets consisting of hundreds of groups of complex specialized equipment, implementing a comprehensive asset management system is difficult. In this section, we propose a set of principles and processes that together make up a comprehensive asset management system for healthcare technology which we call the Healthcare Technology Management System.

In the preceding sections, we have seen that healthcare technology management has a significantly greater scope than just the maintenance or care of medical equipment. It is more closely aligned with the realization of the organization's goals, aiming to ensure that the healthcare technology is managed to support those goals. Thus, to achieve this, healthcare technology management will require the consideration and addressing of often conflicting priorities and the recommendation of optimal solutions.

Ensuring medical equipment and systems are appropriate to support the delivery of care does not necessarily mean that every item of equipment needs to be the most up-to-date technology. Such a solution would be costly to acquire and costly to continuously achieve over time. Rather, the focus should be on the technology being optimum for the task, as well as effective and safe. Of course, where new technologies offer advances in care or new benefits, there may be a compelling case for their adoption. However, there is always a limit to the resources available and so the organization should have a means of managing how groups of equipment are deployed and replaced over time, but always within the context of the organizational goals. When organizations are investing in healthcare technology, a balance must be struck between short-term performance opportunities and long-term sustainability and between capital investments and subsequent operating costs, risks and performance.

Figure 4.11 identifies key principles and attributes of organizational-wide asset management which can be applied to any industry or service, reliant upon technology assets. It proposes that asset management, healthcare technology management in our case, requires

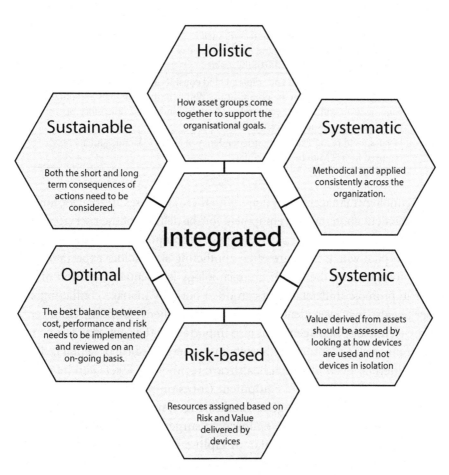

FIGURE 4.11 Key principles and attributes of Healthcare Technology Management. (Adapted from BSI, PAS 55 Parts 1 and 2: 2008, Asset Management, BSI, London, UK, 2008. With permission.)

a holistic view, with different parts of the organization united in its delivery. Note that PAS 55 (BSI 2008) was the source document from which the ISO 55000 series of international standards was developed.

The key principles and attributes of HTM, summarized in Figure 4.11, are important and deserve to be looked at in more detail:

Holistic	HTM should look at how the combined groups of assets come together to support the corporate goals. It should acknowledge that the technology assets do not exist in isolation, but must work synergistically with other assets, such as staff, buildings and information technology. It should identify the functional interdependencies between assets groups rather than thinking about assets in silos.
Systematic	HTM should be methodical and applied consistently across the organization. The approach should be repeatable with the effect of actions capable of being measured.
Systemic	When analyzing the value delivered, assets should be considered in the context of the systems within which they are used and should not be looked at as stand-alone devices in isolation. The systems engineering principles described in Chapter 2 could be used to support this systemic approach.
Risk based	Resources and expenditure should be focused and prioritized based on identified hazardous situations, with the process of reducing risk and delivering value.
Optimal	HTM should acknowledge that factors such as cost, performance, usability (human factors) and risk are interdependent and a best balance or optimal solution must be found. The balance may vary over the life of the assets.
Sustainable	HTM must be developed and implemented considering how both its short- and long-term consequences impact on the organization's ability to meet its goals and obligations. These consequences will include economic and environmental sustainability, operational performance and societal responsibilities.
Integrated	HMT should recognize the interdependency of all these factors and acknowledge all are required for HTM to be successful.

Healthcare technology management systems with these attributes are vital for hospitals which are heavily dependent upon medical equipment for the delivery of their services. Where there are large numbers of equipment, or a diverse range of equipment and technologies used, particularly in a hospital where there are often conflicting stakeholder expectations, a systematic approach to managing the healthcare technology is essential. Inclusion of a variety of stakeholders to promote understanding can often not only manage conflicting stakeholder expectations but also may even identify opportunities for better working. However, appropriate healthcare technology management is also important in the context of less complex heath delivery situations such as community hospitals or general medical or dental facilities.

There are different levels at which healthcare technology assets can be identified and managed – ranging from discrete equipment (infusion pumps, ventilators) to complex functional systems (ICU physiological monitoring system, MRI suite). They may also be grouped by functional type (e.g. endoscopic surgery or resuscitation equipment) or by equipment used within a particular discipline (e.g. audiology or ophthalmology). Grouping assets by clinical discipline helps align the HTM to the aims and objectives of the clinical department and through this to the organization's strategy. Information technology (IT) networks such as clinical information systems must also be considered.

The approach to managing such a diverse healthcare technology asset base will vary. For strategic management, devices might be grouped (anaesthetic equipment, imaging systems), whilst at the same time, for operational management, devices may be managed individually (anaesthetic machine, gamma camera). It does not matter at what level the healthcare technology management system defines an asset or asset group provided that:

- the organization's goals and strategic priorities are directly reflected in the strategic Healthcare Technology Management Plan;

- the asset life cycle costs, risks and performance are considered and optimized. This will usually require the use of clearly defined asset boundaries for measuring performance and life cycle expenditures and for attributing associated risks;

- the aggregation of assets into integrated asset groups and the contribution of value (as part of the organization's portfolio) are both managed in a coordinated and consistent manner;

- all parts of the organization understand and use the same and consistent terminology in relation to the assets, their components and their asset system groupings or aggregation.

In the following chapters, we will look at how an organization might implement an HTM system. It requires coherent direction and guidance from senior management and a governance structure that ensures competent people are given the responsibility and authority to act. One way of doing this is to develop and regularly review a Medical Device Policy (MD Policy) for the organization. It is the MD Policy which drives the HTM system as a whole, and we will look at how the policy might be formulated in Chapter 5.

4.5.1 Introduction to HTM in Practice

As the delivery of healthcare becomes more dependent on the use of healthcare technology, both the increasingly sophisticated and the relatively simplistic, there has emerged a corresponding requirement to ensure that the strategic and operational management of this technology supports the healthcare organization's objectives and delivers value. This management of healthcare technology must be informed by and remain responsive to both the changing strategic plans of the organization and to external regulatory frameworks. The organization must be able to plan for the deployment and ongoing management of healthcare technology to maximize the benefit to patients and to the organization and reduce and control any associated clinical, corporate and financial risk. This can be described as strategically managing the healthcare technology assets to maximize their value, using that word in its broadest meaning.

In Chapters 5 and 6, we describe the detail of an HTM system. Many organizations will have something similar in place but we propose a system whose driving force is embedded in the top management of the organization. We identify and describe the

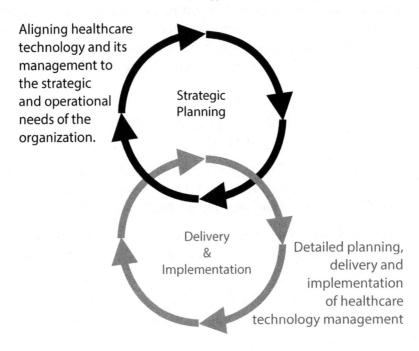

Healthcare Technology Management System

Aligning healthcare technology and its management to the strategic and operational needs of the organization.

Strategic Planning

Delivery & Implementation

Detailed planning, delivery and implementation of healthcare technology management

FIGURE 4.12   The two interlocking cycles of a Healthcare Technology Management System.

structures and processes required to manage medical equipment assets over their life. The HTM system we propose consists of two interlocking management processes as shown in Figure 4.12.

The first strategic management process deals with the responsibility of the Board and the Chief Executive Officer (CEO) to outline the corporate medical devices policy. Think of it as the big picture view of how medical equipment and technology will support the delivery of healthcare to meet the organization's goals over a five to ten year period. The focus is on how the Board and the CEO outline the vision and expectation for the organization and how they have decided to set up structures that deliver strategic asset management such as planned replacements, safety and quality improvement. The establishment of governance structures that assist others in the organization to deliver the scientific and the technical support to medical equipment are also dealt with in this process. Chapter 5 examines these processes and structures that deliver the strategic component of the HTM system.

The second of these two interlocking processes will be familiar to anyone who develops and delivers clinical engineering services. In Chapter 6, we will describe how a typical clinical engineering department (CED) might structure itself to deliver on the organization's aspirations for medical equipment asset management set by the Board. The discussion will focus on the operational delivery of support services for medical equipment management, and how year on year, such a process can be improved through operating as a quality cycle.

The HTM system principles we describe apply regardless of the size of the organization or the context within which the medical equipment is used. They are as relevant for a healthcare organization that places devices into the community to support a 'hospital at home' model of care, as they are for a hospital managing the fleet of equipment within its buildings. Equally, they are as relevant for a small healthcare organization such as a primary care centre as they are for the big tertiary care facility possibly operating across multiple sites. We describe them in terms of a large teaching hospital, as doing so highlights the complexity of medical equipment asset management systems and the governance and structures that need to be put in place. However, where the same principles need to be put in place in other healthcare contexts, the structures and relationships between elements of the system may be reduced and simplified. We will illustrate the application of these principles in different settings with three Case Studies (CS5.1 through CS5.3).

Regardless of the context, the two interlocking processes we propose make up a complete HTM system that ensures the value of the medical devices and equipment are realized and optimized for the organization's stakeholders. Together the two processes should provide a complete 'line of sight' view of medical equipment asset management from government or institutional funders to the healthcare organization's Board and CEO, through to the clinical engineering Head of Department, the clinical engineers and external suppliers who provide services, to the equipment users and last, but by no means least, to the patients.

The intersection between the two cycles needs active management. In a small organization, this could be achieved by identifying that the strategy needs to inform a documented plan for how the assets are practically managed. However, when considering how a large hospital might do this, other challenges emerge such as how the many different people in the organization who have a role to play need to be aligned in their thinking, understand how the money flows in the organization and are empowered so the system as a whole can be effective. Using a large hospital as an exemplar, we will describe how this is achieved by establishing a multidisciplinary team that we refer to as the Medical Device Committee (MDC), which operates at senior management level. This committee plays a pivotal role in translating the top level policy, through a strategic plan into a series of programmes that can be put into action by technical support departments charged with medical equipment asset management. It also seeks reassurance on behalf of the organization that the service is being delivered at the point of care. We will see that the Medical Device Committee is also the route for those who deliver services to be able to escalate risks or communicate opportunities upwards to the organization's Board and CEO.

Describing a complete, generic HTM system requires us to consider strategic issues that perhaps do not feature in the day-to-day management of clinical engineering departments. Yet they are important. As stated previously, the methods by which different healthcare organizations manage their individual HTM system vary greatly and there is no one size fits all; therefore, there is no agreed approach or terminology. Who takes responsibility for each component of the HTM system also varies between similar size organizations. Clinical engineers, who develop and deliver HTM Programmes, also contribute to hospital committees that deal with strategic issues. So it can be complicated to clearly delineate the activity in a generic way. To guide and inform the discussion, we have reviewed a number of standards and approaches to the management of systems and physical assets which are explored in more detail in Chapters 2 and 3. These include the Asset Management suite of standards, ISO 55000 which provides a model for managing assets with the important recognition that assets have a function, an objective, to serve the strategic aims of, and add value to, the organization. ISO 55000 defines asset management as (the) "coordinated activity of an organization to realize value from its assets", translating "the organizations objectives into asset-related decisions, plans and activities, using a risk based approach" (ISO 2014).

In Chapter 1, we presented an overview diagram of such an HTM system (Figure 1.9). It identified the strategic and operational processes and referred to them as the HTM policy and HTM Programme, respectively. In Figure 4.13 we develop Figure 1.9 to include the Medical Device Committee and show that it plays an important role in communicating between those delivering the HTM Programmes and the CEO and the Board. In effect it is the Medical Device Committee which acts as the linking agent between the two management processes. We also identify the different stakeholders who have an interest in HTM and position them within the system, acknowledging that it is not always possible or appropriate to place stakeholders in one place only. The HTM system provides a clear 'line of sight' management system for medical equipment asset management for all these stakeholders.

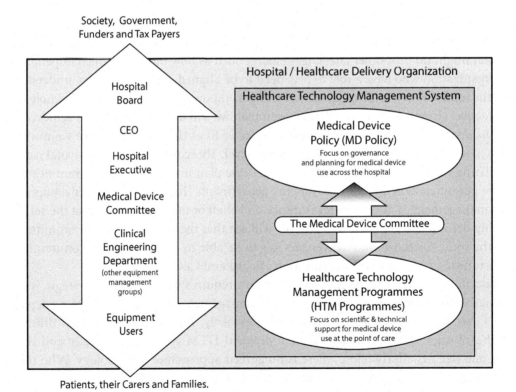

FIGURE 4.13 The HTM system showing a clear 'line of sight' management system from the Government or funders through to patients and their carers.

4.5.2 Definition 1: The Medical Device Policy

A policy is a set of rules and guidelines that ensure consistency and compliance within an organization. They are usually short statements setting out principles and commitments. The Medical Device Policy (MD Policy) is therefore a short statement, which sets out the objectives for healthcare technology management across the whole organization and outlines the business rules and guidelines that ensure HTM is consistent and in compliance with the organization's strategic direction. The policy is owned by the Board who, through the policy, determine the scope of the HTM activity within the organization.

4.5.3 Definition 2: The Medical Device Committee

The Medical Device Committee (MDC) is the committee that turns the MD Policy (guidance, rules and objectives) into actionable plans (an intention to do something). The MDC is multidisciplinary and draws together experts from a number of professional groups at senior management level (typically Finance, Risk Management, Clinical Engineering, Medical, Nursing, Facilities Management, Control of Infection, Procurement, etc.).

The MDC has a number of tasks:

- Draft the organization's MD Policy and seek approval from the Board for its adoption and publication.

- Assist with the further development and regular review of the organization's MD Policy. Whilst the Board ultimately owns the MD Policy, it will seek advice and guidance from the experts within the organization to help define and regularly review the policy.

- Develop and deliver a strategic healthcare technology management plan for the organization. This Strategic HTM Plan (an intention to do something) sets out the actions and arrangements the organization will undertake with regard to meeting the objectives set out in the MD Policy.

- Develop, set into action, regularly review and update an MDC Action Plan. This will be the working document of the MDC.

- Review and monitor the HTM Programmes to ensure that they support the strategic aims of the organization.

- Review and monitor risks associated with the deployment and use of medical devices.

- Review and monitor the delivery of health technology management for the benefit of patients and carers at the point of care.

- Provide a liaison role to the Board and Chief Executive Officer.

So the MDC is a multidisciplinary committee established at corporate management level within a healthcare organization. It is charged with responsibility for the developing, reviewing regularly and implementing the organization's MD Policy and its Strategic HTM Plan.

4.5.4 Definition 3: The Strategic HTM Plan

The Strategic HTM Plan is a plan that sets out the organization's HTM objectives and describes how the structures, roles and responsibilities of the different components of the HTM system come together to meet these objectives. It will set out which department will be responsible for the development and delivery of each of the different HTM Programmes. The objective and expectations for each HTM Programme will be set by the MDC and will usually be assessed by reviewing Key Performance Indicators reported to it regularly and formally reviewed annually. In this way the MDC can ensure compliance and consistency with the MD Policy.

4.5.5 Definition 4: The MDC Action Plan

The MDC Action Plan is a detailed working plan for how the organization manages its medical equipment assets. The main working document of the MDC, it is dynamic and reviewed and updated at every meeting. It records actions agreed, responsibilities allocated and target dates and progress made.

4.5.6 Definition 5: Healthcare Technology Management Programmes

An HTM Programme is a planned series of future events which together make up the scientific and technical support objectives required to manage medical equipment assets. An HTM Programme is developed and owned by and is the responsibility of a department, such as clinical engineering, who deliver HTM services. The details of the HTM Programmes derive from but are not part of the Strategic HTM Plan. Given that there may be several such departments each charged with delivering services to specific areas or groups of equipment, a hospital may have a number of HTM Programmes, each owned by a different department. For example, in a large academic teaching hospital, the responsibility for managing X-ray and other ionizing radiation equipment may be assigned to Radiology through a Radiation Protection Service. The number of different types of medical equipment in a hospital dictates that the planned series of future events that make up an HTM Programme will inevitably need to be further broken down to allow specific plans to be developed for particular types of equipment. So each HTM Programme is likely to consist of a number of specific equipment support plans.

4.5.7 Definition 6: The Equipment Support Plan

A plan is an intention to do a specific thing, a detailed proposal for doing or achieving something. An equipment support plan will contain the details on how to manage specific types or groups of medical equipment, such as renal dialysis, patient monitoring or anaesthetic equipment within the HTM Programme. Equipment support plans will be developed for all devices or groups of devices and will be discussed in detail in Chapter 6.

4.5.8 Summary

Figure 4.14 summarizes the role of the MDC in linking the Board's MD Policy through the MDC's Strategic HTM Plan and its MDC Action Plan to the HTM Programmes

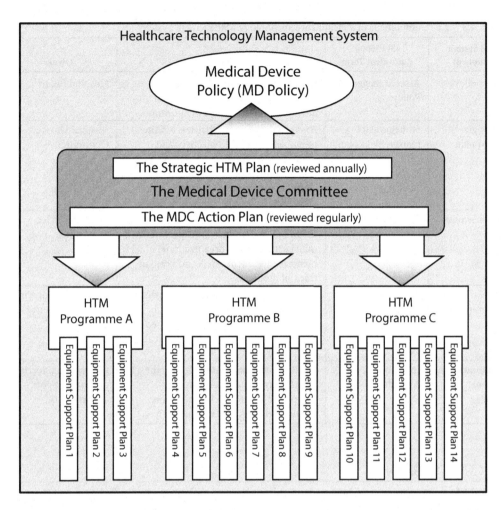

FIGURE 4.14 The HTM system showing how the Strategic HTM Plan and the MDC Action Plan devised by the MDC provide the linkage between the MD Policy and the HTM Programmes. Different HTM Programmes may be developed by a number of departments, each programme consisting of a series of equipment support plans.

and on to detailed equipment support plans. The integration of these Policies and Plans will be further developed and clarified in Section 5.3. The Medical Device Committee and the Heads of each department charged with developing HTM Programmes have a role in ensuring that each is informed by, and consistent with, the MD Policy owned by the Board. Developing specific equipment support plans in line with the department's HTM Programme objectives ensures that the detailed support plans are informed by and consistent with the MD Policy.

4.6 MAPPING THE HTM SYSTEM PROPOSED IN THIS BOOK TO ISO 55000

Table 4.1 sets out the key components of an HTM system mapping them against the ISO 55000 terminology. Figure 4.14 has shown how this can be structured in practice.

TABLE 4.1   Key Components of a Healthcare Technology Management System

HTM System Component	ISO 55000 Equivalent Term	Purpose	Owner
MD policy	Asset Management Policy	A statement, which sets out the objectives for all aspects of medical devices across the whole organization.	Hospital Board
Strategic HTM plan	Strategic asset management plan	An agreed document that sets out the organization's HTM objectives and describes how the structures, roles and responsibilities of the different components of the HTM system contribute to meeting these objectives.	Medical Device Committee
MDC action plan	(No direct equivalent)	A dynamic document that sets out how the MDC is monitoring the active asset management of the organization's medical equipment assets and keeping track of progress.	Medical Device Committee
HTM Programme(s)	Asset management plan(s)	A planned series of future events which together make up the scientific and technical support objectives required to manage specific or groups of medical equipment assets.	Clinical Engineering or other support departments
Equipment support plan(s)	(No direct equivalent)	Detailed proposals outlining how the department will deliver the many equipment specific technical actions required to meet the HTM objectives.	Clinical Engineering or other support departments

4.7 CONCLUSION

The HTM system presented here as a whole delivers a programme that allows for the organization's objectives, set out by top management, to feed into the Strategic HTM Plan and down through the system to the day-to-day equipment support activity. It supports whole life cycle management and the MDC provides a structure for multidisciplinary activity to optimize the value of the healthcare technology assets. At all levels within the system, decision-making is risk based and the reporting structures provide a means for risk escalation. The HTM system also provides a structure for different players in the system to report and audit performance at each level so that assurances can be given that the HTM objectives are being met. These aspects have been illustrated in Figure 4.15.

Whilst traditionally, clinical engineers are central to the design and delivery of the HTM Programmes, they can also take a leadership role in the activities of the Medical Device Committee and indeed in top management.

In Chapter 5, we will focus on the MD Policy and both the Strategic HTM Plan and the MDC Action Plan developed and implemented by the Medical Device Committee. The thinking will be focused on the strategic management perspective and how the supportive policy and plans are developed, implemented and managed. We will see that this strategic activity increases value for the organization but requires investment. We will not describe the HTM Programme activities developed by hospital departments; they will be

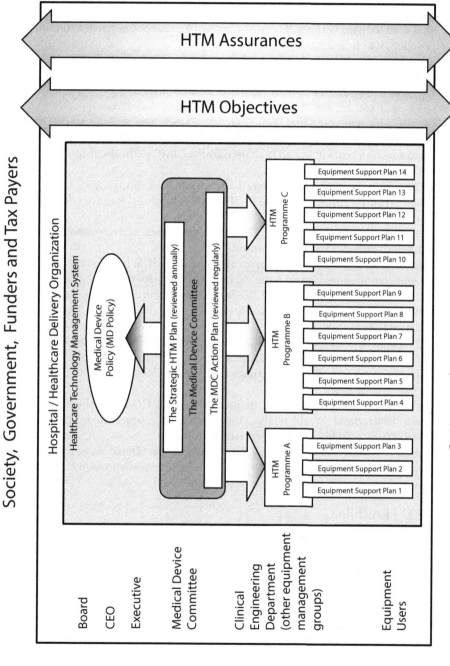

FIGURE 4.15   The HTM system and its stakeholders

discussed in Chapter 6. However, we will look at how the authority and responsibility for these HTM Programmes is assigned and how their effectiveness is assured by the Medical Device Committee.

REFERENCES

Abraham N. 2000. Can we gain without the pain? *Controls Assurance Conference* Hammersmith Hospital, London, UK, June 2000.

Amoore J.N. 2014. A structured approach for investigating the causes of medical device adverse events. *Journal of Medical Engineering and Technology*, 2014: Article ID314138. http://www.hindawi.com/journals/jme/2014/314138/ (accessed 2016-03-01).

BSI. 2008. PAS 55 Parts 1 and 2: 2008, Asset Management. London, UK: BSI.

Chapanis A. 1980. *The Measurement of Safety Performance*, pp. 99–128. New York: Garland STPM Press.

Department of Health. 2000. An organization with a memory: Report of an expert group on learning from adverse events in the NHS chaired by the Chief Medical Officer. London, UK: Stationary Office, 2000.

ECRI Institute. 2014. Top 10 health technology hazards for 2015. https://www.ecri.org/Press/Pages/ECRI-Institute-Announces-Top-10-Health-Technology-Hazards-for-2015.aspx (accessed 2016-03-01).

ISO. 2014. ISO 55000 Asset management – Overview, principles and terminology. Geneva, Switzerland: International Standards Organization.

The Joint Commission. 2013. Medical device alarm safety in hospitals. Sentinel event alert issue 50. http://www.jointcommission.org/sea_issue_50 (accessed 2016-03-01).

Kohn L., Corrigan J. and M. Donaldson (eds). 2000. *To Err is Human: Building a Safer Health System*. Committee on Quality of Health Care in America. Washington, DC: Institute of Medicine, National Academies Press.

McCarthy J.P. 2015. Personal communication. McCarthy adding the phrase 'Dispose of it Right' to the 'Buy it Right, Use it Right, Keep it Right' phrase coined by Abraham 2000.

Reason J. 2000. Human error: models and management. *British Medical Journal* 320: 768–770. doi:10.1136/bmj.320.7237.768.

WHO. 2011. Health technology assessment of medical devices. WHO medical device technical series. Geneva, Switzerland: World Health Organization http://apps.who.int/iris/bitstream/10665/44564/1/9789241501361_eng.pdf (accessed 2016-03-01).

WHO. 2016. Medical devices. Health technology management series. Donation of medical equipment. Geneva, Switzerland: World Health Organization. http://www.who.int/medical_devices/management_use/manage_donations/en (accessed 2016-05-20).

SELF-DIRECTED LEARNING

1. Select examples of medical equipment and analyze their life cycle costs, drawing a graph similar to Figure 4.2 for the total costs of ownership with time. Does the graph show operating costs changing with time? Have you been able to include all the life cycle costs, and if not, which costs are most difficult to identify? How could difficult-to-identify costs be determined?

2. Select examples of medical equipment and analyze in-house and external maintenance costs. Does the analysis point to advantages or disadvantages of either model, or a shared partnership model?

3. How can your management of the medical equipment better benefit patients and carers? Consider the different aspects of HTM from the acquisition phase,

through operational use and disposal. Does consideration of the medical equipment needs in the home for patients and carers have lessons that can be applied for hospital use of medical equipment?

4. How would you respond to a question asking what processes are in place to ensure the safety and effectiveness of medical equipment for patient care in your organization? Consider this from the perspective of an informal question from a clinical colleague, a patient or a fellow clinical engineer. How would you present the answers to a formal meeting, for example, a meeting of your organization's risk management group or corporate governance committee?

CASE STUDIES

CASE STUDY CS4.1: MEDICAL DEVICE CONFIGURATION

Section Links: Chapter 4, Section 4.4.1; Chapter 7, Section 7.2.3

ABSTRACT

The operational characteristics of many medical devices can be personalized to better suit the operational requirements of particular clinical departments. This process is known as configuration and requires careful consideration by clinical and technical staff to ensure that the resulting device is safe and effective. Standardization and control is required to avoid having outwardly identical devices but whose configuration differences produce different functional characteristics (Gibson et al. 1998).

Keywords: Configuration; Personalization; Medical devices

NARRATIVE

Clinical engineers are frequently asked to personalize medical equipment for clinical departments by altering their manufacturer supplied default configuration. A clinical engineer developed a systematic approach for managing the configurations. The objective was for all equipment to be configured to best meet the clinical requirements, backed up by a recording system stored in the Medical Equipment Management System. The configuration would, except for good reason, be standardized for all similar equipment across the healthcare organization. Where there was good reason for different standardizations for specific clinical areas, for example, the neonatal unit, this would be agreed by the appropriate clinical manager and documented. The medical device would be labelled indicating the specific configuration, for example, 'Neonatal Configuration'.

Configuration decisions can significantly affect the operation of medical equipment and its mode of operation. Consequently, configuration development required clinical and clinical engineering cooperation supported, where applicable, by the equipment supplier. Each would be signed off by the appropriate clinical manager and controlled by the Clinical Engineering Department (CED), with only the CED authorized to configure the equipment.

The configuration process, which includes clinical and clinical engineering approval and audit of its completeness for each medical device, is summarized in Table CS4.1A. The process was implemented by ensuring that it was included in the commissioning of all new medical equipment. The configurations of existing medical equipment would be developed on a priority basis using the same process; this might involve changing existing configurations.

TABLE CS4.1A Configuration Process

Objective
Configure the medical equipment to best provide the required clinical service.
Step 1: Decide the Configuration
• Select a medical device type for configuration;   • Convene a short life project team to agree the configuration;   • Membership: clinical staff, clinical engineer, supplier support;   • Develop and test the configuration;   • Finalize the configuration;   • Document the reason for the configuration chosen;   • Approval by the project team;   • Signed off by the clinical manager(s).
Step 2: Document the Configuration
• The Medical Equipment Management System (MEMS) to have a field for each type of medical device indicating whether it requires a configuration and in the record for each device, a field for the configuration reference for that individual device;   • Configuration reference and details and its documented approval to be recorded in the MEMS;   • Link to the equipment support plan (Chapter 6).
Step 3: Configure the Medical Device(s)
• Using the agreed configuration for a particular device type, configure all the appropriate devices;   • Cloning tools facilitate consistent accurate configuration;   • Record in the MEMS the devices configured and the date when configured.
Step 4: Audit the Configuration Process
• Audit the process, reporting on the number configured and awaiting configuration;   • Continue Steps 3 and 4 until all devices are configured;   • Report to the Clinical Manager(s) and the Medical Devices Committee (Chapter 5).

Medical equipment requiring configuration can range from the humble electronic clinical thermometer (oral or ear equivalent temperature) to infusion devices (rate limits, occlusion pressure, alarms), electrosurgery machines (cutting and coagulating currents), critical care monitors (alarms, presentation of monitored waveforms and parameters), ventilators (alarms, ventilation parameters) and imaging systems (imaging protocols, dose reduction processes). In all cases, it is important that careful attention be given to determining the configuration, developing it in conjunction with clinical, technical and supplier staff.

ADDING VALUE

Controlling configuration adds value by ensuring that the medical devices are optimally configured for healthcare delivery in the healthcare organization. In particular alarm configuration is important to reduce unnecessary triggering of alarms. Configuration does cost in terms of staff time to investigate and implement, but it can make the medical equipment easier to use and reduce problems and adverse events. The benefits to patient safety are considerable, so value increases.

Benefits : Cost ∴ Value

SUMMARY

A structured documented approach for configuration planning and implementation is described. The approach benefits the healthcare organization, clinical staff and clinical engineers by ensuring standardized, carefully thought-out configurations are adopted to best match the clinical requirements for patient care.

REFERENCE

Gibson C., McCarthy J.P., Powell A., Roberts D., Spark P. and R.J. Truran. 1998. Minimizing clinical risk in the use of active intravenous infusion devices. *British Journal of Intensive Care*, 8: 114–119.

SELF-DIRECTED LEARNING

1. How does having agreed medical equipment configurations benefit the patient and the clinical user? How do the needs of the patient and the clinical user differ?
2. Does your healthcare organization have a policy on medical equipment configuration? If not, can you suggest an approach to creating one? Who would the stakeholders be in such a project?
3. Take any medical device in your organization. How is it configured? What processes are in place to review its configuration in response to any clinical operational changes?

CASE STUDY CS4.2: MEDICAL EQUIPMENT TRAINING FOR CLINICAL STAFF

Section Links: Chapter 4, Section 4.4.2; Chapter 7, Section 7.4.11

ABSTRACT

Clinical staff are required to be competent with the medical equipment they use for patient care. Training should include the characteristics and limitations of the medical equipment, its technical operation and how to apply it clinically for patient care.

Keywords: Training, Clinical staff, End users, Medical devices

NARRATIVE

The Clinical Engineering Department (CED) was asked to develop a medical device training system as part of a hospital-wide initiative to improve the quality of its healthcare and as part of its undertaking to its Patient Safety Organization partner. The Nursing Director was concerned that the supplier-led infusion device training did not address clinical conditions in the hospital and the Medical Director was aware that the medical staff had limited understanding of basic vital signs monitoring, now all carried out using electronic devices.

The Head of the CED met with the Head of the nurse's clinical professional training department (CPTD) to discuss the medical device training needs. The current infusion device training dealt purely with the technical details of the pumps, with limited consideration of the contexts of their clinical application. Discussions were held with the infusion pump supplier, who currently provided the training, to understand their views on the current training, positive and negative. Most infusion devices were set and managed by nursing staff, but some specialized pain relief pumps were managed by anaesthetists, with no organized training. There was no training for the clinical thermometers, non-invasive blood pressure monitors (NIBP) or pulse oximeters.

A strategy was adopted that addressed the general principles of operation of each device and with a twin focus on its technical and clinical operation. Each department would be responsible for keeping a training record of its staff, using a training record system that the CED sourced. The training record would be coordinated by the Personnel Department and jointly managed by the CED and CPTD. Funding released from the hospital's indemnity insurance scheme by having the training system would pay for a medical device nurse trainer and pay 20% of the salary of a clinical engineer to support device training.

The CED and CPTD presented the proposal to the Medical Device Committee (Chapter 5, Section 5.2) who approved it and passed it to the hospital Board for endorsement. Half-hour general principle training sessions on how to safely deliver infusions would alternate every fortnight with sessions on vital signs monitoring (thermometry, blood pressure and oxygen saturation). The supplier-organized training sessions would be revamped with a twin-track approach, dividing the hour-long session into the technicalities of the pumps (provided by the supplier) and the clinical applications including documentation (provided by the nurse trainer). Attendance at training sessions would be recorded directly into the Training Record Database. Following attendance at training sessions, competency testing would take place at department level, managed by the clinical leader for the department and supported by CED and the nurse trainer. Training records would be reviewed at the annual appraisal of both nursing and medical staff. The clinical engineers would also be expected to attend both of the twin tracks of the infusion device training and the general principles of safe operation of medical devices and would be monitored via their appraisal system. The clinical engineers would gain from a clearer understanding of operational concerns by attending the sessions with clinical colleagues.

The general principles of safe operation of infusion devices, presented in fortnightly half-hour sessions, addressed setting the pump controls, checking with the medication prescription and double-checking the rate setting prior to starting the infusion with a colleague (Amoore and Adamson 2003). Syphonage and its risks were explained with updates on current safety warnings and lessons from adverse events provided. The vital signs monitoring sessions covered the indirect nature of many of the measurements, the differences between different types of electronic thermometers (oral, tympanic and temporal artery, all used in different parts of the hospital), the limitations of automatic blood pressure monitors and when to use manual auscultatory devices and the importance of cuff size. The pulse oximetry training summarized the principles behind the device and conditions, such as smoke inhalation, when the readings might be inaccurate.

The training scheme was audited after 6 months. Training records were examined to summarize attendance at training sessions and the follow-up competency assessments. Records of incidents and near misses showed a small but encouraging decrease. The aim was to have 80% of nurses trained and competency assessed by the end of year 1; after 6 months, 45% of staff had been trained. Questionnaires were handed out during training sessions to gauge reception of the training, with positive comments about how to improve the sessions received which were incorporated into revised training plans. Questionnaire feedback revealed concern about the human usability of certain devices, which would be fed back into enhancing the specifications for procuring future devices; this recognizes that training is not a substitute for poor device design.

Of concern was that only 45% of nursing staff had been trained on infusion devices after the first 6 months and this was reflected in the annual audit with only 75% competency assessed. It was argued that the target was too high, with challenges facing staff attendance, particularly during busy winter and holiday periods. But the twin-track approach was widely welcomed by staff in all groups.

The general sessions on the principles of vital signs monitoring were poorly attended by medical staff and its review, through their appraisal scheme, was not consistent. This would be addressed through the Medical Director. The problems of a diverse range of clinical thermometers would be addressed through the Medical Devices Committee recommending a standardized replacement plan; similarly, the standardization of the blood pressure monitors was recommended.

ADDING VALUE

A year after introduction, the training scheme showed an improved competency of the clinical staff, with the trend of recorded infusion devices incidents decreasing. The twin-track approach of technical plus clinical application training added to the confidence of nursing staff. The costs of administering the scheme were met by the reduction in indemnity insurance, whilst the overall time cost of staff attendance did not increase due to the tighter control on the previous supplier-led sessions. The benefits to patient safety through better medical equipment training and a cost neutral reallocation of resources increased value.

Benefits : Cost ∴ Value

SUMMARY

A structured training scheme for medical devices, operated jointly by clinical engineers and the nurse training department, provided both technical and clinical operational training with competency assessment on infusion devices. Training was also provided on vital signs monitors. The introduction of the scheme focused management attention on both the need for training and for the development of a culture of improved understanding of medical equipment.

REFERENCE

Amoore J. and E. Adamson. 2003. Infusion devices: characteristics, limitations and risk management. *Nursing Standard*, 17(28): 45–52.

SELF-DIRECTED LEARNING

1. What are the deficiencies of medical device training in your organization? How would you address them?
2. How would you assess the competence of clinical staff in the use of medical equipment such as infusion devices?

CASE STUDY CS4.3: FLUID WARMER CABINET INCIDENT; BOUGHT IT WRONG, USED IT WRONG

Section Links: Chapter 4, Sections 4.4.1 and 4.4.2; Chapter 2, Section 2.4.2.

ABSTRACT

Fluid used for irrigation during an arthroscopy procedure had overheated in a fluid warming cabinet and was administered to a patient. The Clinical Engineering Department (CED) had no previous involvement with this equipment but was called upon to investigate. The root cause of the incident was ascertained and recommendations made.

Keywords: Adverse event, Root cause analysis, Inappropriate purchasing, Incorrect use

NARRATIVE

The CED was asked to investigate an incident in which irrigation fluid used in an arthroscopy procedure had overheated in a fluid warming cabinet and had been used on a patient. Interestingly, the technical support and maintenance of this type of equipment was not allocated to the CED, but the immediate response of the clinical staff was to call on the CED for assistance.

The technical reasons for the overheating were quickly established. Two sets of these warming cabinets had been purchased some years previously as part of a capital project to upgrade the orthopaedic operating theatres and the CED had not been involved in the specification of this equipment. They had been purchased from catalogue and were designed to a 1974 Standard for warming bottles of fluids. The near-universal practice is now to use bags rather than bottles of fluids.

The construction of the cabinets had a heater in the base covered by a metal rack with many perforations and a set of low vertical dividers designed to assist the spacing of bottles of fluids. There is then an upper perforated shelf, again with dividers, under which a controlling air temperature sensor is fitted. There is a small circulating fan attached to the right-hand side and a non-self-resetting temperature cut out fitted about half way up, towards the back on the right-hand side.

On examination, the cabinet was full of 2.5 L bags of fluids, about ten bags crammed in on the bottom rack and another ten on the upper shelf. Upon unloading these bags, it was noted that parts of the outer plastic packaging of three of the bags from the bottom rack had become heat welded together and the outer packaging of other bags were 'crisp', showing signs of having been overheated.

On switching on the now empty cabinet, with the temperature set appropriately at 39°C, it reached that temperature (independently measured) and the heater cut out but the displayed air temperature overshot to just under 60°C and then started to fall back.

The cabinet was then operated loaded with ten bags in the top and three in the bottom. It performed as expected with the heater cutting out at a displayed and measured temperature of approximately 39°C, with a similar overshoot of the displayed air temperature.

The root cause of the technical problem was the lack of circulation of the warm air from the heater in the base of the cabinet to the controlling temperature sensor mounted on the underside of the upper shelf due to the space below the upper shelf being very full of bags of fluids. This was preventing the sensor from effectively detecting rising temperature and thus causing the heater to stay on and seriously overheating the bags in the bottom half of the cabinet and clearly affecting those in the top half as well.

However, there is a more fundamental root cause. The purchase of this type of warming cabinet had been inappropriate and without any technical input. These were bottle warming cabinets, manufactured to a Standard titled, *Specification for hospital storage cabinets for bottled fluids (electrically heated)*. The adverse implications of them being used for warming bags of fluids were not appreciated by either the purchasing department or the clinical users or the hospital engineering department who nominally looked after the equipment.

The recommendation from the CED was that all the fluid warming cabinets of this make or similar design be replaced by a more up-to-date design made to the IEC 60601-1 Standard for medical electrical equipment.

ADDING VALUE

This is an example of a situation in which the benefit to patients is clear and the action is necessary for their safety, but the cost is also increased because of the need to purchase new equipment. However, the potential cost, in both reputation and monetary terms, of allowing the equipment to continue in use is considerable, even with staff instruction and training to reduce the risks. A slight mitigating factor is that the running costs of the suggested alternative equipment are significantly lower than those of the inappropriate equipment due to better design. Replacing inappropriate medical equipment can be expensive but the benefits to patient safety far outweigh the costs, increasing value.

Benefits : Cost ∴ Value

CULTURE AND ETHICS

Ethically, the CED had no alternative but recommend replacement of the inappropriate equipment. There was not simply a potential for an adverse incident, one had occurred and a patient had been harmed, albeit not too seriously. The equipment, although operating to specification, was inappropriate for the use to which it was being put and had other out-of-date safety features.

However, consideration had to be given to managing the immediate aftermath. A strict prohibition on further use of the equipment would have resulted in the cancelation of many operations which required the use of irrigation fluids. There is a clear risk/benefit discussion needed. The CED could provide clear written instructions and staff training to help manage risk in the interim situation.

SUMMARY

The purchase of inappropriate equipment without technical input as part of a capital project and the uninformed method of use of the equipment by clinical staff led to an injury to a patient. The CED investigated the technical causes of the problem and made recommendations that had cost implications. Abraham's adage "Buy it right, use it right, keep it right" had not been followed (Abraham 2000).

REFERENCE

Abraham N. 2000. Can we gain without the pain? *Controls Assurance Conference*, Hammersmith Hospital, London, UK, June 2000.

SELF-DIRECTED LEARNING

1. Accepting the fundamental CED recommendation, what immediate steps should the CED suggest or take, following the investigation of this incident?
2. How should the risk/benefit discussions be facilitated and concluded? Who should be involved?

CASE STUDY CS4.4: TEAMWORK AND THE DISPOSAL OF PHYSICALLY LARGE MEDICAL EQUIPMENT

Section Links: Chapter 4, Section 4.4.4; Chapter 2, Section 2.4.5

ABSTRACT

Each stage of the life cycle of medical equipment is important, but the end-of-life disposal often gets overlooked. We describe particular difficulties that arose when disposing of a large MRI scanner located centrally within a hospital and the teamwork required for a successful outcome.

Keywords: Teamwork; Leadership; Disposal; Life cycle; Delegation; Systems approach, Multidisciplinary

NARRATIVE

Deteriorating image quality, caused by imperfections in its 20-year old RF shield, led to the decision to replace an 8-year old MRI scanner. After its installation clinics had been built around radiology, leaving the MRI room centrally sited within a 600 bed hospital without external access. The MRI room led off to a small courtyard; other than that its only access was via long hospital corridors.

The Clinical Engineering Department (CED) was charged with removing and selling the old MRI scanner on the buoyant second-hand market. Several bidders were interested with a mini-auction held to select a preferred buyer.

How to remove the old MRI safely? The head of CED convened a small working group including Facilities and Health and Safety to develop solutions. The magnet's physical size precluded pushing it along the corridors as this would require removal of electricity cables and medical gas pipes above the suspended ceiling. The outer wall of the MRI room leading into the courtyard could be removed and the magnet pushed into the courtyard – but how to remove it from the courtyard? The courtyard had no external access and to the north, east and west was surrounded by medical and surgical wards. Health and Safety regulations precluded craning the 6 ton magnet over areas where there could be people, ruling out access from these directions. Between the MRI room and the southern hospital external wall were offices and clinics. Intensive Care and the Emergency Department (ED), both operating 24 hours a day, were located nearby.

The small working group reported these findings to the larger multidisciplinary project team that included support services, ED and clinical neighbours of radiology. The Head of CED recommended taking a systems approach to develop the solution. The objective was clear – safely remove the MRI, with a crane the preferred method. Elements requiring solutions were the following: the crane location (Facilities, Health and Safety, ED); crane access route between MRI room and external road, ensuring no one was under the route (Fire Officer, Facilities); continuing function of ED and intensive care (clinical leads for those areas and clinical engineer); patient data confidentiality (Information Governance); risk assessments (Facilities and Health and Safety); ensuring no dust from any work spreading to clinical areas (Infection Control); and keeping all informed (Communications).

Communications proved particularly vital when the only feasible route required a crane with a span of 90m to be situated outside the main hospital entrance. Removing the MRI at night allowed continual function of ED, but necessitated closing the main hospital entrance immediately after visiting time at 9 p.m. The crane's size required it to be built on-site (an estimate of 6–8 hours) after which it would have to be parked across half the main car park before being taken down (5–6 hours) the following night. Rerouting vehicular access to the hospital would be required, necessitating consultation with the local authority and bus and taxi companies. The crane's size required that it be reported to the local airport.

The following decision was made: removal by crane, erection of crane from Friday evening 9 p.m., removal of MRI by Saturday 5 a.m., park the crane on car park by 6 a.m., dismantle the crane from Saturday evening 9 p.m. and removal of crane off-site by Sunday morning 6 a.m.

The MRI purchaser would be responsible for safely demagnetizing the MRI and pushing it into the courtyard, the wall having been removed by a build contractor (Facilities). The crane contractor would lift the MRI from the courtyard onto a truck supplied by the MRI purchaser.

The project required the coordinated effort and cooperation of many. No one discipline has all the knowledge, skills, responsibilities and contacts required. The systems approach provides a framework in which each discipline involved takes responsibility for its own area, delegated by the overall project leader. For example, ensuring that no patient data remained on the old MRI scanner (which stores patient data in short-term buffer storage) was delegated to Information Governance with support from the Radiology department and the purchaser. Good communication was vital, with regular planning meetings arranged for airing problems, listening to each other and finding solutions, knowing that the action of one affects others – the interdependency of elements in the systems approach. Leadership was important with crucial decisions having to be made, based on assimilating the collective information from the project team.

The main entrance was closed at 9 a.m., with the crane contractor assembling the crane. Beautiful teamwork saw the crane assembled within 6 hours, the MRI magnet by this time pushed out into the courtyard. Craning the magnet out of the hospital took less than 10 minutes, greatly facilitated by a clear winter night with no wind (prayers answered). Visitors to the hospital on the Saturday were greeted by the sight of an enormous crane dwarfing the hospital entrance, but by Sunday morning the crane was gone.

ADDING VALUE

The teamwork not only led to the successful removal of the MRI but also strengthened bonds for future work. The money raised from the sale of the MRI covered the removal costs with money remaining helping to furnish the new MRI room thus increasing benefit.

$$\text{Benefits}:\text{Cost} \therefore \text{Value}$$

SUMMARY

Plans for medical equipment disposal should be considered during acquisition and reassessed periodically during its lifespan. The systems approach provides the framework for keeping focus on the overall objective whilst giving sufficient attention to all the elements and their interdependencies.

SELF-DIRECTED LEARNING

1. *Scenario*: The removal of a major equipment installation from one of imaging, critical care or theatre. What problems are likely to occur? Who will take responsibility? What disciplines would be required for the removal and how would you develop the multi-disciplinary project team? Why is communication, internal and external, so vital to the success of the process? Why is leadership vital?

CASE STUDY CS4.5: A NEAR-MISS BREACH OF PATIENT DATA CONFIDENTIALITY

Section Links: Chapter 4, Section 4.4.4

ABSTRACT

The storage of patient data and its access by authorized people presents particular problems for certain medical equipment. Returning loaned equipment containing stored patient data raised concern.

Keywords: Information governance; Data; Patient data; IT; Confidentiality

NARRATIVE

As part of evaluating new equipment in a medical equipment replacement programme, respiratory therapists requested the loan of a pulmonary function test analyzer. The equipment consisted of an analyzer and separate computer system that recorded, analyzed and reported on each patient's tests. The patient's hospital identity number, name and date of birth were required for each test set, and these were stored on the internal hard disk, with options for transferring data to DVD, USB or an electronic patient record. Preliminary pre-use testing of the equipment was undertaken by the Clinical Engineering Department (CED), following the appropriate equipment support plan (Chapter 6) for short-term loans.

At the end of the loan period, clinical engineers realized that the computer still contained patient data. Discussions were held with the manufacturer and IT colleagues specializing in information governance (IG). Several options were examined:

- *Straightforward data deletion*: A simple process but leaving the risk of data being recoverable.
- *Deletion and overwriting of data to a 'military' standard*: This would require specialized software and technician time but would ensure the data was not recoverable.
- *Destruction of the disk*: A quick blunt approach but permanent, advocated by IG for hospital computers. Its disadvantage: it requires the supplier to install a new hard disk and operating system software with the associated expense.

The most cost-effective resolution in this case was the physical destruction of the hard disk. This meant the loss of the operating system and proprietary software, and a replacement drive had to be purchased by the hospital and installed by the supplier.

It was realized that patient identifiable data may already be present on equipment received on loan having been previously loaned to other hospitals. This would place those other organizations in breach of information governance guidelines, rendering them liable to large financial fines and bad publicity. It was not clear if the hospital or supplier would be liable but plainly the supplier needs to be aware of the risks. The supplier advised that it would be implementing changes to check for patient identifiable data retained after loans. The supplier considered methods of avoiding this problem, for example, restricting the storage of patient data on loan equipment. Alternatively 'false' patients pseudonyms could be created, identifiable locally but not to anyone without internal access to records. A further possibility could be storing patient data on a separate hard disk or other storage medium.

Clinical engineers took the initiative in meeting with Information Governance (IG) colleagues to produce a set of questions to manage the risks of personally identifiable data stored on medical equipment prior to loan or purchase, including these in the pre-purchase/loan documentation. It was agreed that solving these problems required dialogue with suppliers and IG, arranged through clinical engineering.

Furthermore, the CED modified its internal procedures to check for data that would be stored on incoming loan medical equipment. Changes included:

- Prior to use communicating with suppliers about removal of patient identifiable data;
- Checking incoming equipment for data storage options or functions;
- Checking incoming equipment for existing data from other organizations;
- Investigating the nature of data stored;
- Restricting data storage if possible (switching off USB connections, Hard Disk writing etc.);
- Advising clinical users, where possible, to avoid using *identifiable* data if data must be recorded in order to use the system.

In addition, the Medical Equipment Management Database was modified to include a flag to indicate if equipment stored patient identifiable data.

The CED proactively investigated the storage of data within existing medical equipment. Those equipment storing patient data were flagged on the database. Discussions were held with IG about procedures for controls when sending equipment off-site for repair. CED received assurances from Procurement that national terms and conditions of supply had clauses advising suppliers of their information governance responsibilities. However, it was agreed that not all suppliers might recognize these responsibilities and that each case should be discussed with suppliers before proceeding to external repair.

ADDING VALUE

Although there was a one-off cost in time and effort sorting out the immediate issue that had arisen, there was no long-term cost, and the development and implementation of modified internal procedures led to benefit in both patient confidentiality and organizational risk reduction.

$$\uparrow \qquad \uparrow$$
Benefits : Cost ∴ Value

PATIENT CENTRED

Patients entering hospital for investigations and procedures accept and agree that their personal data can be recorded in order for them to receive the best possible care. However, controls need to be in place to ensure that these records are only seen by those with a justifiable need. Allowing others to see those records, accidentally or not, is a serious breach of trust. CED has a responsibility to ensure that patient identifiable data on the medical equipment it manages is controlled, with access restricted.

CULTURE AND ETHICS

Within hospitals and other healthcare organizations, patient data are used and managed with such frequency that there is a danger that clinical staff regard it as commonplace. Regular briefings on confidentiality help to keep the issue to the fore of all staff.

SUMMARY

This case study highlights another aspect of Health Technology Management requiring vigilance on the part of clinical engineers. In this case the enhanced data storage opportunities of the equipment were appreciated by clinical colleagues but the consequences of the data storage on loaned equipment to be returned were not recognized by clinicians or the supplier.

Both suppliers and healthcare organizations need to be aware of their responsibilities regarding patient identifiable data stored on medical equipment. Procedures to cover the management of equipment containing patient identifiable data need to be developed by clinical engineers in conjunction with IG to minimize risks to the organization. These should be incorporated into the Medical Device Policy.

SELF-DIRECTED LEARNING

1. If the equipment had left the premises without anyone realizing confidential data were still stored on it, what do you think the risks to the organization would be? How would you assess the magnitude of the risk to the patient, service, organization? What remedial actions could be taken to minimize the risks?
2. What are the alternatives to replacing the Hard Disk Drive? What are the limitations and risks of the alternatives and how can the risks be minimized?

CHAPTER 5

The Healthcare Technology Management System

Strategic

CONTENTS

5.1 INTRODUCTION

Chapter 4 introduced a model for a comprehensive Healthcare Technology Management (HTM) system consisting of two interlocking processes, the top one strategic and the other operational. However, these two processes do not happen in isolation – they are interlocked and interdependent. In a large organization, the pinnacle of the operational process will in fact be closely aligned with the strategic planning which is part of the top cycle. In this chapter, we will explore in more detail the top strategic component of the HTM system looking at its relationship and intersection with the operational.

Central to the understanding of the system is the function and structures of the management process that delivers the strategic component and facilitates its interrelationship with the operational process. This linking entity we describe as the Medical Device Committee (MDC). In a small organization, this function might not require a committee *per se* and be vested in an individual. However, as organizations increase in size and complexity, there is a corresponding need for clear functions and processes to be in place as management systems. In this chapter, we will explore in more detail the role of the MDC within the context of a large teaching hospital, in which it is best provided by an actual committee with terms of reference, resources, governance and authority. We have provided a case study based on this sort of larger organization (Case Study CS5.1) and another illustrating some of the complexities of providing community services across a number of different organizations (Case Study CS5.2). Case Study CS5.3 looks at the links between the Medical Device Policy (MD Policy) and the Strategic HTM Plan.

It is an MDC with this wider remit that we discuss in more detail in Section 5.2. The MDC forms part of an ISO 55000–based Asset Management System (ISO 2014a) as discussed in Chapter 4, and summarized in Table 4.1 and Figures 4.12 through 4.14. In Chapter 4 we have defined the asset management terms we have used in this book. Figure 4.14 is particularly worth bearing in mind as you read this chapter because it illustrates the relationship between the different aspects of the HTM system.

5.2 THE MEDICAL DEVICE COMMITTEE

In many healthcare organizations, a multidisciplinary technology committee has been or is being established. The reasons for their development and hence their remit vary dependent on the healthcare organization's management and funding structures but arise out of real needs. Some are driven by the need to ensure safety and effectiveness and to control costs. Some U.S. organizations have set up technology assessment committees or Value Analysis Committees through which clinicians and management work together to assess and select optimal technologies for their organizations (Montgomery and Schneller 2007).

The requirements to address the wider aspects of HTM have led to the recommendation, by the United Kingdom's Medicines and Healthcare Products Regulatory Agency (MHRA), for a multidisciplinary management committee to oversee medical devices and their management in a healthcare organization. The MHRA advises: 'Healthcare organizations should establish a medical devices management group to develop and implement policies across the organization' (MHRA 2015). Such committees are effective and fulfil many aspects of the role of the MDC as defined in Chapter 4.

The MDC is tasked with implementing the Medical Device Policy (MD Policy) and contributing to its annual review. Many stakeholders have important contributions to such a group; Board members, clinicians, both medical and nursing, general managers, finance managers, clinical users of healthcare technology and clinical engineers. However, it is only clinical engineers who have this asset management activity as central to their role and many senior clinical engineers provide leadership in this regard within their organization.

The MDC will report to the Board through a senior Board member, often the Medical Director, Chief Nurse or Chief Executive Officer (CEO). This Board member may lead the MDC or may delegate that task to another senior person. In some organizations, this is the Head of Clinical Engineering but more often it is a senior clinician with an interest in medical equipment and the ability to act in a neutral role in chairing the committee in its job of allocating resources.

The responsibility for formulating and reviewing the MD Policy is likely to be devolved to the MDC, but the policy reflects the view of the organization's Board and must be signed off and owned by the Board. It is only by doing this that the MD Policy reflects top management views and has authority. In this way governance rests with the Board. This arrangement is in line with the requirement in the ISO 55001 Standard, subclause 5.2, 'Top management shall establish an asset management policy that:...' (ISO 2014b).

The MDC should analyze the deployment of devices and systems, reviewing their associated risks and benefits, to ensure that their application supports the organization's clinical, corporate and financial goals. The analysis may require the MDC to group medical equipment by functional type (e.g. resuscitation, renal dialysis), working with subgroups that include the appropriate clinical teams. Systems engineering thinking which we have described in some detail in Chapter 2 will help ensure that all the resultant components work in harmony to support the organization's strategic aims. Having analyzed the deployment, the MDC develops a Strategic Healthcare Technology Management Plan (Strategic HTM Plan), regularly reviewed, which details how the MD Policy will be delivered.

5.3 THE MEDICAL DEVICE POLICY

This is a top-level policy document that focuses on the delivery of appropriate technology management in the context of and support for the organization's overall priorities and strategic objectives. It provides a framework and driver for the Strategic HTM Plan which focuses on operational details and in turn leads to specific programmes and procedures.

The MD Policy is owned by top management and sets out the organization's current position and intentions in relation to medical devices and healthcare technology management. This policy must be aligned with corporate objectives and provide the framework within which managers and staff develop the strategic and operational plans which deliver the healthcare technology management system on the ground.

The mechanisms by which the MD Policy is delivered will vary between organizations, dependent on their local governance arrangements, and also between jurisdictions. They will also vary depending upon the organization's size and maturity. The MD Policy discussed in this book will be that typically implemented by a large hospital or hospital group. However, the principles upon which the model is based are also applicable to organizations that use different governance structures or which operate on a smaller scale.

5.3.1 Overview

There are a number of key aspects, discussed in the following text, that should be built into the MD Policy.

Since HTM plays a significant part in the organization delivering its key objectives, albeit 'behind the scenes', the MD Policy must align with the organization's key strategies. It must address, at the policy level, all aspects of medical devices and their management throughout their life cycle, including facets of technology support such as acquisition planning, risk management and incident investigation. It must provide a means for the organization's top management to communicate their vision for HTM, recognizing HTM's effects on the management and running of the organization. If the MD Policy does not align with the organization's key objectives, it will not be effective in guiding the delivery of an HTM that adds the value that these assets should contribute.

The MD Policy should be appropriate and clear regarding the range of healthcare technology assets under consideration. This should include all types of medical devices: single use; disposable; non-active reusable devices, for example surgical instruments; medical equipment, that is active medical devices; systems of interconnected equipment which may involve IT networks. The policy should consider all such assets. Where the organization's asset management plans and responsibilities for different group of assets are distributed between different departments; the policy should apply to all. The strategy for these different asset groups may be different, but the overall policy principles should be consistent. The policy should apply throughout the organization and meet the reasonable expectations of all stakeholders.

The policy should address at the strategic level the scientific and technical support that goes beyond basic conventional maintenance activity. This includes:

- The requirement for an all-inclusive inventory of assets which supports and enables planning for equipment replacement;

- The processes for the assessment and introduction of new clinical techniques and services which should include early engagement in such projects so as to inform and advise on HTM before decisions are made;

- The planning and coordination requirements for the installation of new equipment;

- The handling of untoward incidents and the necessary follow-up action;

- The provision of advice regarding formal Standards and regulations.

These activities are all important in the wider context of healthcare technology management and add value to the organization's operations.

The policy should also be consistent with other policies within the organization. In particular, it is likely that the organization will have policies on procurement, on risk management and on safety. There will be links to all these in the MD Policy.

Whether or not the organization has a specific environmental policy, aspects of the MD Policy must address environmental issues. This will include consideration of energy usage, the extent to which single-use medical devices and accessories are used, and the management of their disposal. The policy must be very clear on any allowed reprocessing and reuse of single-use devices. The policy should also consider the principles of how medical equipment is managed once withdrawn from service in the organization. Detailed plans for disposal are likely to differ in various circumstances, but the policy will set an overall framework.

5.3.2 Structure and Content of the Policy

Organizations sometimes have a corporate template for policies, and this structure should be followed if one exists. However, as an example, details under the headings below should be included:

- Policy statement and aims;
- Definitions;
- Legislative framework;
- Organizational accountability;
- Review and audit arrangements;
- Systems management:
  - HTM organizational arrangements;
  - Equipment life cycle;
  - Inventory and records;
  - Acquiring new devices and systems and developing new services;
  - Putting new equipment and systems into use;
  - Training;
  - Appropriate prescription and use of devices;
  - Cleaning and decontamination;
  - Maintenance and repair;
  - Removal from service;

- o Disposal of withdrawn equipment;
- o Managing adverse events;
- o Dealing with safety alerts and field safety notices (FSNs).
- References;
- Appendices.

The Policy statement and aims should be short, concise and clear, for example:

> It is the policy of this organization that all healthcare technology is acquired, used, supported, managed and maintained in such a way as to maximise the safe and effective use of medical technologies for patient care, minimizing risk to patients, staff and visitors, and providing best value to the organization in delivering its strategic objectives.
>
> The aim of this policy is to form the basis for detailed strategies and plans that will ensure that whenever healthcare technology is used, it is:
>
> - suitable for its intended purpose;
> - used according to its 'intended use' by suitably trained persons;
> - supported, managed and maintained in a safe and reliable condition;
> - disposed of appropriately at the end of its useful life.

Terms such as *healthcare technology, medical device, medical equipment laboratory equipment* are likely to be used in the policy and should be defined clearly so that there is no ambiguity as to what the policy applies to. It may be helpful to include examples. Also the statutory and regulatory frameworks that apply in the national or state jurisdiction should be referenced in outline. It is important that the lines of accountability and responsibility at Board level are defined in the policy. Also, the responsibility for consultation on and drafting the policy should be spelt out. This is usually through and to the MDC. The policy should state clearly the arrangements for its review and for the operational audit of its application.

The policy will usually close by citing other documents, certainly legislation, regulations and formal guidance, but may well also cite peer-reviewed papers and textbooks. These should all be formally referenced. One style of presenting a policy of this nature is to write the policy statements as clear requirements without detailed explanation and then to include an explanation of each requirement in an appendix. In this way the policy is sharp, explicit and readily accessible, but the background and explanation for each requirement is available for more detailed reading. This has the added advantage that the policy itself concentrates on key core principles, leaving details that are subject to variation arising from organizational and regulatory changes in the appendices.

5.3.2.1 HTM Organizational Arrangements

The policy should make it clear that detailed, strategic organizational arrangements for HTM will be decided by the MDC. It is better that the MD Policy, owned by the Board, does not go into such detail, so that the necessary flexibility is available without high-level change of policy.

5.3.2.2 A Life Cycle Approach to Equipment Management

The concept of managing technology by considering its complete life cycle and life cycle costs has been discussed in Chapter 4. The policy must make it clear that such an approach must be followed in the best interests of the organization. For example:

> 'A key feature of all effective equipment management is that a life cycle approach, including whole life cost of ownership, is taken. This involves careful strategic consideration of needs and options before equipment is acquired, appropriate operational management of that equipment while it is in use, and the legal and economic disposal of the equipment when the decision is taken to remove it from service. This brings the cycle back to the starting point again.'

A diagram may be helpful in discussing the policy, with Figure 5.1 summarizing aspects of the strategic and operational HTM cycles.

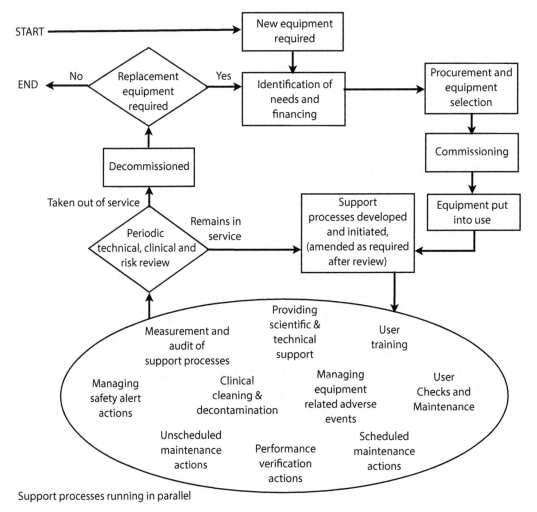

FIGURE 5.1 Equipment life cycle.

5.3.2.3 Inventory and Records

The policy should mandate that a comprehensive inventory and medical equipment management record keeping system is in place for the whole organization. Such a system will be software based, and the policy may also mandate which department is responsible for this database.

5.3.2.4 Acquiring New Devices and Systems, Developing New Services

The policy should clearly state the responsibilities and principles to be followed when new healthcare technology, be it devices, equipment or systems, are planned for and acquired. Clinical need, clinical effectiveness, value for money and the organization's overall plans and objectives must be key drivers. This will include ensuring that the MD Policy addresses issues regarding the procurement of healthcare technologies, with appropriate links to the Procurement department. A more detailed list of issues to be considered which should be included in the policy is set out in Section 5.6.4. Detailed procedures will be developed based on the MD Policy.

5.3.2.5 Putting New Equipment and Systems into Use

The policy must set the overall requirements such that once acquired, new devices or systems must be put into use in a planned and safe way. Detailed plans will flow from this policy requirement.

5.3.2.6 Training

The policy must state that appropriate training for both clinical users and staff who maintain these assets is required. Users may in some circumstances include patients or lay carers some of whom may be at home, and this must be acknowledged. Detailed training requirements will not be part of the policy but will be developed elsewhere as part of equipment support plans (ESPs) (Chapter 6).

5.3.2.7 Appropriate Prescription and Use of Devices

The prescription of devices is the selection of the most appropriate device to use for a given clinical situation. The policy should set an overall requirement that this will only be made by staff with the appropriate professional qualifications. Competency to prescribe must be assessed, recorded and audited to ensure consistency and accuracy of prescribing procedures.

5.3.2.8 Cleaning and Decontamination

Cleaning and decontamination of medical equipment is a vital aspect of infection control. The policy should set a requirement that equipment is clean and decontaminated before use and mandate the development and responsibilities for detailed procedures and the necessary training.

5.3.2.9 Maintenance and Repair

Appropriate maintenance and repair ensures the continuing safety and effectiveness of equipment. The policy should set the overall requirement and responsibilities. If, within the organization, different asset groups are managed by different technology management departments, the overall policy must be applied consistently. It may be useful to include as an annex to the policy a table of the various asset groups against the departments responsible for their technical support and maintenance. The policy may also name a 'medical devices officer' to coordinate between various departments and act as a single point of contact for clinical users.

5.3.2.10 Removal from Service

The policy should state a list of criteria that will be used to determine whether an item of equipment should be withdrawn from service. This is likely to include:

- Worn out beyond economic repair;

- Damaged beyond economic repair;

- Unreliability (based on service history);

- Clinically or technically obsolete;

- Spare parts (manufacturer's support) no longer available;

- More cost-effective or clinically effective devices have become available;

- Unable to be decontaminated effectively;

- Advice or requirements from regulatory authorities;

- The availability of safer alternatives as required by new regulations.

5.3.2.11 Disposal of Withdrawn Equipment

There are a variety of ways of disposing of withdrawn equipment: it can be scrapped, in which case waste legislation applies; it can be sold or given to another organization, in which case issues of transfer of liability and copies of service records must be considered; or it can be kept as spare equipment within the organization, in which case control and management must continue.

The policy should flag up these various alternatives and mandate that procedures are in place to deal with each one.

5.3.2.12 Managing Adverse Events

The MD Policy should reference the organization's incident reporting policy and include a clear statement that adverse incidents are to be reported in a timely and open fashion, detailing to whom they are reported and an outline of the actions to be taken at the time. This should include retaining equipment and all accessories and consumables in use at the time and taking photographic records where appropriate. Responsibilities for investigating incidents and reporting them to regulatory authorities should also be explicit.

5.3.2.13 *Dealing with Safety Alerts and Field Safety Notices*

Safety alerts which warn of problems that have been identified with medical devices are issued either directly from manufacturers (often called field safety notices) or by national regulatory authorities. The mechanisms involved will be detailed later in Section 5.6.6, but each healthcare organization has a responsibility to ensure that it has a system in place to receive and act as appropriate on the safety alerts. The policy should detail the arrangements, ensuring compatibility with any appropriate processes developed by the organization's risk management committee.

5.3.3 Communicating the Policy

The MDC should consider how the MD Policy is communicated within the organization and its public availability.

It is usual for such policies to be available to all staff via the organization's intranet. It is likely that the whole policy, or sections of it, will have a specific impact on particular departments within the organization, and this should be made clear in the policy. For example, the whole policy will be relevant to the Clinical Engineering Department (CED) for whom it will form the baseline document from which strategies and plans will be developed.

The extent to which the full policy is put into the public domain, for example by being available on the organization's website, will depend on the nature of the particular healthcare organization. In principle, it would seem reasonable that a publicly funded organization should be completely open in this area. A 'for-profit' healthcare organization or a 'non-profit' one working in a commercially competitive healthcare system might choose to put an executive summary in the public domain but not the detailed policy.

5.4 FOLLOWING ON FROM THE MD POLICY: THE STRATEGIC HTM PLAN

The MD Policy is the organization's top-level document dealing with the management of its healthcare technology and ensuring alignment with and contribution to its overall strategic plans and objectives. The policy will require regular review to ensure this but is unlikely to require radical change from year to year unless strategic objectives change substantially.

However, such a top-level document does not detail the practical, operational activities that are necessary to implement the policy. These are detailed in a strategic Healthcare Technology Management Plan (Strategic HTM Plan), reviewed regularly, at least annually. This is considered in detail in the remainder of this chapter.

5.4.1 Developing the Strategic HTM Plan as a Quality Cycle: Increasing Value

The challenges involved in establishing and implementing a Strategic HTM Plan are great. It is unlikely that a perfect system will ever be reached, rather that the Strategic HTM Plan is the subject of a continuous quality improvement cycle. This allows for the plan to be influenced by changes in corporate and operational policy or by changes in external governance, healthcare policy and accreditation requirements. Figure 5.2 shows the work of the MDC in this respect, a Strategic HTM Plan imagined as a quality cycle taking the familiar Plan–Do–Check–Act (PDCA) format. We will see that through the Strategic HTM Plan the organization can assure safe and effective service delivery and improve patient care and satisfaction. The Plan also supports the organization to build processes which are robust, documented and evidence based. In doing so it promotes effective service provision, in compliance with standards and regulation, and acts as a means of adding value and controlling cost. Case Study CS5.4 illustrates the value added by linking the Strategic HTM Plan with a strategic clinical aim of the organization.

5.4.1.1 Plan

We have already seen that, although the MD Policy is a Board level document, it is likely that its drafting and refining will be delegated to the MDC. Through their multidisciplinary membership, they have the knowledge and expertise to draft this policy and present it to the Board for approval.

The Strategic HTM Plan belongs to the MDC. It is based on the MD Policy but is a more detailed and flexible document concerned with the detail as to how the policy will be implemented in practice. It is strategic because it deals with the top-level details as to how the corporate MD Policy will be implemented. Issues to be addressed in the Plan include:

- The arrangements for having available a complete inventory of medical devices;

- The establishment, updating and review of a risk register;

- The allocation of responsibility and authority for management and maintenance of different types of medical devices to different departments who deliver healthcare technology management programmes;

- The arrangements to be in place for the allocation of finance for new and replacement devices.

As with the MD Policy, this Strategic HTM Plan requires regular review, probably more frequently than the Policy, to ensure it remains consistent with corporate strategy and policy, remaining relevant as circumstances change. The Plan needs to be backed up by an implementation plan which might take the form of a familiar action plan table with tasks, outcome criteria, allocated responsibility and target dates, reviewed regularly at each meeting of the MDC. We refer to this as the MDC Action Plan.

Case Study CS5.5 illustrates how strategic planning for a new clinical service reduces clinical and financial risks.

5.4.1.1.1 Medical Device Inventory: Capital and Revenue Equipment.   Meaningful equipment planning requires a clear and comprehensive understanding of the healthcare

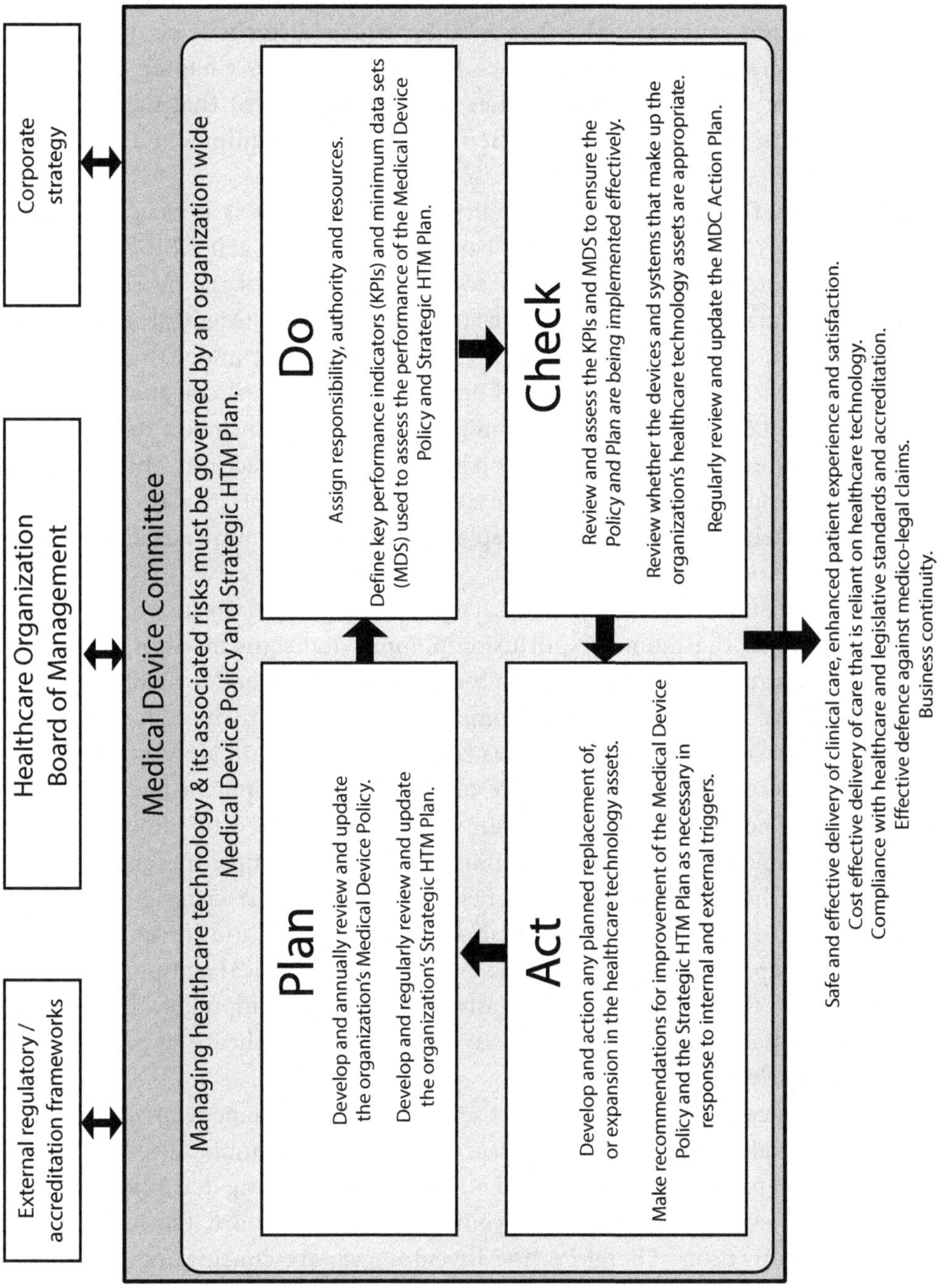

External regulatory / accreditation frameworks

Healthcare Organization Board of Management

Corporate strategy

Medical Device Committee

Managing healthcare technology & its associated risks must be governed by an organization wide Medical Device Policy and Strategic HTM Plan.

Plan

Develop and annually review and update the organization's Medical Device Policy.

Develop and regularly review and update the organization's Strategic HTM Plan.

Do

Assign responsibility, authority and resources.

Define key performance indicators (KPIs) and minimum data sets (MDS) used to assess the performance of the Medical Device Policy and Strategic HTM Plan.

Act

Develop and action any planned replacement of, or expansion in the healthcare technology assets.

Make recommendations for improvement of the Medical Device Policy and the Strategic HTM Plan as necessary in response to internal and external triggers.

Check

Review and assess the KPIs and MDS to ensure the Policy and Plan are being implemented effectively.

Review whether the devices and systems that make up the organization's healthcare technology assets are appropriate.

Regularly review and update the MDC Action Plan.

Safe and effective delivery of clinical care, enhanced patient experience and satisfaction.
Cost effective delivery of care that is reliant on healthcare technology.
Compliance with healthcare and legislative standards and accreditation.
Effective defence against medico-legal claims.
Business continuity.

FIGURE 5.2   The PDCA cycle for the MDC in relation to the MD Policy and the Strategic HTM Plan.

assets that the organization owns or uses. Therefore, a comprehensive asset register is vital. Finance departments usually require an asset register for accounting purposes and are concerned about equipment owned by the organization with a value above a defined amount or representing a significant investment in the organization's capabilities. The inclusion criteria of the accountant's asset register will be determined by the organization's accounting rules. We have discussed in Chapter 4, Section 4.2.2 the concept of capital expenditure that acquires equipment that increases the organization's assets. Such equipment is often called 'capital' equipment and we will use that term.

Figure 5.3 shows the distribution of medical equipment that is typical of a mid-size UK healthcare organization, in value bands starting at <£250 GBP through to >£250,000 GBP. There should be a database to record this and all the necessary asset management information. We have called this database the Medical Equipment Management System (MEMS). This should include all equipment under the control of the organization, whether purchased, leased or acquired through a 'consumables deal' (Chapter 4, Section 4.3.2). The organization might choose not to include a full database entry of items that are on short-term hire or on loan from a manufacturer whilst a repair is being carried out but such items need to be tracked in an appropriate way. The MEMS database will definitely include equipment deployed into the community as described in Case Studies CS5.2 and CS5.5.

Figure 5.3 demonstrates that much medical equipment costs relatively small sums of money with electronic thermometers, infusion pumps, vital signs monitors and even many defibrillators ranging from less than a £1000 ($1400) to under £5000 ($7000). Whereas the Finance Department may in some circumstances only be concerned with 'capital' equipment which they may define as costing over a certain amount, the clinical engineer will recognize the importance of all medical equipment because of its clinical function. The data in Figure 5.3 can be further analyzed to show the number of items in various cost bands. This is illustrated as an example in Table 5.1. It is important to note that equipment costing below £5000, the cost criteria commonly used in the United Kingdom to distinguish between 'capital' and 'revenue' equipment, accounts for approximately 93% of the organization's medical equipment assets, though only 30% of the cost base. Such relatively lower-cost equipment, deployed in large numbers, is vital for the safe and effective delivery of healthcare to patients and needs to be effectively managed.

The MDC is concerned with the impact and proper management of *ALL* equipment whatever the value, whether owned, leased, borrowed or manufactured in-house. Many of these, vital to patient care, will fall outside the accounting definition of capital equipment (however that is defined in your organization) or are items not legally owned by the organization. Therefore, the inventory we are considering here must be comprehensive and the database must be capable of recording the 'cost' of each item, that is what was paid for it, which may be zero, and the value, often recorded as the list price at the time of acquisition and referred to as the Replacement Asset Cost

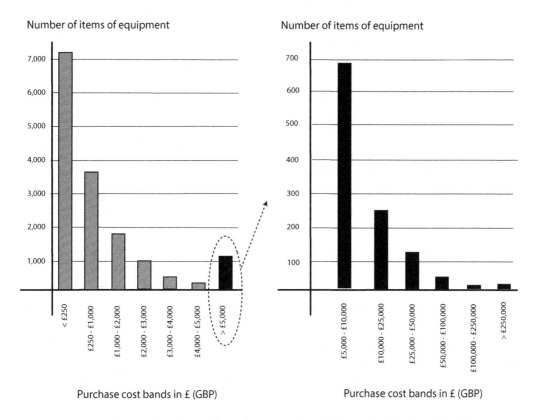

FIGURE 5.3   Distribution by the number of items in defined cost bands of a medical equipment inventory typical of a midsize UK healthcare organization.

TABLE 5.1   Distribution of Medical Equipment Assets Detailing the Number of Items by Cost Band and the Estimated Total Cost within Each Band

Medical Equipment Assets Data				
Cost Bands	Number of Items	% of Items	Total Cost in Band	% of Total Cost
Up to £250	7,147	46.6	£893,375	2.5
£250 to £1,000	3,540	23.1	£2,212,500	6.1
£1,000 to £2,000	1,795	11.7	£2,692,500	7.4
£2,000 to £3,000	1,004	6.6	£2,510,000	6.9
£3,000 to £4,000	471	3.1	£1,648,500	4.5
£4,000 to £5,000	238	1.6	£1,071,000	2.9
£5,000 to £10,000	690	4.5	£5,175,000	14.2
£10,000 to £25,000	250	1.6	£4,375,000	12.0
£25,000 to £50,000	130	0.85	4,875,000	13.5
£50,000 to £100,000	39	0.25	£2,925,000	8.0
£100,000 and £250,000	12	0.08	£2,100,000	5.8
Greater than £250,000	14	0.09	£5,880,000	16.2
Totals	15,330	100	£36,357,875	100

(Chapter 4, Section 4.2.4). It is also strongly preferable that all this information is held on a single database, the Medical Equipment Management System; the formal capital asset register, which the accountants require, can be a subset of this.

5.4.1.1.2 Risk Register.   The planning and development of a register of risks associated with the use of healthcare technology is key to many of the actions the MDC will take. Hazards or hazardous situations may have been identified as a result of incidents which occurred or from information based on staff imaginatively foreseeing problems and bringing them to the attention of the MDC. Alternatively, risks may be calculated through the implementation of the HTM Programmes and by extracting data from the inventory system. In most cases, such risks will be assessed and managed as part of those HTM Programmes. For risks that cannot be managed 'locally', the MDC is the escalation route used to bring the risk to the attention of senior management. It may be useful in large organizations to establish a subcommittee of the MDC to analyze risks and incidents associated with medical devices.

The risk register will be an important input to the planning of a healthcare technology asset acquisition list. Items may be placed on this acquisition list arising from changes in corporate strategy, for example a new service development. Alternatively, existing devices may be considered for replacement as a result of issues highlighted by internal or external risk management or as a means of cost control. Similarly, there may be recommendations to replace or increase devices to comply with best practice or to support delivery of care improvements. The MDC may need to seek Board approval for new acquisitions and should only proceed to the acquisition phase once the sources of funding for acquisition and whole life operation and eventual disposal have been identified.

5.4.1.1.3 Allocation of HTM Support Activities.   It is unusual, though not unprecedented, for a single scientific and technical department to be responsible for supporting all types of healthcare technology assets. An example of an arrangement in which a single organization provides comprehensive support can be seen in the public healthcare system in Hong Kong where the Electrical and Mechanical Services Department of the Government of the Hong Kong Special Administrative Region provides such a service. See http://www.emsd.gov.hk/en/engineering_services/hospitals_clinics/scope_of_services/index.html.

More typically, the MDC has the responsibility for agreeing the allocation of responsibility for scientific and technical support to more than one department, such as Clinical Engineering, Medical Physics, Facilities Management, Information and Computer Technology, Radiology, Pharmacy, Laboratory Medicine and Sterile Services. In most cases the Clinical Engineering Department carries the largest load both in terms of number of devices and in total value. Where an external service provider is contracted to deliver support to some of the medical equipment, the MDC must allocate clear responsibility to one of the technical support departments for managing and monitoring that contract. Experience shows that it is ineffective and costly to leave this task to the clinical departments concerned. It is not their primary role and usually not within their expertise. Procurement departments are also not usually well placed to manage technical support contracts.

5.4.1.1.4 Allocation of Finance. The MD Policy will include robust and defendable arrangements for allocating funding for medical equipment procurement. The organization's Board may delegate this task to the MDC or to a funding allocation committee. The funding allocations must support the corporate objectives, address medical equipment risks and promote clinical services and their developments (Case Study CS5.5).

Requests for funding will include requests for both additional and replacement equipment. Clinical engineers can assist in informing these decisions through their HTM knowledge and experience and from careful use of information extracted from the MEMS database.

5.4.1.2 Do

The ongoing work of the MDC will flow from the Strategic HTM Plan and the MDC Action Plan.

Where risk reduction actions have been approved and funding assigned, the MDC should clearly task one of the technical groups with the responsibility for their implementation. If these actions are related to a change or increase in the ongoing equipment management processes, the action will be assigned to the department delivering the relevant HTM Programme. Sometimes risk reduction actions require the convening of a multidisciplinary 'task and finish' group to implement a specific risk reduction project, for example the establishment of a pooled equipment loan service, sometimes called an 'equipment library' (Keay et al. 2015). The MDC should establish such a project group and give it the authority to implement the action.

The MDC should ensure that planned replacement programmes are in place. These support the corporate financial planning of the organization and orderly and cost-effective equipment replacement. These replacement plans may typically cover 5–10 years, ensuring forward planning, whilst being flexible enough to respond to changes in clinical priorities. In this way procurement decisions are not made in a rush without adequate prior analysis. These are sometimes referred to as rolling replacement plans (Case Study CS5.6).

When new devices are approved to be acquired, and the funding for these has been secured, the MDC should establish an appropriate multidisciplinary acquisition group, which must include the Procurement Department, to execute the acquisition project. This group may well be a standing subcommittee of the MDC which works in the background to analyze and prioritize equipment procurement requests and is brought into full action when funds become available.

The planning of new acquisition projects should consider not only the devices themselves but also how they are to be used over their useful life. In doing so, the capability of the organization to use, manage and maintain the devices effectively needs to be considered and planned for. Only by having in place and following a corporate policy can the range of risks and the conflicting demands on financial resources that all healthcare organizations face be effectively managed. This principle is no different from that which applies to a non-healthcare organization. The unique feature is that inadequate healthcare technology management can result in serious risks to patients.

Another task for the MDC to 'Do' is to agree key performance indicators (KPIs) and any applicable minimum data sets to be applied to each of the technical support groups (CMS 2015; HSCIC 2015). These will significantly assist the MDC in the CHECK stage of its work.

5.4.1.3 Check

The 'Check' work of the MDC within the PDCA cycle reviews the HTM activities. The driver for this review process will be the MDC Action Plan. This will be a dynamic document with items falling off as they are resolved or completed and new items being added as identified. Some 'Checks', such as KPIs, may have scheduled review times.

KPIs that are agreed with service providing departments should not be regarded as targets but as measures. The clue is in the term – they are indicators. Goodhart's law is popularly formulated as 'When a measure becomes a target, it ceases to be a good measure' (McIntyre 2001). If a KPI is getting worse, the review process should concentrate on asking the following questions:

- Why?

- Is it a real problem or an anomaly of the KPI?

- Are there short-term explanatory factors?

- What are the long-term trends?

- Is the KPI meaningful?

This approach will encourage honest and meaningful reporting rather than encouraging performance to be skewed to maximize the target, probably to the detriment of the service's real need.

Other factors that require review will be checking the effectiveness of risk control measures taken and checking the progress of equipment procurement projects and service improvement projects.

On a longer timescale, probably yearly, the MD Policy will be reviewed and any suggestions for change made to the Board. The Strategic HTM Plan will be reviewed particularly for continued alignment with the MD Policy. As this is an MDC document, changes will be discussed and agreed by the MDC as necessary.

5.4.1.4 Act

The check/review processes will inevitably lead into a need for actions. Again the MDC Action Plan can support this. Actions will be allocated to appropriate MDC members. Target dates will be set and planning will proceed based on the actions agreed.

The MDC is accountable to the Board and/or the CEO and thus will be required to report regularly to the Board. The Board may require summary statements highlighting any concerns at its routine meetings but may also require a more formal report annually.

So the PDCA cycle continues.

5.5 LEADERSHIP OF AND FROM THE MDC

5.5.1 Leadership of the MDC

Given the importance of the MDC in the delivery of HTM to the healthcare organization, it must be led by an appropriate person with the necessary authority, expertise and commitment to healthcare technology. Such a person needs to have sufficient understanding of the impact of technology on clinical care, an understanding of the clinical implications of the use of various types of medical devices and an understanding of clinical, financial and reputational risk. Additionally, because of the wide ranging and diverse impacts that HTM can have on the organization and the role of the MDC in drafting then implementing the MD Policy, the leader of the MDC, the Chairman (male or female), needs sufficient seniority to have ready access to the CEO, the Director of Finance and the Board.

An important leadership role for the Chairman is to bring together and facilitate the collaborative working of the range of expertises that must be represented on the MDC. Therefore, the Chairman should have good interpersonal and team building skills.

In many organizations, the leader of the MDC is either the Board member who has been given specific responsibility for medical devices or a person directly representing that Board member. In some organizations the Head of Clinical Engineering is tasked with this role.

5.5.2 Membership of the MDC

The range of expertise required by the MDC for it to deliver its function is worth considering.

Clearly, the leaders or their representatives from the various scientific and technical support departments exampled in Section 5.4.1.1.3 need to be members. There is a need for medical and nursing input at a level of seniority, high enough for actions agreed to be taken forward in those professions. Input from the Finance Department and from the Procurement Department is important. Practical and useful input can be contributed by risk managers and control of infection specialists. The important thing is to have available on the committee a broad range of relevant experts who can contribute in a symbiotic way to the governance and effectiveness of HTM.

5.5.3 Leadership from the MDC

The MDC, and particularly its Chairman, is an important leadership resource for the heads of the technical support departments. These heads may well sit in other managerial structures. For example, a Clinical Engineering Department may be part of a Medical Physics and Clinical Engineering Division within a Clinical Support Services group in the organizational structure.

The line manager of the head of the Clinical Engineering Department may be a radiation physicist and not the most appropriate person to sit on the MDC. The Head of Clinical Engineering has organizational-wide responsibilities in respect of HTM. Therefore, the chairman of the MDC, in the senior position outlined earlier, has a leadership role for the Head of Clinical Engineering and the other heads of technical support departments despite not being their direct line managers. Rigid hierarchical management structures are not appropriate for the effective delivery of HTM. A form of matrix management as outlined by Sy and D'Annunzio (2005) and by Tonn (2007) is likely to be more appropriate.

5.5.4 Actions for the MDC to Lead On

In the next five sections, we will look in more detail at the activities that the MDC must engage in. We will look at:

Section 5.6   Actions to Address Risk, including Actions to Address Procurement;

Section 5.7   Actions to Realize Opportunities;

Section 5.8   Communication and Documentation Requirements of the MDC;

Section 5.9   Performance Evaluation;

Section 5.10   Continual Improvement.

In all of these activities, we will see that the clinical engineer has an important part to play.

5.6 ACTIONS TO ADDRESS RISK

5.6.1 Fundamental Corporate Decisions

We will look first at three fundamental decisions that the MDC will have to take which have organization-wide implications:

- Should all medical devices be considered or only certain types, for example, only medical equipment and not disposable devices?

- Which technical support department has responsibility for which devices?

- How should an inventory of medical devices be managed?

Decisions on each of these will form part of the Strategic HTM Plan and be reviewed on a regular basis.

A challenge for the organization when developing the Strategic HTM Plan is the question of which medical devices to include within the scope.

At present, there is considerable variation between organizations, within individual countries and internationally. If one starts with an idealistic view of the world, then the answer is simple – the strategy and plan must include all medical devices, from the simplest bandage to the most complex radiological scanner, from the wheelchair to the sophisticated anaesthesia workstation. If one thinks from the patient perspective, as indeed we should, whatever device is currently being utilized to provide diagnosis, treatment or assistance is an important, possibly critical, device in their care at that instant, and there is an organizational responsibility for it to be safe and effective. This prompts the use of a risk assessment process which has at its core the impact that the use of the device has on the patient's safety and well-being. In the real world of finite healthcare budgets, choices have to be made and risk assessments will assist in the debate about where to prioritize service delivery. It is far better to be in position where risks have been identified and considered and some steps taken to mitigate them and a residual risk accepted, than that in which no assessment has been made.

A sensible, practical approach is for the MDC to have an overarching remit for all medical devices but, in terms of inventory, not to include all, but to deal with devices that are low risk and non-powered or single use such as bandages, syringes and catheters through a materials management inventory system.

Once the MDC has clarified which devices are included, it should organize these into logical grouping and assign them to specific technical support departments. In many healthcare organizations, the responsibility for supporting medical devices is split between different departments. A typical scenario is, for example, that the majority of powered medical equipment (monitors, infusion pumps, HF surgery equipment, operating microscopes, endoscopy equipment, non-invasive blood pressure (NIBP) machines, electronic clinical thermometers, etc.) are looked after by the Clinical Engineering Department, but beds and patient hoists are looked after by the Facilities department. Patient weighing scales are on a contract managed by the Procurement department, and no one takes proper care of patient transport wheelchairs.

These situations have often grown up through historical precedent and one major task for the MDC is to bring clarity and rationality to what can be a confusing situation for clinical staff on wards and in clinical departments. When they have problems, they want to be able to contact one person who will take ownership of the issue, which may cut across more than one type of equipment, and point them in the right direction and give appropriate advice. The appointment of a 'Medical Devices Coordinating Officer' or a 'Medical Devices Safety Officer', as discussed in Section 5.6.2, may address this issue and help to reduce organizational risk.

Some organizations may find it helpful to develop an Equipment Responsibility Matrix identifying different types of medical devices (e.g. fixed equipment such as operating lights, pendants, wheelchairs, specialist seating, consumables) and which group is responsible for their life cycle processes (specification, commissioning, maintenance, safety management, reporting to the MDC). The MDC coordinates and clearly sets out the Strategic HTM Plan, which is the mechanism by which the MD Policy is to be implemented. From the Strategic HTM Plan, support departments then develop detailed HTM Programmes which enable them to deliver their services.

Historically, clinical engineering departments developed inventory systems to help them manage the devices allocated to them. These started as card index systems but are now powerful software databases that not only provide the basic data about each item but keep records of equipment history, time and money spent on maintenance, schedules of preventive maintenance, technical staff training and competency, etc. Invaluable as these database systems are, they have often covered only those items managed by clinical engineering, with other technical support departments running different non-compatible systems or relying on spreadsheets. Without a single comprehensive database of powered and higher risk medical devices, the MDC will find it difficult to have a complete and comprehensive picture of the task that has been assigned to it. The MDC must mandate the development of such a database and allocate responsibility for its management and security. Many commercially available databases designed for this purpose can be divided into sections, so allowing different support departments to have appropriate day-to-day control over their own section. However, since clinical engineering usually has responsibility for the greatest proportion of the total inventory, it is usual for them to be allocated overall responsibility for the systems management of the database.

5.6.2 Organizational Arrangements by the MDC to Address Risk

Within the MDC, there are a number of arrangements and issues that will need to be given consideration and appropriately implemented:

- Are all medical devices which the organization owns or has control over covered by the management systems in place?

- Would it reduce risk to have in place a subgroup with a specific remit to consider and advise on issues relating to single-use medical devices?

- Would the appointment of a Medical Devices Safety Officer, with a wide coordinating role and acting as a 'single point of contact', help to reduce risk?

In many healthcare organizations, the majority of devices are on the inventory and are being managed in a systematic way, but in some organizations, there is equipment that is on external service contract or is hired in as needed which is not managed by any of the technical support departments. Worse still, there may be equipment that is not being serviced at all. An extension of this problem is where a particular type of equipment in use by one clinical department is on service contract managed by that department, only for the same type of equipment in another clinical department to be on a separate service contract with the same external organization, both departments paying for service and call-out visits at the maximum rate.

This sort of situation has risk implications in the clinical, financial and potentially reputational arenas. Clinically, if equipment is not being maintained at all, there are clear clinical risks as well as reputational and financial risks if patients are harmed. If the clinical engineering or another appropriate technical support department knows about this equipment, they may well decide, as part of their HTM Programme, to have an external service contract but will manage this and provide a first-line response to any problems that arise. They will also coordinate a single contract for all equipment of the same type across a number of departments with economies of scale and cost, based on their first-line input. They cannot do this if they do not know about the equipment and it is not 'in the system'. So again there are clinical risks because equipment failure, however minor, will lead to delayed patient diagnosis or treatment and financial risks because the suboptimal arrangements are more expensive.

The question of how to deal with single-use devices has been raised earlier. The MDC should take overall responsibility for such devices but, particularly in a large organization, should consider whether a subgroup should be established to manage the details. Such a subgroup could be nurse led but control of infection specialists would have a prominent role. Clinical engineers could contribute through their scientific and engineering expertise and, perhaps more importantly, through their knowledge of the use of technology in healthcare. Furthermore, many single-use devices are accessories to the medical equipment which they look after, so supporting the application of these single-use devices is part of their holistic support for the application of technology for patient care.

To help provide a coordinated service to clinical areas, the MDC should consider the appointment of a person, perhaps titled a Medical Devices Coordinator or a Medical Devices Safety Officer. The need for a role with this unifying function has been identified in some healthcare organizations and jurisdictions. In particular, there is a need for a single point of contact dealing with the dissemination and action on safety warnings and support for adverse incident investigation. The role may be assigned to a member of the Clinical Engineering Department or the Risk Management department.

One model has been established in the National Health Service (NHS) in England in 2014. The role has been formulated to address and improve the coordination of incoming hazard warning notices and manufacturers' field safety notices, and the reporting of incidents to the MHRA who are the competent authority in this jurisdiction. The Medical Devices Safety Officer title has been used in this initiative (NHS England 2014a,b). NHS England state, "This person will support local medical device incident reporting and learning, act as the main contact for NHS England and the MHRA and medical device manufacturers and be a member of the new National Medical Devices Safety Network". The requirement is that healthcare organizations "… identify an existing or new multi-professional group to regularly review medical device incident reports, improve reporting and learning and take local action to improve the safety of medical devices".

5.6.3 Planned Replacement List

An important role for clinical engineers in their advisory role to the MDC is to be planning ahead in respect of the equipment they have responsibility for. We have already seen, illustrated in Figure 5.1, that all equipment has a life cycle and one identified step in this cycle is a review of each item or group of items in respect of their risk status and continuing clinical use. Clinical engineers will need to consider and provide advice to the MDC in respect of:

- the short-term drivers that may lead to advice that replacement is required;
- long-term drivers that may lead to similar advice;
- the difficult issue of funding for both so-called capital equipment and non-capital (revenue) equipment. (The distinction has been discussed in Section 5.4.1.1.)

There are many factors that need to be taken into account when advising that medical equipment ought to be replaced. Some are predictable and can be foreseen some years ahead. Manufacturers will often notify their customers of an 'end-of-life' date (EOL) for a type of equipment beyond which they cannot guarantee support. These dates should be entered into the MEMS and clinical engineers should be planning for those. Look again at Case Study CS5.6. Other factors are unpredictable or arise at relatively short notice. Issues such as damage to or failure of an item that is uneconomic to repair but which otherwise would remain in use should be considered. The previous service history of the equipment and the impact of its absence on clinical service delivery are also important. Making the judgement on the economies of repair is itself multi-factorial. Suppose the equipment requires expensive replacement parts and the expertise to fit these and recalibrate can only be provided by the manufacturer, then a cost of repair that exceeds, for example, 50% of the replacement cost might be considered uneconomic. Successful negotiation of discounts might well confirm the replacement course of action. On the other hand, if the repair can be carried out appropriately by the clinical engineering team, the potential full cost of the job should be estimated including the true internal labour cost, but a higher threshold might be considered since internal labour costs are an opportunity cost (doing this means something else does not get done at that time) rather than a 'cash' cost. Doing the repair might also be justified to buy time whilst a properly considered replacement project is progressed. Factors to consider are the workload and priorities within the Clinical Engineering Department and the urgency of getting the equipment back into use.

Another short-term driver occurs when the regulatory authority declares that a particular equipment type should be withdrawn from service or places the device under restricted use on safety grounds. It is possible that this safety advice may have been issued despite no problems having been experienced in your hospital with these particular devices. This is a time when the comprehensive inventory is indispensable. The question, 'How many do we have and where are they?' can quickly be answered. The medical equipment management facilities in the database should also be able to answer the question, 'What problems have we had with this device?' A risk assessment in close collaboration with clinical users will be necessary to cover continued use until a replacement can be procured. This can be a protracted process as we can see in Case Study CS5.7.

Clinical engineers, together with their clinical colleagues, also need to consider the impact of service delivery or combined technological and clinical developments that render existing technology less than effective, perhaps even clinically obsolete. Examples include the following:

- The clinical evidence became available towards the end of the twentieth century supporting the use of biphasic rather than monophasic electrical shocks for treating cardiac fibrillation. This spurred on the replacement of the monophasic defibrillators with biphasic defibrillators (Bardy et al. 1996).

- The increasing clarity of high-definition surgical endoscopic systems has led to clinical demand to replace existing surgical endoscopic systems with those providing better images for the surgeons performing delicate endoscopic procedures.

- Concerns about mercury poisoning and its environmental hazards have led some jurisdictions to outlaw the use of mercury thermometers and mercury sphygmomanometers, requiring mercury-containing devices to be removed from service, although still functional.

For longer-term replacement planning, the equipment inventory held within the equipment management database is essential. This will give equipment ages and should also have a field for recording 'withdrawal of support' or EOL dates for each equipment type. Many manufacturers will give 1–3 year's notice of the date beyond which they will be unable to guarantee the availability of certain critical spare parts. This is a crucial input to drawing up a planned equipment replacement list which needs to be dynamic, prioritized and up to date. Throughout the year, a rich source of intelligence will have been accruing as HTM Programmes have been ongoing, risks identified, safety alerts addressed, incidents investigated, new devices acquired, changing clinical practices observed and new technologies identified. The clinical engineer is well positioned to collate, analyze, interpret and ultimately recommend priority areas for investment. It is a daunting task as the information does not necessarily all reside in one place, but once alerted to the importance of such data, it can be gradually gathered throughout the year. Case Study CS5.8 illustrates a systematic way of involving clinical staff in contributing to the prioritization process.

A commonly experienced scenario is that, towards the end of a financial year, money suddenly becomes available to purchase medical equipment. Unexpected announcements of funding available for equipment replacement can be reduced by continual dialogue. The Head of Clinical Engineering should work with the Director of Finance to discuss healthcare technology replacement planning, including the funding of 5–10 year replacement plans. If replacement lists are in place and some prior planning has been done, appropriate and effective action can be taken at short notice to add value and support patient care. Conversely, purchase decisions taken in a rush, without prior planning, often lead to less than optimal acquisitions and poorer value.

In some jurisdictions and healthcare organizations, the rigid separation between funding allocated to capital or revenue spending can cause problems when sourcing the funds to replace 'capital' and 'revenue' equipment (see Section 5.4.1.1.1). Where they exist, these separate medical equipment funding allocations have to be managed separately. The MDC should, in these cases, work with the Director of Finance to ensure a holistic approach, with coordination of equipment procurements whatever the funding source. This coordination should extend to all funding sources, including charity funding and donations.

Where these separate 'capital' and 'revenue' funding allocations exist, a healthcare organization may sometimes find itself in the position that it has capital, but no revenue funding available when faced with an urgent need to replace a large number of revenue equipment. An example would be the need to replace a fleet of infusion pumps, each individually costing less than the capital threshold, but taken together falling well above that threshold. In these circumstances, collaborative working with the Director of Finance or their staff may provide solutions within the accounting rules such as grouping together items whose individual costs are less than the threshold, but whose cumulative costs are above the threshold. It may then be possible to treat the grouped items as 'capital' and the resulting capital asset depreciated over the appropriate time period.

The important point is that clinical engineers need to have an understanding of the finance rules that apply in their jurisdiction and organization and be able to work constructively with colleagues from the Finance department to achieve the desired outcome for the benefit of patients.

5.6.4 Acquisition

The introduction of new technologies requires understanding of healthcare technologies and how they can contribute to and improve healthcare. Healthcare is facing increasing expectations by the public and patients within a context of restricted resources. Technology in many areas of life is seen as the solution to increasing the quality of life whilst containing costs: healthcare technologies face the same challenge and their introduction requires skills and understanding of both technology and healthcare, the key role of the clinical engineer.

The MD Policy, endorsed at the organization's Board level, is needed so that all acquisitions of healthcare technology are made with full analysis of the risks and benefits; clinical, corporate and financial. The days of individual clinicians arranging the purchase of expensive equipment on the basis of experience in another organization or a trip abroad, without proper organizational scrutiny, should be long gone. The Value Analysis Committees set up within many U.S. healthcare organizations can help to tackle what can otherwise be a scenario of conflict between clinicians' demands and organizational strategic planning (Montgomery and Schneller 2007). In this book, we suggest that the MDC can play an important part in this cooperative planning.

The MD Policy should state that, as a minimum, the following issues are considered prior to any medical equipment acquisition:

- Alignment with the organization's corporate objectives;

- Fitness for purpose as judged against a duly considered specification;

- Standardization with other similar types of equipment already in use;

- Whole life costs to include consumables and disposables, staffing implications, maintenance and training;

- Usability engineering/human factors issues;

- The training needs of users of the equipment;

- Availability of appropriate staff who can operate or be trained to operate the equipment;

- Maintenance implications, that is cost of maintenance, warranty terms, availability and training of in-house technical support and quality of support from the supplier;

- Reliability, based on experience in this and other comparable organizations;

- Any need for decontamination of the equipment and the availability of suitable decontamination facilities;

- Possible delivery and installation complications;

- Any implications for the healthcare organizations physical infrastructure, for example buildings, floor loadings, power and other utility supplies;

- Any implications for the healthcare organization's IT systems and infrastructure, for example systems compatibility, integration with electronic patient record or PACS systems;

- Statutory medical devices regulations, safety standards, health and safety regulations;

- Risks to the healthcare organization both clinical and corporate;

- The most advantageous financial arrangements, for example outright purchase, lease and managed service contract;

- Disposal implications and possible costs.

Many of these considerations are illustrated in practice in Case Study CS5.9, and specific issues applicable to clinical IT systems are illustrated in Case Study CS5.10.

Only by having in place and following the principles set out in the MD Policy on device acquisition can the range of risks and the conflicting call on financial resources that all healthcare organizations face be effectively managed. A strategic approach, documented in a policy, ensures that all medical devices and equipment proposed for acquisition are considered in a consistent way and will contribute effectively to the organization's short- and long-term objectives. This reduces both clinical and financial risks. To give just two examples: adherence to the policy ensures that equipment is not introduced without adequate staff training or appropriate maintenance provision and expensive adaptations to buildings or facilities do not unexpectedly have to be met after purchase. A strategic policy also contributes towards meeting corporate and clinical governance standards and regulatory requirements. In most jurisdictions, there are statutory quality and governance requirements that all healthcare organizations must follow and against which they are inspected.

Ensuring that appropriate mechanisms are in place for procurement of healthcare technology is an essential part of the MD Policy. Whilst there are many triggers that lead to acquisition projects, the need for corporate oversight and direction is common to all. Well-established efficient and speedy arrangements should be in place for smaller acquisitions, particularly for spare parts, but for large projects once approved, the MDC should convene a multidisciplinary team to deliver the acquisition. Management of the acquisition project is a specialist task which requires particular expertise. Central involvement in the procurement process by the Procurement and Finance departments should ensure that a standard approach to acquisition is taken for each project but neither of these departments should have absolute control. Rather Procurement's remit is to advise on and ensure that procedures comply with best procurement practice in accordance with procurement regulations. This will involve competitive tendering in many cases. Finance will want to ensure that the funding is available, both for the initial acquisition costs including any associated installation and commissioning costs and for the ongoing whole life year-by-year cost of ownership made up of running costs including accessories, consumables, maintenance and staff training. Familiarity and understanding of these commercial issues is part of the competencies required by a professional clinical engineer.

The predominant role for the clinical engineer in the procurement process is to interpret and put into unambiguous terms the technical specifications required by the clinical user and those required for the effective management of the equipment such as arrangements for maintenance and user and technical training. The clinical engineer is best qualified to contribute this part of the tender documentation. They may be best placed in some circumstances to lead on the tender evaluation process as described in Case Study CS5.11.

We use the term 'acquisition' in the heading of this section because outright purchase is not always the most cost-effective arrangement for acquiring the use of devices (see Chapter 4, Section 4.3). Other methods that may be appropriate are leasing, both short term and long term, or agreeing with manufacturers of medical equipment which require dedicated single-use disposables that they will supply the equipment free of charge against an agreement to purchase a minimum number of the disposable items per annum, at an agreed price. Such agreements may also include service and maintenance of the equipment or flexibility on the number of items of equipment provided.

Acquiring medical equipment through alternative pricing arrangements has been successfully applied to equipment acquisitions with predictable revenue costs, for example infusion devices, patient warming devices, surgical diathermy and laboratory equipment. In each case, the supplier can plan for and take into account assured revenue streams from the healthcare organization for the supply of consumable items to cover their costs incurred in supplying the equipment. However, good commercial knowledge and robust negotiating skills are needed within the healthcare organization to ensure best value.

Once an equipping proposal is approved, it becomes a project within the MDC Action Plan. The acquisition project has distinct phases and these are: (1) planning, (2) procurement and (3) commissioning. Regardless of the size of the project, consideration should be given to each of these steps, but the depth of work involved in any one step will vary considerably between procurements.

5.6.5 Internally Identified Hazards, Risk Management and Mitigation

In many cases, the management of risk associated with particular technologies requires cooperation between departments. The MDC provides a framework for identifying these needs and facilitating the establishment of cross-department working groups to solve particular problems. As a multidisciplinary group, the MDC is well placed to review all aspects of the deployment of medical devices and may identify actions which would reduce risk. This might include recommendations for change of practice and increase in education and training in some area, or campaigns to raise awareness of safety issues. In some cases, this process may recommend the replacement of devices if they have led to adverse events or if new technology offers the possibility of providing the same functionality more effectively and safely.

Once hazards and hazardous situations have been identified, associated risks can be scored using the standard likelihood of occurrence combined with the severity of the consequence method that objectively and numerically scores risk. Organizations will have their own risk management policy and risk registers identifying key corporate risks. Clinical engineering staff should be familiar with their organization's approach to risk management and ensure that healthcare technology risks are identified, analyzed and escalated by way of the organization's management structures if mitigating actions are insufficient to achieve acceptable residual risk.

Risks will essentially form a spectrum, namely those that can be addressed directly by actions of the end user, ones that local clinical management teams can resolve, ones that can be resolved by the clinical engineering team and others that can be appropriately mitigated by changes to clinical guidelines, with some requiring a corporate response and perhaps investment.

Unfortunately, though rare in proportion to the total number of care episodes, adverse incidents in healthcare do occur that harm or, without mitigating factors including timely staff actions, could potentially harm patients or staff (Kohn et al. 2000). Healthcare organizations should have in place a policy for reporting internally and dealing with such situations. Where such incidents involving medical devices occur, the MDC has a role in reviewing them and acting to change the Strategic HTM Plan or in some cases even the MD Policy, to reduce and control the associated risks. So in its reactive role, the MDC provides a forum where risk incident investigation can be discussed by a multidisciplinary group. Where appropriate the MDC may undertake the investigation of risk occurrences directly or it may delegate a task group to do so, keeping a record on the MDC Action Plan. An outcome of a well-organized MD Policy and Strategic HTM Plan should be a reduction in adverse incidents. The monitoring of the rate, nature and locations of incidents can be used as a metric of effective management.

5.6.6 Externally Identified Hazards, Risk Management and Mitigation

Most jurisdictions have in place a system for providing health organizations with technical or safety alerts. In the United Kingdom, MHRA Medical Device Alerts provide this function and the FDA in the United States has similar arrangements as does the

Therapeutic Goods Administration of the Australian Government Department of Health. All such safety alerts need to be actively considered against the inventory of medical equipment held by the healthcare organization – hence one reason amongst many for having a comprehensive database of all equipment 'on the books'. Many of these medical device alerts concern non-active devices such as disposable accessories. The MDC needs to consider how these devices are recorded and therefore be traced and appropriate action taken on the alerts.

The MDC should ensure that an effective dissemination system for these medical device alerts is in operation within the organization, liaising as appropriate with the organization's risk management team. Action on medical device alerts must not be delayed until the next meeting of the MDC, but it will act as an advisory group concerning the management of any actions that arise as a result of device alerts. Responses may vary from determining that the healthcare organization has none of the device type implicated in the alert and therefore no further action is required, to a realization that a major programme of upgrades over a large number of items is required; this will require careful planning and execution as a specific project. Actions will form part of the MDC's dynamic MDC Action Plan and progress will be reviewed through that.

Manufacturers' Field Safety Notices (FSNs), which are far greater in number, should be dealt with in the same way. These are not always cascaded in an effective and timely manner to the relevant people who need to be aware of their contents and who are responsible for ensuring that the required actions are carried out. Lack of responses to these FSNs by healthcare organizations to the manufacturers can lead to the regulatory authority having to send out a medical device alert reminding users about a manufacturer's field safety notice. This is part of the post-market surveillance responsibilities of manufacturers, supported by regulatory authorities. It is suggested that the Medical Devices Safety Officer (Section 5.6.2) may support effective responses to FSNs.

The MDC should also review all internal adverse incidents to see whether the healthcare organization should report the incident to the national or regional medical device vigilance reporting systems. This will be discussed in more detail in Chapter 9. There should be a standard route for external incident reporting from the organization and a designated point of contact for responses from the regulatory authority. It is less than helpful for incident reports to be going out from the organization to the regulatory authority or to manufacturers from multiple points within the organization. For example, without a coordinated reporting system mandated by the MDC, a single incident might be reported multiple times, say from the ward where the incident happened, from nursing administration and from the Clinical Engineering Department, giving the impression of multiple similar incidents. Again the Medical Devices Safety Officer may help and support.

5.7 ACTIONS TO REALIZE OPPORTUNITIES

A strategic and forward-looking role of the MDC is to identify and consider the opportunities for improving the HTM system. Suggestions may come from issues that the MDC has to deal with or from proposals brought to it from members based on their own experiences and local departmental discussions. The multidisciplinary membership of the MDC is a good place to air such ideas, get useful authoritative feedback and judge the practicality of and support that there would be for a new idea. The MDC should have sufficient reputation and authority that any proposals to senior management that are backed by the MDC are taken seriously.

Some possible opportunities worth considering are discussed next.

5.7.1 Flexible Deployment of Assets

An aspect of devices management which needs to be centrally considered by the MDC and clearly documented is the deployment approach taken for each group of devices. A sensible starting point in addressing deployment issues is first to categorize items into those devices that have the potential to be shared and those that should permanently reside in a particular department. Clearly, for example, the ophthalmology department and the urology department will each have equipment and devices specific to their own speciality, not required in other departments. Some widely used equipment may need to be configured specifically for a particular department, for example neonatal infusion pumps, and these will need to be separated out and identified. But very many departments will have regular use for general infusion pumps and blood pressure measuring devices and access to a defibrillator.

If the policy is for each area to be equipped with sufficient of these widely used devices to meet its peak demand, then for much of the time some of the assets will sit idle. An alternative is to have available sufficient to meet the totality of the organization's demand, but to dynamically allocate these according to need. Potentially, fewer devices need to be owned, but a policy and processes that support sharing is required. This has resulted in the establishment of so-called medical equipment libraries. This is often an initiative that is led by the Clinical Engineering Department with the support of the MDC. The most common devices managed in this way are infusion pumps but the system can be extended to other types of equipment (Keay et al. 2015). Treating equipment such as beds or pressure-relieving mattresses as organization-wide assets means departments are always guaranteed availability, at the time of need, of the right number and type. Appropriate spare equipment will be held centrally for deployment rather than each area keeping a spare for use if there is a failure (Newton 2001). Planning and managing the appropriate deployment of healthcare technology is important: oversupply is financially wasteful and can lead to crowding of clinical areas with equipment not immediately required; undersupply can adversely impact on patient care. Studies have shown that a significant cause of adverse events is lack of equipment at the point of clinical need (Weerakkody et al. 2013).

The deployment of devices to ensure that clinical, corporate and financial requirements are met is usually considered as part of the acquisition process. Clinical engineers contribute significantly to planning deployment of devices. They are well placed to do so as a result of them having an in-depth appreciation of the impact of different deployment approaches in clinical practice and their particular expertise in designing maintenance programmes supporting the devices in practice.

5.7.2 Identifying Opportunities for Promoting Quality

Quality does not have a simple single meaning. Some words given as synonyms are *excellence, superiority, value, worth*, all of which have elements of applicability to HTM.

Dr. Robert Burney, a U.S. anaesthetist, has an interesting blog entry (Burney 2013) in which he writes about the definition of quality specifically in healthcare and quotes the Lexus slogan 'The Relentless Pursuit of Perfection'.

For a Clinical Engineering Department, quality is about the systematic organization of work in order to meet the expectations and needs of the department's stakeholders. This must include not just the expressed and documented needs/requirements but also the implied needs. There are times when the Clinical Engineering Department, because of its particular knowledge, must provide services that the clinical departments do not realize that they need. The ISO 9000 Quality Management Systems standards provide a formal framework around which a QMS can be structured. This is based on the familiar Plan–Do–Check–Act cycle which we have already mentioned.

The role of the MDC is to require, or at least strongly encourage, the application of QMSs in all of the technical support departments. The MDC can be a forum for the sharing of good practice in HTM across departments and might consider requiring a degree of cross auditing between departments to bring this about. The aim is to ensure that experiences gained in one group which are proven to be beneficial can inform and facilitate the development of similar practices in other groups.

5.7.3 Identifying Opportunities for Promoting Safety

We have discussed already the role that the MDC should play in monitoring the processes that deal with both incoming safety alerts and outgoing incident reporting. Beyond these obvious contributions to device safety, the MDC must ensure that the processes that are in place for device procurement, based on well-thought-out procurement specifications, and the processes for training of clinical user staff are robust and effective. A key contribution that the MDC can make arises from its multidisciplinary composition, with relevant people from all those departments that contribute to HTM convened into a forum where activity is reviewed and good practice shared.

The MDC is also the appropriate forum where organization-wide safety issues that significantly involve medical devices can be discussed and resolved. A relevant example is the thorny issue of the use of mobile communication devices (e.g. phones, two-way radios, emergency service phones/radios) in close proximity to medical devices that may be susceptible to interference. The MDC can be the driver for developing consistent, evidence-based, organization-wide policy by commissioning literature reviews (Ettelt et al. 2006) and, if necessary, tasking groups, such as the Clinical Engineering Department, to carry out appropriate tests.

5.7.4 Identifying Opportunities for Promoting Effectiveness

Horizon scanning is a term used to describe the activity where hospital staff look to the market to identify new products and technologies which might play a role in meeting the clinical, corporate or financial goals of the organization. Horizon scanning is likely to be undertaken by all members of the MDC motivated by different needs. Clinicians are looking for technology to deliver better outcomes, finance departments for systems which can deliver cost containment, whilst the CEO might be interested in systems which reduce length of stay in hospital for patients. Horizon scanning is a process which feeds into and supports the development of planned equipment replacement lists. Sources of information include trade journals from suppliers, formal and informal links with colleagues, meetings and conferences run by professional organizations, commercial exhibitions and review of scientific and technical literature. The involvement of clinical engineers in research, innovation and Standards development discussed in Chapters 1 and 3 also promotes awareness of new technologies and the benefits they might bring. As part of the MDC, the clinical engineer must keep up to date on new and emerging technologies, reviewing the state of the art.

Horizon scanning leads to the requirement to evaluate and select medical devices most suitable for the local organization. As briefly discussed in Chapter 2, health technology assessments and evaluations are valuable activities that investigate the strengths, weaknesses and effectiveness of healthcare technologies. They are carried out by dedicated organizations and groups and their outputs are often in the public domain. Two such internationally known organizations are as follows:

- The ECRI Institute, a U.S. non-profit organization "...dedicated to bringing the discipline of applied scientific research to discover which medical procedures, devices, drugs and processes are best, all to enable you to improve patient care" (ECRI 2016a). Their home URL is https://www.ecri.org/Pages/default.aspx and the link to the health technology assessment page is https://www.ecri.org/components/HTAIS/Pages/default.aspx.

- The National Institute for Health and Care Excellence (NICE) (http://www.nice.org.uk/) is a UK government–funded body established on a statutory basis by an Act of Parliament, accountable to the Department of Health, but operationally independent of government. Their guidance and other recommendations are made by independent committees. Within NICE, the Centre for Health Technology Evaluation develops guidance, including technology appraisals, on the use of new and existing treatments and procedures within the UK NHS, such as medicines, medical devices, diagnostic techniques and surgical and other interventional procedures. Much of the work is carried out by contracted External Assessment Centre such as Cedar in Wales, UK (http://www.cedar.wales.nhs.uk/home).

Regardless of whether a technology delivers a required outcome in another organization, there is a requirement to look carefully at the technology and how it might affect the local care pathway before considering it for acquisition. Typically, the clinical engineer will develop specific expertise in different areas, including aspects of health economics, as clinical requirements and local developments dictate.

The issue of training is dealt with more extensively in Chapter 7, Section 7.4.11. Staff who are competent in applying the devices they have to use will enhance the effectiveness and safety of their work. The MDC has a role in overviewing the training systems that are in place and asking for review and evidence of effective training, including analysis of training records showing how many staff in each area have been trained and competency assessed.

Therefore overall, the MDC provides a forum within the organization to promote and coordinate a holistic approach to HTM for the benefit of safe and effective patient care and in alignment with the organization's corporate objectives.

5.8 COMMUNICATION AND DOCUMENTATION REQUIREMENTS OF THE MDC

5.8.1 Overview

We have seen in Section 5.4.1.1.3 that one role of the MDC is to agree and allocate responsibility between different technical support departments and communicate this information throughout the organization with the objective of ensuring that all types of medical devices have a technology management programme associated with them. This is an important central role as it ensures that all devices are covered with no gaps, overlaps or inconsistencies. For example, for historic reasons, it can be that a particular type of device, say patient transport wheelchairs, is looked after by one technical support department in one hospital and by another in a different hospital within the same healthcare organization. The result is that no one has the big picture, there is no consistency of level of service, and clinical staff moving between hospitals face potential confusion when requesting support.

Having decided the allocation of responsibilities, there are several ways in which the MDC can and must support the technical support departments. This is in line with the requirements in Clause 7 of the ISO 55001 Standard which deals with the support the organization must provide towards asset management. There will not be one common pattern across all healthcare organizations. Local regulations and funding regimes will lead different healthcare organizations to adopt different approaches to HTM and the role of the multidisciplinary MDC. However, aspects of the issues discussed in the following text will feature in the Strategic HTM Plan and fit into the corporate structure of all healthcare organizations.

5.8.2 Awareness

It is important that all staff involved in the delivery of the HTM Programmes are aware of the organization's MD Policy and Strategic HTM Plan. Clearly, the leaders of the technical support departments will require a detailed knowledge and understanding of these documents and may well have contributed to their drafting and approval. However, their operational staff must be clear where their role contributes to the overall plan and helps deliver high-quality services that benefit patients.

It is also important that clinical staff are aware of the MD Policy and the Strategic HTM Plan. In Section 5.6.2, we have made the point that clinical staff need clarity as to where to turn and who to contact when technical problems arise. Clinical users have daily oversight of their medical devices and have a major responsibility to bringing problems to the attention of the relevant technical support departments. A coordinating role for the MDC is to spread the awareness of the HTM system through, for example, briefings, newsletters, internal web pages, and through formal inclusion on the agendas of corporate meetings. In the United States, the Association for the Advancement of Medical Instrumentation (AAMI) sponsor an HTM Week with an award for the HTM department that is judged to best promote its work within its host organization. In the United Kingdom, some clinical engineering departments have run successful 'open days', inviting staff from the CEO down to visit their department and see the facilities and work done.

All such activities help spread awareness of the importance of HTM to the safe delivery of patient care and help to spread understanding of the systems in place that are designed to best deliver that HTM support.

5.8.3 Communication

Spreading the awareness of the MD Policy and Strategic HTM Plan described earlier is part of a communications exercise. However, as part of the Check phase of the MDC's Plan–Do–Check–Act cycle, the MDC should review from time to time how the MD Policy and Strategic HTM Plan is being communicated throughout the organization in a formal way, and whether it is being communicated effectively. The best policy and plans in the world will not be effective if the staff whom they affect do not know about them. Patients too need to be made aware of those aspects of the implementation of the Strategic HTM Plan that impinge directly on them. This is likely to be a particular issue where medical devices are issued to patients for use at home, for example: have they been adequately trained as a user; do they know how to and who to contact for technical support?

5.8.4 Information Requirements

In Section 5.3.2.3, we propose that the MD Policy should mandate that the organization has a comprehensive database of all medical devices. This is the fundamental basis of the asset information requirements of the organization. The ISO 55001 Standard in subclause 7.5 requires that the organization determines its information requirements, and this is further explained in the equivalent section in the ISO 55002 guidelines document. Determining the information requirements is a role for the MDC. Clearly, more information is required than a simple list of assets. Consideration of the wider information requirements will influence the specification of the asset management database. Properly specified to include dates of purchase, service history, etc., the database will provide information on a range of factors such as age and reliability that will help in determining equipment replacement priorities. A complicating factor is that 'medical devices' includes large numbers of non-powered consumable devices such as syringes and cannulae. The database suitable for medical equipment will not be appropriate for such devices. As we have noted in Section 5.6.1, the MDC will need to consider how to have such devices on record, perhaps through a 'materials management' database.

5.8.5 Documentation

It is important that the formal approved policies and plans that have been discussed are made available and appropriately archived. In large organizations, it is likely that the archiving will be done digitally on servers that are backed up in a systematic way. In small organizations, to which the principles outlined can be applied in a proportionate way, archiving could be done by appropriate storage of paper copies.

Databases can only be useful if they are software based, even in small organizations. Again, backup and archiving of databases is essential. For large organizations, this should be straightforward because they will have ICT departments charged with these tasks. For small organizations, it is important that the issue of backup of databases is thought through and plans made and implemented.

Clause 7.6 of ISO 55001 and the associated guidance in the same sections in ISO 55002 deal with this issue.

5.9 PERFORMANCE EVALUATION

Since the MDC allocates responsibility for the management and maintenance of types of medical devices to appropriate technical support departments, it has the responsibility for receiving reports and reviewing their performance. The MDC might well mandate a consistent format for reports to make the review process less onerous and more comparable between technical support departments.

It also has a responsibility for reviewing quality and safety programmes and in doing so being assured that they are working and are effective. Further, the MDC will keep an ongoing list of identified risks associated with medical devices. This will include a risk-prioritized equipment replacement plan, with device-related field safety notices and safety alerts notified to it. These are also issues for review and evaluation.

Clause 9 of the ISO 55001 standard and the corresponding guidance in ISO 55002 cover this issue and have some useful and appropriate requirements.

5.9.1 Review the Performance of HTM Programmes

A set of KPIs developed for the work undertaken by the departments delivering scientific and technical support programmes is a valuable and clear method for monitoring the HTM Programmes. The management of the organization and the members of the departments will have a keen interest in how well their department is working.

In developing KPIs, it is useful to adopt a balanced scorecard approach, where several aspects of the management programme arc taken into account, rather than present an overwhelming list of technical data. There is a mass of literature on the balanced scorecard approach, but the initial concept came from Kaplan and Norton (1996). We discuss KPIs further in Chapter 6, Section 6.5.3, in the context of implementing and auditing the delivery of ESPs.

There are no fixed rules as to what should be included, but reference to the service specification agreed would be a good start in KPI development. KPIs prove a useful tool when looking at trends over a long period of time. Changes in work practice and external influences on the department can be quantified. It has earlier been noted in Section 5.4.1.3 that KPIs should be regarded as measures which guide and inform decisions around resource allocation, and not targets.

KPIs should be determined and formulated by the leadership of the technical support department in consultation with their staff. They should not be thought up and imposed by the MDC, though discussion in the multidisciplinary forum of the MDC may well inform and influence departmental KPIs which should be agreed and signed off by the MDC as part of continual improvement.

KPIs from each support department should be reviewed periodically, say at quarterly meetings of the MDC, but exception reporting should be the norm at most meetings with a more thorough annual review.

5.9.2 Reviewing Quality and Safety Programmes

Much of the review of quality and safety programmes will result from the review of the MDC Action Plan at every meeting of the MDC. However, where a technical support department works within a registered, externally audited quality management system, the results of periodic inspections should be reported back to the MDC. A similar report and review should take place where quality management systems are not externally registered and audited but a system of internal cross audit, as suggested in Section 5.7.2, is in place.

With regard to responding to safety alerts from regulatory authorities or from manufacturers, action on these must start in the relevant technical support department as soon as they arrive, but review of all such alerts, consideration of their applicability to the organization and progress on actions in hand should be on the MDC's agenda. For example, if user training issues are clearly becoming a cause of problems, then the technical support department may have proposals for addressing the issues, but it is through a review at the MDC that agreed action can be moved forward.

Benchmarking with other similar organizations is also a useful method of performance review. The problem is often being sure that the comparisons are being made against similar data sets and similar operational practices. One-off benchmarking exercises are often inconclusive but larger benchmarking groups who work together over a number of years and refine their techniques can be useful once mature.

Two good examples from the United Kingdom are the London Clinical Engineering Benchmark Group (http://www.lcebg.co.uk) which has been operating since 1999 with a current membership of clinical engineering departments from some 15 hospitals in and around London. Each department agrees to three external independent audits of a number of ward-based KPIs per year. The National Performance Advisory Group (http://www.npag.org.uk/) facilitates two clinical engineering benchmarking groups that have been meeting for over 10 years and have developed common clinical engineering KPI definitions (NPAG 2016).

There are also benchmarking tools available through membership of two U.S. organizations:

1. The ECRI Institute offers a membership "BiomedicalBenchmark" service, "... with inspection and preventive maintenance data and best practices to help you gauge efficiency and improve effectiveness within your clinical engineering department" (ECRI 2016b).

2. AAMI offer their "Benchmarking Solutions—Healthcare Technology Management". This is described as "... a popular web-based tool designed to help clinical engineering departments measure their budgets, personnel, practices, and policies against similar departments at other facilities" (AAMI 2016).

5.9.3 Risk Register

We have seen in 5.4.1.1.2 that two activities prompt the development of a register of healthcare technology risk issues. The organization is likely to have a high-level risk committee which keeps a register and decides on all serious risks which can conveniently be grouped into four categories: safety risks (in their broadest sense including device issues), financial risks, operational risks (which may include device issues) and reputational risks. The MDC will have to decide which of the risks relating to devices should be escalated to the corporate risk register. A good example would be if the fleet of anaesthetic machines was at the end of their projected life and reliability was beginning to deteriorate. This would be an issue that would feature in all four categories and should be flagged up at the highest level.

The risk management activity may identify processes or ways of working which can be improved through changes to the Strategic HTM Plan. If a risk occurrence highlights the need for a group of devices to be brought into the Strategic HTM Plan and actively managed, then this would be noted on the MDC risk register for evaluation and in the MDC Action Plan for resolution. It may be that during a risk investigation a number of recommendations are made which would change the way devices within the existing Strategic HTM Plan are managed. The HTM Programmes of each technical support department may also suggest changes to the Strategic HTM Plan to better control risk. Even if no adverse events occur, clinical engineers who work in the delivery of HTM Programmes can often identify actions which improve practice and control risk. For example, if a clinical engineering team identifies a cluster of requests for assistance from users where the cause of the problem is lack of familiarity with the use of a device, it may highlight the need for more training. In many cases, the solutions can be effected within their HTM Programme, and noted at the next MDC meeting, but where resources or governance does not allow this, the department will escalate the issue up to the MDC by identifying the risk and placing it on the risk register for evaluation. Case Study CS5.12 describes a situation in which an HTM risk had to be escalated to the MDC for wider consideration and placing on the risk register.

5.10 CONTINUAL IMPROVEMENT

A consistent feature of quality management systems is a requirement to strive for continual improvement of the product or service delivered. This concept is present in both ISO 9001 and ISO 55001. Since we propose that the MDC and the HTM system should work to a Plan–Do–Check–Act quality cycle and that, where possible, technical support departments are formally QMS registered or certified to a relevant international standard, it is appropriate that the MDC also embraces the continual improvement concept.

Continual improvement can be reactive or proactive. Both are equally valuable. If for example a series of 'no fault found' problems occurred with a particular type of equipment, pointing to a clinical staff training issue, this could be discussed at the MDC and a new or revised training programme introduced. That is an example of a reactive improvement. A proposal brought to the MDC by a technical support department to seek support for some space close to a clinical area in order to carry out preventive maintenance and thus reduce downtime would be an example of a proactive improvement.

The MD Policy and the Strategic HTM Plan through which it is implemented should be reviewed regularly. Organization policy may well dictate the review cycle for formal Board-level policies, but as advice to the Board on the MD Policy will come from the MDC, it would be prudent for the MDC to review this policy annually. The MDC can then make recommendations to the Board in a timely manner and ahead of the formal review time if circumstances warrant it.

The Strategic HTM Plan belongs to the MDC. It is a strategic document and plan, so it should not require frequent amendment. It should be reviewed at least annually.

The MDC Action Plan is a dynamic document that is the means of keeping track of the work of the MDC, setting objectives and tracking progress. This document will be reviewed at every meeting of the MDC. The MDC Action Plan is probably the best source of and evidence for continual improvement. It is through a culture of striving for continual improvement that positive change can come about.

5.11 CONCLUSION

In this chapter, we have explored how the development and continual review of an organization-wide MD Policy guides and enables the MDC to develop and keep up to date a Strategic HTM Plan, managed and monitored through the MDC Action Plan. Through this process, the MD Policy can assist a complex organization in achieving its clinical, corporate and financial goals. In doing so, the organization increases the value of its healthcare technology assets for all of the organization's shareholders.

It is important to remember that the MD Policy and the Strategic HTM Plan as described in this chapter are discussed within the context of a large modern, perhaps multisite, healthcare organization. The principles described would apply to a smaller hospital, a care home or a primary care facility. However, it is likely that in smaller organization the functions and processes may be grouped and structured differently to be sustainable and practicable for that particular organization. However, the principles described are the same and equally applicable.

Implementing the MD Policy through developing the Strategic HTM Plan and the HTM Programmes is one of the key ways to maximize the benefit to the patient and to the organization and reduce and control risk, including financial risk. Whilst the whole HTM system and its implementation involves many stakeholders, it is only clinical engineers that contribute to all elements. Clinical engineers contribute significantly to the development and review of the hospital-wide strategic policy. The links that the Clinical Engineering Department has with senior management of the organization and the clinical leads for the different specialities will support the necessary dialogue to ensure effective equipment prioritization decisions, whether these involve the transfer of existing equipment or procurement of additional or replacement equipment. The involvement of the Clinical Engineering Department and individual clinical engineers as described in this chapter is collaborative and interdisciplinary.

In Chapter 6, we will look at how clinical engineers develop and deliver the device-specific HTM Programmes. It is important to note at this stage that the involvement of clinical engineers in HTM Programme delivery informs every aspect of their contribution to the elements of the MD Policy, and vice versa. As we will see many times in the case studies in this and other parts of this book, the 'advancing and supporting care' role provided by clinical engineers is interweaved with the equipment management role.

REFERENCES

AAMI. 2016. Benchmarking solutions – HTM. http://www.aami.org/productspublications/ProductDetail.aspx?ItemNumber=1062 (accessed 2016-05-21).

Bardy G.H., Marchlinski F.E., Sharma A.D. et al. 1996. Multicenter comparison of truncated biphasic shocks and standard damped sine wave monophasic shocks for transthoracic ventricular defibrillation. *Circulation* 94: 2507–2514. http://circ.ahajournals.org/content/94/10/2507.full (accessed 2016-05-10).

Burney R.G. 2013. Healthcare costs – New thoughts. Available at http://hcfocus.blogspot.co.uk/2013/01/new-thoughts.html (accessed 2016-05-21).

CMS. 2015. Quality measures: What's new. Baltimore, MD: Centers for Medicare and Medicaid Services. https://www.cms.gov/Medicare/Quality-Initiatives-Patient-Assessment-instruments/NursingHomeQualityInits/NHQIQualityMeasures.html (accessed 2016-05-10).

ECRI. 2016a. About ECRI. https://www.ecri.org/Pages/default.aspx (accessed 2016-05-21).

ECRI. 2016b. Biomedical benchmark. https://www.ecri.org/components/BiomedicalbenchMark/Pages/default.aspx (accessed 2016-05-21).

Ettelt S., Nolte E., McKee M. et al. December 2006. Evidence-based policy? The use of mobile phones in hospital. *Journal of Public Health*, 28(4): 299–303. http://jpubhealth.oxfordjournals.org/content/28/4/299.full (accessed 2016-05-10).

HSCIC. 2015. Data sets. Leeds, UK: Health and Social Care Information Centre. http://www.hscic.gov.uk/datasets (accessed 2016-05-10).

ISO. 2014a. ISO 55000 series – Asset management. Geneva, Switzerland: International Standards Organization.

ISO. 2014b. ISO 55001 Asset management: Management systems – Requirements. Geneva, Switzerland: International Standards Organization.

Kaplan R.S. and D.P. Norton. 1996. *The Balanced Scorecard: Translating Strategy into Action.* Boston, MA: Harvard Business School Press.

Keay S., McCarthy J.P. and B. Carey-Smith. 2015. Medical equipment libraries – Implementation, experience and user satisfaction. *Journal of Medical Engineering and Technology*, 39(6): 354–362.

Kohn L.T., Corrigan J.M. and M.S. Donaldson (Ed.). 2000. *To Err is Human: Building a Safer Health System*. Washington, DC: The National Academies Press.

McIntyre M.E. 2001. Goodhart's law. http://www.atm.damtp.cam.ac.uk/mcintyre/papers/LHCE/goodhart.html (accessed 2016-05-10).

MHRA. 2015. Managing Medical Devices – Guidance for healthcare and social services organizations. Ed1.1 https://www.gov.uk/government/publications/managing-medical-devices (accessed 2016-05-10).

Montgomery K. and E.S. Schneller. 2007. Hospital strategies for orchestrating selection of physician preference items. *Milbank Quarterly*, 85(2): 307–335.

Newton H. 2001. Developing an equipment library: The solution to increasing demand. *British Journal of Nursing*, 10(Suppl 5): S59–S64.

NHS England. 2014a. Patient safety alert to improve reporting and learning of medication and medical devices incidents. http://www.england.nhs.uk/2014/03/20/med-devices/ (accessed 2016-05-10).

NHS England. 2014b. Improving medical device incident reporting and learning. Available as a PDF from http://www.england.nhs.uk/2014/03/20/med-devices/ (accessed 2016-01-25).

NPAG. 2016. Benchmarking activity. http://www.npag.org.uk/our-services/benchmarking-clubs-best-value-groups/benchmarking/ (accessed 2016-01-27).

Sy, T. and L.S. D'Annunzio. 2005. Challenges and strategies of matrix organizations: Top-level and mid-level managers' perspectives. *Human Resource Planning*, 28(1): 39–48.

Tonn G. 2007. Matrix management: A scoping review. Province of British Columbia. Ministry of Labour and Citizens' Services.

Weerakkody R.A., Cheshire N.J., Riga C. et al. 2013. Surgical technology and operating-room safety failures: A systematic review of quantitative studies. *BMJ Quality and Safety*, 22: 710–718.

SELF-DIRECTED LEARNING

1. Develop in outline an MD Policy suitable for the healthcare organization in which you work.

2. Prepare a talk for your colleagues on how the MD Policy finds expression in the Strategic HTM Plan, and how the Medical Device Committee works with its MDC Action Plan to ensure delivery of the Strategic HTM Plan.

3. If you have previously attempted Self-Directed Learning exercise (3) in Chapter 3, revisit that exercise in the light of what you have learned from this chapter.

CASE STUDIES

CASE STUDY CS5.1: A PRACTICAL IMPLEMENTATION OF A MEDICAL DEVICE COMMITTEE

Section Link: Chapter 5, Section 5.1

ABSTRACT

Healthcare Technology Management (HTM) requires a structured systematic approach. From an organizational perspective, this approach is developed and embodied in a Medical Device Committee. This case study describes the formation and working practice of such a committee.

Keywords: Healthcare Technology Management; Medical Device Committee; Structured systematic approach

NARRATIVE

An important component of any HTM system is to have an organization-wide group, which we call a Medical Device Committee (MDC), charged with specific governance responsibilities for medical equipment. The makeup and exact terms of reference will vary between organizations but key features will usually encompass:

- Oversight of the Medical Device Policy (MD Policy);
- Developing a strategic, organization-wide HTM plan;
- Monitoring the performance of the technical support departments that deliver the HTM Programmes and equipment support plans (ESPs);
- Supporting equipment replacement planning;
- Being the focal point for discussions relating to all types of medical devices and their management.

We describe the implementation of a medical device committee in a district general hospital operating across three sites. It serves a population of 420,000 with 700 beds and 18,000 medical devices valued at $70million. The committee is called the Medical Device and Equipment Group (MDEG). The group dates back to 1997, originally designed as a multidisciplinary forum bringing together key leaders of medical device management with clinical users of medical equipment and management representatives. The group's responsibilities and membership have evolved over the years and the current representation includes the following:

- Clinical engineering – three representatives, including the Head of Clinical Engineering who chairs the group;
- Finance;
- Procurement;
- Risk Manager;
- Health and Safety Manager;
- Organizational lead for Information Services;
- Organizational lead for decontamination;
- Head of Estates;
- Strategic Planning Lead;
- Clinical Management Representatives – attendance is mandated from each of the five clinical divisions within the organization;
- Patient liaison representative;
- Specialist clinical advisors:
    o Infection prevention and control team;
    o Medical device training facilitator;
    o Moving and handling coordinator.

The group has four main functions:

1. Oversight of the MD Policy, including implementation issues, monitoring of policy efficacy and monitoring of equipment support plans;
2. Reviewing medical device safety issues to inform risk management;
3. Developing and monitoring equipment replacement programmes;
4. Identification of activities to promote more effective HTM.

The group is empowered by the organization to take key HTM decisions, making it a respected and effective forum. Those decisions that are outside the group's remit are appropriately routed through the organizational hierarchy. A key operating principle is that no medical device will be acquired without the group's approval, ensuring that best practice is adhered to in terms of standardization and rationalization. The group is empowered to co-opt members as necessary and to create specific project teams.

Meetings are held monthly to ensure its responsiveness and regular dialogue between key individuals. Issues are regularly reviewed and emerging issues and priorities aired. The comprehensive makeup of the group enables almost all aspects of HTM to be discussed with informed input from the relevant experts, eliminating the need to take issues to other forums.

Meetings follow a standard agenda which includes review of:

- Policy effectiveness;
- Medical device safety alerts;
- Medical device risks;
- Replacement programmes;
- Urgent replacement needs;
- Medical device projects.

Equipment replacement planning is approached by grouping medical equipment as either specialist to particular departments or hospital wide. Hospital-wide equipment is that which can be used interchangeably across the organization, examples being beds, infusion pumps, pressure relieving mattresses and oxygen flowmeters. These are proactively managed by the group with an allocated budget for their replacement across the organization. This frees up clinical divisions to focus on their specialist equipment.

The key to the success is the group's reporting links to various hospital forums including the following:

- Capital Management group – Reporting on capital replacement priorities and procurement progress;
- Risk committee – Identifying and escalating device risks and unresolved safety alerts;
- Commercial Development Group – Advising on new developments as opposed to replacements;
- Decontamination committee – Ensuring a robust link on medical device decontamination issues.

The MDEG chair has a seat on all these organizational forums and is the budget holder for the medical equipment capital programme and the replacement budget allocated to replacing smaller hospital-wide items. The chair meets weekly with the Risk Manager to discuss safety issues, useful for identifying and progressing specific emerging and unresolved issues.

From a practical perspective, it is quite onerous to meet monthly, to ensure that minutes and papers are distributed in a timely fashion. The group has also faced challenges where it has not been quorate, with colleagues perhaps not seeing the importance of attending when all the money has been allocated. The group has been proactive in keeping senior hospital executives abreast of such issues and have gained their support in mandating all equipment issues are presented prior to progressing to purchase. At the time of writing, the group is experimenting with setting aside specific meetings each year for major reviews of future expenditure planning, reviewing device incidents and strategic reviews of equipment. This leads to an annual work programme, with meetings in between taking on a monitoring nature and identification of urgent issues. What has been interesting is that the forum has remained a key decision making group despite numerous organizational restructuring programmes.

ADDING VALUE

The group has aided the organization to understand healthcare technology management issues and resulted in more coherent decision making.

$$\uparrow \qquad \uparrow$$
$$\text{Benefits : Cost} \therefore \text{Value}$$

PATIENT CENTRED

There are a large number of practicing clinical staff on the group who have a clear focus on doing the best for patients. The concept of 'organization-wide' assets underlines the commitment to ensuring equipment is deployed as required rather than benefitting one clinical area.

CULTURE AND ETHICS

A positive culture has developed where the group takes collective responsibility for ensuring that the highest risks are addressed as opposed to each individual's own priorities. The group has often been seen to take a longer term view than usual in-year budgetary constraints allow. This has given clinical colleagues confidence their requirements will be addressed albeit not immediately. The consensus approach has aided executives in addressing conflicting requirements and affordability issues.

SUMMARY

An example of the structure, roles and reporting mechanisms of a medical device committee is outlined.

SELF-DIRECTED LEARNING

1. What are the existing decision making forums that review device safety issues and equipment replacement priorities in your organization?
2. How are unresolved issues escalated?

CASE STUDY CS5.2: HTM IN SMALLER HEALTHCARE ORGANIZATIONS

Section Link: Chapter 5, Section 5.1

ABSTRACT

Healthcare Technology Management in the community setting is complex. Devices are prescribed to meet user needs and, in the case of wheelchairs, are individually customized. There is a likelihood of failure from a number of causes and the outcome can be catastrophic. With thousands of service users out in the community using their devices on a daily basis, we must ensure their needs are met and their safety is maintained, all within the constraints of limited resources.

Keywords: Wheelchair service; Community healthcare organizations; Electrically powered wheelchair; Medical devices.

NARRATIVE

In the NHS in England, money for universal healthcare from general taxation is allocated to Clinical Commissioning Groups (CCGs) in each area. They then commission healthcare services from a variety of providers, mostly but not exclusively public sector organizations usually called NHS Trusts. They deliver services. There is a split between Trusts providing home/community care services and those providing secondary and tertiary care (see Chapter 1, Section 1.2).

In a big city like London, the geographical boundaries of each CCG are clear but many providers can be providing care into a single CCG area so coordination is vital for good patient care. The diverse community healthcare services are provided by a range of organizations such as community Trusts, acute Trusts, social enterprise, charities and the private sector (Foot et al. 2014).

This case study discusses the approach of a consortium of four community Wheelchair Services (Services) working across multiple sites, in responding jointly to equipment failure and the resulting patient incidents. The case study examines the part that Healthcare Technology Management Systems have to play in medical device management and maintaining patient safety in a community healthcare setting.

The four Wheelchair Services in this case study were part of three separate Trusts and served eight Clinical Commissioning Groups (CCGs). When referring to the Healthcare Technology Management System illustrated in Chapter 4, Figure 4.13, the structure in this case becomes quite complex. The healthcare organizations are separated across the three Trusts, all with individual Medical Device Policies, Medical Device Committees and Healthcare Technology Management Programmes.

The Trusts' Medical Device Committees oversee medical devices within their Trusts. The Trusts manage custom devices such as splints, posture management, e.g. static seating and pressure cushions, and mobility devices from low-tech walking frames to Functional Electrical Stimulation (FES). Wheelchair Services provide seating, accessories, pressure care, custom-made seating devices and manual and electrically powered wheelchairs. Electrically powered wheelchairs are one of the more complex medical devices managed by the Trusts.

The subcontracting and reconditioning model that Trusts use for managing the wide range of low-tech devices issued out into the community is not appropriate for wheelchairs due to the lower turnover of equipment and the low percentage of equipment returned which is able to be reconditioned and still be safe and economically viable and the complexity of the powered wheelchairs.

Wheelchair Services manage powered wheelchairs either in-house or through Approved Repairer subcontractors. The day-to-day management of this equipment is generally undertaken by Rehabilitation Engineers who ensure that best practice according to the regulations and infection control and decontamination guidance is followed and that quality system standards and manufacturer's guidelines are applied. Medical Device Committees within Trusts are involved in pre-contract reviews to ensure the contractors are able to meet the requirements of the regulations, and regular reviews of agreed KPIs are undertaken.

The consortium of Wheelchair Services had jointly subcontracted a single Approved Repairer. This enabled the Services to pool expertise in managing the contract and provided advantages in the shared procurement and reconditioning of equipment, giving financial benefits to the consortium and quality of service delivery to our patients.

To ensure joint working and collaboration of approach between the four Services, a joint Technical Group, chaired by a senior clinical engineer, has been set up to oversee the management of their medical devices across the consortium. This is a multidisciplinary group consisting of Wheelchair Service senior representatives including both Therapists and clinical engineers, a representative from the contracted Rehabilitation Engineering organization and a representative on behalf of the consortium of CCGs and the Approved repairer. The representatives from each Service then reported to their individual Trust's Medical Device Committee and up through the Trust to the respective Boards. The equipment users are separated by the four Services and four separate budgets. However, there is a joint procurement strategy across the Services to ensure economies of scale across the consortium; therefore, all four Services have the same types of equipment procured through the same suppliers as agreed by the Technical Group.

The Wheelchair Services reported via the Technical Group that there had been a significant number of failures of one model of powered wheelchair. Some failures had occurred whilst patients were out in the community using their wheelchairs, one whilst crossing a road which had put the patient at significant risk of serious injury. Through investigation with peer networks such as the National Wheelchair Managers Forum and the South Coast Rehabilitation Engineers Group, it was found that failures had also occurred in other parts of the country.

To ensure that the approach was holistic, discussions needed to include all key stakeholders. This involved not only the member of the Services representing their patients but also MDCs within the Trusts and stakeholders within the CCGs to ensure the approach met with organizational goals. Regulatory bodies, in this case the MHRA, were also involved in reporting adverse incidents to them, and them reporting back what action if any had been taken with the manufacturer to date. The contracted Rehabilitation Engineers and Approved Repairers were involved as they would need to potentially perform the remedial action and take advice from the manufacturers.

A joint strategy and objectives were formed at the Technical Group, the main objective being to ensure patient safety and prevent any further incidents. The strategy was to perform a risk analysis in the first instance to establish what risks there were to patients and then form an action plan based on the analysis. The Technical Group gathered evidence to analyze the hazards and potential harm to patients. The MEMS database provided the total number of failures that had occurred and the total number of the specific model of chair being used across the four Services. This gave a local picture of the potential likelihood of failure and therefore the potential risk. Evidence was also gathered from reports to MHRA and discussions with patients as to how the incidents had occurred and if there were any common factors.

Discussions with the manufacturer established what the failure rate had been nationwide. Following enquiry from the Technical Group, the manufacturers disclosed that no changes had been made to the build, material or manufacture of the wheelchair. Failures had been occurring over a number of years from as early as 3 months after issue to five plus years after issue with no clear evidence as to why, although patient misuse of equipment was discussed as a potential factor. The manufacturer had devised guidance for engineers in the field to inspect the chairs for potential failure at two sites on the frame and had released an updated frame design in response to the failures. The manufacturer agreed to have a number of these at the Approved Repairer on a call-off basis to ensure availability and to reduce the impact to patients.

The immediate action was to stop prescribing the power chair affected. Patients who were at potential risk were identified by reviewing their repair records and assessment records on the MEMS to see if there was frequent outdoor use or if they had a high level of repairs and therefore may be more prone to potential failure. Patients deemed to be at risk either had their frames replaced with the new design or were given alternative power chairs which suited their needs.

The MHRA were contacted to establish if a UK-wide device alert had been raised, if they were aware of the issue and if they had recommended actions. The MHRA had not issued an alert due to the small percentage of failures across the United Kingdom compared to the total number of these wheelchairs in use. The MHRA at that point were happy with the manufacturer's response to provide replacement frames to resolve the issue.

To prevent further incidence of failures, the Technical Group agreed that two rather than one Planned Preventative Maintenance (PPM) should be performed per year and to perform the frame checks in line with manufacturer's guidance as part of the PPMs.

The individual Trusts and CCGs were informed of the decisions that had been made at the Technical Group. Risk assessments were performed by each Trust's risk lead as to whether the risk reduction measures of frame replacement, repair strategy and extra PPMs reduced the risk to acceptable levels. Each Trust agreed that the risk reduction was acceptable with these measures in place. The CCGs were also in agreement with the risk reduction measures. The financial implications also had to be considered as this potentially affected a large number of power chairs. The financial outlay was considered to be acceptable to the Trusts, CCGs, Approved Repairers and manufacturer. The Services accepted that there would be an increase in contacts with patients for the Services to take remedial action.

As a result of the actions, the number of frame failures has reduced significantly. Discussions are still ongoing to monitor and review the situation within the localities and also to keep the MHRA and manufacturer informed of any new incidents. Communication and agreed shared objectives are key where a large number of stakeholders are involved.

ADDING VALUE

An integrated approach as shown in Chapter 4, Figure 4.11 was essential for the success of this project. Various organizations were involved in the process including four Wheelchair Services, one contracted NHS clinical engineering organization, three NHS Trusts, eight CCGs and two private companies. In order to achieve the optimal outcome, a holistic approach needs to be taken with support from all organizations. Although costs went up, the benefit to service users went up so overall value was increased.

Benefits : Cost ∴ Value

SYSTEMS APPROACH

A systematic approach was taken, ensuring all stakeholders were addressing the issue in the same manner, with the same objectives, to ensure the agreed actions and outcomes were consistent and to agreed time frames across the North West London consortium. Good communication between the stakeholders was essential to keep the project on track and to ensure the buy-in from all involved.

PATIENT CENTRED

NHS England national focus is on maintaining care at home and reducing hospital admissions (Foot et al. 2014). The community sector is integral to achieving this goal.

SUMMARY

In Community Trusts where a number of individual services are being managed over a large geographical area, with medical equipment being provided out in the community, joint working is essential to ensure services can be delivered to the required standards and patient safety is maintained.

REFERENCE

Foot C., Sonola L., Bennett L., Fitzsimons B., Raleigh V. and S. Gregory. 2014. Managing quality in community health care services. London, UK: Kings Fund. http://www.kingsfund.org.uk/publications/managing-quality-community-health-care-services (accessed 2016–01–29).

SELF-DIRECTED LEARNING

1. How would you ensure the key stakeholders in the many organizations involved in the management of Healthcare Technology work together to ensure what is planned is delivered and implemented?

2. Think about and list some of the challenges in managing technology in support of health services in a community healthcare setting.

CASE STUDY CS5.3: MD POLICY SUPPORTS BOARD LEVEL QUERY LEADING TO ACTION

Section Links: Chapter 5, Section 5.1

ABSTRACT

The Board and the Chief Executive Officer (CEO) must be able to exercise leadership of the healthcare technology and its management. The Medical Device (MD) Policy provides the process: the Board working through the Medical Device Committee (MDC) with its Strategic Healthcare Technology Management (HTM) Plan and action delivered through its MDC Action plan.

Keywords: MD Policy; Strategic HTM Plan; Medical Device Committee; Action Plan; Clinician training

NARRATIVE

A non-executive member tabled a question about medical equipment training for the Board's theme meeting on staff training. This meeting was a follow-up to a serious adverse incident strongly linked to poor staff understanding of the safe procedure for use of a particular type of medical equipment which had led to adverse publicity. The CEO referred to the Medical Device (MD) Policy: staff must be competent when using medical equipment, assessed at annual appraisals. A Policy annex had details: the personal professional responsibility of doctors and nurses to be competent in applying medical equipment, the hospital to provide training, staff competency to be assessed and training records kept. However, with no details or statistics on staff training available, the CEO was not immediately able to respond and asked the MDC to report.

The MDC reviewed the details on medical device training in its Strategic HTM Plan which linked to the MD Policy. The Plan was vague on how training would be provided and the MDC discussed how time and cost pressures had caused training initiatives to slip. The supplier of infusion pumps provided training; the pumps were procured under a consumables purchase agreement that included continuing staff training. However, no training records were kept and anecdotes suggested poor attendance without follow-up competency assessments. Other than at the commissioning of new equipment, no other medical devices training was known to take place.

The MDC asked the Head of Clinical Engineering to convene a project team to investigate. Improving medical device training was added to the MDC Action Plan which called for a summary of the current provision, clarifying the views of the Medical Director and Director of Nursing and their training and development departments and development of a practical and implementable medical device training programme.

The medical and nursing heads agreed that medical device understanding was inadequate, but pointed to the crowded working week and stretched staff. Gradually, working with the training departments and senior medical and nursing staff, opportunities for training were revealed. There was scope for some separate general training for medical and nursing staff. A 3-monthly cycle of general principles of safe application of commonly used medical devices would be offered, one at a monthly 45 min session, with the cycle repeated three times annually but avoiding the busy winter time. Targeted training was required. The head of surgery asked for electrosurgery training for surgeons, covering how the technology worked and its risks; this could be timetabled with their update programme. The theatre manager wanted theatre nurses and assistants to understand the surgical and medical equipment they prepare for theatre use, to be incorporated in their monthly development programme. Resuscitation training was good, but clinical engineering was asked to deliver an annual resuscitation technology update. The anaesthetists had their own training programme, covering theatre and critical care equipment, but asked for help in training the critical care nurses. Maternity and neonatal services had a number of requests: cardiotocograph training for maternity staff, an improved understanding of blood gas measurements to overcome the limited knowledge of problems with pulse oximetry and how to achieve accurate transcutaneous blood gas measurements and when these needed to be supplemented by laboratory gas measurements. Other departments had their own speciality-specific requests. The project team reflected on the unmet needs throughout the hospital.

The company trainer for the infusion pump programme was frustrated by the lack of uptake on the training. Reviewing and updating the training material, and with support from the Director of Nursing, the Head of Clinical Engineering stressed the importance of infusion pump training to the senior charge nurse forum. But it is not only infusion pumps we are concerned about, the senior charge nurses responded – vital signs monitoring has changed so much over the past decade. The opportunity was taken to explain the planned 3-monthly cycle of general principles training which would cover this. Clinical thermometers were also supplied on a consumables purchase agreement; update training would be included in the upcoming contract renewal.

A training records scheme was proposed, with commercially available software installed on the hospital network. Registration at training sessions would be linked to this, with hospital departments taking responsibility for reviewing their staffs' training and competency assessments. Clinical engineering would manage it, providing 6-monthly reports to the MDC to be included in its annual Board report.

The project team compiled a report for the MDC. Following discussion and revision, it was submitted to the Board, summarizing the current position and immediate and medium term development plans. The report recommended changes to the Strategic HTM Plan and to the MD Policy. Clinical engineering was asked if departmental efficiencies could free up the 0.5 WTE staffing required to support training or to report back requesting funding.

ADDING VALUE

The HTM process, starting with its MD Policy, provided the mechanism for the Board to understand aspects of how its medical equipment was managed. The process, with its policies and plans, incurs a cost but reduces risk and ensures that clear answers can be given. The process added value, with benefits greater than costs.

Benefits : Cost ∴ Value

SUMMARY

The HTM process enabled concerns raised at the Board to be addressed. The non-executive director noted: "Congratulations to the CEO and MDC for having a simple but effective process for responding to Board questions about healthcare technology management".

SELF-DIRECTED LEARNING

1. Discuss with clinical engineering colleagues the advantages and disadvantages of the HTM process described in this book in which the Board manages its medical technology through its MD Policy and its MDC committee. Continue by discussing whether the MDC having a Strategic HTM Plan and an MDC Action Plan can help improve the management of medical equipment.

CASE STUDY CS5.4: LINKING HTM STRATEGY TO ORGANIZATION'S STRATEGY ADDS VALUE

Section Links: Chapter 5, Section 5.4.1

ABSTRACT

A need to replace most of the defibrillators in a healthcare organization on grounds of obsolescence (an HTM issue) coincided with an evidence-based clinical push to move to biphasic defibrillation waveforms and Automated External Defibrillation (AED) technology for better patient outcome. Clinical engineers worked with clinical colleagues to achieve a total replacement of the fleet.

Keywords: Defibrillator; Biphasic; AED

NARRATIVE

A healthcare organization has a Resuscitation Committee with links through the management structure up to the Board. Its remit is mostly clinical – discussing and approving resuscitation policy and protocols and resuscitation training. The Resuscitation Training Department (RTD) reports to this committee. The Clinical Engineering Department (CED) has representation on this committee and good working relationships with the RTD.

The organization had 146 defibrillators, 129 of which were of one of two types of mains/battery-powered devices delivering monophasic waveforms. The remaining 17 were of three different types and not in the main hospital. The organization's defibrillators were 'owned' by individual wards and clinical areas but all were looked after by the CED.

The CED was aware that 42% of the defibrillators were obsolete (no longer manufactured and/or with no availability of some spare parts) and 46% were obsolescent (no longer manufactured but still supported). The manufacturer had notified the CED of an impending cessation of support. These two categories accounted for the 129 noted earlier.

CED brought these technical issues to the attention of the Resuscitation Committee and the MDC, the latter asking the Resuscitation Committee to draft an action plan. Because of their involvement with the Resuscitation Committee and the RTD, CED was aware of the current clinical evidence: first, the clinical evidence for and the international resuscitation guidelines recommending the use of biphasic defibrillation waveforms and, second, the clinical benefits of rapid defibrillation. Both pointed to the deployment of biphasic AED technology. Most manufacturers were offering this combination. The Resuscitation

Committee agreed that this should be the strategic aim. So the technical need to replace old equipment converged with a clinical strategic aim which would provide clinical benefit to patients.

The CED member of the Resuscitation Committee suggested that a 'Defibrillator Working Group' should be set up to manage the replacement. This was agreed and was chaired by the head of the RTD with the CED representative as secretary, that is, doing most of the drafting. A strategic proposal and business case was drawn up which proposed that the 146 defibrillators be replaced by 204. These would be an appropriate combination of mains/battery defibrillator/monitors, all with AED capability, and simple 'shock box' AED machines placed in clinical areas with low usage of defibrillators. This plan provided for much wider distribution of defibrillators throughout the organization in order to meet the desired 'time to shock' following a cardiac arrest, which published evidence showed to be critical in improving survival rates.

The other two strategic proposals were that the Resuscitation Committee be given the authority, after consultation, to decide on the distribution of defibrillators throughout the organization and that defibrillators would be 'owned' and managed corporately rather than by individual wards and clinical areas.

The business case was endorsed by the Resuscitation Committee, passed through the MDC to the Board and accepted. A very successful tender exercise was carried out, the CED working closely with the Procurement department and the RTD, resulting in very advantageous prices. The implementation of the plan was phased over a number of years, with the obsolete equipment being replaced as a matter of urgency, followed by replacement of the non-standard types.

ADDING VALUE

The replacement incurred costs; you cannot have a major acute hospital or its associated smaller hospitals relying on obsolete defibrillators. The oldest equipment could have been replaced individually, but this might have led to diversification of equipment types with different clinical protocols depending on which type of defibrillator was being used and would certainly have led to higher overall cost of replacement. Clinical user training would have been less than optimally effective. This would have increased patient risk. The coordinated, organization-wide approach resulted in a lower cost than would have otherwise been incurred and in greater benefit to patients.

Benefits : Cost ∴ Value

PATIENT CENTRED

The fundamental driver for this project was the understanding that availability of defibrillation to a patient in cardiac failure had been less that optimal, given new published evidence, both in terms of delays and the improved clinical benefit from biphasic waveforms.

CULTURE AND ETHICS

The presence of a clinical engineer on the Resuscitation Committee and the close working relationship between the CED and the RTD promoted an easier and focused approach to solving the problem.

SUMMARY

A fairly complex procurement exercise was developed out of a strategic clinical objective approved at the Board level. A significant clinical engineering input was to propose and argue for a strategic change of 'internal ownership' arrangements which provide for better management of this vital type of healthcare technology.

SELF-DIRECTED LEARNING

1. In the end, because of financial constraints, the whole fleet of defibrillators could not be replaced in one hit but the project was implemented over a number of years. Consider the implications and risks of this approach.
2. Consider what other types of equipment might be suitable for a similar approach.

CASE STUDY CS5.5: ISSUING MEDICAL DEVICES TO PATIENTS FOR HOME USE: POLICY AND RISKS

Section Links: Chapter 5, Section 5.4, Section 5.4.1.1

ABSTRACT

Increasingly, there is a trend to deliver care closer to, or even in the home, to give patients more control over their healthcare in a more familiar environment. This presents challenges for secondary care providers who increasingly need to rethink delivery models as they support patients with chronic conditions who need ongoing community-based care as well as occasional episodes of inpatient support. There is a growth in the deployment from secondary care organizations of medical devices for patient use in the homecare environment. This case study describes the challenges of managing healthcare technologies located in the homecare environment and the steps organizations must consider to ensure risks associated with the use of equipment are minimized.

Keywords: Community healthcare; Managing homecare equipment

NARRATIVE

Patients and their carers are increasingly taking responsibility for and contributing to decisions about their own healthcare, as society comes to terms with an ageing population living longer and coping with often multiple co-morbidities. Hospitals have limited bed capacity and the flow of critically ill patients who need specialist inpatient care is often hampered due to reduced bed availability with medically fit patients not being able to be discharged for a variety of social and logistic reasons. Many hospitals are attempting to speed up discharges and contribute to community care initiatives to minimize admissions, thus preserving vital specialist care for those who need it. Increasingly, this requires considering sending patients home with medical devices for their personal use. Sometimes, this may be a short-term loan until other arrangements are made or alternatively may be a permanent issue of technology to provide continuing care. Any healthcare technology deployed in this way needs to be effectively managed, with the risks and management arrangements carefully considered.

So what are the things that can go wrong? Well, consider a patient who has been discharged from a respiratory ward with a nebulizer to dispense bronchodilator therapy at home. Late at night on the day of discharge, therapy is attempted and the device will not turn on. The patient phones the ward and asks for help, but help over the phone doesn't resolve the issue;

the patient becomes increasingly distressed due to lack of treatment and is advised to present at the Emergency Department. The following day clinical engineering discover the device had a faulty mains lead. Another example of the risks associated with the long-term issue of electromedical equipment is around periodic maintenance; where does the responsibility lie if a patient does not bring a device back to the issuing institution as agreed at the point of issue – what if they physically can't? In the event of an injury to such a patient as a result of failure to maintain a device, the responsibility is likely to lie with the issuing institution. The use of devices in the homecare environment creates a myriad of challenges – devices are not being used by healthcare professionals and the usual support mechanisms, such as ready access to clinical engineering workshops or spare equipment, are not available. It is inappropriate to issue a medical device for home use unless done so with appropriate consideration of the risks and the development of necessary technical and clinical support systems.

The focus of this case study is to present a generic approach to guide clinicians as to the essential considerations and steps to take if considering issue of medical devices for patient use in the non-hospital environment. As articulated in this chapter, it is essential that all technology used in the hospital is done so within the organizational HTM framework. So an essential first step if the organization plans to issue equipment for patient use is that a strategic decision is made to do so. There needs to be a clear line of sight from the Board to the clinical front line so executive management supports the development of the necessary systems and resources to manage technologies deployed directly to patients. In practice this means the MDC must take in hand the development of appropriate advice and control mechanisms, ensuring that policies reflect the special factors such deployments require. A recurrent theme throughout this book is that systematic planning and effective consideration of risks prior to issuing of technologies will yield significant increase in value. The time spent to rectify an incident that has occurred in a patient's home is costly in manpower, potential medicolegal cases and organizational reputation. But most importantly, effective planning will improve the intended clinical outcome and enhance patient experience.

So having aligned the intent to undertake such activity with organizational policy, attention can be turned to the practical advice that clinical users can be given prior to issuing the technology. This is where clinical engineering can provide valuable guidance and insight to clinical colleagues, leading on developing an appropriate equipment support plan (ESP) for this type of deployment (ESPs are discussed in Chapter 6). Existing ESPs for deployment within the hospital can be a useful starting point from which to consider what additional risks and challenges may be encountered with home deployment. The following steps are designed to enable a structured approach to guide clinicians through the thought processes prior to commencing device issue:

- Identification of the clinical pathway, patient group and technologies involved, together with a rough estimate of numbers. In effect a statement of the outcome and aims to be achieved, including whether this is a permanent issue of the device or short-term allocation.
- Engagement with key stakeholders who may be able to contribute directly or indirectly to the project, for example Physicians, Nurses, Therapists, clinical engineers and Infection Prevention and Control. This step leads to an embryonic ESP being developed, capturing the risks associated with the device deployment.
- Operational factors to be considered:
  - Equipment acquisition: It is important that such deployments do not destabilize equipment availability elsewhere, so it needs to be established where the equipment is acquired from, funding arrangements established, etc.
  - Roles and responsibilities: Who will issue and receive returned equipment? What are the arrangements for issuing spare devices in the event of failure for example? What will maintenance arrangements be? It is essential to establish points of contact to advise patients. Out of hours support is key.

- o Guidance for users: Within the hospital, clinical staff will have received training in operating the device; what information is passed to the patient and their carers to ensure that they can use the equipment safely? Care needs to be taken to ensure that training material is in a form that is readily comprehensible. Should demonstrating the equipment to patients be part of the deployment process? Guidance and perhaps demonstration needs to be given on key user maintenance, such as cleaning and battery management. If consumables and/or accessories are used, advice on when and how to change them will be required. Requirements for ongoing maintenance need to be established: how will the patient get their device serviced?
  - o Reprocessing on return to the organization: It is essential that when a device is returned following use outside the hospital, it is brought back into clinical use in a controlled way. This may entail all devices going through clinical engineering workshops to ensure devices are cleaned and fully serviced and in suitable condition for the device to be redeployed.
- Record keeping: It is essential to identify where devices have been deployed to; how is equipment to be tracked via the MEMS database? Tracking enables records to be kept demonstrating effective medical device management and importantly the ability to respond to safety alerts.
- Feedback and communication: It is essential that all stakeholders keep in touch to monitor and review how these arrangements are working, enabling the fine tuning of arrangements. A key area to consider is capturing feedback from patients to inform system improvements.
- Monitoring: The MDC will keep a record of devices which have been authorized for use outside the hospital and will periodically seek assurance that those areas are auditing practice, identifying and managing risks as required.

ADDING VALUE

Clinical engineers should develop the knowledge and skills to objectively determine how their activities affect costs and benefits. In this case study for example, managing equipment systematically should not increase costs, but if there is a need to deploy extra equipment, costs will increase.

Managing equipment systematically is likely to cost no more than managing it badly, but benefit is increased.

Deploying extra equipment into a homecare environment has a cost but brings organizational-wide benefit.

SYSTEMS APPROACH

The approach discussed demonstrates systems thinking in practice, taking the approach of considering stakeholders and the aims and objectives. Putting emphasis on the roles, responsibilities and mechanisms ensures the organization is acting responsibly.

PATIENT CENTRED

By addressing issues in advance and ensuring key issues are thought through, there is a better patient experience and less risk of clinical service disruption. This work paves the way for more significant change to service delivery outside the hospital environment.

SUMMARY

The clinical engineering function can ensure robust HTM arrangements are put in place associated with issuing of medical equipment for use at home. The case study describes a proactive approach which minimizes risk and problems which could otherwise occur.

SELF-DIRECTED LEARNING

1. Review and comment on the arrangements for issuing equipment for home use in your organization.
2. Do guidelines and controls exist, with the MDC having oversight? How could these ideas be introduced into your organization?

CASE STUDY CS5.6: REPLACEMENT PLANNING: A CONTINUING JOURNEY FROM WHERE WE ARE TO WHERE WE NEED TO BE

Section Links: Chapter 5, Sections 5.4.1 and 5.6.3; Chapter 7, Section 7.2.1

ABSTRACT

Medical equipment replacement is a continuing journey, travelling from the current asset status to that of the required equipment availability, taking into account the healthcare organization's strategic plans. It is a continuous journey because where we are and where we want to be change over time. These moving start and end points define the equipment planning strategy, setting the scene for prioritizing needs against available resources on a year-by-year basis. We describe a structured approach covering up to 10 years, detailed for the first few years, more general for the longer-term planning, that levels out annual replacement and investment costs whilst supporting standardization.

Keywords: Medical equipment replacement; Planning; Structured approach; Current equipment status; Desired equipment status; Lifespan; Decision making; Rolling replacement.

NARRATIVE

A strategic medical equipment procurement approach was developed by clinical engineering, the budget holder for medical equipment, in conjunction with the Director of Finance who developed similar strategic planning for other technical support departments, for example estates. Each group had a budget holder, with quarterly joint meetings to discuss expenditure against the year's procurement plan. Finance requested a 10-year plan to support their future projections and agreed to negotiate forecast budgets.

The approach was based on the following:

1. Objective: To provide, within the available finances, the most clinically appropriate and economically advantageous medical equipment for patient care.
2. Realistic: Recognition of where we are, taking account of the current status of medical equipment.
3. Forward looking: Clinical leaders, hospital management and clinical engineering identifying short-term and forecasting long-term medical equipment needs.

These three summarize the objective and the start and end points of the procurement journey. The start point was determined from the inventory of medical equipment held and maintained by clinical engineering as part of their HTM due diligence. To clarify the start point, the current equipment status, the inventory was divided into groups, based on the equipment functionality. For simplicity in describing the approach in this case study, the equipment is divided into 10 groups as shown in Table CS5.6A. In reality there were more groups, chosen to help facilitate discussions with the clinical users.

Discussions were held with Finance to clarify the funding available in relation to estimated needs. The hospital's medical equipment had an estimated replacement value of a little over $100 million. Finance had set aside $10 million annually over the next 3 years for medical equipment procurement but was prepared to consider future increased funding allocations based on a robust 10-year procurement plan.

Two goals should dictate the procurement planning: the organization's strategic plans and ensuring the availability of safe, reliable medical equipment that provides clinicians with the tools for healthcare delivery. The duration of support from manufacturers introduces equipment lifespan, nominally 10 years for most, but shorter where reduced by technological developments or wear and tear. For example, for imaging and endoscopic equipment, quoted lifespans are closer to 7 years, with medical professional groups often endorsing these shorter recommendations.

Clinical users were consulted as to their immediate and long-term plans. They provided good guidance on the immediate equipment needs (year 1 planning) but their focus on the immediate challenges of care provision often inhibited longer-term planning. However,

TABLE CS5.6A   Summary of 10-Year Medical Equipment Procurement Plan

		Year									
		1	2	3	4	5	6	7	8	9	10
Budget		$10	$10	$10	$12.5	$12.8	$13.0	$13.3	$13.5	$13.8	$14.0
Allocated		$9.0	$9.1	$9.0	$11.7	$11.9	$12.1	$12.6	$12.6	$12.7	$12.9
Contingency		$1.0	$0.9	$1.0	$0.8	$0.9	$0.9	$0.7	$0.9	$1.1	$1.1
	Asset Value	Allocation by Medical Equipment Type									
Imaging	$28.2	$2.3	$2.5	$3.1	$3.8	$3.3	$3.3	$3.3	$3.3	$4.4	$4.4
Endoscopy	$25.5	$3.4	$1.4	$2.5	0	0	$3.4	$3.5	$3.5	$3.9	$3.9
Theatre and critical care	$18.6	$2.6	$3.4	$2.2	$2.1	$2.1	$1.8	$2.1	$1.9	$0.3	$0.3
General ward	$8.5	$0.2	$0.2	$0.2	$1.2	$1.2	$1.2	$1.2	$1.2	$1.2	$1.2
Infusion systems	$5.0	0	$0.1	0	$2.0	$3.0	0	0	0	0	0
Maternity and neonatal	$3.5	0	$1.0	0	$0.4	$0.4	$0.4	$0.4	$0.4	$0.4	$0.4
Dialysis	$2.9	$0.1	$0.1	0	0	$0.5	$0.5	$0.5	$0.5	$0.5	$0.5
Cardiac defibrillators	$1.6	0	0	$0.6	$0.9	0	0	0	0	0	0
Clinics	$7.4	$0.2	$0.3	$0.3	$0.9	$0.9	$0.9	$0.9	$0.9	$0.9	$0.9
Community	$2.6	$0.2	$0.1	$0.1	$0.4	$0.5	$0.6	$0.7	$0.9	$1.1	$1.3

Notes: Values are in million dollars. The budget was $10 million for each of years 1–3, with increases estimated for years 4–10 following negotiations with Finance.

some predicted changes in clinical care practices, for example increased day surgery, shorter but more intense hospital care and enhanced community care. A small community outreach service had been started, with expected 20% annual growth that the plan should support.

Clinical engineering had a long history of 'emergency' demands for equipment procurement during any financial year; consequently, it was agreed with Finance that about 10% of the allocated budget should be set aside as a contingency fund. If not required to fund emergency procurements, it would be used to bring forward procurement plans.

Equipment should be standardized to facilitate user competence, particularly for devices in widespread use throughout the organization (thermometers, general vital signs monitors, infusion devices and defibrillators). The procurement plan should facilitate this.

Comments on the draft were sought from the forum of clinical leads (medical and nursing) and the MDC, leading to refinements, after which it was presented to the Board. Negotiations with Finance for long-term budgetary estimates followed. Finance shared the procurement objective (point 1 in the list) and requested that the Board endorse 'in principle' the budget allocations, to be adjusted later for inflation, for the 10-year period.

Table CS5.6A summarizes the asset value and the replacement plan by equipment type. Standardization is promoted by grouping for replacement, into 1 or 2 years, equipment of the same functionality. Detailed discussions with major equipment users such as Imaging and Endoscopy led to their plans levelling out the projected required finances between years, for example not replacing the MRI, CT and nuclear medicine imaging machines in the same year. Their year-by-year allocations would also vary to accommodate peaks in, for example, critical care and theatre procurement planning. Growth in community care was accommodated for by allocating a 20% increase from year 4.

The first 3 years funded only $30 million of the requested $45 million, but year on year for the next 7 years, the allocation increased. Lifespans of major imaging and endoscopy equipment were estimated at 8 years with procurement phasing programmed to meet clinical requirements and level off the annual financial spend over the period.

The journey is not static; the starting point changing as medical equipment is acquired and old medical equipment disposed. The ending point will change, perhaps not as rapidly, but with clinical and technical developments and with the evolution of the organization's strategy. Developing the plan requires close cooperation between clinical engineers and clinicians, each understanding each other. Finance must be part of the team; recognizing the importance of ensuring the availability of the appropriate medical equipment at the point of clinical need, Finance can strive to incorporate the replacement planning in the organization's overall financial management and planning.

The clinicians were delighted that attention was being paid to future medical equipment procurements, welcoming the prospects of future increases in funding but concerned about the immediate short term, a concern partly addressed by the contingency reserve. Finance and the Board were reassured that this aspect of their corporate planning had a more robust platform, rather than *ad hoc* annual demands to which they had become accustomed. All recognized that this was just a start with the planning needing refinement and improvement, but that the journey had been started.

The medical equipment procurement planning should not exclude low-cost items – indeed relatively low-cost items such as infusion devices and defibrillators are included (Table CS5.6A). This case study deliberately does not address finance distinctions that may arise from different funding streams for different equipment types. Rather it focuses holistically on the need for procurement planning whether the medical device concerned is an electronic thermometer or an MRI scanner.

ADDING VALUE

A planned rolling replacement that does not rely on keeping old equipment in service will increase costs, but less so than those associated with unplanned procurements that cost even more without providing benefits. The cost of a structured and coordinated forward plan may be a little more but provides significant benefit, thus value is increased.

Benefits : Cost ∴ Value

PATIENT CENTRED

By focusing on the clinical benefits provided by medical equipment and recognizing procurement as a journey towards an ideal equipment status, the procurement planning described delivered an approach that met with the approval of clinicians and the hospital Board. The plan accommodated flexibility through its contingency allocation and ensured detailed focus on the immediate few years with a general outline covering a 10-year period.

SUMMARY

A 10-year medical equipment procurement plan was developed through consultation with Finance and clinical users that was based on a clear patient-centred focus. The procurement journey from current status to where it was desired to be in 10 years mirrored the strategic 10-year planning of the Board. The plan accommodated flexibility through its contingency allocation.

SELF-DIRECTED LEARNING

1. Have you discussed with Finance long-term procurement planning? If yes, what were your experiences of it? If no, how would you approach Finance?
2. Define and describe the current status of the medical equipment in your organization.
3. Describe your goals for medical equipment procurement planning.
4. Develop a 3- to 10-year equipment procurement plan. What key factors need to be considered?
5. How would you develop replacement plans to standardize medical equipment in widespread use in the organization? Consider scenarios where the starting point is a lack of standardization and also where standardization has already been achieved.

CASE STUDY CS5.7: INVESTIGATING THE RISK OF CONTINUING TO USE EQUIPMENT THAT IS SUBJECT TO A SAFETY WARNING NOTICE

Section Link: Chapter 5, Section 5.6.3

ABSTRACT

An unexpected safety alert is received for a range of syringe pumps advising that they require field updates that will not be available for a considerable time. This case study weighs the risks involved in deciding whether to continue to use the equipment or not.

Keywords: Safety alert; Warnings; Risk; Field updates

NARRATIVE

A fleet of syringe pumps used by a hospital was subject to a safety warning issued by the manufacturer. It was advised that a combination of hardware and software changes needed to be made to address a potential risk that patients may not get the correct drug delivery.

The manufacturer advised that the pumps could still be used in the interim with heightened vigilance, and additional training would be provided for staff until the fixes were available. Clinical engineers along with clinical colleagues assessed the risk from two points of view: the risk to the patient of continuing to use the equipment and the risk to the patient of not continuing and the ensuing issues.

The risk to the patient of continuing to use the equipment was not clear. If the clinical staff had been trained in the use of the pump and now received further training, then they would be able to continue to use the pump. However, there was a possibility that they might omit the further checks required, being so used to previous operational ways. There was also the aura of the equipment being the subject of a safety issue which also raised concerns as to its suitability, even though these were being addressed. The issue of replacing the pumps as soon as possible was voiced.

The option of withdrawing the fleet and replacing with new equipment was an option that seemed attractive but there were several drawbacks. Funding would be a problem as this had not been foreseen or planned for and the capital outlay would be considerable and financially damaging as the pumps were only a few years old. Even if alternatives could be sourced, then there would still be delays in obtaining and deploying the replacements and were the existing pumps to be withdrawn immediately? What impact would this have had on patient safety and activity? After all, the patients and clinical users had been using these pumps for several years and there were no local incidents to alert of an issue.

In the end, a risk assessment was completed jointly by clinical engineering and clinical colleagues that put forward scores for both courses of action and concluded that the *status quo* should remain with heightened training. There was, however, the proviso that the risk assessment should be revised regularly in the light of any incidents or updates from the company. In time, the field service updates were completed.

ADDING VALUE

Clinical engineers added value in this situation by having an understanding of the technology of the pumps, good records of service performance and the ability to contribute this to a multidisciplinary project group. The unexpected additional workload was a cost but the benefit was continuity of clinical service at acceptable risk.

Benefits : Cost ∴ Value

SYSTEMS APPROACH

This is a clear case of having to assess the risk of different possible solutions to a problem and weighing up the balance of risk between them. The systematic approach involves clarity of the objective (we need to continue to provide patient service), involving all stakeholders in the development of solutions and then assessment of options and regular review of progress.

PATIENT CENTRED

The safety of the patient is central to all those involved in their care. Occasionally, situations like this arise that seem to place the patient in a 'no-win' position. Use a 'bad' pump or use no pump at all. Whilst this is an extreme view, the reality is that sometimes risks have to be weighed and the best course chosen. It happens with direct medical care and it can happen with medical equipment. The main concern is that the care of the patient is kept in focus.

CULTURE AND ETHICS

How long the manufacturer had been aware of the safety issue is not known but there is a responsibility under the vigilance system for manufacturers to disclose as soon as possible to the regulatory authorities if there is any change in risk. This information is in the public domain and so the staff should be candid with the patient if they are concerned about equipment. Patients rely upon clinical staff to make the right choices for them.

SUMMARY

Weighing up risks is part of the clinical engineers' world, but when it directly influences patient care, as in this case, it is a more sobering experience. Reducing the situation to mere numbers in order to analyze the risk may seem cavalier and impersonal, but it is only through such quantitative methods that a meaningful conclusion can be arrived at.

SELF-DIRECTED LEARNING

1. Can you think of any other options other than the two mentioned earlier? Think about the perspectives of all stakeholders involved.
2. What would you feel would be a maximum timescale for the manufacturer to sort out the problems?
3. What would you do if they exceeded that time?
4. Could/should a patient representative have been involved in the discussions leading to a decision?

CASE STUDY CS5.8: GATHERING DATA TO EVALUATE AND PRIORITIZE MEDICAL EQUIPMENT REPLACEMENT

Section Links: Chapter 5, Section 5.6.3; Chapter 7, Section 7.2.3

ABSTRACT

In an ideal world, the proposal to replace a medical device or system would be planned and anticipated long before its eventual demise. However, circumstances frequently change and this case study looks at a methodology that considers both technical and clinical user input to provide a method of prioritization within a replacement programme.

Keywords: Replacement; Prioritization; Clinical opinion; Planning; Methodology

NARRATIVE

Although the clinical engineer can plan for the eventual replacement and disposal of equipment contained within the asset inventory, there are occasions where problems occur unexpectedly that can have an impact on this planning, especially within a tight financial budget. The process of managing this is discussed around the example of an EEG recording system not anticipated to require replacement for a few years.

An EEG recording system was scheduled for replacement in a few years time, but unexpectedly, the manufacturer ceased trading leaving no supplier service expertise and support and no source of spare parts. To compound the issue, the system developed some very intermittent problems that the clinical users found to be a problem, but managed with some difficulty to work around to maintain the clinical service. However, they had initially failed to log the issue with clinical engineering as they knew that the company had ceased training and presumed that help would not be available.

Clinical engineering were trying to find alternative sources of service support and escalated the issue of no system support within the medical equipment replacement programme. Those responsible for the medical equipment replacement programme use several factors in determining the expected replacement date of equipment. These are as follows:

- Expected or actual, End-of-Life (EOL) date from the manufacturer or supplier. For larger and more expensive medical equipment, this is usually available from the manufacturer and can be recorded as part of the MEMS database.
- Technical life. This is a clinical engineering estimation based upon usage and environment. For example, small handheld equipment is more likely to suffer damage than desktop type equipment, which in turn is more likely to need replacing before a major fixed installation does. Sometimes, the technical life can be far in excess of the EOL date through use of third-party expertise and equivalent spare parts, whilst at other times the technical life is reduced by damage in use.
- Current condition. Simply a good, satisfactory or poor summary by clinical engineers.
- Current level of reported faults. An increase in faults is usually an indication of unreliability.

Some of these are more variable than others but none take into account the actual experience of the clinical user in prioritizing replacement. To include this real-life experience, a methodology was created that was based on an extended 5 × 5 risk scoring method as shown in Table CS5.8A.

The clinical users are provided with a copy of the scoring method and asked to complete their two parts, considering the impact factor ('I') and the clinical consequences factor ('C') that each has the highest effect, and then enter the corresponding scores.

The clinical impact factors ('I') are given under the following headings:

- Confidence: Are there performance issues with the equipment? Is there an issue surrounding how the equipment is expected to operate? Will it work today, or will it not? Does it need frequent resetting or power cycling for example?
- Quality: Is the equipment producing usable data, results, treatments, etc., on a regular basis? Are there issues, for example, with tests having to be re-run to verify results?
- Efficiency: The equipment might operate and produce the results correctly and reliably but runs so slowly that it is inefficient compared to alternative means.
- Compliance with professional recommendations: Some professions such as ultrasound imaging services have provided guidance for when they think equipment should be replaced to ensure appropriate technology continues to be used.

They can then judge and score the clinical consequences factors ('C').

These two scores are multiplied to give a score between 1 and 25. This is then multiplied by the technical score allocated by clinical engineering to give a final score between 1 and 125. It is this final score that is used to allocate priority for replacement.

The clinical users were delighted to be able to have an input into the replacement process that did not rely solely upon measurable and quantifiable data alone but rather allowed for softer input.

ADDING VALUE

It costs no more to use a systematic approach to planning equipment replacement, but the value is increased because all parties benefit.

Benefits : Cost ∴ Value

TABLE CS5.8A   Priority Matrix

Quality and Performance Risk Matrix

Clinical Impact I

		1	2	3	4	5
Confidence		No issues	Occasional minor concerns but manageable	Occasional issues giving concern	Regular issues giving concern or requiring intervention	Regular (weekly or greater) performance issues
Quality		No issues	Occasional minor concerns but manageable	Infrequent/minor issues with data/results/tests	Regular queries on, or poor quality of data/results/tests	No confidence in data/results/tests; unusable outputs
Efficiency		No issues	Occasional downtime but service unaffected	Infrequent, inconvenient breaks in service	Regular or prolonged breaks in service	Downtime unacceptable; unable to provide service
Compliance		Within compliance recommendations	At limit of compliance recommendations	>1 year out of compliance recommendations	>2 years out of compliance recommendations	>3 years out of compliance recommendations
		1	2	3	4	5
Clinical Consequence C	Minimal injury requiring no/minimal intervention or treatment — 1	1	2	3	4	5
	Minor injury or illness requiring minor intervention; increase in length of hospital stay by 1–3 days — 2	2	4	6	8	10
	Moderate injury requiring professional intervention; increase in length of hospital stay by 4–15 days; an event which impacts on a small number of patients — 3	3	6	9	12	15
	Major injury leading to long-term incapacity disability; increase in length of hospital stay by >15 days; mismanagement of patient care with long-term effects — 4	4	8	12	16	20
	Incident leading to death; multiple permanent injuries or irreversible health effects; an event which impacts on a large number of patients — 5	5	10	15	20	25
	Impact on the safety of patients staff or public					

Performance P

	Planned replacement	Condition and support by company	Reliability
1	≥ Year of estimated replacement	Good condition and supported	0 reported faults per annum
2	Replacement >1 year overdue	Satisfactory condition and supported	1–3 reported faults per annum
3	Replacement >2 years overdue	Good/satisfactory condition but not supported	4–5 reported faults per annum
4	Replacement >3 years overdue	Poor condition but Supported	6–7 reported faults per annum
5	Replacement >5 years overdue	Poor condition and not supported	≥8 reported faults per annum

X

Source: This table is copyright Paul Blackett; used with permission.

SYSTEMS APPROACH

Having a logical approach to equipment replacement is essential in order to prioritize equipment. The approach taken that includes data from multiple sources, including the real-life quality aspect of the equipment in use, uniquely allows the position of hard and soft data to be seen and included (separately if need be) in the replacement planning.

SUMMARY

This case study shows the importance of taking into account the experiences of medical equipment users and not to solely rely on the data obtained from manufacturers and servicing organizations. In fact the operation and clinical consequences of the equipment are of paramount importance, influencing as they do the users' confidence in its operation, reliability and outputs. Perception as to how well equipment is operating influences care, in the extremes through lack of trust and confidence in equipment regarded as unreliable.

The measurable technical data are obtained and scored to produce a technical ranking but, when multiplied by the users' quality aspects, enable further stratification and finer detail for replacement prioritization.

SELF-DIRECTED LEARNING

1. Is it reasonable to take into account the possibly subjective, user opinion of the operation of medical equipment in the planning of its replacement? Is it not preferable to leave it to purely technical scores? Discuss.
2. Do you think this approach leaves the replacement programme open to abuse or manipulation? Think about the drivers and pressures in replacement planning.

CASE STUDY CS5.9: IMPLICATIONS WHEN PROCURING ADDITIONAL MEDICAL EQUIPMENT

Section Links: Chapter 5, Section 5.6.4; Chapter 7, Section 7.2.3

ABSTRACT

The procurement of additional medical equipment may require additional resources including staff, premises, utilities (electric power, medical gases), general support services and consumables and accessories required directly by the equipment. The implications of procuring an additional CT scanner are described.

Keywords: Procurement; Resources; Teamwork; Equipment life cycle; Finance; Procurement costs; Operating costs; Clinical guidelines; Clinical efficacy

NARRATIVE

CT scanning demands in a 600-bed general hospital had steadily increased, resulting in calls by radiologists, supported by surgeons and Emergency Department consultants, for a second CT scanner to cope with the rising clinical demands. The single CT scanner was not able to meet current demands, with costly (financial and reputational) transfer of patients to neighbouring hospitals for scans.

An outline business case summarized the anticipated costs (acquisition and operating) offset against the anticipated benefits (improved patient care and income from the additional CT scanning time). It summarized the option of doing nothing, the inability to meet current and rising demands and the resultant costs of buying CT scans from neighbouring healthcare organizations. The hospital's Board approved the acquisition in principle, provided that appropriate space could be found and all the increased operating costs (staff, facilities, consumables and the CT scanner itself) could be identified and justified against the increased revenue anticipated from an additional CT scanner.

Radiology had budgetary responsibility for all costs except the CT scanner itself for which Clinical Engineering Department (CED) had responsibility (procurement costs and ongoing maintenance). Consequently, the heads of radiology and CED were asked to lead a procurement project. They formed a multidisciplinary team comprising radiologists and radiographers, a consultant medical physicist (primary responsibility; specification, evaluation and selection of the CT scanner), Facilities, general hospital management, hospital hotel services, health and safety, infection control, fire safety officer, communications, patient liaison officer and clinical engineering.

Resource implications of an additional CT scanner are significant. Space and accommodation are required together with additional staff, radiologists, radiographers, nurses and support staff. Annual consumable costs including injector dyes were estimated at $100,000. Additional electrical supply costs as well as air-conditioning costs would be required as would housekeeping and cleaning costs.

Clinical engineering would require an estimated $60,000 for a comprehensive maintenance contract and radiation protection charges but would not require additional personnel. Finance offered to fund, from accrued operating surpluses, the acquisition costs for the CT scanner and its accommodation build costs, but would require radiology, through its income stream from the additional scanner, to fund the ongoing operating costs and an additional $100,000 per year depreciation costs to ensure funding was available for its future replacement.

The head of Radiology developed, in consultation with radiologists and radiographers, a bold accommodation plan, redesigning CT around a twin facility with common control room, patient reception and waiting areas. It made better use of existing space but would require an extension of radiology's outer boundary into landscaped and parking areas. The proposal had been discussed with the hospital's patient liaison officer who was delighted to hear of the proposed improvements to the CT patient areas and the prospect of not requiring regular transfer of patients for scans. Facilities supported the proposal and helped gain approval from the Board at the full business case stage.

Implementing the approved proposal required structured teamwork that was organized around fortnightly meetings of the multidisciplinary team, with each team member taking responsibility for specific elements of the project. For example, Facilities arranged local authority planning approval for the hospital building changes. Radiology, Communications, Facilities and clinical engineering planned the build construction phasing to minimize disruption to patient care within Radiology and adjacent areas.

Building started by extending the department's outer wall to accommodate the additional CT scanner and shared control room which enabled the additional CT scanner to be installed and commissioned before embarking on changes to the existing radiology department. A specialist imaging build contractor was employed, with Infection Control guiding the process to ensure that dust and other hazards from the build work would be contained within the build site.

Whilst Facilities managed the build work, the CT scanner selection panel chose the preferred CT scanner, with clinical engineering negotiating favourable acquisition and maintenance costs for the eventual replacement of the existing scanner. Staff recruitment challenges were a major hurdle, exacerbated by national shortages of available skilled radiology staff. The business case envisaged a phased opening, starting with two staffed-CT scanners available for 40 h per week and one for the rest of the week. This met current clinical demands and generated sufficient income to cover the increased costs; it would be reviewed annually with a view to full operation within 2 years.

The fortnightly team meetings, with update reports from the leads of the various elements, kept the team focused on the objective whilst allowing each to share aspects of concern in their areas of responsibility, adding to the project's team spirit. Problems that the team could not resolve were investigated by the joint leads, working with senior management and clinicians to find solutions. Communications proved vital in keeping the hospital informed of developments.

ADDING VALUE

Additional medical equipment procurements have operational resource implications which must be carefully investigated. No one hospital department has knowledge of all aspects. Consequently, multidisciplinary teams with technical, clinical, procurement and financial elements must work together, mindful of the improving patient care objective. Clinical Engineering can add value by guiding the project team.

Benefits : Cost ∴ Value

SUMMARY

Medical equipment procurements have resource implications. Recognizing that no one group or department has all the answers, clinical engineering can lead multidisciplinary project teams of clinicians, finance and the organization's Board to identify the resource implications and develop robust plans for meeting them, keeping focus on patient care and the strategic aims of the organization.

SELF-DIRECTED LEARNING

1. Teamwork is required when planning and implementing medical equipment installations. What benefits are ensured by including multidisciplinary staff such as Infection Control, Fire Officer, or Communications in major projects such as described here? What are the implications of not including some or all of them?
2. You are asked to guide the increase in number of endoscopy suites. Describe how you tackle the task.

CASE STUDY CS5.10: VALIDATING A DATA INTERFACE BETWEEN TWO CLINICAL SYSTEMS

Section Links: Chapter 5, Section 5.6.4; Chapter 7, Section 7.3.4

ABSTRACT

As part of the commissioning of a new ICU clinical information system, a clinical engineer was tasked with performing a risk management exercise to assure a key interface asset. The approach taken added value by developing a more rigorous testing approach than that suggested by the supplier and by examining the way the people in the organization communicated around the management of the interface.

Keywords: Interface validation; Black Box testing; Grey Box testing; IEC/ISO 80001

NARRATIVE

A new clinical information system was to be installed in the ICU. This system was to integrate data from medical equipment such as monitors and ventilators as well as data from other clinical systems including the radiology information system and the laboratory results reporting system. A component of the system is the interface engine which connects the lab system to the ICU clinical information system. During commissioning and in the testing phase, before the new interface was put online, the supplier of the new interface engine was able to demonstrate that test results entering the interface from the lab system were being appropriately passed through the interface to the clinical ICU system. Some lab results are communicated along with acceptable reference ranges for that result. These reference ranges change over time. Where a result is outside of the communicated reference range, the result should be presented as highlighted for review in the end system, the clinical information system, for the doctor's review.

The clinical engineer who contributed to the management of the clinical information system in the ICU was also the system's Risk Manager. Whilst confirming that black box testing of the new interface indicated it was working, the clinical engineer felt that further work was required before the interface was put online. As a black box tester, the clinical engineer was unaware of the internal structure of the interface being tested. The clinical engineer judged that whilst the testing indicated that the interface was working for the test data set, it did little to test whether it would work for a real-world data set which changes over time.

To add to the confidence in this critical system, the clinical engineer undertook a validation of the interface. This required the suppliers of the new interface, the existing lab system and the clinical engineer to sit together and document the code in the interface that passed and restructured the lab data stream into the ICU system-ready data stream. During this process the emphasis shifted from 'is it working' to identifying 'how could this go wrong'. Documentation of the interface did not identify any failings, but did identify the need to test how the interface handled results that were outside of their associated reference range, a function not checked during the black box testing. A number of test data sets were created. Each of them tested all possible combinations of results being, higher, within and lower than a particular reference range. Different data sets used different reference ranges. These data were run through the interface and the data checked by the clinical engineer at a number of different points in the process, to assure the algorithms in the interface engine were manipulating the data as expected.

ADDING VALUE

The exercise resulted in the hospital-based clinical engineer having a much more developed understanding of the interface and a greater ability to independently assure it. The exercise did not highlight any technical deficiency in the interface; however, it did highlight an otherwise unidentified weakness. The testing required the clinical engineer to go to the lab to verify the data stream leaving the lab system and generate test data. In conversation with the lab system manager, it became obvious that the lab staff were unaware of the fact that the data from their system were being used within another system. The lab staff subsequently changed the communication log for the lab system to include the ICU system staff so that any planned upgrade or downtime would, in future, be notified to ICU. The validation did identify a weakness in the culture of how distributed systems are managed by the people in the organization, with people working in silos whilst data were being distributed across systems.

The clinical engineer was tasked with being the system's risk manager so no extra cost was involved in taking the systematic approach described. This resulted in greater assurance of the robustness of the whole system to the benefit of patients and clinicians.

Benefits : Cost ∴ Value

SYSTEMS APPROACH

When the clinical engineer was tasked with validating the interface, they researched current thinking on testing software and rightly identified Grey Box testing as a useful approach. Grey Box testing was beneficial in this case as it combines the straightforward technique of Black Box testing with the presentation of the range of conditions the interface was likely to encounter and, in doing so, rendered the interface understandable and allowed the engineer to verify it. The clinical engineer also referred to the IEC/ISO 800001 suite of standards which are concerned with risk management of Medical IT Networks. It was familiarity with these Standards that prompted the clinical engineer to also look for potential failing in the social aspects of managing sociotechnical systems.

SUMMARY

The involvement of a clinical engineer with appropriate expertise in the deployment of an ICU clinical information system led to a more rigorous risk assessment and testing protocol when another remote IT system was to be connected. This went further than the original testing protocol and drew out additional potential risks which were then resolved.

SELF-DIRECTED LEARNING

1. Are you familiar with the concepts of Black Box and Grey Box testing of software systems? If not, investigate these.
2. Can you identify a process or system in your own work which is tested using a Black Box approach? Briefly discuss whether there would be merit in testing the process or system you identified using a Grey Box approach, and explain why?
3. In relation to the situation described in this case study, in your opinion which of the following was more like happen if the clinical engineer had not validated the system:
   a. The interface would have passed data but introduced an error.
   b. The interface would have passed data but not identified it was outside the reference range.
   c. The lab staff would have turned off the data stream going to the ICU for system upgrades without warning the ICU staff.

Justify your answer with a brief explanation of why you chose the answer you did.

CASE STUDY CS5.11: PROCUREMENT: TENDER EVALUATION CRITERIA, JUDGING TENDERS, EVALUATION OF THE USABILITY OF ALTERNATIVES

Section Links: Chapter 5, Sections 5.6.4

ABSTRACT

A replacement fleet of vital signs monitors was to be purchased. The weighting of various factors such as cost, usability and manufacturer support had to be decided as part of the tender process. A method of assessing user preferences had to be devised.

Keywords: Tender specification; Evaluation criteria; User evaluation; Human usability

NARRATIVE

Our healthcare organization decided to replace its aging fleet of vital signs monitors consisting of a variety of makes and models with a single make and a model (though some would have additional functionality) to be used in all clinical areas.

The benefits foreseen were the following: up-to-date equipment using better algorithms and with improved reliability, standardization of equipment facilitating clinical user training, flexibility in use because equipment moved from area to area would be familiar to all, selection to prioritize human usability leading to less use errors, lower purchase cost from bulk purchase discount and financial accountancy benefits because the fleet replacement would allow the expenditure to be capitalized which suited the organization's financial position.

Outright purchase was decided on for two main reasons: the equipment does not have any associated single-use disposables so there was no prospect of a 'consumables deal' as described in Chapter 4, Section 4.3.2; the tender specification included a requirement for further one-off purchases of additional equipment at the agreed price over the following 5 years. Furthermore, the Clinical Engineering Department (CED) was experienced and able to support equipment of this general type, with in-house support enhanced by standardization; standardization makes an equipment loan service model of management of the equipment in clinical areas possible.

CED convened a project team which included clinical users, Procurement and Finance staff to draw up a manufacturer neutral tender specification. One key decision to be made was the weighting to be given to the various factors against which tender bids would be judged. The following were agreed:

- Cost – 35%
- Usability, judged as described below – 25%
- Cost of accessories such as NIBP cuffs and SpO_2 probes which have a lifetime less than the equipment – 20%
- Supplier provided training, clinical and technical – 15%
- CED judgement on build quality including that of the accessories and on after-sales support – 5%.

Because the project highlighted the importance of the equipment's usability, user evaluation was very important. The option of placing examples of each manufacturer's devices in various wards requires significant resources for training nursing staff to competently use the equipment and was thus considered not an appropriate method. Over ten responses to the tender were received. Consequently, it was decided to conduct an initial short list of the tender returns, evaluating the initial responses without including the usability criteria. This left a short list of five.

The CED then organized a demonstration and evaluation of the devices on the short list; demonstrations during week one followed by evaluations during week two. Each manufacturer presented a series of demonstrations during the course of a day, a different day for each manufacturer. For the evaluations, two devices of each type were available in the same evaluation room for users to handle, use and compare, with a questionnaire, drawn up by CED, for recording the evaluation results including human usability. A clinical engineer or clinical member of the project team was present at all times during the open access to the equipment to observe and provide assistance when necessary. The programme was widely publicized to clinical staff and was also attended by experienced clinical engineers.

The project team met after the evaluation exercise to analyze the questionnaires and discuss observations. From this a usability criteria score was agreed for each device.

Table CS5.11A shows the unweighted and Table CS5.11B the weighted scores.

TABLE CS5.11A   Summary of Evaluation Scores (Max 100) for Each Element

Supplier	Cost	Human Usability	Accessory Costs	Training	CED Judgement	Total Score
A	50	85	95	95	95	420
B	100	25	100	95	95	415
C	98	74	97	100	100	469
D	95	93	95	90	90	463
E	98	84	91	80	75	428

TABLE CS5.11B   Summary of Weighted Evaluation Scores for Each Element

Supplier	Cost	Human Usability	Accessory Costs	Training	CED Judgement	Total Score
Weighting factor %	35	25	20	15	5	100
A	18	21	19	14	4.8	77
B	35	6	20	14	4.8	80
C	34	19	19	15	5.0	92
D	33	23	19	14	4.5	94
E	34	21	18	12	3.8	89

The tables highlight the importance of considering all the factors, with due consideration to the weight of each. Based on cost alone, product B would be selected, with the lowest acquisition and accessory costs. However, its human usability was very poor. Users commented that basic functions such as measuring the blood pressure was very difficult, perhaps even dangerous as the machine always initially inflated the cuff to 200 mmHg; this could not be changed and would be very painful for children, leading to possible bruising. The alarm settings could not be configured and were always ON, leading to numerous false alarms when tested by the focus group. Consequently, 'B' scored very low on Human Usability.

The non-weighted score put product C slightly ahead (Table CS5.11A), but when the weighted criteria were included, product D came first, largely because of its best Human Usability. Clinical engineering agreed that their slight reservations were outweighed by the usability factors.

Finance accepted that medical equipment should not be judged solely on cost and were satisfied that product D's roughly 5% higher cost was justified by its overall clinical performance. The users were delighted that the equipment's usability was taken into account.

ADDING VALUE

With any procurement project, obtaining best value is important, but when replacing a whole fleet of widely used equipment, it is vital that value is not judged solely on upfront price. All costs including accessories and support must be taken into consideration. Consideration must also be given to ease of use with lower likelihood of use errors. Rigorously assessing and weighting each criterion contributes to obtaining best value.

Whilst the cost of the preferred device was not the lowest, the bulk purchase project yielded an acceptable overall price, and one lower than would have been obtained by multiple purchases of smaller numbers from individual budgets. The coordinated project realized an improved device with good ease of use.

Benefits : Cost ∴ Value

SYSTEMS APPROACH

Clinical engineering was able to put input into setting up and organizing the demonstrations and the availability of the equipment for comparison. This process was an example of a systems approach to achieving the goal of establishing an appropriate measure of the usability of the equipment under consideration.

SUMMARY

Usability of medical equipment is a key factor in its safety, evidenced by its inclusion in the IEC medical electrical equipment safety Standard IEC 60601-1. Comparing this and other factors of similar devices as part of a procurement exercise is difficult. It is not always practical to put candidate equipment into clinical use in an ethically acceptable way because of the time and effort required for pre-evaluation user training. The 'focus group' approach described is an alternative method of gathering useful opinion. Procurement projects and tender evaluation criteria must include input from the intended clinical users as well as from the CED.

SELF-DIRECTED LEARNING

1. Consider alternative methods of obtaining a range of clinical user opinion of the usability of equipment proposed for purchase.
2. What contribution can a clinical engineering input make to judging usability?
3. What impact can the choosing of the selection criteria make to the medical equipment chosen? Consider different selection criteria, and how the type of medical equipment might affect determining the criteria and the weight applied to each. How should the CED determine criteria weighting?

CASE STUDY CS5.12: ESCALATING IDENTIFIED RISKS ASSOCIATED WITH MEDICAL EQUIPMENT TO THE MDC

Section Links: Chapter 5, Section 5.9.3; Chapter 7, Section 7.4.3

ABSTRACT

Maintenance on a range of newly delivered medical devices cannot be put into place because they have not been identified sufficiently accurately to enable this to happen. What are the risks and who owns them?

Keywords: Risk; MDC; Ownership; Governance

NARRATIVE

The Clinical Engineering Department (CED) was given the responsibility of organizing and managing the contract maintenance of a new range of electrically powered surgical tools and accessories that had been delivered direct to the Sterile Services Department (SSD), bypassing the CED, and had now been in use for many months. The project aim was to implement regular servicing of these as soon as possible. However, as with all surgical tools, they spend much of their lives stored in a sterile environment and only emerge from it when put into use. Identification of the equipment was going to be an issue. The project team which was established with the Sterile Services Department (SSD) identified several barriers:

- The time necessary to catalogue all the equipment was going to be significant.
- The impact on availability of equipment might be an issue if it was being sterilized after inspection and needed urgently.
- The cost of an additional sterilization cycle would be a factor.

The barriers worked against the team's main objective and this was considered a risk. Options to take the project forward were investigated and evaluated. Each had varying levels of benefits and risks and these are listed in the following text:

- Request Operating Theatre staff to record contents when opening the sterile trays in order to use them.
  - o Advantages: Positive identification, no additional sterilization cycle, accurate costing.
  - o Disadvantages: A slow process with no guarantee that all trays would be used/cycled quickly. They could be on a shelf for up to a year. Maintenance delayed until completion of survey.
- Attack the situation with as many people as possible. Open all trays, record and re-sterilize.
  - o Advantages: Quick result, accurate costing, complete inventory check. Maintenance implemented quickly.
  - o Disadvantages: SSD may not be able to process quickly enough to avoid impact on clinical services, additional cycles at cost and loss of surgical time.
- Request the maintenance company to service all equipment over a fixed timescale, identifying equipment and invoicing afterwards.
  - o Advantages: Sensible timescale, accurate check, minimal delay for maintenance to be completed.
  - o Disadvantages: Unknown financial outlay, additional sterilization cycle (but avoids one after maintenance).

The options were taken to the Operating Theatre management for financial authorization to proceed. The first option was chosen as this offered the most cost-effective method of checking the inventory. However, the risks involved were higher for this option as it delayed addressing the overall aim of dealing with the regular service due on the surgical equipment.

A risk assessment was undertaken jointly between the Operating Theatre management and Clinical Engineering. This assessment was tabled at one of the regular Medical Device Committee meetings and was discussed with clinical and technical colleagues present. Having been assured that the risk was correctly identified and scored, it was passed on for inclusion in the hospital's Risk Register. The risks revolved around the delays in completing necessary servicing whilst equipment was catalogued, and the potential risks to patient safety because of it.

The risk was reviewed at each meeting until such time as the inventory check had been completed and the equipment had been serviced.

ADDING VALUE

The starting point of this project was a procurement process and maintenance situation that was not well addressed and therefore put the organization at un-assessed risk. The agreed course of action clarified the situation but had an associated cost and a known risk. Therefore, there was benefit in carrying out the project, but the potential value is deferred until the maintenance can be put in place.

$$\uparrow \quad \uparrow$$

Benefits : Cost ∴ Value

SYSTEMS APPROACH

A strong and clear route for all members of healthcare staff should exist for the reporting, escalation and recording of risks. Risk assessments are to be encouraged as a means for the organization to understand its risks, and it can then decide if reallocation of resources should take place. Review is key, as a stagnant equipment register is of no use; evidence of review needs to be included.

CULTURE AND ETHICS

The demands placed upon individuals, teams and departments sometimes place them in a moral dilemma. In this case, the team had clear aims to get the equipment serviced but the decision was taken, because of the uncertain financial commitment, to catalogue them first.

SUMMARY

This case study shows how the demands placed upon a project can conflict; money versus completion and service. The risks that were identified as a consequence were escalated to the Medical Device Committee, making the organization aware and, importantly, allowing it to decide whether it accepted that risk or should take an alternative course of action. The lesson to be learned here is one of corporate responsibility. If a risk is identified but not flagged up to those who need to know, the responsibility will remain with that individual or department.

SELF-DIRECTED LEARNING

1. Which option would have been better if funding for maintenance had not been an issue? In what other ways would the options impact the clinical users and support departments?
2. Can you think of any other methods that could have been considered to manage the project?
3. Is clinical engineering best placed to manage this equipment?

CHAPTER 6

Developing Equipment Support Plans in the Context of the HTM Programme

CONTENTS

6.1 INTRODUCTION

Chapter 4 described a healthcare technology management (HTM) system consisting of two interlocking processes, namely the strategic planning and the delivery and implementation process as summarized in Figure 4.12. Chapter 5 discussed the strategic planning process. In this chapter, we will describe the operational process which an organization uses to implement its strategic equipment management plans for healthcare technology. As discussed in Chapter 5, the Medical Device Committee (MDC) is the construct that facilitates the strategic and operational interaction. In this chapter, we will describe how a Clinical Engineering

Department (CED) charged by the organization's senior management with managing medical equipment through the working of the MDC goes about turning policy into action.

A key concept in this chapter is that of the 'equipment support plan' (ESP). It will be used to describe the proposed actions that are planned for a particular group of medical equipment. Given that the HTM system as a whole is to be broad in its scope, it should be no surprise that ESPs go beyond technical considerations and include procedures for ensuring the equipment is used optimally and in a way that supports person-centred care and the goals of the organization. The ESP sets out what needs to be put in place from the asset management perspective to unlock the value of the equipment.

The organization will have many ESPs, each tailored for a different group of equipment. There might be an Anaesthetic Workstation ESP or a Defibrillator ESP or a Dialysis Machine ESP, for example, each developed and delivered to meet the device-specific technical and clinical requirements of that equipment group, recognizing the clinical context in which the equipment is used. The ESP is where the objectives set by the hospital's Medical Device Policy (Chapter 5, Section 5.3) become realized in a way that is appropriate and specific to that equipment group. The Medical Device Policy and the Strategic HTM Plan (Section 5.4) discussed in Chapter 5 are pan-hospital and strategic. The ESPs are particular to the equipment they address, and support their day-to-day use, and so could be characterized as operational.

This chapter will clarify and describe the ESP in more detail. However, it is perhaps worth summarizing key fundamentals of the ESP:

- The equipment support plan is a package of procedures whose goal is to ensure the safe and effective application of medical equipment for the benefit of patients.

- The ESP recognizes the shared role of technical and clinical staff and, particularly in community and home situations, of the patient and carer in achieving the goal of safe and effective medical equipment.

- Following on from this second principle, the shared responsibility between technical, clinical and patient groups is recognized in developing the ESP, albeit it is anticipated that the clinical engineer will be expected to lead the process.

- The ESP is not a static support plan, but one whose effectiveness and applicability will be checked and evaluated, and where appropriate, refined and improved.

We express explicitly the need for dialogue in the development of the ESP, but you should remember that this is not a new concept. Standard Operating Procedures (SOPs) for the routine maintenance of defibrillators have often incorporated aspects of clinical user maintenance, for example, check that devices are kept on charge and, for manual defibrillators, perhaps carry out daily discharge checks which both check functionality and ensure familiarity with defibrillator use. Similarly, clinical engineers expect theatre staff to check operating-time clocks on ventilators to ensure adherence to routine scheduled inspection programmes and, more actively, to check and clean air filters on equipment such as capnography monitors.

The ESP builds on processes that clinical engineers have long followed in caring for and supporting the medical equipment they are responsible for. In simple terms, the ESP provides a coherent structure for the various procedures. Clinical engineers will be aware of Standard Operating Procedures (SOPs), and we have briefly discussed them in Chapter 3, Section 3.4.2. A brief summary of an ESP is that it is a package, a collection, of SOPs relevant to a particular equipment type or group of equipment types. We will see that the ESP will include procedures for quality and verification checks and for scheduled and unscheduled maintenance and for helping ensure the safe and effective use of the equipment. We will keep in mind the need for dialogue between technical, clinical and user groups in the development and evaluation of the ESP. It is not the intention that the ESP should reproduce material that already exists within SOPs or manufacturers' service manuals, but the ESP may take the form of a high-level document that brings together SOPs.

We will discuss the development and delivery of the ESPs based on the life cycle management approach discussed in Chapter 4. Each ESP should be tailored to ensure that it delivers optimum value for its particular group of equipment, and so it will be no surprise that we will take a holistic view of how the equipment should be managed, to deliver benefits in a way that also controls cost.

A Clinical Engineering Department will develop and deliver a number of ESPs all running concurrently. Together, they make up the complete healthcare technology management activity delivered by the department. We will use the term 'Healthcare Technology Management Programme' (HTM Programme) to describe this collection of ESPs, owned and delivered by a single department.

Many of the routine services delivered by the CED are often invisible at Chief Executive Officer level and those that report to that level. A formal HTM Programme with its ESPs for the individual equipment types can provide a readily understood process of documented support that can be promoted at executive level and used as an argument for the necessary resources to carry out this work. ESPs also constitute a detailed part of a quality management system (QMS).

Dependent on a healthcare organization's size and structure, the medical equipment may be managed by a single or by several different departments. Where, as is often the case, the responsibility is divided between different departments, the division will typically reflect the nature of specialized equipment. Thus radiology and radiotherapy equipment may be managed by Medical Physics; rehabilitation equipment by Rehabilitation Engineering; renal dialysis equipment by specialized Dialysis Technical Services working within Renal Units; pathology, biochemistry and other clinical laboratory equipment by the Laboratories themselves; or sundry medical equipment including beds, patient-lifting devices and weighing scales looked after by Facilities. As discussed earlier, in this book, we use the term Clinical Engineering Department (CED) to describe the department managing medical equipment, whilst recognizing that different terms are also used (e.g. Medical Engineering). Whatever the term used to name the department and whether medical equipment is managed by one or more departments, the principles outlined here are applicable.

Where multiple departments are charged with managing different groups of the medical equipment inventory, each will develop ESPs focused on supporting the equipment for which they have responsibility. The activities of each of the departments will need to be coordinated and, as discussed earlier, we are using the term Medical Device Committee (MDC) to describe the corporate body that acts to assign responsibility and authority for management of specific groups of equipment to particular departments. The MDC also reviews the performance of the HTM Programmes delivered by different departments to ensure each meets the hospital's Medical Device Policy. In doing so, the MDC takes an organization-wide view. However, within each department, the responsibility and authority for each ESP are assigned and their performance measured and controlled within the department. Like their design and delivery, the measures used to assess the performance of a particular ESP are tailored to the equipment being supported.

In this chapter, we will be looking at how an individual department devises, develops and delivers an HTM Programme, made up of ESPs running concurrently. The discussion is based around what happens within a Clinical Engineering Department. Of course this is by way of example, and the processes described apply equally to any other departments who deliver HTM Programmes. Figure 6.1 illustrates the system.

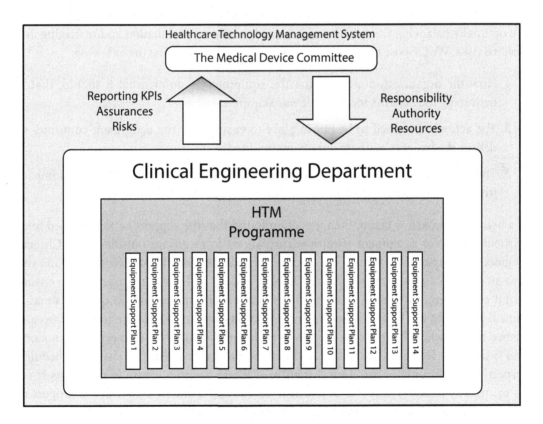

FIGURE 6.1 The HTM Programme consisting of concurrent ESPs, developed and managed by a Clinical Engineering Department.

6.2 ESP

6.2.1 Holistic ESP Planning Which Is Patient Focused

Despite emphasizing the key principle of shared technical, clinical and patient roles in equipment support in the introduction to this chapter, the technical content of an ESP could lead the clinical engineer to fall into the trap of thinking of it as purely a technical maintenance plan for medical equipment. Traditionally, technical maintenance was the mainstay of and remains the most visible activity undertaken by a Clinical Engineering Department. Unfortunately, many of the more complex and valuable support and advice activities delivered by clinical engineers are invisible to those who visit clinical engineering departments. The existence of physical workshops, test equipment, service manuals, etc., adds to the impression that equipment support is only concerned with technical maintenance. However, to ensure that a medical equipment asset is used appropriately, safely and effectively requires planning and actions that involve a wide range of people, including those who use the equipment. A holistic ESP will include the traditional technical maintenance activity but extended further to include those aspects of the support for the use of the equipment that ensure its benefits are delivered. Such an approach requires us to look at the use of the device in a broader context and consider and clarify what needs to be done, who needs to do it, what resources and training do they need and how to plan and action all of those mentioned. Given the resource constraints we all have, this analysis needs to be done within a businesslike framework, to ensure that we are optimally balancing the benefits and cost side of the value equation and managing any residual risks. We propose that each ESP should aim to clearly define the following:

1. How the organization assures that the equipment is doing what it should, that is delivering the benefits for which it was acquired.

2. The actions that need to be put in place to ensure that the equipment continues to deliver the benefits, with its quality maintained.

3. The preventive actions that should be put in place to reduce the likelihood and impact (the risk) associated with unforeseen failures of the medical equipment.

If a holistic approach is taken, then it is often found that the support actions needed for a particular piece of equipment require participations from groups outside of the Clinical Engineering Department (CED). Let us consider a defibrillator located in an ICU and use it as an example to illustrate the point. Clearly the CED will put in place actions to ensure that it continues to deliver the benefits for which it was acquired and its quality is maintained. No doubt the CED will set up an inspection schedule where the technical performance of the defibrillator is checked. Such a plan goes some way to meeting points one and two earlier. However, given the resources and demands on the CED, such a scheduled inspection might only occur every 3 months. Perhaps another schedule of actions is set up to ensure that every year the battery will be changed. A holistic view of the support of the defibrillator would also identify that the nurses in the unit perform a daily check of the defibrillator, including discharging it into its internal load as part of the manufacturer's self-check process. So on a daily basis, the nurses are also assuring that the device is doing what it should and that it is capable of delivering the benefits for which the equipment was acquired. Now, it could be said that the checks done by the nurses are not of interest to the Clinical Engineering Department, and it does not need to concern itself with what the nurses do, other than ensure if they find something untoward they know to call clinical engineering. However, it is not that simple when you are taking a holistic view of equipment support. The adequacy of the clinical engineering plan for performing detailed technical checks every three months is only appropriate within the context of the nurses performing the daily checks. Imagine how the level of confidence in the quality of the defibrillator would change if the nurses stopped doing the daily checks, and the quality was only assured by the clinical engineering three monthly check. Confidence would decrease significantly. The point is that the adequacy of the Clinical Engineering Department's contribution to the support of the equipment has to be reviewed in a wider context. Continuing the example, the Clinical Engineering Department might go further and contract with the equipment supplier's technical support department to ask them to perform an independent assessment of the defibrillator once a year and at the same time perform any software upgrades. Whilst this increases the cost of the defibrillator support plan, it makes it more robust as it brings in a level of independent assessment of the work of the Clinical Engineering Department and brings added benefits associated with the adoption of new software. So in this high-level example, we quickly identify that three groups are involved in the defibrillator support plan, the clinical users, the Clinical Engineering Department and the supplier's technical support department. This example also emphasizes the important principle that the responsibility for the care and the maintenance of medical equipment is a shared responsibility of clinical and technical teams, not simply delegated to the technical department.

You would think that the actions that should be put in place to reduce the impact or risk associated with unforeseen failures of the defibrillator would more closely align with the Clinical Engineering Department. However, once again the users and supplier have a role. Where a defibrillator fails during a routine daily check by the nurse, it is important to have processes in place to ensure that this is reported to the Clinical Engineering Department in a timely manner. So the support plan should identify this and ensure that these processes are understood and effective. In this example, we imagine that the clinical engineers will undertake the repair, but this may entail purchasing parts or expert service support from the supplying company. Again all three groups could be involved. Yet there needs to be clarity as to who does what and how decisions are made. The governance needs to be clear as to who has responsibility and authority to call in the supplier's technical support. The ESP might include a provision to ensure that a loan defibrillator is made available to the ICU during the time the repair is being affected, to reduce the impact of the unforeseen failure. So the availability of loan devices, held and managed by the Clinical Engineering Department for such an eventuality, would need to be planned and resourced.

In the event of the device failing in clinical use, there is the added issue of how the support plan identifies the appropriate management of the consequences of that failure. Such a failure could result in an incident which might impact seriously on patient care and could result in patient injury. If such an unfortunate incident occurs, the plan for dealing with that failure might differ given the need for there to be some investigation of the incident and the contribution the technical performance of the defibrillators made to that incident. If the incident became the subject of a formal review, the ESP would no doubt be reviewed to ensure it was appropriate for the device and its intended function, so the plan itself should be documented and there should be evidence of it being actioned. The log of the nurse's daily check, the documented results of the clinical engineering three month inspection and the documented annual independent check by the supplier technical department would all be reviewed.

This example highlights that taking a hospital-wide holistic approach to defining an ESP requires leadership by the Clinical Engineering Department whilst acknowledging that many others outside that department play a part. Keeping the focus on how the organization as a whole works in concert to support the application of the device is important. Ultimately the guiding principle for any ESP is to ensure patient care is optimal, and whilst this obviously involves technical considerations, the 'technical' in isolation should not drive the process. Case Study CS6.1 illustrates a further example of how holistic working can be beneficial to clinical engineering, clinical users and patients.

In summary, developing a holistic ESP is in effect developing a system aimed at providing optimal value (in its widest meaning) out of the equipment in question. It is therefore very relevant to keep in mind the principles of a systems engineering methodology as set out in Chapter 2, Section 2.3:

- Define the system's objective.

- Identify the system's constituent elements.

- Identify the relationships between constituent elements.

- Monitor and improve the system using a Plan–Do–Check–Act (PDCA) approach.

- Measure the performance of the system.

6.2.2 ESP Planning Principles

The example earlier identified those who are concerned with implementing a plan to ensure that the equipment is doing what it should, it continues to deliver the benefits and its quality is maintained and the consequence of any unexpected failure is minimized. The support actions that need to be included in each ESP can be grouped under four headings:

1. *Performance verification (PV) Actions*: Scheduled actions which are put in place to assure that equipment which appears to be working is performing to specification. These processes both assure the quality of the equipment and identify hidden failures like calibration drift. Activities include quality assurance programmes to assess that equipment is performing within specification and safety assessment such as electrical safety testing.

2. *Scheduled Actions*: Actions undertaken to minimize the risk of failure in service. These include battery management programmes, cleaning or replacing filters and scheduled replacement of parts which have a known expected lifetime. Such actions are sometimes referred to as 'Planned Preventative Maintenance', abbreviated as PPM; we use the term 'Scheduled' to emphasize that they are actions undertaken which are timetabled.

3. *Unscheduled (reactive) Actions*: Actions undertaken to address the consequences of unexpected technical faults or loss of function for any other reason. The origins of the unexpected loss of function may be component failure, wear and tear, accidental damage, or user difficulty with operating the equipment, etc. This will include actions required in response to adverse events, linking in to the adverse events procedures of the healthcare organization.

4. *Training and user support*: Actions undertaken to support the clinical use of the equipment including the training of end users. The role of clinical engineering in ensuring user training and operational support will vary with the equipment type and its deployment. Including this aspect in the ESP emphasizes both its holistic objective and the shared roles of technical, clinical and, where appropriate, patient and carer in supporting the equipment.

Remember some of these actions may not be recognizable as traditional technical maintenance. A nurse changing the batteries in an electronic thermometer every month is a scheduled action designed to prevent a failure. Swapping an anaesthetic workstation with another if the first fails is a contingency plan put in place to reduce the impact of an unexpected failure. It might well be carried out immediately by a competent user. As an unscheduled action, it is part of the plan of how the failed device will be maintained. Once you start to take this holistic view that focuses on the impact these four action types have on the patient and their care, your locus of the support activity should move away from pure technical work to be more broadly located around optimizing the use of the equipment, for patient benefit. So when developing or reviewing any particular ESP, it is valuable to consider who delivers which action and how are they interdependent as illustrated in Figure 6.2.

We should acknowledge that often the patient or their carer is in fact the equipment user. For example, glucometers are pieces of medical equipment often used by patients in their own home. The patient's training will include how to use the equipment, carry out some routine maintenance and recognize signs of faults and action to take when faults are found. User routine maintenance in this case is simple, changing the batteries regularly and cleaning the sample window if it gets stained with blood. Often users are asked to perform basic performance verification by regularly inserting a test strip into the device, to ensure it is reading correctly. As more equipment moves into the community setting, there is a need for careful consideration as to how performance verification, scheduled maintenance (SM) actions and unexpected failures will be managed. Clearly, training of end users who use equipment in the home and community environments will become a feature of an ESP. Whilst beyond the immediate scope of this section, it is worth pointing out that there is growing literature available on steps to ensure the safe use of medical devices in the home environment. In particular, the reader is pointed to the FDA 'Home Use Devices Initiative' (FDA 2016).

6.2.3 ESP as a Process That Is Structured as a Quality Cycle

The HTM Programme is the sum of the ESPs that run within it and it operates as a quality cycle. Therefore, it should be no surprise that each ESP is itself recognizable as a quality cycle. The first step in this cycle is the initial development of the ESP. Analysis needs

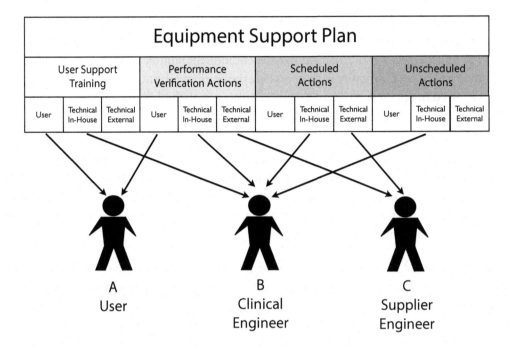

FIGURE 6.2 Examples of stakeholders who contribute directly to the delivery of different components that make up a holistic ESP. 'User' is the end user of the equipment, for example, a nurse, but could also be the patient or their carer in home healthcare.

to be undertaken as to the needs of each group of equipment, being mindful of the clinical context within which it is used and the consequence of failure on patients. This is the first step in a 'systems' approach. A template based on the principle discussed previously is a useful starting point and is shown in Figure 6.3. Such a template challenges the planners to consider the four types of activity that need to be considered, performance verification, scheduled actions, unscheduled actions and training and user support, as well as who is going to be tasked with performing them. Such a plan should be developed in consultation with all relevant stakeholders which could include patients, staff and external maintenance suppliers. This is one way of ensuring clarity as to who is taking responsibility for what. The plan should also clearly identify how the actions will be recorded and what systems will be used to monitor performance. Clinical users and ward and theatre managers should be able to view relevant parts of Medical Equipment Management System (MEMS) database, viewing equipment lists and status of equipment, helping develop partnership between users and the CED in carrying out the ESPs.

6.2.3.1 Developing the ESP

The manufacturer's service documentation is the starting point for developing an ESP but local experience in supporting the equipment in its local context and conditions of use must also be taken into consideration. There is no standard approach; each department must work out what is right for the particular equipment they support within the local context of its use. An ESP is a bringing together of many equipment and non-equipment

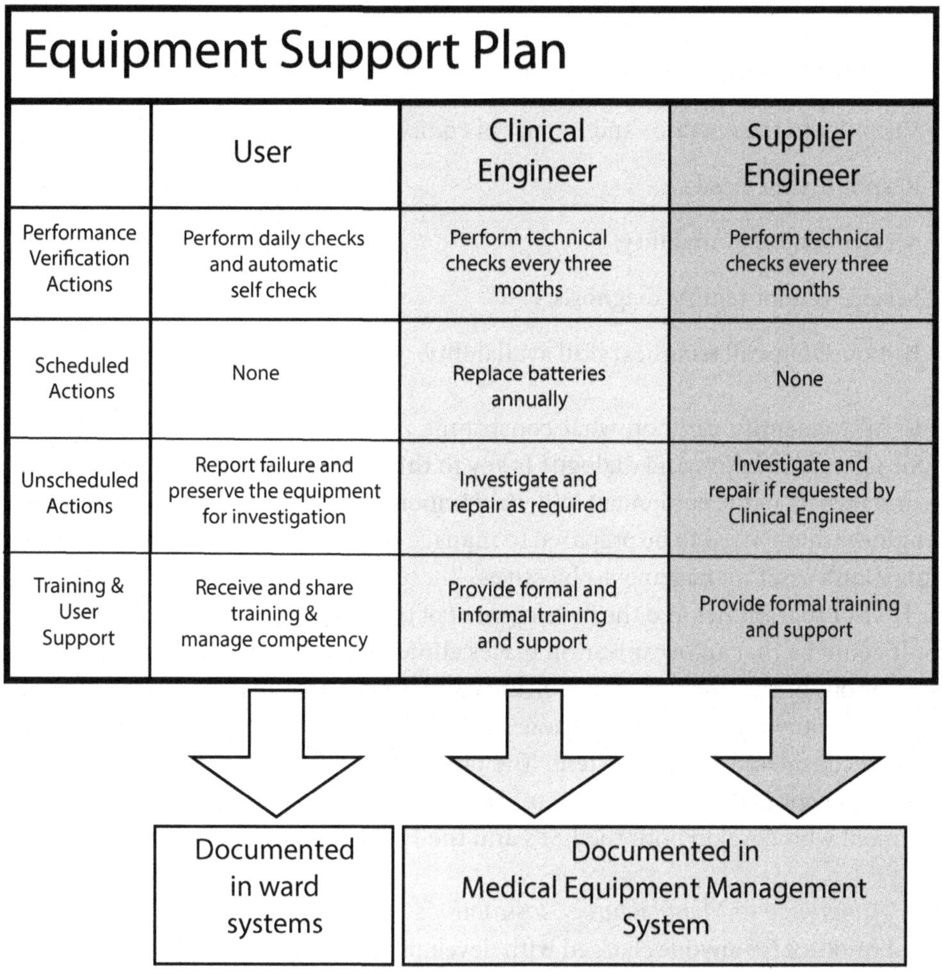

Equipment Support Plan

	User	Clinical Engineer	Supplier Engineer
Performance Verification Actions	Perform daily checks and automatic self check	Perform technical checks every three months	Perform technical checks every three months
Scheduled Actions	None	Replace batteries annually	None
Unscheduled Actions	Report failure and preserve the equipment for investigation	Investigate and repair as required	Investigate and repair if requested by Clinical Engineer
Training & User Support	Receive and share training & manage competency	Provide formal and informal training and support	Provide formal training and support

Documented in ward systems	Documented in Medical Equipment Management System

FIGURE 6.3 Example of a template used to determine the responsibilities of different stakeholders in the ESP.

requirements with the aim of ensuring the equipment receives the appropriate support at the appropriate time and at the appropriate level. Overall, the ESP must reflect the organization's asset objectives, be in keeping with the MD Policy, and also be informed by many other aspects such as national standards, local clinical need, patient activity, in-house skills and experience and financial risk, all of which all play a part in the determination of the ESP. The following list includes a number of considerations that inform, constrain or shape the ESP design:

- Organization's asset management objectives;
- Frequency of use;
- Availability of alternatives;
- Frequency of maintenance recommended;

- Mandatory, statutory or professional;

- Criticality to the patient (Risk level);

- Visibility to the operator and impact of equipment failure;

- Response time needed;

- Service vendor availability;

- Service vendor remote diagnosis;

- In-house/On-call technical skill availability.

It is useful to identify early on what constraints there are likely to be in the creation of the ESP. Communication and dialogue is key in this process; discussions with the clinical users and owners of the equipment will yield important clinical perspectives, but the clinical engineer might need to be prepared to manage their expectations keeping in mind the organization's asset management objectives. These objectives will influence the running of the HTM Programmes and the development of the ESPs by placing emphasis on certain areas. It could be that an organization places clinical preference and need above all other considerations or that the total expenditure on maintenance must not exceed an allocated budget under any circumstances. These extremes are unlikely but are aspects that will need to be considered to some extent. The task of prioritization and management of the ESP development is the responsibility of the leadership within the Clinical Engineering Department who develop both the ESPs and the HTM Programme as a whole.

6.2.3.2 Adherence to Manufacturers' Instructions

A useful resource for anyone charged with developing an ESP is the standard ANSI/AAMI EQ89:2015 titled 'Guidance for the use of medical equipment maintenance strategies and procedures' (AAMI 2015). This document aims to provide basic information to healthcare technology management professionals by describing and identifying, in general, various maintenance strategies and methods for efficient, effective and timely maintenance of medical equipment in healthcare facilities.

It deals, amongst other things, with the issue of the extent to which maintenance activities must follow exactly the instructions provided by the manufacturer. This issue is one that causes much worldwide debate and difficulty within the clinical engineering community. We understand that in some jurisdictions there is a legal requirement to follow manufacturers' instructions 'to the letter'. However, even in those jurisdictions, the resources necessary to do this are not always made available. This places the clinical engineer developing an ESP in a very difficult position. The lowest personal risk course of action for the clinical engineer is to concentrate resources into a plan that follows the manufacturers' instructions at the expense of a more holistic, patient-centric approach based on the actual circumstances of use of the equipment. The former course of action would not necessarily provide the best value for either the patients or for the organization. Some organizations have misapplied scarce resources in pursuit of rigid adherence to manufacturers' instructions, leading, for example, to very extended downtimes for broken equipment or the inability to engage in user training, both to the detriment of patients.

A more professional approach is to bring to the attention of top management the lack of resources needed to follow the manufacturers' instructions to the letter; propose alternative courses of action firmly based on the manufacturers' instructions, backed up by risk/benefit arguments; and request the organization to take responsibility for these alternatives. All this would need to be carefully documented.

Fortunately, there is evidence that 'authorities having jurisdiction' are beginning to understand that a rigid adherence to manufacturers' instructions is not necessarily the optimum arrangement. In the United States, the Centers for Medicare & Medicaid Services (CMS), a Federal agency which administers the Medicare and Medicaid programmes, have from 2014 adjusted their 'Hospital Equipment Maintenance Requirements' to allow "alternative equipment maintenance frequency or activities" under certain defined circumstances (CMS 2013).

In the United Kingdom, the guidance document from the MHRA 'Managing Medical Devices – Guidance for healthcare and social services organizations' (MHRA 2015), referenced in Chapter 5, was circulated in draft form for consultation in 2013. This draft required maintenance of medical equipment to be carried out 'in accordance with manufacturers' instructions', but this phrase was not used consistently in the draft. In some places, 'in line with' was used, in others, 'based on' or 'taking account of'. Following comments, the final Ed 1.0 of this MHRA document used 'in line with' consistently throughout; this phrase has been retained in Ed 1.1 referenced. Although the difference between 'in accordance with' and 'in line with' is very subtle and has not been tested in a court of law, we believe it is a slightly less strict requirement. It would allow, for example, an ESP to specify a greater level of maintenance activity if that had proved necessary, which would not be 'in accordance with' the manufacturer's instructions. MHRA have also acknowledged in Section 8.2 (Chapter 8) the possibility of a risk/benefit-based approach when "finalizing the specification of any maintenance and repair services".

Finally, in the second edition of the IEC Standard 62353 'Medical electrical equipment – Recurrent tests and test after repair of medical electrical equipment' (IEC 2014), a proposal was made by one of the authors (JM) in his capacity as a UK member of the relevant standards committee which resulted, after discussion, in the following note being included:

> "NOTE: A RESPONSIBLE ORGANIZATION having appropriate expertise can also take responsibility for modifying MANUFACTURER's proposals based on local conditions of use and risk assessment."

None of those experts around the table who were employed by medical equipment manufacturers objected to this note which was accepted and appears in IEC 62353:2014.

In conclusion, ESPs must be firmly based on manufacturers' instructions. If equipment is new to you, or you only have a single item of that type in your organization, and you therefore have little or no experience of it, it would be very difficult to justify not following the manufacturer's instruction on servicing to the letter. As you gain experience and are perhaps managing multiple examples of the same type of equipment, for example, a fleet of anaesthetic work stations, you will build up experience and understanding that will enable you to review your ESP and make adjustments based on maintenance data and risk assessment. As already stated, documenting your thought processes, a formal risk assessment and your conclusions are vital. These all form part of 'continual improvement' within the quality cycle.

6.2.3.3 Implementing the ESP

Once the ESP is developed, its implementation must be organized. The groups involved in its delivery must each be tasked with implementing the parts of the plan for which they are responsible. So for the defibrillator example discussed earlier in this chapter, the user's role in daily performance verifications needs to be communicated and perhaps training provided where required. The importance of documentation should be part of this activity. Similarly the supplier responsibility should be detailed in the form of a written service contract which includes all the deliverables including the documentation and means of auditing the work performed by them, and of course costs. The in-house clinical engineering team also need to be given authority and responsibility to set the plan into action and also need to give clear instructions as to the objectives of the in-house technical programme and the resources to carry out and document their actions as shown in Figure 6.4.

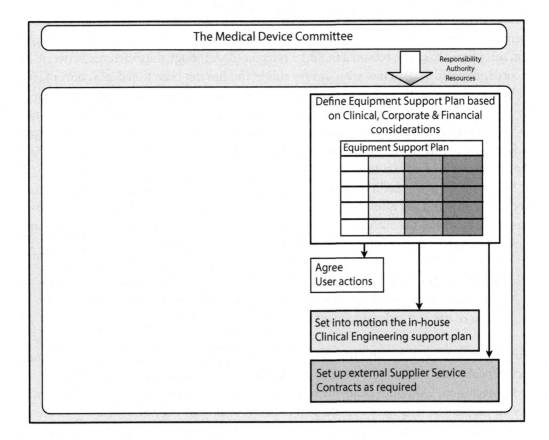

FIGURE 6.4   Implementing an ESP, initial planning and setting into action.

Once the equipment is defined in the ESP and the resources are assigned, teams are set into action. The scheduling of work is an important task for the team leader. They have to develop and implement processes that include routine scheduled actions and performance verification, whilst also having sufficient resources to respond to breakdowns and other unscheduled support request. Also remember that ongoing user support and training might be as important in supporting the clinical work as routine performance verification.

Typically equipment requires some form of inspection on a regular basis. This performance verification activity is intended to assure that the device is performing to specification and is safe. Common practice suggests a device to be seen at least once per year and manufacturers usually provide a schedule of actions to be performed. Clinical engineering departments use these manufacturer's instructions, but, as described earlier, these can be moderated by local experience and evidence of the performance of devices in the particular environment of use. Experience and evidence may point to doing more than the manufacturer recommends. Performance verification and scheduled maintenance activities can be aligned to increase efficiencies. Moderating manufacturer's instructions to decrease or increase scheduled maintenance are illustrated in Case Studies CS6.2 and CS6.3, respectively.

Unscheduled actions such as corrective maintenance are unpredictable and the department should put in place a system to ensure that resources are available to deal with unscheduled requests for corrective maintenance or user support. Lack of resources can lead to unsatisfactory support for clinical users in two ways, either requests for unscheduled support are prioritized and scheduled preventive maintenance suffers or unscheduled requests take a very long time to be resolved because scheduled maintenance on working equipment takes precedence. In practice, a risk-based balance has to be struck when allocating workload priorities.

Contracting an external service supplier requires careful consideration. Doing so does not remove all responsibility for the delivery of the support actions from the Clinical Engineering Department. The contract needs to include provision for the supplier to provide assurance that actions have been appropriately carried out. There must be technical oversight of the provision of external services and mechanisms to ensure that the service records are consolidated within the organization's Medical Equipment Management System (MEMS) database. It is not satisfactory for a Procurement Department to manage external service contracts for medical equipment.

Where an external supplier is contracted, the nature and level of the service agreement needs to be decided. This ranges from fully comprehensive cover that includes all the required inspections and any repairs necessary to a more basic cover that only includes the recommended scheduled service or performance verification actions. Many variations exist in between, parts excluded, on-site or off-site servicing, accidental damage included, parts included to a monetary value, only one repair visit per annum included, etc. It is usually the Clinical Engineering Department's decision, in consultation with the clinical users, as to which level of cover is best suited to the organization.

Shared service models can also be effective. In such a model, a service contract for particular actions is put in place with the external contractor, and the in-house Clinical Engineering Department performs 'first-line response'. These or similar contracts are often

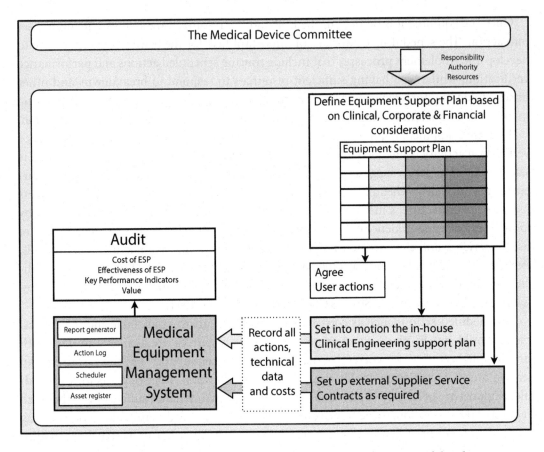

FIGURE 6.5 Implementing an ESP, recording and auditing the performance of the plan.

called partnership contracts. With such a contract, when a fault is reported, the Clinical Engineering Department takes the initiative and initially looks to see what the problem is with the equipment and liaises with the contractor if necessary. If repairs can be undertaken by the Clinical Engineering Department or a usability problem resolved, then this will be done, but if not, the contractor at least knows the general condition of the equipment when they attend on-site. This solution allows the Clinical Engineering Department to retain involvement with the equipment and provide support to the clinical user whilst having the manufacturer's technical backup easily to hand and at known cost.

Whichever method of external contract is made, records need to be kept of the contract and period of cover, along with service reports. The scheduled and unscheduled maintenance delivery by the external service supplier must also be documented within the MEMS database as shown in Figure 6.5. This will help ensure that the information in the database can be used to determine the whole life cost of ownership of the equipment (Chapter 4).

6.2.3.4 MEMS

A Medical Equipment Management System (MEMS) database is a software system used to schedule and record actions performed on individual pieces of equipment. At its simplest, it consists of an asset register or table of medical equipment within the organization, a schedule of actions to be performed and a table of all actions performed. The requirement to document actions and record them in the MEMS should be part of the planning and organizing process. Combined with a flexible report generator, such a system is invaluable in helping control the implementation of the ESP. It allows the effectiveness of the plan to be monitored in real time and supports those implementing it to get helpful data to support day-to-day decision-making focused on ensuring the programme delivers its goal.

6.2.3.5 Auditing the ESP and the Quality Cycle

We will discuss the checks and audit of the ESP in much greater detail later in this chapter, but in closing this section on the development of the ESP as a process or system and structured as a quality cycle, some general comments should be made.

The audit process, supported by the MEMS, both reassures that the process is working and operating as planned and identifies any deficiencies that might arise. Evidence-based audit greatly improves the ability of leaders to make good decisions to mitigate problems that arise and to act to revise the plan so that such problems are less likely to occur in the future.

Of course we must remember that the ESP's objective is in turn determined by the objectives set out by the MDC in the organization's hospital-wide Strategic HTM Plan. The MDC enables the department to develop and implement their HTM Programme and ESPs by assigning it responsibility, local authority and resources. The MDC also requires the department to report on the performance of the HTM Programme, including individual ESPs. High-level KPIs, such as percentages of schedule actions completed, are useful ways of doing this. If the ESPs are well planned and implemented, but cannot meet the objectives set by the MDC due to extenuating circumstances or lack of resources, then there is a duty for the leaders in the department to report this to the MDC, who then need to consider how to mitigate the deficit at an organization level.

Figure 6.6 shows the implementation and ongoing development of an ESP as a quality cycle managed as a process within a department. The leadership within the department must exercise good management to ensure that this specific plan is effective. So they must organize and control it and measure its performance to optimize it in real time and to report its success or failings to the MDC. Of course this ESP is just one of many, and in Section 6.5, we will look at how the departmental leadership manages the HTM Programme as a whole.

6.2.4 Financial Analysis of the ESP

All the actions identified as part of each ESP need to be resourced, set into action and controlled. Controlling costs is an important part of the process and so some means of measuring resource utilization needs to be established.

The optimum mix of in-house and externally contracted maintenance services needs to be decided, and the primary driver for this decision should be optimizing the support based on its impact on care of the patient. It is important to remember that the ESP must also support the hospital's financial goals. Regardless of the service support mix

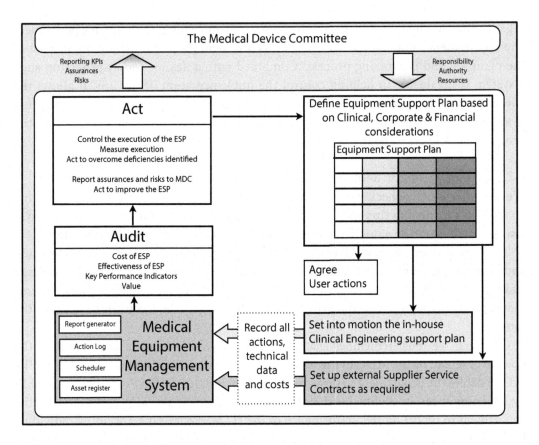

FIGURE 6.6   A complete ESP operating as a quality cycle.

chosen, there is a need to estimate the costs associated with it. Where the service solution is fully outsourced, the costs associated with external contractors will be readily available as service contracts will be procured and their cost clearly identified. However, in such situations, it is not uncommon for the Clinical Engineering Department to also provide some degree of front-line support. As a minimum, this usually includes managing communications between the clinical staff and the service company, but may extend to having a quick first look to rule out user error or difficulty with operational situations. Where service is delivered in-house, it can be difficult to assess all the costs associated with its provision. It should include as a minimum an estimate of staff costs, external service support that is purchased where no service contract is in place (non-contract call outs), spare parts and cost of staff training. Other overhead costs such as workshop space and energy costs can be less visible in the hospital context. Regardless, the Clinical Engineering Department should make all reasonable attempts to construct a complete financial analysis of its function and include financial factors in the decision-making when developing the ESP.

Being able to estimate the total cost of each ESP allows analysis of its cost-effectiveness. An estimate of the cost of the plan can be achieved by looking at the internal and external costs associated with each element. Table 6.1 shows a simple template which can be used to

TABLE 6.1   Template Used to Identify the Costs Associated with an Equipment Support Plan

Equipment Support Actions	CED Internal Costs		CED External Costs	
	Staff	Parts	Contract	Non-contract
Performance verification				
Scheduled				
Unscheduled				
End-user support				
Subtotals				
Total cost				
Cost of internal support				
Cost of external support				

Source: From McCarthy, J.P. et al. 2014. Healthcare technology management, in *Clinical Engineering – A Handbook for Clinical and Biomedical Engineers*, eds. A. Taktak, P. Ganney, D. Long and P. White, pp. 43–57, Academic Press, Oxford, UK. With permission.

get an estimate of the support costs associated with ESPs. An example of how this table can be used to assess the cost and cost-effectiveness of an ESP can be found in Case Study CS6.4.

Expressing this total cost as a ratio to another metric which is based on the contribution and complexity of equipment supported can be useful to get a sense of how cost-effective the ESP is. So expressing the ESP as cost per device, or per hospital bed, or its cost contribution to supplying a clinical procedure can help in understanding and appreciating how the resources are being assigned.

It can also be helpful to measure and contrast ESPs weighted towards in-house or external supplier support models. Clinical engineering departments can no longer assume that the approach to supporting the equipment developed years in the past is still optimal. Within any quality cycle, it is important that decisions are evidence based, and so there is merit in looking at the in-house versus external model costs for each plan to ensure that financial, as well as corporate and clinical, goals are being met year by year. Doing so may challenge accepted norms in the Clinical Engineering Department and may prompt a change of approach with some devices moving to an external service model and others being incorporated into in-house models. It can prompt a Clinical Engineering Department to find cost savings within its own delivery system or prompt it to negotiate more cost-effective external support contracts.

6.2.4.1 Cost-Effectiveness Measures

There are a number of measures used to assess the cost-effectiveness of an ESP. The ESP should record a 'cost' or 'value' to the equipment, and on the face of it, this seems simple – it is the amount the organization has paid for the equipment. However, what if the organization pays nothing for infusion pumps which are funded through an arrangement against the cost of the disposables? Or alternatively, what if a healthcare organization is able to negotiate significant discounts from the list price of equipment when they are purchasing multiple items or complete systems such as equipment for an intensive care unit? Alternatively, what if a manufacturer is prepared to sell equipment at a very advantageous price to a healthcare organization with a national or international reputation, as part of a marketing strategy? In all such cases, measures which use the cost or value of the equipment under consideration could be seriously distorted unless a consistent approach is used.

There is no universal set 'cost price' of an item of medical equipment; the price that is paid is dependent on negotiations, quantities, discounts, service agreements and warranties. The most logical measure to choose is the list price of the equipment rather than the actual price paid. This will ensure consistency of measurement of the cost-effectiveness of ESPs – and comparison between different clinical engineering departments. The design of the Medical Equipment Management System should include a field for list price and a field for the price paid. Developing from this, a useful Cost-effectiveness Index (CEI) of an ESP can be calculated:

$$CEI_{ESP} = \frac{\text{Annual cost of equipment support plan}}{\text{List price of equipment}} \times 100\%$$

This allows the Clinical Engineering Department to compare and analyze how resources are being distributed to the different equipment groups. The CEI should be applied to low-value items as well as expensive ones, especially where there are large numbers requiring some regular maintenance, for example, flowmeters and suction controllers.

Various values for the expected CEI are quoted. Wilson et al. (2014) reported that, in the United Kingdom, manufacturers typically charge 8%–10% of the value of the equipment for a comprehensive annual service contract. They went on to write that, for a large teaching hospital, it is possible to reduce this to 6%–8% by reducing or varying the level of cover and augmenting it with in-house support. Further, they suggest that by predominantly using an in-house support model, it is possible to reduce costs by a further 2%–3%, especially where evidence is used to match the level of support to the risk of failure and in doing so simultaneously reduce equipment downtime.

This cost-effectiveness index is helpful for replacement planning. If the ESP expenditure in any year is 20% of the cost of replacing the equipment, it means that the organization spends as much on supporting the item in 5 years as it would in replacing it with a new piece of equipment. This is, in general, too high for medical equipment, but may be appropriate for equipment that is delicate and in normal use subject to significant wear and tear. Some equipment, such as optical endoscopes do suffer significant physical and chemical wear and tear. Nevertheless, the high ratio may also indicate that there are poor operating practices associated with the endoscope, perhaps the handling and reprocessing of the devices do not consider the delicate nature of the technology. Alternatively, it might indicate that the endoscopes themselves are not of a high enough quality. On the other hand, a CEI of 20% would be extremely high for a physiological monitor which consists mostly of electronics. The index for such devices is more likely to be 3%–4%; an index of over 20% for an electronic device merits detailed investigation into the cost-effectiveness of the ESP or the reliability of the device or both.

Low CEI percentages can also give rise to concerns. They might indicate that equipment is under used, or that not enough proactive maintenance is being undertaken which in turn gives rise to questions as to the safety, reliability and quality of the device in clinical use.

Looking for peaks and troughs in the CEI percentages is a useful way to assess the performance of the ESPs. Review can prompt the Clinical Engineering Department to find quality improvements and ways of controlling cost.

6.2.5 Analyzing the Benefits Delivered from an ESP

The overall goal of an ESP is to ensure that the equipment functions as intended in a safe and effective manner and to ensure that the equipment is available for use when needed. Following a review of the support required for an equipment group, the Clinical Engineering Department will develop a considered holistic and coherent strategy as discussed earlier. The complexity and extent of ESPs will vary dependent on the nature of the equipment and its clinical use. Some items, like the defibrillator discussed as an example earlier in this chapter, will require more comprehensive support plans than that of a battery-operated ophthalmoscope used in an outpatient's department. On the other hand, technological defibrillator developments, in particular as in certain AEDs, will require less verification and scheduled maintenance, with manufacturers recommending only minimal scheduled maintenance. In these cases, the ESP will predominantly focus on the unscheduled actions required, with minimal performance verification and scheduled actions required (Figure 6.7). Thus the clinical engineer will review the weight put on verification, scheduled and unscheduled actions in the ESP, dependent on manufacturer's recommendations, experiential evidence and risk management.

It is not necessary for every ESP to be as comprehensive as that for some of the more complex and clinical critical equipment. However, it is important that the ESP requirements for each group of equipment are carefully assessed and considered in the light of manufacturer's recommendations and the local experience of its use in the organization. The ESP which arises from that exercise must be both documented and regularly reviewed. The scope of ESPs will change over time in response to these regular reviews and as other ESPs develop, demanding more of the finite resources the organization has available to allocate to HTM.

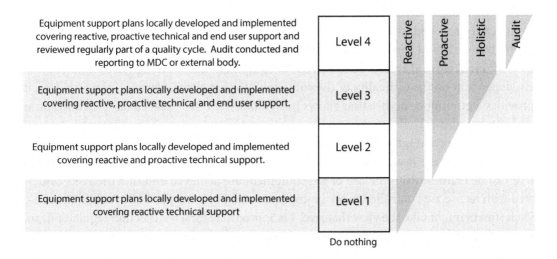

FIGURE 6.7 Four different levels of ESPs.

Accepting that at any given time the designed ESP is judged to be appropriate, we need some method to assess the scope of the plan so we can compare and contrast the approach with other plans. Whilst the cost of a support strategy can be quantified in financial currency, be that dollars, yen, euros or pounds, there is no unit of measure for complexity of an ESP. So we have to take a qualitative approach.

We propose a qualitative scale which consists of four levels and suggest that any ESP can be placed on this scale. As you move from level 1 to level 4, the ESPs move from being reactive to including proactive actions, from being technically focused to being more holistic and at level 4, subject to external review and audit. These levels are differentiated by the scope and complexity of support activities as shown in Figure 6.7.

Level 1 is used to characterize a support plan that is purely reactive. Here we would expect that equipment failures or observed deterioration by the users would be reported, dealt with and documented. These ESPs, because of their reactive nature, do not require performance verification or scheduled actions to be undertaken. Level 1 is often referred to as a 'run-to-failure' or 'maintain-on-failure' approach. It will be used for medical equipment whose failure will both be obvious to clinical staff and not impact in the short term on patient care.

Level 2 is used to characterize a support plan that is proactive as well as reactive, and where all these actions are documented, but the focus is on traditional technical maintenance support for the devices which includes scheduled actions, sometimes called 'planned preventive maintenance'.

Level 3 covers plans where a locally defined and agreed support plan extends further than the technical maintenance activities to include those suggested by taking a holistic view of an equipment support strategy. So it might include provision of loan equipment, perhaps through an equipment library (Keay et al. 2015) or include an ongoing training activity for users both of which are part of the support programme.

Level 4 describes an ESP locally developed by the Clinical Engineering Department in consultation with the appropriate clinical departments. In addition to the services described for level 3, this level also includes audit or reporting that goes outside the Clinical Engineering Department, either to the MDC or to an external organization.

In reality the boundaries between these levels can be blurred. Remember this is a qualitative scale and some degree of subjectivity is to be expected. Flexibility is required as in different clinical areas, the same type of equipment may be assigned to different levels, dependent on the criticality of the equipment to that clinical service. Nevertheless, it provides a guiding structure that allows individual ESPs to be placed on one continuum and compared.

What is important is that the support plan is appropriate to that equipment type and how it is used within the healthcare organization. It should be at a level such that the benefits that can be realized from the use of the equipment are achieved and that the risks associated with its use are controlled to an appropriate level. For example, a Clinical Engineering Department might take the view that level 1 is appropriate for a wall suction regulator. If so, and a level 1 ESP is put in place, then that is all that is required for this equipment group. However, if an Anaesthetic Workstation was supported with a level 1 plan that would rightly be considered inappropriate; such life-support equipment demands higher-level scheduled maintenance and performance verification actions for it to function as intended in a safe and effective manner and to ensure that it is available for use when needed.

Individual items of equipment supported by plans classified to the higher levels are likely to require more resources to implement than those classified to lower levels. So it is not necessary to have all the equipment supported by level 4 plans. Classifying the support for all medical equipment as level 4 might not provide the best benefits for patients and clinical staff and is unlikely to be cost-effective or appropriate for the organization.

6.2.6 Assessing the Value Delivered by an ESP

Value in healthcare is the ratio of benefits delivered to cost. The value of an ESP is therefore the ratio of the level of benefits it delivers to its cost.

Once an ESP is designed that is appropriate for the equipment it relates to, then we can say that, if implemented fully, the expected benefit that can be derived from the plan will be achieved. In practice, if the actual implementation falls short of the ideal expectation, then the benefits of that support plan are not fully realized. The actual delivery of the plan can be assessed qualitatively during its implementation and when completed by reviewing the KPIs measured as part of the internal reporting and controlling process within the Clinical Engineering Department. We can use the qualitative scale to compare the ideal designed plan to the actual delivered plan. Figure 6.8 shows the ideal designed ESP as the white bar reaching level 4, and the measured delivered ESP as the black bar only achieving

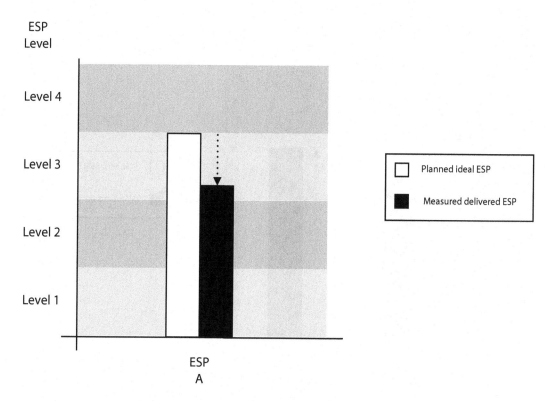

FIGURE 6.8 Diagram comparing the ideal designed ESP to the actual delivered ESP.

level 3. Therefore, the implemented ESP is performing less than that planned, and so all the benefits that should be derived from the ideal plan are not being achieved.

Where an ESP falls short of the ideal, optimal value is not being delivered, and the plan and its implementation should be reviewed to identify the actions needed to improve the process of its implementation. Of course it could happen that the implemented ESP level exceeds the required ideal ESP level as in ESP B in Figure 6.9. This can occur when traditional approaches to support are not reviewed when new devices, often built to a higher specification, are deployed.

We can see that the implementation of ESP B exceeds that of its ideal, suggesting that resources might be over assigned to this group of equipment. In this case the Value delivered is not optimized, as the deployment of resources and consequently cost exceeds that necessary to achieve the benefits of an ideal plan. The delivered ESP A however did not meet the ideal requirements. In this case, value is also not optimized as the benefits required are not being delivered. In managing the HTM Programme, the Clinical Engineering Department might act to improve the value of the programme by moving resources from one ESP to another. In this example, if the implementation of ESP B was reviewed to ensure that the benefits were delivered with fewer resources, then those extra resources could be applied to improving the implementation of ESP A (Figure 6.10). This is one of the reasons why ESPs need to be critically reviewed on a regular basis, reviewing the ESP blueprint, the required plan, and how it is actually delivered. Often practices which were once relevant

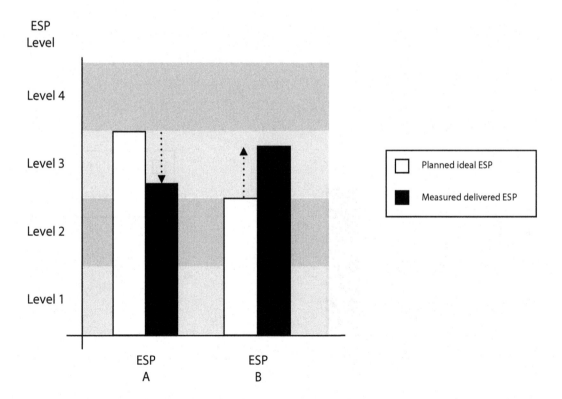

FIGURE 6.9 Comparison of two ESPs showing the ideal and actual delivered level for each.

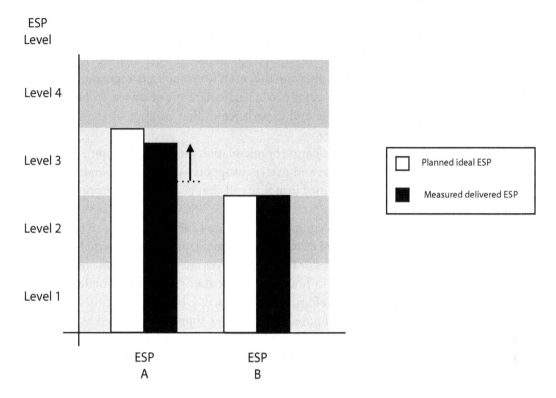

FIGURE 6.10   Diagram showing how the delivered ESP of the two ESPs shown in Figure 6.9 might change as a result of reallocating resources to each ESP.

such as safety testing become less required due to advances in materials, equipment design or improved use. At the same time new technologies being implemented may require a whole new type of support (Case Study CS6.5).

Whilst we assign discrete numbers 1–4 for the levels of service support (Figure 6.7) when we measure the delivered support, we may judge it as lying between two levels. Hence, as shown in Figure 6.8, the measured level developed may not equate to one of the discrete levels.

Whenever the implemented plan exceeds the ideal level, then careful consideration should be given as to whether resources used to exceed the level would be better deployed elsewhere within the HTM Programme.

In designing and implementing ESPs, different approaches can be taken. One example would be the mix between in-house and external support. In-house delivered ESPs nearly always have the advantage of proximity and flexibility, which together with local knowledge and contacts gives staff the ability to respond quickly to problems that arise and perform front-line troubleshooting well before any external organization could get a person on-site. For more specialized and complex equipment, the detailed technical knowledge available to in-house staff will usually be inferior to that of the manufacturer who have the advantage of seeing a given type of equipment perform in a variety of settings and who can access extra resources such as specialized test equipment and access to expensive spare parts. In-house delivered ESPs have the advantage of greater speed for front-line troubleshooting and can reduce cost: manufacturer-delivered ESPs have the advantage of technical expertise for more complex equipment requiring specialist device–specific knowledge. Paradoxically, it is sometimes the case even for complex equipment that when a healthcare organization has a large number of a particular type of device, in-house technical staff may have more knowledge and experience of that equipment than a service engineer sent out by the manufacturer.

It is therefore not possible to set a clear recommendation for specific equipment groups, as local needs and supply of in-house and external support vary. The boundary between what can and what cannot be maintained in-house will change for each institution. Rather, the boundary should be critically reviewed and should be adjusted to produce the best mix of support approaches for individual equipment types, and the effectiveness of the chosen mix should be evaluated and reviewed as part of an ongoing quality improvement process.

Equipment such as patient monitors or infusion devices are present in large numbers in hospitals and are often looked after in-house. Equipment of reasonable complexity present in moderate numbers such as defibrillators, endoscopy imaging systems and ventilators might also be maintained in-house, with supplier support purchased as required. Some more complex equipment such as Anaesthetic Machines and X-Ray rooms might be managed through a mix of in-house, providing front-line support, supported by manufacturer with partial service contracts. Some equipment, including some of the more complex imaging systems, may not be managed in-house at all because specialist expertise is essential and minimizing downtime is a critical factor. In some cases equipment is put on service contract as a means of reducing the financial risk associated with possible expensive repairs (e.g. associated with expensive spare parts) rather than as a result of a lack of competency in the in-house team.

In deciding on the best approach, those defining the ESP must look at the impact of failure on clinical care, business continuity and cost and optimize the approach to deliver value for the organization's stakeholders. The proportion of technical support between in-house and external suppliers can vary but neither the organization nor the manufacturer can avoid some level of responsibility for maintenance activities (Wilson et al. 2014). Each Clinical Engineering Department must determine the most appropriate mix of service options for each equipment grouping, to meet clinical, corporate and financial needs. This is done as part of the design of the ideal ESP, but must also include consideration of the cost. So optimizing value is the objective. By regularly reviewing each ESP as part of a quality cycle, the Clinical Engineering Department can ensure the optimal mix of support options for the diverse range of equipment in its care.

The value ratio can be useful in determining the optimal approach. It allows two different ESPs to be designed, each taking a different approach yet meeting the appropriate level, ensuring that each delivers appropriate and comparable benefits. Each plan can then be costed, and the cost used to guide decision-making. Figure 6.11 shows how the level of the ESP and its cost expressed using the CEI can be plotted on the same graph. By representing benefits and cost on the same graph, we can assess value delivered. In this example both ESPs deliver the same level of benefits, but ESP 'A' Version 1 is more cost-effective

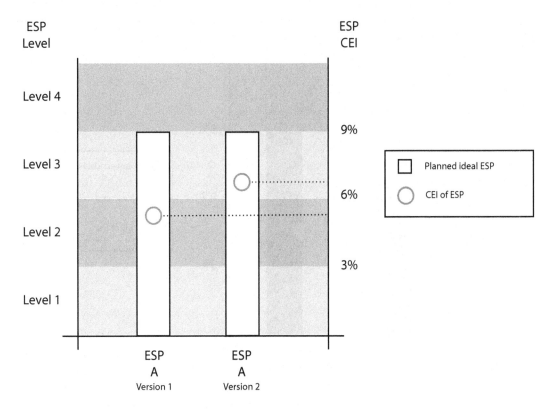

FIGURE 6.11   The planned Level of two ESPs plotted with their CEIs on the same graph.

than Version 2. Therefore Version 1 offers better value. Where cost is used in this way, it is important to use the appropriate cost-effectiveness index as described in Section 6.2.4.1, so that it corrects for the number and cost of the assets being supported by the plan.

The value delivered by an ESP can be reviewed in many ways using diagrams such as that shown in Figure 6.11 to indicate best value.

Figure 6.12 shows the planned and delivered support level and the associated cost index (CEI) for two ESP versions, 1 and 2, both planned to deliver level 4 support. Prior to assessing the delivery of the plans, version 1 with its lower cost index would have been judged to be better value than version 2. However, a review of the actual support delivered shows that version 1 failed to reach the required support level, whilst version 2 did achieve the aspired level 4, albeit for a higher CEI. In this case, you might say that whilst version 2 was more expensive, its value was higher than the less costly version.

Clinical engineers as a group are focused on optimizing value in their daily work. They both critically review equipment and its use and act to foresee and address problems before they manifest. Where failures occur, they not only work to solve the problem appropriately, but also cost-effectively. In this way clinical engineers are increasing value for the organization all the time. However, clinical engineers need to get better at demonstrating this and measuring our effectiveness in this regard.

The diagrams indicate value because they indicate the level of benefits derived from an ESP and also its costs. By plotting the performance and cost of many ESPs side by side,

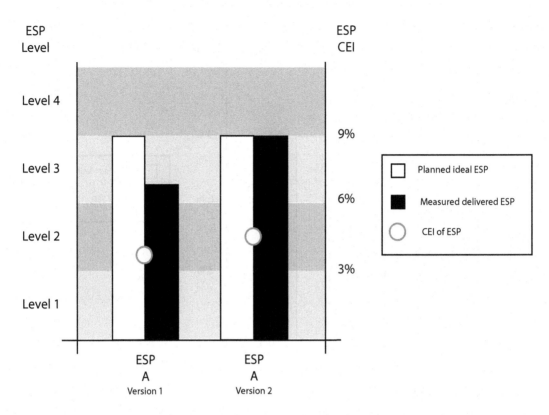

FIGURE 6.12   Diagram indicating the level of support delivered and the costs associated with two versions of an ESP.

the Clinical Engineering Department can get a snap shot of the performance of the HTM Programme as a whole. Such a diagram can be used to identify areas for improvement. Figure 6.13 shows the value plots for a number of illustrative ESPs, and we will use it to demonstrate how the analysis of the relative performance of the different ESPs can be a useful tool. In this example ESP B and ESP C show a balance between ideal and delivered ESP level. However, the CEI of ESP B is high at 9% as compared to 6% for ESP C suggesting that there might be opportunity for better cost control if ESP B were re-examined. Looking at ESP A, we see the delivered programme falls short of that required and its CEI is low at 3%. This suggests that perhaps not enough resources are assigned to this programme. Conversely, ESP D indicates that the level of the ESP implemented exceeds that required and its CEI is high at 8%. Perhaps resources from this plan could be reassigned to ESP A, resulting in both plans achieving their target levels.

6.3 EXAMPLES OF EQUIPMENT SUPPORT PLANS

Later in this chapter, we will consider the grouping of equipment support plans into the HTM Programme and the audit and review of the operation of the package of ESPs within the context of the HTM Programme. In this section, we take a more detailed look at individual equipment support plans. In Case Studies CS6.6 through CS6.9, we will explore the value gained through carefully designed ESPs.

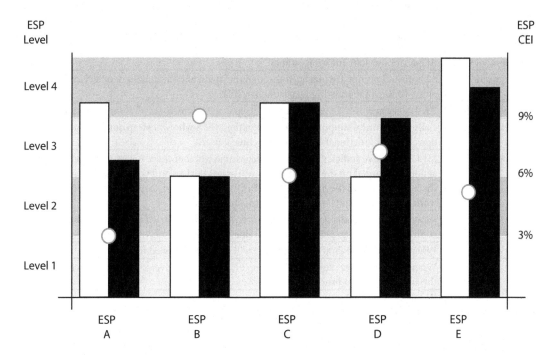

FIGURE 6.13 Diagram showing level of support delivered by and cost of a number of ESPs. The white bars show the planned level of support, the black bars the delivered level.

An ESP does not need to contain all the technical detail regarding the service and maintenance of medical equipment. That is most likely held in 'standard operating procedures', also sometimes called 'work instructions' held as part of the Medical Equipment Management System. These will be controlled documents that can only be changed after discussion, review and authorization. They will provide references to the relevant service manual, configuration settings where necessary and performance and safety checks to be carried out when equipment is serviced or maintained.

The work instruction will be linked to a particular model of equipment and may in some circumstances be linked to the serial number of individual items. For example, the healthcare organization may have standardized on a particular model of infusion pump. Most will have a common configuration for general-purpose use and a work instruction will be written for that. Some will be configured differently for neonatal use, and the serial numbers of those will be linked to a modified work instruction.

The ESP should be a high-level document that provides the overall plan for managing equipment which, for technical details, refers to these work instructions. Each ESP will cover an equipment type or a group of equipment types, grouped together based on their support needs. Each ESP should set out which one of the four levels of support described in Figure 6.7 is appropriate for a particular group or individual equipment and should summarize the outline of the support at each level. Some examples of how ESPs might look are given in Tables 6.2 through 6.5. Table 6.2 covers infusion pumps, used in clinical areas throughout the healthcare organization. In this example, because of the large number of

TABLE 6.2   Example of Equipment Support Plan for Infusion Pumps

ESP reference	ESP 1.			
Equipment type	ACME model 4001 infusion pump.			
Background	All infusion pumps are managed through the Medical Equipment Loan Service (MELS) of the Clinical Engineering Department (CED).			
Support level	1. Reactive	2. Proactive	3. Holistic	4. Audit ✓
Reactive	Calls for technical support are directed to the MELS who will swap out the reported device with a functional device for the clinical users.			
Unscheduled	MELS staff have sufficient training to determine whether there is a technical fault. The action is recorded on the MEMS. They will assess devices and where a technical fault is suspected will send the device to the CED for detailed work, using the appropriate work instruction which includes performance verification.			
	Where technical faults are identified, the repairs will be carried out, followed by configuration setting, performance and safety checks.			
	Where no technical fault is found, CED will consider the need for training.			
	Where the device was suspected of being involved in an adverse event, CED will follow the adverse incident procedures.			
	Repaired infusion devices are returned to the MELS base.			
Proactive	As per standard operating procedure in the MEMS.			
Scheduled maintenance	Each year, the MEMS will be interrogated by CED staff, and any device that has not been through the CED lab for 2 years will be recalled to the department for checking.			
Performance verification	Performance verification is part of the check work instruction.			
Holistic	A comprehensive clinical staff training programme is in place that covers this type of equipment. This is managed and provided from the CED by the Medical Equipment Training Officer.			
Training and user support	The Medical Equipment Training Officer or experienced CED staff provide problem-solving support to clinical areas on request.			
Audit	The Medical Equipment Training Officer provides senior nursing staff with a review of levels of training and a review of incidents, actions and outcomes involving equipment of this type.			
	File reference	Date		Issue no.

Note: The support level refers to Figure 6.7.

these devices, the Clinical Engineering Department manages them through its Medical Equipment Loan Service (MELS), often called an equipment library, which manages the deployment of these devices centrally for all departments.

6.4 DEVELOPING CONCURRENT ESPs AS PART OF A COMPLETE HTM PROGRAMME

To recap, the HTM Programme is a collection of ESP processes delivered by a competent provider, designed to provide appropriate scientific and technical support. As noted in Section 6.1, the process is applicable to any department in a healthcare organization charged with managing medical equipment, although for convenience we use the term Clinical Engineering Department (CED). The following sections expand on the typical considerations and steps involved in creating a suite of ESPs and how these form part of the HTM Programme.

TABLE 6.3   Example of an Equipment Support Plan for Multi-Parameter Patient Monitoring

ESP reference	ESP 2			
Equipment type	ITU multi-parameter patient monitoring, central station and server.			
Background	This equipment is all from the same manufacturer.			
Support level	1. Reactive	2. Proactive	3. Holistic	4. Audit ✓
Reactive Unscheduled	Most unscheduled calls for support will be of high priority and most will require a visit to the ITU.			
Proactive	Instances of equipment out of action may require assistance to clinical staff to swap over equipment.			
Technical support	Preventive maintenance including the cleaning of filters on the central station server will be carried out in accordance with the work instructions available from the MEMS for each of the various types and model of equipment.			
Performance verification	Scheduled service and intervals are held on the MEMS and flagged up as a job request when required.			
	Performance verification checks and electrical safety tests are included in the scheduled servicing.			
Holistic	Support for clinical staff, for example, problems with poor signal quality, is to be provided on request.			
Clinical support	Requests for changes of screen layout or colours must be approved by the ward manager before implementation to ensure consistency and understanding by all staff. No ad hoc changes to be made.			
Audit	Six-monthly reports back to the ward manager to report on the agreed KPI and highlight any particular issues.			
	KPI 1	Resolve time for unscheduled support calls.		
	KPI 2	Percentage of scheduled services completed within scheduled interval +10%.		
	File reference	Date		Issue no.

Each of the ESPs that make up the HTM Programme should be developed, put into action, checked that they are working and regularly reviewed. In general the checking and review of the experience of operating the ESP will suggest amendments that can be used to deliver improvements. In short, we propose that the HTM Programme be run as a quality cycle within a department.

This quality cycle approach will require planning to develop the ESPs and set out their proposed delivery methods. This in turn involves reviewing the requirements for different equipment types and, most importantly, the context within which the equipment is used and within which the ESP will be applied. A plan is useful but means nothing if it is not implemented.

Once the ESPs are developed, the department will need to organize itself so that it can put the plan into action. Organizing involves matching the tasks to be done to the resources available. This is when specific teams might be convened to manage particular equipment based on technology type or clinical area supported. The head of department will also have to clarify what aspects of the plan are to be actioned by clinical users or outsourced to external service suppliers. The plan is set into action by formally assigning responsibility and authority to individuals, teams or agencies to carry out the actions. In assigning responsibility, the head of department will need to be mindful that resources

TABLE 6.4   Equipment Support Plan Example for Customized Special Seating for Wheelchair-Bound Clients

ESP reference	ESP 3			
Equipment type	Customized special seating for wheelchair-bound clients			
Background	This clinical service is provided to individual patients referred to the Rehabilitation Engineering Unit.			
Support level	1. Reactive	2. Proactive	3. Holistic	4. Audit ✓
Reactive Unscheduled	Once issued with a seating system, clients will be given contact details to report any issues and request support.			
Proactive	Once referred and assessed by the multidisciplinary team as requiring special seating, clients will be allocated to a clinical engineer depending on case load and seen as soon as possible.			
Scheduled	A custom seating system will be designed and constructed in accordance with the work instructions that are part of the QMS. Clinical data will be kept on the patient record system.			
	A custom seating system fitted to an appropriate wheelchair will be issued to the client with normally no more than three clinic visits.			
	Once the seating system is issued, clients will be recalled for a review visit every 12 months for adult clients and every 6 months for child clients.			
Holistic	The Rehabilitation Engineering service will work closely with the clients and their carers and other associated clinical professionals involved.			
Clinical and technical support	Particular attention will be given to ensure that adequate information and training is given to the client and their carers and that written instructions for use are issued.			
	Particular attention and suitable engineering steps will be taken and training and written instructions given if the client is to be transported in their wheelchair in a vehicle.			
Audit	Six-monthly reports back to the service manager to report on the agreed KPI and highlight any particular issues.			
	KPI 1	Referral to issue time; 80% <17 weeks.		
	KPI 2	Percentage of review visits undertaken within 57 weeks for adults and 29 weeks for children.		
	File reference	Date		Issue no.

TABLE 6.5   Equipment Support Plan Example for Wall-Mounted Suction, Medical Gas Regulators and Flowmeters

ESP reference	ESP 4.			
Equipment type	Regulators and flowmeters for piped medical gases and suction.			
Background	Piped medical O_2 and suction are provided at all bed positions on all wards.			
Support level	1. Reactive	2. Proactive ✓	3. Holistic	4. Audit
Reactive	Repair on failure. No preventive maintenance required.			
Unscheduled	Non-functioning flowmeters will be replaced. Non-functioning or damaged regulators will be replaced. Regulator manufacturers provide a free replacement service for those less than 5 years old.			
Proactive	On a rolling replacement programme, all flowmeters, regulators and associated hoses will be replaced every 10 years.			
Scheduled	If a ward or clinical area is to undergo an upgrade, replacement of regulators and flowmeters will be included in the planned cost of the upgrade project. Replaced items that are less than 5 years old will be retained for reuse.			
Holistic	Not applicable			
Audit	Not applicable			
	File reference	Date		Issue No

are finite and be conscious of the resources that have been allocated to the department for these targets, that is be mindful of the budgetary considerations. Therefore, planning and organizing must be done within a context that ensures finite resources are appropriately allocated to meet the goals of the HTM Programme as a whole, balancing the demands of the various ESPs; the tools outlined in Section 6.2.6 can help this process, optimizing value across the ESPs. Resources include staff time and skills, costs of parts and accessories as well as travelling costs, leadership and administrative support for the teams and information and computer technology to help manage implementing the HTM Programme. Once assigned, there is continual need for leadership to manage communications and motivate all involved to work to meet the objectives of the HTM Programme as a whole and of its constituent ESPs.

During the implementation of the ESPs, it is inevitable that problems will arise. Part of the role of the leaders is to identify these and act to overcome them. Feedback from staff delivering the ESPs is also valuable, so a culture of constructive review is important. ESPs need to be controlled to ensure that each meets its ideal designed objectives. Reviewing the performance of the programme regularly, initially weekly, is a powerful way to identify early any deficiencies and gives enough time to act to control any deficits which occur. The performance review process includes having agreed measures and key performance indicators (KPIs) in place that allow the performance to be assessed in a systematic way. Through these and the evidence gained from the experience of implementing them, the ESPs are adjusted in a controlled and risk assessed way to continually improve performance.

On top of the local regular reviews of the implementation of the ESPs, a formal regular review, usually annually, will be undertaken. This formal review will involve assessment of the ESPs as constituent components of the overall HTM Programme, an assessment where the structured objective-driven systems approach outlined in Chapter 2 can assist. This review will thus include a more direct comparison of the levels achieved by each of the ESPs together with their associated Cost-Effectiveness Indices (CEIs). In the context of the HTM Programme as a whole, the review may move resources from one area to another in response to how effective different ESPs were in meeting the goals of the programme. Figure 6.14 shows how the process management of the HTM Programme functions and delivers an HTM Programme that operates as a quality cycle. In the following sections, we discuss the steps that can be taken to deliver such a programme.

It is not the intention of this book to instruct the reader on how to set up a quality management system (QMS), but there are certainly topics that are so relevant it is necessary to include them. The key QMS Standard, ISO 9001, has been discussed in Chapter 3. The reason for having a quality management system is that it brings discipline into the Clinical Engineering Department. The QMS consists of a collection of processes that work together for the Clinical Engineering Department to meet its quality objectives and which are audited both internally and externally to verify that they are working correctly. As part of the QMS, there are logs that record corrective actions and improvement activities. Corrective actions are those taken in response to something not happening as it should, sometimes called a non-conformance. Improvement actions are those which are generally suggesting a better

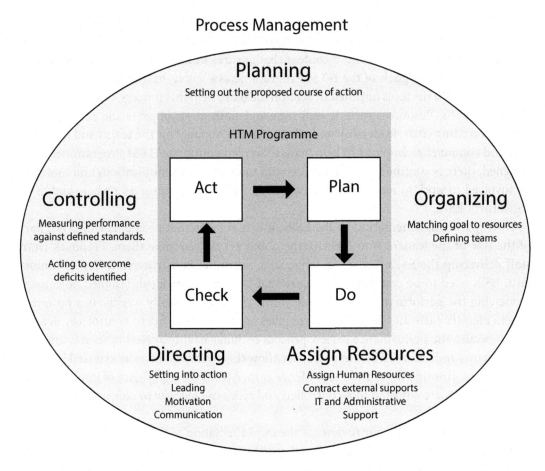

FIGURE 6.14   Processes that support the delivery of an HTM Programme constructed as a quality cycle.

way of working, perhaps more efficiently or effectively. Preventive actions are those taken in response to a risk-based approach, perhaps a situation being uncovered as a result of an audit, with the intention of preventing its recurrence. The value lies in finding out early on in a process that something is adrift and making alterations to the process, rather than finding out too late that things have gone wrong and have led to losses or harm. The Clinical Engineering Department leadership, by looking at these logs on a regular basis, will help form a picture of how vigilantly its department is performing and how proactive it is in making regular improvements.

6.4.1 Step 1: Group Equipment into Support Groups

With typical medical equipment asset databases containing thousands of equipment items and with a wide variety of equipment types and models, there needs to be some attempt to group assets together to make planning their ESPs efficient. At a high level, asset names offer a straightforward method to group equipment, but seasoned clinical engineers will understand that there are pitfalls. For example, an ECG monitor could be a simple low acuity monitor for basic monitoring and local transport or a high-end system with additional features such as analysis, trending, printing and networking capability. Both have similar roots but their support requirements differ. The objective of a grouping exercise is to identify equipment that has aspects of their support in common. Approaches differ and include grouping by clinical use (ventilation, anaesthesia, dialysis) or by the area where they are used (Critical Care Unit, Endoscopy Unit) or by their technology (Optical Devices, Electronics, Mechanical, Pneumatic). There are several methods that can be of help in the grouping of assets, and they can be nested within each other. They are not exclusive and can be mixed as required.

A further method of grouping assets can be made by the type of maintenance that the item requires. For certain equipment, the manufacturer advises that no regular maintenance is required at all, as failure would be readily apparent and only an inconvenience to the user. In this case the ESP could state that the equipment is only maintained 'On Request', sometimes called 'Run to Failure'. Remember though that mains-operated equipment may be subject to national health and safety guidance and that a regular electrical safety test may be recommended. Where equipment requires routine maintenance or inspection and is usually located in some form of system or assembly, it may make sense to create an ESP for the whole system. For example, endoscopic video systems comprising processors, displays and light sources are frequently assembled together and used together semi-permanently. In this case testing would very likely be carried out on the whole system in order to ensure interconnectivity was performing as expected.

Finally, when an organization completes a standardization project on say defibrillators or infusion pumps, the opportunity exists to group these assets together in a very logical fashion. Obviously this is the most preferable option as the ESP would contain identical equipment, but in the practical world it does not happen as often as the clinical engineer would wish. Nevertheless, for the purposes of illustration in the rest of this chapter, we will describe an HTM Programme where devices are grouped by equipment types (Infusion, Physiological Monitors, etc.) accepting that within each grouping the particular ESP might vary from model to model.

6.4.2 Step 2: Estimate Impact of Equipment Failure

The development of all ESPs will be guided by assessing the value they deliver and each will include an appropriate process to reduce the risk associated with failure. We propose that the grouping be prioritized for review on the basis of the harm that their failure might cause for clinical care, business continuity or financial control (Case Study CS6.10).

Logical systems have been developed for classifying the level of risk to patient care and the healthcare organization that failure of a particular device would entail. These structured systems are useful in providing an objective method of comparing the risks of different types of equipment. The risk will depend on the functionality of the equipment, the environment in which it is used and the impact of equipment failure (or lack of its availability) on patient care. Assessing the level of risk will involve a dialogue with clinical users. An example of a general risk classification offering definitions for low-, medium- and high-risk equipment is that published by ECRI (ECRI 2008). Typically the classification systems require the clinical engineer to combine the various factors and critically assess the resulting level. They offer a high-level method of grouping assets together, perhaps most useful in prioritizing maintenance when sufficient resources are not available. Where resources are limited, for example, an ESP could be developed for low-risk equipment stating that scheduled maintenance for this equipment is at a lower frequency than for medical equipment rated at higher-risk classifications. However, careful and holistic thinking is required to establish appropriate ESP as is illustrated in Case Study CS6.11.

6.4.3 Step 3: Prioritize Equipment Groups for Review

Having grouped the asset register of medical equipment into logical grouping, the intention is to develop ideal ESPs for each of these groupings and in doing so develop the HTM Programme. However, this takes time and resources in itself. It is unlikely that a comprehensive HTM Programme can be established or reviewed in a short period of time. Remember each ESP requires research, gathering of evidence of device performance and development of a means of delivery. It is much more likely that the HTM Programme grows and evolves over time as part of the quality cycle.

The Clinical Engineering Department needs to prioritize the planning of its ESPs and HTM Programme, being mindful of the strategic aims of the organization and the HTM Policy developed through the Medical Device Committee. Working within constraints, be they budget, premises, staff or lack of skills and experience, can inevitably lead to a pressure to prioritize work within the development of the HTM Programme and ESPs. It is important that a logical and demonstrable method is adopted to ensure that the appropriate consideration is given to the task and that it is evidenced based.

6.4.4 Step 4: Develop the Ideal ESP for Each Equipment Grouping

Once stratified, the department can start the development of the ESPs using the factors described in Sections 6.2.5 and 6.2.6.

6.4.5 Step 5: Assign Available Resources to ESP Delivery Teams

The Clinical Engineering Department (CED), when planning the implementation of the ESPs that form the HTM Programme, will need to assign resources to ensure that the ESPs can be delivered effectively. We have already seen that the planning will have involved classification of the risk levels of different medical equipment and this will inform the prioritization of the ESPs.

The CED will typically be structured with different skill sets and skill levels. In larger departments, the CED may employ staff to carry out routine, less technically intensive work, and the CED leadership will review the skill levels, assigning to individual clinical engineers' tasks appropriate to their skill level. This may result in the assignment of the procedures in ESPs that involve largely routine 'mechanistic' tasks to staff with less skill, whilst assigning the more complex procedures to more skilled and experienced staff. Such planning will lead to the development of ESPs that deliver the desired level of service and have an appropriate cost-effectiveness index as described in Section 6.2.6. The CED leadership will typically provide a continuous professional development programme that will seek to enhance the skills and experience of all staff members to enable them to carry out their assigned procedures competently.

The implementation of ESPs in the context of the HTM Programme requires administrative tasks that could perhaps be carried out by support staff within the CED. The CED leadership will consider the administrative burden, how it can be minimized by appropriate methodologies and technologies (including IT support and handheld recording systems) and also how administrative support staff can free the clinical engineers from administrative tasks to concentrate on technical tasks.

We have emphasized earlier that carrying out the ESPs involves a team approach, recognizing the shared roles of clinical, technical in-house staff and external medical equipment suppliers, as well as where appropriate patients and carers (Figure 6.2). The ESPs will have been designed recognizing these shared roles, and in planning the implementation, the CED leadership will need to ensure that tasks, as outlined in the ESP, are assigned to the most appropriate staff. The CED leadership may need to review the shared roles in light of changing resources available within the department in relation to the overall task of implementing the HTM Programme.

An important leadership task for the head of the Clinical Engineering Department is to assign the available resources to action the ESPs. This is most likely going to be a risk management exercise as typically there will not be enough resources to implement all the ideal ESPs developed. In step 3, we discussed the prioritization of equipment grouping for review using risk assessment. Now that the demands and resources are known, the leadership must assign the resources to best mitigate the risk and optimize value. This will be an ongoing and important activity within the quality cycle. We suggest that clinical risk be given the highest priority.

6.4.6 Step 6: Identify and Communicate Any Residual Risk

A consequence of step five will be that there will be some risks which the HTM Programme cannot control. These should be documented in a risk register and communicated to the MDC, so that the MDC has the information it needs to manage the allocation of resources and risk across all groups in the organization who are implementing HTM Programmes.

6.5 IMPLEMENTING AND AUDITING THE ESPs WITHIN THE HTM PROGRAMME

6.5.1 Set ESPs into Action

The leadership in the department needs to set the ESPs into action. Usually, this is done by assigning their implementation to individuals or staff groups and encouraging and mentoring those individuals to deliver the ESPs. But with such a wide variety of medical equipment and applications, it is not going to be possible for each clinical engineer to be an expert on all the equipment supported by the department. To solve this issue, larger departments have chosen to allow clinical engineers to specialize, in the same way as clinical staff, in that they have a basic knowledge across the range but a deeper knowledge of a speciality such as cardio or imaging or anaesthetic equipment. Although this solves one problem, it can create others should that person leave or be busy elsewhere when needed. There may also be challenges in providing out-of-normal hours 'on-call' services. So a balance needs to be made with specialization. However, we must remember that others in the organization, such as equipment users, may have a role in the ESPs, and they also need to be empowered to act. Where external agencies are part of the ESP, then service level agreements need to be agreed and formally put in place. The secret of success is good planning.

6.5.2 Continual Management of the ESP Delivery

Problems will arise with the practical delivery of the HTM Programme and its ESPs. These could be caused by failures of multiple devices at the same time, by product recalls of equipment deployed in large quantities or by corrective actions required due to manufacturer's field safety notices, by external factors such as mains power failures or flooding or by sudden acute staff shortages within the CED. Hence the process will need careful management.

The management will start by recognizing the professionalism of the clinical engineers tasked with implementing the ESPs, and the individual clinical engineers need to recognize their professional standards and ethics. Management of the process involves team management, with its emphasis on listening, dialogue and recognizing the contribution of all team members.

Where unforeseen problems arise or difficulties are encountered, the clinical engineers delivering the ESP will attempt to solve the problems as they arise. Where problems are beyond the capability of the individuals or teams to solve, they should be escalated to the department leadership.

Working within constraints is a day-to-day pressure on the Clinical Engineering Department and the skill of finding a way of understanding, acknowledging and then working to these limits in the most effective way possible needs to be acquired. Very often it is a balancing act, trading off one aspect against another in order to find a balance that provides a good service to the clinical user and a safe and effective service to the patient, and does both in the most cost-effective way.

6.5.3 KPIs: Seeking Assurance That the HTM Programme Is Being Delivered

The HTM Programme, having been planned and set in motion, can progress, but after some time, a check will need to be made to see if it is carrying out what it is supposed to be doing and if objectives are being met. The leadership in the Clinical Engineering Department will seek to provide an assurance that processes and outputs are in line with expectations and targets set by the MDC and will flag this up if it is not the case. Assurance is not to be confused with reassurance. Assurance is factually based with evidence to back it up whilst reassurance only seeks to comfort. Assurance is freedom from doubt, even if the outcome is uncomfortable.

Key performance indicators that measure the delivery of the ESPs can be useful in helping those tasked with their delivery. A key performance indicator (KPI) sometimes referred to as a performance measure can be defined as 'a quantitative tool (e.g. rate, ratio, index, percentage) that provides an indication of an organization's performance in relation to a specified process or outcome' (JCAHO 2005, p. 100). Parmenter has written two instructive books on KPIs, the second aimed at 'Government and non-profit agencies' which clinical engineers might find helpful (Parmenter 2007, 2012). Summary reports on the performance of the HTM Programme may include relevant KPIs.

A paper, 'On target – The practice of performance indicators', published in 2000 by the UK Audit Commission (Audit Commission 2000) suggested key components of successful KPIs:

Relevant: The key is to focus on the right topics, think about what is important – what the quality objectives are and the Clinical Engineering Department's aims. Relevant topics could be to do with economy (cost of repairs this year), efficiency (cost per repair) and effectiveness (are the users happy with the cost of the repair).

Understandable: KPIs should be easily understood with a minimum of jargon. The source of the data should be clearly defined and verifiable with evidence to back it up.

Comparable: There is an old adage about not being able to compare apples with oranges because they are completely different fruits. That may be so but they are both fruit and both round, so there are some common denominators already. It has to be acknowledged, however, that organizations differ in how they work and manage assets but that still does not exclude some functions being compared between hospitals; an example is 'resolve time', that is the time from when a problem is first brought to the attention of the CED to when the issue is resolved, for example a repaired item of equipment is returned to service. Benchmarking which we have discussed in more detail in Chapter 5, Section 5.9.2, is a useful tool for CEDs seeking to improve their services comparing their performance in relation to other comparable departments. There is also another slant on comparability and that is comparing KPIs over time. This is sometimes called 'self assessment' and it is a valuable process as it is comparing like with like over time. However, in order to show change over time within the same organization, it is important to maintain the same definition of indicators.

Timely: A responsive indicator is one which provides a result within a useful timescale. If a weekly KPI is required, but takes a week to gather and calculate, it is probably too late and not useful.

Action focused: Beware of measuring for measuring sake. Is the data useful or not? If there is nothing to be done with the result, good or bad, then is it worth measuring? Care needs to be taken in developing indicators to avoid any 'perverse incentives'. These are unwanted side effects of, for example, the decision to monitor repairs completed within a target time, but which results in a lowering of quality because the job is rushed. We have already made the point in Chapter 5 (Section 5.4.1.3) that when a measure becomes a target, it ceases to be a good measure.

Tailored: There is a need to understand the recipient of the KPI data. Who is the data for? Does the data provide good and useful information? There will be a different set of data requirements depending upon whether it is for the public, technical staff or the Medical Device Committee. Think carefully about their needs.

Balanced: A typical KPI dashboard would be considered balanced if it had KPIs from four perspectives:

- Service users;
- Internal Management;
- Continuous Improvement;
- Financial.

TABLE 6.6   Template for a Balanced Scorecard Assessment Methodology for Equipment Support Plans and the Healthcare Technology Management Programme as a Whole

Service User	Internal Management
Response time – from the requesting phone call to first action	Scheduled maintenance completion – percentage attempted in a 12-month period
Time to completion – from the requesting phone call to completed job	Repair jobs open for more than 3 months
Continuous Improvement	**Financial**
Annual 'Customer Satisfaction Survey'	Average annual maintenance cost per item of medical equipment
Time spent on engineer training – hours per year	Average annual maintenance cost per bed-space

Balancing KPIs is important because rarely can the full picture of the operation of an HTM Programme, particularly one which aims to take a holistic approach to equipment support, be obtained by looking at one single aspect of it. A typical balanced 'scorecard' of KPIs is suggested in Table 6.6.

The ability of KPIs to help manage ESPs within the context of the HTM Programmes is discussed in Case Study CS6.12.

The balanced scorecard can and should be defined for individual organizations with the principles set out earlier used as a guide. The evaluation of the balanced scorecard could well be used to measure the effectiveness and appropriateness of the delivered ESP. So the evaluated scorecard for each ESP could be used as the measure of benefits delivered when comparing ESPs and optimizing value.

6.5.4 Reviewing the ESPs within HTM Programme

The leadership of the Clinical Engineering Department will review all the ESPs regularly to ensure each is appropriate and delivering and also to ensure that the resources assigned to the department are being distributed across the various ESPs to manage risk and optimize value. The outcome of the analysis of the impact of failure on clinical care will result in a hierarchy of ESPs with some judged to have a high risk associated with failure, others with medium risk and another group with low risk. By grouping the ESPs by risk and plotting their ideal and delivered ESP level, as measured using the balanced scorecard, the clinical engineering leadership can get an overview of the performance of the HTM Programme that includes risk and performance level. Figure 6.15 is presented to illustrate what such a plot would look like.

Reviewing Figure 6.15, we can immediately see that neither ESP A or ESP C are delivering to the required level even though they are judged to be in the high-clinical-risk equipment group. The delivered level of ESP E on the other hand exceeds requirements, suggesting perhaps resources could be reassigned from it to improve the delivered performance of ESP A and ESP C.

Such a plot could be developed to also give an indication of value by including the CEI of each ESP (Figure 6.16).

The addition of the financial index gives more information and helps with developing the plan for further investigation. Continuing the example, we see that of the two high-risk

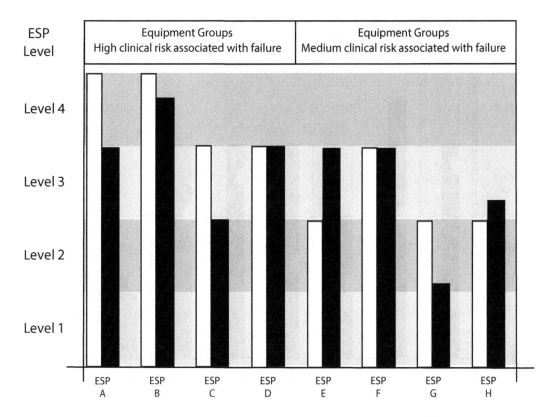

FIGURE 6.15   Comparison of planned and delivered support from a cohort of ESPs that make up an HTM Programme. The white bars are the planned support levels, the black bars, the measured delivered support levels.

plans which are not delivering as required, ESP C has a low CEI, perhaps indicating that there are simply not enough resources assigned to it, further suggesting it might best benefit from any reassigning of resources from ESP E. ESP A on the other hand has a high CEI and yet is not performing well; this would suggest that the design or managed delivery of this plan is not optimized and the quality improvement action might be a careful review of its implementation to discover the reason why and then act to rectify issues identified.

A final note on KPIs is that clinical engineering departments are not professionally in competition with each other. There is a temptation to withhold information as it could be commercially sensitive, and there will be some information in certain CEDs to which this applies, particularly in commercial for-profit or not-for-profit departments. However, sharing information between departments in safe environments such as benchmarking clubs is mutually beneficial. We must remember the core focus of our profession is to increase value to the patient. Lack of sharing between CEDs is likely to obstruct that goal.

6.5.5 Auditing to Ensure That All Equipment Has Been Checked

Audit will include assessment of whether all the equipment in the healthcare organization has been checked as part of the scheduled maintenance process: audits may also involve physical asset audits in the clinical environment.

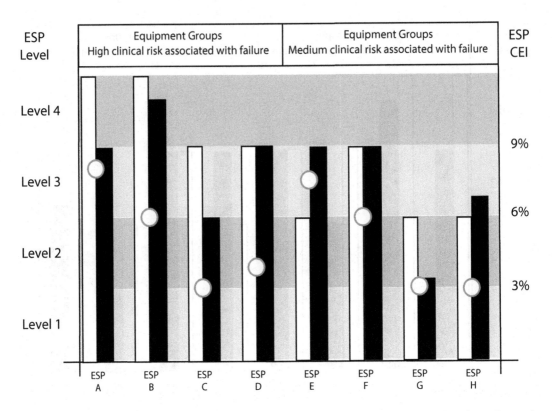

FIGURE 6.16 Comparison of planned and delivered support and costs associated with a cohort of ESPs that make up an HTM Programme.

Processes need to be adopted to ensure that equipment is not overlooked or disposed of incorrectly. The processes will include communication with clinical departments, informing them if equipment has missed a maintenance visit, is faulty and is beyond economic repair or not able to be found, potentially lost. Medical equipment may have been loaned to a clinical department by a supplier without having gone through the CED, posing a risk as it may be unknown to the organization and not supported properly. Missing equipment can be a real problem in a large organization, particularly when dealing with easily transportable or shared equipment. Several solutions exist for tracking medical equipment, such as active radio-frequency (RF) identification (RFID) tags, but these can be expensive to install. As usual, there is a trade-off, this time between the ease of finding equipment and cost of looking for it.

6.6 QUALITY IMPROVEMENT OF THE ESPs AND HTM PROGRAMME

6.6.1 Act to Improve the Equipment Process

The review process and the KPI data may suggest improvements in the HTM Programme. This could involve reassigning resources as suggested by data analysis (e.g. see Figure 6.16).

The equipment types and models included in a particular ESP may need to be reviewed. Perhaps the groupings could be improved, allowing for better resource allocation and improved ESP definition. Critical review of the groupings is useful.

The medical equipment asset inventory is not a static list but is constantly changing. With equipment being added, replaced, scrapped on a daily basis and occasionally lost, equipment will from time to time be added that is not covered by existing ESPs. It may be possible to fit this new equipment into a modification of an existing ESP, or it might be necessary to develop a new ESP.

New medical equipment types or variations on the old ones create work on the medical equipment database, and in a busy, developing hospital, there can be wards and departments opening, closing and changing speciality often. Keeping the MEMS database up to date is obviously important but is quite a time-consuming and difficult task, and priority resources for the management of this key piece of information infrastructure are required. The importance of an accurate inventory of the medical equipment within the organization, essential for replacement and maintenance planning, cannot be overemphasized and it must not be overlooked.

6.6.2 Review and Revise the Estimate of Impact of Equipment Failure

Just as implementing the HTM Programme may prompt a review of the equipment grouping, it may also suggest that a different prioritization for review and resource allocation be considered. Experience in implementing the programme or changes in the organization may affect the evaluation of clinical care, business continuity and financial risks. So reviewing and revising the estimates of the risks associated with grouping is prudent.

6.6.3 Develop a Quality Improvement Plan for the HTM Programmed

The final subject to tackle, and to complete the Plan–Do–Check–Act cycle, is to act on the results from the evaluations and implement improvements to the programme making it more efficient and effective in attaining its quality and asset objectives. Changes identified in Sections 6.6.1 and 6.6.2 prompt revision of the programme to improve its quality. Reviewing the programme from other perspectives can help identify a list of actions that can also form part of a quality improvement plan.

Both internal and external sources of information can inform improvements to the HTM Programme. It is helpful to direct ones thoughts from time to time in each direction to provide a differing perspective on the HTM Programme. There is often a temptation to become too inward looking, trying to invent solutions to issues when the solution may already be developed and available from elsewhere. Why reinvent the wheel when clinical engineering colleagues are happy to share best practice? Benchmarking can assist sharing best practice. Remember, the patient is at the focus of all we do.

6.6.3.1 Looking Inwards for Improvements

An active HTM Programme will produce much data on how it is performing and where the weak areas are in its implementation. KPIs, audits and customer feedback will all inform the leadership of the Clinical Engineering Department who then need to take the appropriate action.

The assessment of the implemented ESPs from a value perspective can be useful in illuminating where resource reallocation is merited. Where implemented ESPs fail to meet appropriate levels, further investigation is warranted, particularly if the equipment concerned poses a high risk to the patient if it fails. Similarly, groups of equipment where the cost-effectiveness index exceeds expectations merit review from a financial risk perspective. Case Study CS6.13 gives a practical example. The leadership of the department should suggest and document changes to the HTM Programme to maximize value across concurrent ESPs.

It is often assumed that improvements in reliability may be achieved if ESPs are resourced to deliver more scheduled actions, but conversely evidence suggests that a reduction in the frequency of scheduled actions might not have any impact on the safety or reliability of some medical equipment. There is also evidence that unnecessary maintenance interventions can cause failures that may not otherwise occur. The topic of reliability-centred maintenance (RCM) (Moubray 2000) is debated regularly and widely amongst clinical engineering colleagues with heated reasons for and against its adoption. We feel that it is up to the organization to consider whether this approach is acceptable or not, but would encourage critical ongoing review of all ESPs based on a risk management approach that prioritizes patient impact and uses a value-based assessment that balances both benefits delivered and cost. RCM can be easier to do in a large organization that can effectively acquire all the data set needed to properly support the process. In smaller healthcare organizations, there will be merit to working with other organizations. Pooling of data between healthcare organizations can provide valuable insights into service support strategies and also equipment reliability (Gandillon 2013).

Some questions that might provoke a more holistic review of the ESPs include the following:

- What scientific data support the recommended frequency for scheduled actions?

- What is the worst-case scenario should the equipment develop a fault in any way?

- How does this vary, dependent on equipment type – for example life support, diagnosis, treatment or alleviation?

- What responsibilities rest with the user/operator of the equipment and does this have any impact?

- What legal and ethical responsibilities need to be considered?

Clinical engineers should lead and be proactive in seeking out improvements. This responsibility distils down into communication between fellow engineers, between engineers and clinical users and also with patients where appropriate and practical. The statistics themselves are not to be taken simply at face value and some interpretation is often required. Remember our assessment of value is qualitative and subjective; consequently, triangulating the value measure with risk assessment and other information sources is warranted.

6.6.3.2 Looking Outwards for Improvements

Looking outwards from your own Clinical Engineering Department to other departments or services should be encouraged as a learning exercise to clarify your current performance and identify where improvements can be made; the process is not aimed at scoring performance on a sort of league table. Many clinical engineering departments, organizations and services carry out the same tasks as each other but perhaps may have developed different approaches. Investigating and understanding the strengths and weaknesses of the various approaches can reveal processes that can be adapted and changed to implement improvements. The opportunities for improvement come from several directions.

Exemplars: Having already said that there should be no league table, there will often be an exemplar Clinical Engineering Department that is considered by many in your country or region to be a reference department. It may be that one department is particularly good at something compared to another. These strengths should be recognized, whilst not expecting to find one department that is far stronger than all others, nor using this as a pointer on a league table. A well-managed, well-led and resourced department can be inspiring to clinical engineers who desire to develop and improve their own service delivery.

Benchmarking, which we have discussed in Chapter 5, Section 5.9.2, is the process of formally comparing services or parts of services directly with others in a similar position. It might be that one Clinical Engineering Department agrees with another to benchmark the time taken to carry out a particular scheduled maintenance task. The results might differ enough for the departments to investigate the reasons for the differences. Processes, KPIs and financial indices can all be tools for benchmarking with other departments. Regional benchmarking groups have been set up in several areas of the world, for example, between the clinical engineering departments in large cities such as London in the United Kingdom where many different hospitals provide healthcare. Benchmarking tools and support systems have been developed. For example, in the United States, AAMI provide three benchmarking software tools to subscribing members with training support and advice (http://www.aami.org/productspublications/content.aspx?ItemNumber=911&navItemNumber=679).

6.6.3.3 Best Practice

Whenever groups of clinical engineers meet, there is always a desire to improve or learn from each other. After a time, and sometimes with the input of institutes or national governing bodies, a recommendation emerges that can be considered best practice amongst the profession. It might not be a statutory requirement but rather recommended practice. It is important that clinical engineers have the opportunity to network with colleagues from other healthcare organizations through attendance or presenting at conferences and seminars. Much useful information and perspective can be learned from these activities, and clinical engineering leadership should include appropriate resources in their planning to allow for this type of continuing professional development. It brings benefit. Healthcare organizations should encourage CED leadership to attend these types of events and should provide the opportunity for organizational listening and learning available to them when their internal experts network and learn from best practice elsewhere.

There is also benefit in sharing failure rates and failure modes across different centres. Whilst one healthcare organization may have 10–20 of a particular medical device, the shared experience of many organizations could provide experience of over 100 of the devices. The U.S. Veterans department has published interesting work analyzing the reliability of different infusion pumps from their experience of equipment across many of their facilities (Gandillon 2013).

6.7 REPORT ON THE STATUS OF THE HTM PROGRAMME TO THE MDC

The Clinical Engineering Department should be given responsibility, authority to act and resources from the appropriate authorities within the healthcare organization. It should be able to undertake tasks commissioned by the MDC or other departments reporting to the MDC. There is a balancing requirement for the CED to report assurances and residual risks to the MDC and appropriate summary reports on the performance of the HTM Programme are required and may include relevant KPIs.

The quality improvement plan proposed for the next cycle of the programme should be reported for a number of reasons. The proposals themselves are evidence that the plan is active. It may be that implementing it requires extra resources and the MDC needs sight of this for consideration. Learning and improvements identified within one department could be shared across the organization, and the MDC has a role in identifying and facilitating this. Finally, it may be that a change to the HTM Programme proposed by the Clinical Engineering Department is contrary to evidence identified in another department or does not line up with changes in policy and organizational goals, so the MDC has a role in reviewing and endorsing the quality improvement plan.

A special type of risk register is usually developed to identify equipment that needs to be replaced; this may be called the Equipment Replacement Register. It will inform medical equipment procurement planning, helping develop priorities that may modify a baseline rolling equipment replacement plan. Equipment which gives rise to a clinical, financial or business continuity risk that can only be mitigated by its replacement is identified for consideration for replacement by the MDC. Old equipment that has become excessively costly to maintain might be placed on the Equipment Replacement Register to mitigate against rising costs. Old equipment for which service and spare parts are difficult to obtain might be identified for replacement because failure will result in a loss of service and long repair time which will impact negatively on business continuity. Other equipment might be recommended for replacement because it has been superseded by a new technology that delivers better clinical outcomes or reduced clinical risk. In this case, replacing the device mitigates against the delivery of less than optimal care.

6.8 MANAGING THE HTM PROGRAMME FROM A RISK MANAGEMENT PERSPECTIVE

We have discussed risk management from the perspective of individual items of equipment and their interactions with patients elsewhere in this book. At the core of risk management is the identification of hazards and hazardous situations, and here the clinical engineer is perhaps uniquely placed to be the focal point for gathering 'risk intelligence' relating to medical equipment. When managing HTM Programmes, clinical engineers must be aware of the different risks associated with different groups of equipment. In assessing the different risks associated with equipment groups, we need to consider not only how the equipment functions but the clinical context within which it is used. Clinical engineering departments that implement holistic HTM Programmes will not only be familiar with the engineering aspects of equipment but also how they are used and the characteristics of the environments in which they are deployed.

It is almost always the case that resources to implement HTM Programmes are limited and require those who deliver them to prioritize resources to different groups. We suggest this be done with an emphasis on assessing possible risks to the person being treated should an equipment failure occur.

The aim of risk management in the context of healthcare technology management is to identify risks associated with the use of medical devices and ensure that they are minimized enabling benefits accruing to outweigh any harm that may be caused. This is also the essence of a good HTM Programme. The clinical engineer should be able to identify hazards and quantify risks, drawing on intimate knowledge of the inherent risks associated with the use of medical devices and importantly the context in which they are used. The risk management process, as outlined in the ISO 31000 (ISO 2009) standard, is shown in Figure 6.17 and clearly highlights the need to establish the context within which the risk can occur.

The management of the HTM Programme can also be considered as a risk management exercise. However, in HTM Programme management, we are managing the risks of many different pieces of equipment. The challenge is to evaluate the risk associated with each group, develop and implement ESPs for each and measure the effectiveness of the solution.

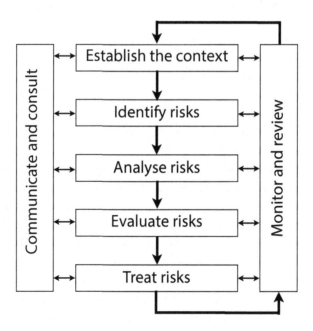

FIGURE 6.17 Risk Management process. (Adapted from ISO 31000 Risk Management – Principles and Guidelines, International Standards Organization, Geneva, Switzerland, 2009. With permission.)

If unacceptable risk in one area is manifest, it may be necessary to change the ESPs to reassign existing resources to mitigate that risk or escalate the issue to the MDC who in turn might assign more resources to mitigate the risk. It is not uncommon for clinical engineers to have to balance resources as best they can to control risk. However, it is vitally important that the risks be considered in terms of the operation of the equipment in context, and from the perspective of the impact on the person receiving care.

When managing risk, it is essential that all actions to address risks are recorded and monitored and that the clinical engineer understands the workings of their host organization. For risks that cannot be managed 'locally' within the department by making adjustments to the HTM Programme, the escalation route, most likely through the MDC, must be known and used. In practice, this often requires an understanding of organizational committees and reporting structures, together with the knowledge of the level of detail and format of reports each group requires.

6.9 CONCLUSION

In this chapter, we described how a Clinical Engineering Department manages the development and delivery of an HTM Programme within a PDCA quality management cycle. Figure 6.18 illustrates the whole process.

The fact that the ESPs and the HTM Programme as a whole is managed as a quality cycle and documented goes a long way to ensuring good risk management practices are built into the HTM system. The processes that support escalation of risk and inclusion of equipment-related risks on the hospital's risk register also add to this, and by keeping the assessment of risk person and care focused, it improves further still.

We have now identified the linkages between the MDC and the hospital-wide Medical Device Policy, the Strategic HTM Plan and the specific HTM Programme developed in response to it by the Clinical Engineering Department. In doing so, we have explored how the design, continuous review and implementation of equipment-specific support plans are the mechanism by which the aim and objectives of the Medical Device Policy are implemented.

The ESPs are best developed and reviewed with an eye to how they deliver value for the organization, and so this requires analysis both of the benefits each delivers and the associated costs. This activity, which takes a holistic approach, requires individuals with a broad range of engineering, management and communications skills. Even where much of the support is outsourced, there remains a need for the organization to resource clinical engineers to manage the HTM Programme and provide clinical user support not available through external contractors. The presence of such a Clinical Engineering Department promotes best practice and delivers value through optimizing the application of technology in the provision of care. This includes the traditional maintenance and equipment management roles but today extends further to management of the assets in a wider context, focusing on other actions discussed in Chapter 7, Section 7.3, which through specific project work advance patient care.

The HTM Programme must also provide the opportunity for managing the relationship between the supplier and the healthcare organization. The selection of equipment from a

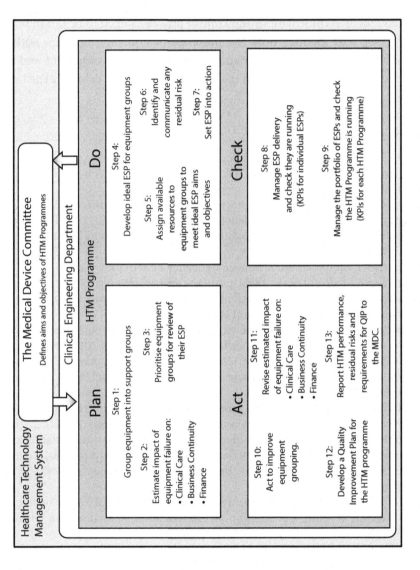

FIGURE 6.18 The steps that make up an HTM Programme structured as a Plan–Do–Check–Act quality cycle.

particular supplier makes that supplier, to varying extents, a partner in the delivery of care. Managing this relationship is particularly important when the supplier supports the equipment remotely, 'dialling in' to work on the equipment. Remote working should not be done without control nor with the appropriate personnel in the healthcare organization unaware of the details. The healthcare organization should always be given advance knowledge of software changes, which may require to be approved by its IT department, with the CED providing important support with this. Remote working can have implications for the security of patient confidential information, and control processes need to be put in place.

In Chapter 7, we will identify how the presence of clinical engineers with experience in both strategic and operational management of medical equipment and involved actively in supporting its use at the point of care can add further value for the healthcare organization.

REFERENCES

AAMI. 2015. ANSI/AAMI EQ89:2015 Guidance for the use of medical equipment maintenance strategies and procedures. Arlington, VA: Association for the Advancement of Medical Instrumentation.

Audit Commission. 2000. On target – The practice of performance indicators. London, UK: UK Audit Commission. http://webarchive.nationalarchives.gov.uk/20150421134146/http://archive.audit-commission.gov.uk/auditcommission/subwebs/publications/studies/studyPDF/1398.pdf (accessed 2016-05-11).

CMS. 2013. Alternative equipment maintenance frequency or activities. Baltimore, MD: Centers for Medicare and Medicaid Services. https://www.cms.gov/Medicare/Provider-Enrollment-and-Certification/SurveyCertificationGenInfo/Policy-and-Memos-to-States-and-Regions-Items/Survey-and-Cert-Letter-14-07.html (accessed 2016-05-11).

ECRI. 2008. Assessing scheduled support of medical equipment. Plymouth Meeting, PA: ECRI. http://www.smbe.asn.au/SMBE_Hot/TimRitterECRI.pdf (accessed 2016-04-11).

FDA. 2016. Medical Devices—Home Use Devices Initiative. http://www.fda.gov/MedicalDevices/ProductsandMedicalProcedures/HomeHealthandConsumer/HomeUseDevices/ucm208268.htm (accessed 2016-05-21).

Gandillon R. 2013. Infusion pump reliability and usability: Veterans health administration examines differences between manufacturers. *Journal of Clinical Engineering*, 38(1): 27–31.

IEC. 2014. IEC 62353:2014 Medical electrical equipment – Recurrent test and test after repair of medical electrical equipment. Geneva, Switzerland: International Electrotechnical Commission.

ISO. 2009. ISO 31000 Risk Management – Principles and Guidelines. Geneva, Switzerland: International Standards Organization.

JCAHO. 2005. Tools for performance measurement in health care: A quick reference guide. Oakbrook Terrace, IL: Joint Commission on Accreditation of Health Care Organizations. http://bit.ly/1qKHOMR (accessed 2016-05-11).

Keay S., McCarthy J.P. and B. Carey-Smith. 2015. Medical equipment libraries – Implementation, experience and user satisfaction. *Journal of Medical Engineering and Technology*, 39(6): 354–362.

McCarthy J.P., Scott R., Blackett P., Amoore J. and F.J. Hegarty. 2014. Healthcare technology management. In *Clinical Engineering – A Handbook for Clinical and Biomedical Engineers*, eds. A. Taktak, P. Ganney, D. Long and P. White, pp. 43–57. Oxford, UK: Academic Press.

MHRA. 2015. Managing medical devices – Guidance for healthcare and social services organizations. Ed1.1. https://www.gov.uk/government/publications/managing-medical-devices (accessed 2016-04-11).

Moubray J. 2000. *Reliability-Centered Maintenance*, Edition 2.2. New York: Industrial Press, Inc.

Parmenter D. 2007. *Key Performance Indicators: Developing, Implementing, and Using Winning KPIs*. Hoboken, NJ: Wiley.

Parmenter D. 2012. *Key Performance Indicators for Government and Non Profit Agencies: Implementing Winning KPIs*. Hoboken, NJ: Wiley.

Wilson K., Ison K. and S. Tabakov. 2014. *Medical Equipment Management*. Boca Raton, FL: CRC Press.

SELF-DIRECTED LEARNING

1. *Clinical user involvement in scheduled maintenance.*

   How important is clinical user involvement in the routine scheduled maintenance of medical equipment? Discuss with technical and clinical staff. The clinical staff could be the nurse in charge of a ward who might be responsible for the use of general medical equipment such as ward monitors and infusion devices. The clinical staff might also be support staff in an operating theatre responsible for a range of theatre equipment including anaesthetic machines, ventilators, patient monitors and surgical instruments.

2. *Continual improvement of an ESP.*

   Reflect on the support plan that you use for medical equipment and choose one which you think could be improved. What are its weaknesses and strengths? What evidence do you have that you could use to improve the support plan? Evidence could come from experience of failure rates, from manufacturer information and from colleagues, perhaps in other healthcare organizations. How could you use this to inform and document the reasons for improving the ESP?

3. *Continual improvement, learning from others.*

   How can you improve the services that you deliver by learning from other CEDs? Does your CED undertake benchmarking exercises with other departments or do you meet colleagues from other departments? Clinical engineering journals and magazines often have articles describing particular aspects of clinical engineering services. Consider an area of your service provision, reflecting on its positive and negative aspects and, with colleagues, carry out a 'Strength–Weakness–Opportunity–Threat' analysis. What evidence from other CEDs or from journal articles could help you improve the service?

4. *The clinical user procedure in an ESP.*

   Following on from Sections 6.2 and 6.6.1, draft an ESP for the clinical user maintenance of an item of medical equipment.

5. *The procedure for an ESP for a home user caring for medical equipment.*

   Develop an ESP for an item of medical equipment that is being used in the community.

6. *Home dialysis.*

The provision of home dialysis can have many advantages for patients and may also have financial advantages for the healthcare organization. What procedures should be incorporated into an ESP to help guide home users of dialysis machines, either the patient or the patient's carer or both? How should the ESP clarify the shared roles and responsibilities in ensuring the continuing safe and effective operation of the dialysis machine, considering the roles undertaken by the home user or carer, the community nurse, the technical support organization and the supplier of the dialysis machine?

CASE STUDIES

CASE STUDY CS6.1: DEVELOPING AN ESP FOR AN MRI SCANNER

Section Links: Chapter 6, Sections 6.2.1; Chapter 4, Section 4.4.1

ABSTRACT

Developing a comprehensive equipment support plan (ESP) for major medical equipment installations such as MRI scanners requires a holistic approach that incorporates technical, MRI safety, clinical, patient and infrastructure service critical support systems.

Keywords: Equipment Support Plan: MRI scanner; Imaging; Holistic; Safety; Patient

NARRATIVE

The Radiology Manager, as part of the planning for the installation of a new MRI scanner, convened a core project team to develop an ESP for it. The Head of the Clinical Engineering Department (CED) was asked to lead on the service contract support and the chief MRI radiographer to lead on the clinical and patient aspects. The Head of the CED noted the importance of infrastructure support and was asked to invite Facilities officers into a wider working group.

The MRI's powerful magnet requires that the risks associated with the magnetic and RF fields are managed through the ESP, thereby protecting staff and patients and preventing unauthorized access to the MRI suite. The Radiology Manager was keen to have in place a comprehensive support package, with the project team suggesting that it encompasses the technical support for the scanner, clinical support, infrastructure support and patient support.

TECHNICAL SUPPORT

The core of the technical support would be based around a comprehensive service contract with 4 h response time during the day (7-day week) and 8 h response time between 18h00 to the following 07h00. A preferred MRI scanner supplier had been selected, and options for original equipment manufacturer (OEM) or 3rd-party service support were considered. The OEM supplier offered an attractive 7-year comprehensive service package with remote support and software upgrades and with refresh of the computer systems after year 3 and year 5. Full labour, parts and RF coils were included as were 6-monthly safety and performance checks that also tested the magnetic field and integrity of the RF shield. Periodic independent image quality assurance checks were also incorporated in the service support package.

The MRI system itself continually carries out system checks, for example, of the helium level, with any problems flagged via the remote Internet support and actioned as appropriate.

The hospital's IT department was consulted to ensure that the remote Internet support, provided through a protective firewall, met with their approval. Including the service support contract as part of the acquisition procurement offered significant financial advantages.

MRI SAFETY SUPPORT

The MRI Safety Officer appointed for the previous MRI scanner would continue in this role. Working with the hospital's Health and Safety Department, the MRI Safety Officer was responsible for the MRI safety policy and for access controls to the MRI suite. These included processes that ensured that anyone who entered the suite would be scanned for metallic objects which would be removed prior to allowing entry.

The MRI Safety Officer was asked to revise the existing MRI safety policy to take into account the new MRI scanner, the policy to be reviewed annually. The MRI safety policy stipulated the controls in place for ensuring that only MR Safe equipment and, under control, MR Conditional equipment are brought into the MRI scanner room.

Update training for the MRI Safety Officer was arranged as was strengthening of liaison with MRI Safety Officers in neighbouring hospitals. Annual MRI safety training would be compulsory for all MRI radiographers and radiologists.

CLINICAL SUPPORT

Applications training to enable radiographers and radiologists to effectively use the functions of the imaging packages in an MRI scanner were included in the ESP, both for installation of software upgrades and regularly during the operating life of the MRI scanner as part of the comprehensive contract. The contract also included software upgrades (excluding release of additional functionalities), with training and support to make the most of them. Applications support specialists from the MRI supplier would work with the MRI radiographers to install and take advantage of improvements to imaging protocols and algorithms. The chief MRI radiographer would maintain training records of the MRI radiographers ensuring that all were trained to use the system effectively.

PATIENT SUPPORT

Careful support for patients entering the MRI suite is required. The process starts with referrer checks and then detailed checks carried out through a departmental questionnaire. Patients are reminded that metallic objects must not be taken into the MRI room and the questionnaire and screening both check for metallic objects including implants. Access to the MRI room is strictly controlled. The hospital's patient liaison officer was asked to advise on updating patient information leaflets, processes for patient screening and patient support including headphones to protect against the high noise levels generated by the magnetic and RF fields. Choices of music were provided, with the ESP including regular review and update of the music selections.

The ESP included annual review and updates of the patient support package.

INFRASTRUCTURE SUPPORT

The installed MRI scanner had a superconducting magnet reliant on helium cooling. The MRI scanner continually checks the helium temperature and will warn if it rises. The Facilities department had been asked to oversee the installation of the quench pipe (which is used to vent helium in an emergency) and to check it annually. Failure to ensure its correct operation could result in helium venting into the MRI room with associated risk of asphyxiation. Facilities were also required to ensure that all support systems including air conditioning are regularly maintained. LED lighting had been installed to minimize the need for changing light bulbs. The Facilities staff responsible for MRI support had received specialist training in MRI safety and this would be updated annually.

The Fire Officer had advised on the design of the MRI facility to ensure that it complied with fire regulations. Whilst the risk of fire in MRI suites is low, it was recognized that procedures need to be in place for the safe management of fires within the area including the controlled access of firemen. Procedures were included in the MRI safety policy, with annual update meetings scheduled with local Fire Brigade officers to ensure that the firemen could advise on and were acquainted with the procedures.

The hospital's hotel services were also contacted to discuss cleaning arrangements and to ensure that only cleaning staff with knowledge of MRI safety were involved in the cleaning of the suite.

Compilation of the ESP

The Radiology Manager convened a meeting of all those involved to discuss and agree the comprehensive MRI ESP. The ESP was developed as a compilation of various sections, each section referring out where appropriate to specific documents and procedures. Thus the ESP referred to the MRI safety policy, noting who was responsible for it, and referred to the MRI service contract and Facilities management arrangements, with clear lines of responsibility and accountability.

ADDING VALUE

Developing a holistic ESP that covers technical maintenance, safety controls for access to the MRI suite, clinical support including applications training and update of clinical protocols, patient advice and support and infrastructure support gave the Radiology Manager confidence that the MRI scanner would provide safe and effective clinical care. The comprehensive nature of the package did not reduce costs (though arranging service contract support upfront did reduce those costs), but added benefits of ensuring safety and efficacy.

Benefits : Cost ∴ Value

SUMMARY

Support for complex medical equipment such as MRI scanners is not restricted to its technical service support contract. It requires a multidisciplinary holistic approach developed by a team that covers technical support for the scanner, procedures to ensure safety and minimize the risks associated with the magnetic and RF fields, support for the clinicians using the system, patient considerations and Facilities management support, with update training provided for all involved.

SELF-DIRECTED LEARNING

1. If elements of the support package are missing, what risks would that bring for (a) the patient, (b) radiology staff, (c) facilities and service staff and (d) operational integrity of the MRI system?
2. Review the support packages available for the MRI scanners in your healthcare organization. What provisions are made for the various support aspects? How is MRI safety assured? Map out who is responsible for the different aspects. And are their knowledge and skills regularly reviewed and updated?
3. This Case Study developed an ESP for an MRI scanner. Develop an ESP for a CT scanner showing the infrastructure, clinical and patient aspects required.
4. Could this approach to an ESP be applied to supporting an endoscopy suite containing three endoscopy rooms? Discuss the benefits of developing an ESP for the complete suite as compared to developing ESPs for the individual endoscopy medical equipment.

CASE STUDY CS6.2: ESP – DOING LESS THAN THE MANUFACTURER'S RECOMMENDATIONS

Section Links: Chapter 6, Section 6.2.3; Chapter 7, Section 7.4.9

ABSTRACT

Sometimes it is justifiable to design an equipment support plan that does less than the manufacture recommends. This must always be based on knowledge, experience and a risk assessment.

Keywords: Defibrillator, Battery conditioning, Patient risk/benefit, Cost benefit.

NARRATIVE

A whole fleet of new mains/battery defibrillators was purchased following an extensive and generally very thorough procurement process. However, after purchase and the considerable task of commissioning, training users and deploying the new equipment, some small print in the technical manual was noticed that mandated a 'battery conditioning procedure' to be carried out every 3 months. It had been known from the start and taken into consideration that a battery change was mandated every 18 months.

The battery conditioning procedure involved fully charging the battery, completely discharging it, and then recharging it, followed by an operational test procedure. It was estimated that the defibrillator would be unavailable for clinical use for about 24 h. There were 24 of these defibrillators deployed across multiple sites and one spare machine held by clinical engineering to cover breakdowns and other short-term emergency situations. If the conditioning procedures were carried out on two defibrillators every week, it would take 12 weeks to get through the whole fleet, by which time the cycle would start again. Withdrawing equipment without providing a replacement was considered to be an unacceptable course of action. Two additional spare defibrillators would be required to ensure clinical availability.

Clinical Engineering considered whether swapping the batteries on-site for conditioned ones rather than withdrawing equipment and providing a spare would be a viable alternative. This would require very careful tracking of batteries independent of the equipment in which they were installed and the purchase of a laboratory battery reconditioning device. The logistics of this plan would not save any time or effort.

Both courses of action would involve purchase of additional equipment that had not been included in the original bid.

When this issue came to light and the analysis mentioned earlier was done, the first course of action was to contact the manufacturer and ask whether this procedure was necessary in the conditions of use that prevailed – ward based, so no extremes of temperature and always on charge. Their response was to agree in writing that the procedure could be carried out every 6 months. However, by that time, and with no battery conditioning procedures having been carried out, the defibrillators that had been deployed early in the replacement programme were due for a battery change and that was being done. The medical equipment management database was carefully searched, and it was found that there had been no issues with any of these defibrillators linked in any way with battery failure. It was decided to continue not carrying out the conditioning procedure, to continue to change batteries as specified by the manufacturer and to carefully monitor the performance of the defibrillators. The head of the CED (a consultant clinical engineer) took responsibility for this decision.

The reasoning and risk assessment were carefully documented and shared with the Resuscitation Committee.

ADDING VALUE

In this case, the right decision was reached, all be it following a less than ideal path. Continuity of availability of the equipment was assured without having to purchase additional equipment and without incurring additional staff time and resources which would have had to be diverted from other tasks. Value was added by ensuring that the patients still had the benefit of equipment support without incurring additional cost.

$$\uparrow \qquad \uparrow$$
Benefits : Cost $\therefore$ Value

CULTURE AND ETHICS: LESSONS LEARNED

It might be argued that, given the high clinical profile of defibrillators, the battery conditioning programme should be carried out in accordance with the manufacturer's instructions despite evidence that it was unnecessary in the circumstances prevailing. The lessons learned were twofold:

1. Read the technical manual carefully prior to purchase.
2. Be prepared to challenge the manufacturer on service and maintenance issues that seem excessive. Insist on any agreed changes to procedure being confirmed in writing.

SUMMARY

By serendipity, evidence was accumulated that omitting a logistically difficult and time-consuming maintenance procedure was having no detrimental effect on performance of a fleet of mains/battery defibrillators. The evidence was obtained from a retrospective study of the failure and maintenance records held on the medical equipment management database.

This led to a documented risk assessment that supported a course of action that modified the equipment support plan away from strict adherence to the manufacturer's instructions in the conditions of use prevailing.

SELF-DIRECTED LEARNING

1. Were there any other courses of action that could/should have been considered?
2. Was the decision reached a justifiable one?

CASE STUDY CS6.3: ESP – DOING MORE THAN MANUFACTURERS' INSTRUCTIONS

Section Links: Chapter 6, Section 6.2.3; Chapter 7, Section 7.4.3

ABSTRACT

Experience and service records showed that a replaceable sensor on an anaesthetic machine was failing either between routine services or before its 'replace by' date. The manufacturer recommended replacement at the next service after its replace by date. The equipment support plan was modified to include a scheduled replacement irrespective of the replace by date.

Keywords: Anaesthetic machine, O_2 sensor, Expiry date

NARRATIVE

Anaesthetic machines are fitted with an oxygen sensor which in this case was a fuel-cell type with a 'replace by' date. The anaesthetic machine manufacturer's instructions were to replace the O_2 sensors on the scheduled service visit following the expiry date.

The anaesthetic maintenance team in a Clinical Engineering Department had responsibility for service and maintenance of the anaesthetic machines in their own healthcare organization over two major and a number of smaller sites, plus they had service contracts with neighbouring healthcare organizations up to 40 km away.

Careful monitoring of the records from the medical equipment management database resulted in a realization that O_2 sensors were failing between scheduled services. In some cases, this was happening before their expiry date, possibly due to low usage. Unscheduled services, especially to the more remote locations, were costly in technician time and travel costs. Failure in service of an anaesthetic machine also carries patient risk because of the need to swap out a machine during an operation.

A decision was taken to replace O_2 sensors irrespective of expiry date at every 6 months scheduled service. Analysis showed that the cost of early replacement of the O_2 sensors was far less than the costs being incurred by unscheduled service for failed sensors. The reduction in patient risk was an added benefit.

ADDING VALUE

The main value added results from a lowering of cost due to a reduction in unscheduled service calls, especially to more remote locations. There is also an increase in patient benefit as described in the following text.

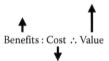

$$\text{Benefits} : \text{Cost} \therefore \text{Value}$$

PATIENT CENTRED

The patient-centred aspect of this change was that it reduced the likelihood of failure in service of the anaesthetic machine. Swapping out an anaesthetic machine during an operation has clinical risks. These are avoided.

SUMMARY

There are times when doing more than the manufacture specifies can add value. Sometimes this will increase cost and therefore the patient benefit must be clear for there to be added value. In this case, doing more actually cost less overall and added patient benefit, so value was clearly added.

SELF-DIRECTED LEARNING

1. Consider any circumstance in which you think that an ESP that does more than the manufacturer specifies would be appropriate and make out a case for this change.

CASE STUDY CS6.4: CALCULATING THE ANNUAL COST OF AN ESP FOR A FLEET OF INFUSION PUMPS

Section Links: Chapter 6, Sections 6.2.4 and 6.2.6; Chapter 4, Section 4.2

ABSTRACT

As part of the annual review of the equipment support plan for infusion pumps, the costs associated with it were determined by mining the data in the medical equipment management system (MEMS) database.

Keywords: Cost-effectiveness index; Infusion pumps

NARRATIVE

The hospital's MDC had identified the volumetric infusion pumps as an equipment type whose use was associated with high clinical risk. Any confusion associated with the use of this equipment or error in their accuracy could result in a patient medication error and possibly injury. The hospital in question had standardized on one volumetric infusion pump for the whole site to make training easier and ensure the competency of the staff. There were 420 of the pumps in use. The equipment support plan consisted of a mix of support activity from both the internal Clinical Engineering Department (CED) and the external supplier service department.

Given the concern over the risk of under or over infusion using this equipment, the hospital CED had a proactive performance verification programme where every pump's fluid delivery rate and electrical safety was checked on the bench once a year. Each pump was set up to run at a rate typically used in clinical practice for at least 12 h (usually overnight). This activity required one full-time equivalent (FTE) clinical engineering post to deliver the programme (but was delivered in practice by a number of individuals). The FTE cost was €40,000 per annum.

The nurse practice development unit also ran an ongoing training programme for this equipment and all infusion pump users were obligated to attend once a year. The CED supported this training programme by having clinical engineers present to demonstrate the equipment and simulate possible error conditions. The cost of supporting this activity was €10,000 per annum. The supplying company also supported the training through the provision of training material and work books, at no incremental cost to the hospital, the costs being covered as part of the purchase of giving sets for the pumps.

A review of the database system showed that there had been one hundred and seventy-six unscheduled service requests in the previous year. One hundred and twelve of these were the result of users having difficulties with operating the device and in each of which no technical error was identified. The remaining sixty-four were the result of technical failures or accidental damage. These were repaired by the CED staff and the staff costs associated with this work was estimated to be €17,000 and the parts cost €5,600. Six pumps were so badly damaged that they were service exchanged to the supplier maintenance department for reconditioned pumps, at a cost of €8,000.

The MDC wanted the CED to implement a conservative battery management programme and follow the manufacturer's advice to change the battery once a year. This scheduled work was outsourced to the equipment supplier's maintenance department. The cost of the batteries was fixed at €8,000; however, the cost of the work was dependent upon the supplier getting access to the devices and was billed separately and came to €16,700.

The costs were tabulated as shown in Table CS6.4A.

The equipment support plan cost €105,300 to implement. Analysis shows that 77% of the costs associated with the equipment support plan were internal with only €24,700 being spent on external service support. The overall cost-effectiveness index (CEI) for the equipment support plan was 11%.

REVIEW OF KPI'S AND OTHER DATA

The KPI for battery replacement showed that only 76% of the pumps had their battery replaced, and the supplier could credibly demonstrate that the gap was a consequence of difficulty getting access to the equipment. So the unfitted batteries were stored in the Clinical Engineering Department.

In contrast, the performance verification programme undertaken by the Clinical Engineering Department had resulted in 94% of pumps being checked. In all cases, there was no pump found to be delivering at a rate outside of the equipment specification. Routine electrical safety tests did not highlight any equipment safety issues; however, visual inspection of the mains leads highlighted minor damage and wear and tear and resulted in 82 of these being replaced (costs were included in the unscheduled parts calculation).

TABLE CS6.4A   Financial Summary of the Equipment Support Plan for the Volumetric Infusion Equipment for 1 Year

Equipment Support Actions	CED Internal Costs		CED External Costs	
	Staff	Parts	Contract	Non-contract
Performance verification	€40,000			
Scheduled		€8,000		€16,700
Unscheduled	€17,000	€5,600		€8,000
End-user support	€10,000			
Subtotals	€67,000	€13,600		€24,700
Total cost	€105,300			
Cost of internal support	€80,600	77%		
Cost of external support	€24,700	23%		
Number of items of equipment	420			
List price of each item	€2,200			
List price of the equipment	€924,000			
Cost-effectiveness index (CEI)	11%			

Opportunities for Improvement

The Clinical Engineering Department proposed two changes. First, there was little evidence that the calibration of the fleet of pumps was drifting and so the performance verification effort spent on this check was considered excessive. They recommended that the routine inspection be continued as it identified wear and tear issues, but the onerous task of running the pump overnight at a typical rate be abandoned and would only be done if a pump came to the laboratory for repair. Second, as configured, the equipment support plan had two independent schedules where pumps were taken out of service, the internal performance verification and the supplier battery change programme. The resources freed up by reducing the need for the rate check could be redeployed to have the clinical engineers change the batteries at the same time as the visual inspection was performed. The in-house team had a better relationship with the users and were more effective at getting access to the pumps.

ADDING VALUE

The proposed changes to the ESP would deliver a better service as the battery management would be more complete, and the costs associated with the external supplier coming on-site to change the batteries could be reduced so value is increased. The costs of the equipment support plan should also reduce, adding further value.

Benefits : Cost ∴ Value

SELF-DIRECTED LEARNING

1. Assuming that the battery replacement programme can be combined with the new visual inspection and a revised performance verification programme and that the staff cost associated with this reduce to €32,000 per annum, what would the cost-effectiveness index of the new equipment support plan be?
2. How would the percentage of internal to external costs change?

CASE STUDY CS6.5: REVISING AND REBALANCING THE RESOURCES ACROSS TWO ESPs

Section Links: Chapter 6, Section 6.2.6; Chapter 7, Section 7.2.8

ABSTRACT

Two equipment support plans which had been developed and run successfully for many years without review were revisited. Changes in the equipment and how it is used in the period between the previous and current reviews, prompted changes to each ESP and required a redistribution of resources to implement the improved ESPs.

Keywords: Suction units; Non-invasive blood pressure (NIBP) monitors; Reassigning clinical engineering resources

NARRATIVE

The equipment support plans for the theatre suction units and non-invasive oscillometric blood pressure monitors used on the general wards were reviewed.

ESP FOR THE THEATRE SUCTION UNITS

The theatre suction unit's equipment support plan was closely aligned to the manufacturer's recommendations to service the motor and vacuum pump once a year. This was an activity that was predominantly a preventive maintenance exercise, changing oil, hydrophobic filters and gaskets, etc. The equipment is uncomplicated to use and no special training programmes or user support were required or in place. A review of data in the medical equipment management system (MEMS) database revealed a trend in the previous 2 years of these units failing before they were due for service. Initially, this was assumed to be due to ageing of the equipment, but the staff who serviced the equipment pointed out that after the scheduled maintenance (SM) service, they were effectively reconditioned. The root cause was traced to a change in the use of the theatres. Since the original equipment support plan was developed, three of the theatres had been designated to support regional trauma services and now ran double shifts every day and often into the weekend. Other changes in theatre list usage revealed that the facilities were operating much longer hours when compared to the schedule when the equipment support plan was established. In consultation with the theatre manager, the Clinical Engineering Department established that the workload on the suction units had increased by on average 40% in 4 years. The Clinical Engineering Department decided to increase the frequency of the SM to every 9 months instead of every 12 months. So the scope of work detailed in the current delivered equipment support plan (black in Figure CS6.5A) was increased for the revised new planned ESP (white in Figure CS6.5A), and this would require more human resources.

ESP FOR THE WARD-USE OSCILLOMETRIC NIBP MONITORS

The equipment support plan that was in place for the NIBP monitors was designed for a fleet of equipment made up of a mix of older models from different manufacturers. At the time it was devised, there was concern about the accuracy of the different models. The old version of the equipment support plan, still being implemented, included a check to put each item of equipment on an NIBP simulator and test its accuracy at a number of simulated blood pressures. This was time consuming. In the intervening years since the ESP was written, all the NIBP monitors had been changed as part of a rolling replacement programme funded by each clinical directorate. The NIBP monitors were now standardized. The manufacturer's recommendation for performance verification (PV) of the new equipment was to perform a static accuracy check on the pressure transducer only, a procedure that made considerable less demands on the clinical engineer's time. A review

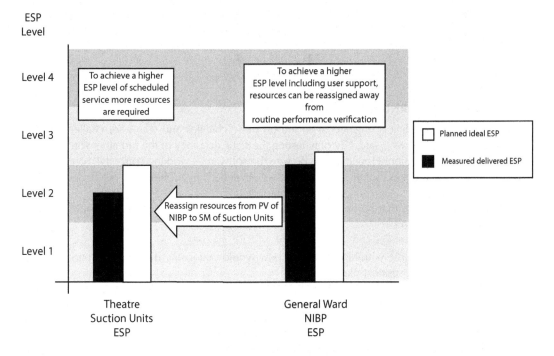

FIGURE CS6.5A   Revising and rebalancing the resources across two equipment support plans.

of the data stored in the MEMS database highlighted that in the previous 2-year period, in 68% of PV actions undertaken, the equipment in question had only a standard size cuff and no large cuff. In discussion with nurses on the wards, the clinical engineering team identified that there was a lack of awareness as to the importance of using the correct size cuff for each patient to ensure an accurate reading. The Clinical Engineering Department decided to abandon the time-consuming NIBP simulation approach and adopt the simpler static accuracy checks as recommended by the manufacturer. The resources this freed up were used to implement a better user support plan. It included an informal training programme where once every 6 months the clinical engineering team would walk the wards and discuss the importance of using the right cuff with staff on duty (this process included night staff). A poster with the same information was put up in staff areas. This elevated the support plan to a level 3 plan including planned user support, even though it was using up less resources than the previous ESP. The net result was that the revised equipment support plan (white in Figure CS6.5A) was to a higher level yet less demanding of resources than that which was currently being implemented (black in Figure CS6.5A).

The resources freed up in the NIBP equipment resource plan redesign allowed for the extra work required to improve the suction unit equipment support plan to be implemented at no extra cost.

ADDING VALUE

As a result of the review, both equipment support plans improved. The benefits for patients and staff were a reduction in failures of suction units and improved accuracy due to the availability and understanding of the need to use larger cuffs for some patients. The changes required to both plans, extra technical work for the suction units and extra training and information for the NIBP users, were cost neutral. The resources used to simulate the blood pressure for routine testing of the NIBP monitors was wasteful and unnecessary now that newer NIBP monitors were in use, and this effort could be redeployed. The redistribution of support resources increased benefit without increasing costs and therefore improved value.

$$\uparrow \qquad \uparrow$$
Benefits : Cost ∴ Value

SELF-DIRECTED LEARNING

Imagine you are the clinical engineer who led on these changes and you were challenged by the Senior Cardiologist who said, "By only testing the static accuracy of the monitors you are not checking their full function! How can you be sure the software is working correctly? Your predecessor always undertook elaborate checking and we never had any problems. Why change now?"

1. Would you agree that he has a point and if so how would you modify the NIBP equipment support plan?
   Or
2. Would you disagree with him and if so, how would you justify the changes made to the NIBP equipment support plan?

CASE STUDY CS6.6: EXAMINING THE VALUE DELIVERED BY AN ESP FOR A CO$_2$ SURGICAL LASER

Section Links: Chapter 6, Section 6.2.5; Chapter 7, Section 7.2.8

ABSTRACT

By reviewing the equipment support plan for a CO$_2$ laser, the Clinical Engineering Department were able to increase value by both reducing costs and adding extra benefits. To do so, they had to take on new roles and negotiate and assure clinicians that the proposed changes would not result in a decrease in the quality or availability of service support.

Keywords: Surgical laser; Service contract negotiation; Safety training

NARRATIVE

A clinical engineer was tasked with reviewing the equipment support plan for a surgical CO$_2$ laser to determine if it was delivering optimal value.

The clinical engineer reviewed the plan. There is only one CO$_2$ surgical laser in the hospital, and when it was purchased, the decision was made to support it through the purchase of an annual comprehensive support contract as availability of the device was crucial to meeting hospital commitments regarding patient throughput targets. This contract covered all performance verification, scheduled and unscheduled action for a fixed cost of £12,000 per annum. There was very little information in the medical equipment management system (MEMS) database as to the support activity associated with the laser. When the clinical engineer discussed the issue with the theatre manager, they were able to confirm that the consultant surgeon would call in the company representative himself if anything was required, and there was no perceived requirement to document or control this since the equipment was on a fixed price contract. The theatre manager was able to confirm that the representative 'did something with the laser every month or so' but no records existed. An email response from the consultant surgeon stated that he was very happy with the support arrangements and he would like them to continue as they were.

TABLE CS6.6A Financial Summary of the CO_2 Surgical Laser's Equipment Support Plan; Year 1

Equipment Support Actions	CED Internal Costs		CED External Costs	
	Staff	**Parts**	**Contract**	**Non-contract**
Performance verification				
Scheduled			£12,000	
Unscheduled				
End-user support				
Subtotals			*£12,000*	
Total cost	£12,000			
Cost of internal support				
Cost of external support	£12,000	100%		
Number of items of equipment	1			
List price of each item	£95,000			
List price of the equipment	£95,000			
Cost-effectiveness index (CEI)	13%			

The clinical engineer calculated the cost-effectiveness index for the existing support plan and estimated it to be 13% (see Table CS6.6A). There was no involvement by the clinical engineering department in the support plan and given the specialist nature of the equipment and the fact that there was only one such item in the organization, it was felt unlikely that the department could train up sufficiently to support this single item.

The clinical engineer met with the company representative to get better information on what services were being delivered for the £12,000 contract. Most of the visits mentioned by the theatre manager and confirmed by the company representative were requested by the consultant surgeon to check that the visible helium neon guide laser and the invisible CO_2 treatment lasers were aligned. The surgeon was concerned that with the movement of the laser in and out of the theatre, the optics could get misaligned, and since there was no increase in cost associated with calling in the engineer, he did so once a month. The contract also covered the cleaning and realignment of the optics by a specialist engineer which took about 3 h but was only required once a year. The contract also covered all breakdowns not associated with misuse or accidental damage including replacement of the laser tube.

The clinical engineer looked at how the value of the support plan could be increased by first looking at how costs could be reduced without decreasing any of the existing benefits and then by looking to see what added benefits could be delivered within a new cost structure.

In discussion with the company representative, the clinical engineer was able to confirm that there were other contract options available to support the equipment. A 'preventive maintenance' contract was available which covered the annual cleaning and realignment service visit, any software upgrades and gave the hospital priority for emergency response call outs (which would be charged for separately). This contract cost £3,500 which represented a considerable saving, but it would not support the regular visits to check the beam alignment which gave the consultant so much comfort. The clinical engineer was confident that this simple check, which needed no specialist test equipment, could be performed by the Clinical Engineering Department. Once he had assured the consultant surgeon that this was the case, the surgeon agreed that in future he would call the Clinical Engineering Department to check the alignment. The costs associated with this check, which only takes 15 min to do, were found by reassigning resources within the Clinical Engineering Department. The clinical engineer also explained to the finance department that by changing the contract type they were reducing annual cost, but exposing themselves to an occasional high call out cost (every few years) should the tube in the laser fail. The finance department agreed to fund the occasional high spend in favour of getting better year on year costs.

In discussion with the consultant surgeon and the theatre manager, it became clear to the clinical engineer that there was confusion and unnecessary fear in many of the theatre staff around the safety procedures associated with the use of the laser. Whilst all recommended safety systems were in place and operating well, the use of the laser and control of the risks was poorly understood. The clinical engineer and a medical physics colleague together developed and ran a series of talks and demonstrations on safe laser use for key clinical and nursing staff as part of the theatre in service education programme.

ADDING VALUE

The revised equipment support plan was put into action the following year. The costs associated with it are shown in Table CS6.6B. The main saving was associated with the change of contract type. Whilst the costs of the beam alignment checks and the training programme undertaken by the Clinical Engineering Department are new, as is the call out charge for a repair required in year 2 (£1700), these costs are considerably less that the saving achieved by switching contract type. The support costs are still predominantly going to the external company (76%), and given the specialist nature of the equipment, this is unlikely to change further. However, the overall costs have reduced by £5200 a 43% saving, and the cost-effectiveness index fell from 13% in year 1 to 7% in year 2. It is unlikely that this level of saving will be sustained when averaged over several years. It is anticipated that the occasional large repair bill, which previously would have been covered by the comprehensive contract, will arise and erode the savings achieved. Nevertheless it is envisaged that, over the equipment's life-time, costs will be reduced while keeping the same level of deliverables from the ESP, thus adding value. The fact that the clinical engineering team were now actively involved in the laser management meant that the documentation of all actions including the monthly beam alignment check improved. The initiation of the training programme for staff on laser safety is a new benefit delivered by the new plan which adds further value.

Benefits : Cost ∴ Value

TABLE CS6.6B   Financial Summary of the CO_2 Surgical Laser's Equipment Support Plan; Year 2

Equipment Support Actions	CED Internal Costs		CED External Costs	
	Staff	Parts	Contract	Non-contract
Performance verification	£600		£3,500	
Scheduled				
Unscheduled				£1,700
End-user support	£1,000			
Subtotals	£1,600		£3,500	£1,700
Total cost	£6,800			
Cost of internal support	£1,600	24%		
Cost of external support	£5,200	76%		
Number of items of equipment	1			
List price of each item	£95,000			
List price of the equipment	£95,000			
Cost-effectiveness index (CEI)	7%			

SELF-DIRECTED LEARNING

The following information gives the non-contract call out charges associated with calling in the company for specialist repairs over the 5-year period that followed the narrative set out in this case study. Assuming that all other costs are set out as in the case study and remain unchanged, calculate and compare the 5-year cost-effectiveness index of both the original and the revised equipment support plans.

CALL OUT CHARGES

Repair costs associated with new ESP and which would have been covered by the contract if the old ESP was left unchanged.

Year 1	£1,700
Year 2	£2,300
Year 3	£1,956
Year 4	£19,500 (tube replacement)
Year 5	£2,700

CASE STUDY CS6.7: DEVELOPING A PARTNERSHIP ESP WITH THE EQUIPMENT SUPPLIER TO MAXIMIZE VALUE DERIVED FROM AN ESP

Section Links: Chapter 6, Section 6.2.3

ABSTRACT

Through a negotiation process, the Clinical Engineering Department of a hospital and the technical support department of an anaesthetic equipment supplier developed a shared equipment support plan that delivered value.

Keywords: Anaesthetic workstations; Partnership with industry

NARRATIVE

As part of its planned equipment replacement programme, a large university teaching hospital replaced all its anaesthetic workstations in one procurement exercise. The workstations for 12 theatres and their associated induction rooms and for four specialist imaging rooms were replaced, resulting in a total of twenty-eight new workstations. As part of the tender process, the winning bidder had offered a comprehensive service contract for €200,000 per annum or 9% of the list price of the equipment purchased (see Table CS6.7A).

When the procurement contract was awarded, the Clinical Engineering Department suggested that they and the supplying company explore the development of a partnership support model for this equipment. The supplier was aware that the complexity of anaesthetic workstations was such that they give rise to the need for significant front-line support to deal with minor user-related issues. The supplier was also aware that the hospital's clinical engineering team had experience in supporting the previous anaesthetic workstations and that they had negotiated service training on the new equipment for two clinical engineers as part of the procurement.

Over a series of meeting, the strengths and weaknesses of both the clinical engineering team and the supplier team in supporting the equipment were identified, and a shared equipment support plan was developed that maximized the benefits of both for the hospital and the supplier. One of the key objectives of the Clinical Engineering Department was to maximize the uptime and availability of the anaesthetic machines as any delay at the start of a surgical case had impact on the effectiveness of the theatre complex as a whole. One of the key objectives of the company

TABLE CS6.7A   Financial Summary of the Proposed Supplier Equipment Support Plan

Equipment Support Actions	CED Internal Costs		CED External Costs	
	Staff	Parts	Contract	Non-contract
Performance verification			€200,000	
Scheduled				
Unscheduled				
End-user support				
Subtotals			€200,000	
Total cost	€200,000			
Cost of internal support				
Cost of external support	€200,000	100%		
Number of items of equipment	28			
List price of each item	€80,000			
List price of the equipment	€2,240,000			
Cost-effectiveness index (CEI)	9%			

was for the equipment to be well received and highly regarded by the Anaesthetists. The hospital was a strategic institution, and if the equipment was perceived to work well here, the company could expect that to positively influence future sales in other institutions. The hospital was a teaching hospital and consequently many junior anaesthetists would be trained on the supplier's equipment; for the reputation of the equipment it was important that the equipment was recognized as being reliable and of a high standard.

The strengths of the clinical engineering team were their ability to respond quickly to any technical difficulty that arose. They were an experienced team who could provide fast and reliable user support and effective front-line repairs and carry out performance verification. They also had the competency and capacity to take on major repairs but only if supported by the supplier's engineers. Thankfully, major problems were rare and so it was difficult for the in-house team to develop a complete expertise in this regard, since they did not support as many devices as the supplier's technical team.

The strength of the supplier's technical team was their depth of training to deal with complex problems, and the availability of expert knowledge and updates from the manufacturing company. However, they could not respond quickly to minor issues or user support calls without basing a staff member full-time in the hospital which would erode the profitability of the proposed support contract.

The solution arrived at through negotiation was that the Clinical Engineering Department would handle user support, front-line maintenance and one annual proactive performance verification on each workstation. The supplier would perform the annual scheduled maintenance for each workstation, including any software upgrades and at the same time perform a full performance verification on that workstation. The scheduling of the work was aligned so that the two performance verifications, independently undertaken by the Clinical Engineering Department and the supplier's technical team, were 6 months apart. Unscheduled maintenance, that is repairs, was shared. The clinical engineering team would always respond first but would escalate to the supplier as soon as it was clear that the repair required the supplier expertise. An agreed sum of €20,000 was paid upfront at the start of the year to cover the costs of any parts required or call out charges. In the event of an equipment failure, this gave clinical engineering and the supplier the freedom to act quickly to order parts and call in engineers without the delays often associated with raising purchase orders. This greatly speeded up the ability of the partnerships to effect

TABLE CS6.7B   Financial Summary of the Implemented Shared Equipment Support Plan

Equipment Support Actions	CED Internal Costs		CED External Costs	
	Staff	Parts	Contract	Non-contract
Performance verification	€10,000			
Scheduled			€75,000	
Unscheduled	€20,000	€20,000		
End-user support	€10,000			
Subtotals	*€40,000*	*€20,000*	*€75,000*	

Total cost	€135,000	
Cost of internal support	€60,000	44%
Cost of external support	€75,000	56%

Number of items of equipment	28
List price of each item	€80,000
List price of the equipment	€2,240,000
Cost-effectiveness index (CEI)	6%

repairs and built a culture of shared responsibility to sort problems effectively and in a timely manner. At the end of the year, the actual costs were reconciled with the €20,000 provision and any extra payments required agreed or balance carried forwards into the next year. This financial management was led by the Clinical Engineering Department with input from the finance department.

The costs associated with the partnership model of equipment support plan developed are presented in Table CS6.7B.

ADDING VALUE

Comparison of the costs for the comprehensive contract and the partnership model shows that the partnership model reduces costs for the hospital. The cost-effectiveness index falls from 9% for the comprehensive contract to 6% for the partnership model. The model was judged to be sustainable by the supplier even though they did not get as much revenue as if the comprehensive contract was in place. The supplier judged that, not having to respond to the many minor calls which for them would be associated with travel time and opportunity costs, added value for them. Above all, the supplier judged, the partnership model, with the in-house team rapidly resolving minor issues, promoted the image of their product in the prestigious teaching hospital environment. The Clinical Engineering Department and the anaesthetists judged that this rapid response to all issues and speedy resolution of minor ones brought benefit to the delivery of patient services and meeting performance targets. The partnership reduces cost and adds benefits so increasing value.

Benefits : Cost ∴ Value

CULTURE

Very quickly the in-house and external engineers developed a culture of helping each other out and this added a lot of efficiency. For example, if a major repair was beyond the competency of the in-house team, then the supplier engineer often had to travel for a few hours before he could get on-site to effect the repair. However, both teams worked together to reduce the impact of the failure in such circumstances. The in-house team would usually have the equipment in the workshop and partially disassembled to the company engineer's instructions before he could get on-site. Sometimes the in-house team would make measurements under instruction from the supplier engineer whilst they were elsewhere, and this allowed for required parts to be ordered with confidence and couriered to the hospital before the supplier engineer had come on-site.

The benefits of the partnership model are that there is a larger pool of available talent to resolve problems that arise with these complex electromechanical items of equipment. The availability of the in-house team to support user difficulty with checking and calibrating the equipment at the start of lists was highly valued by the department of Anaesthetics. The fact that these same individuals were regularly performing minor repairs and detailed performance verification of the equipment resulted in an in-house team with a commanding expertise in the use of the equipment and this contributed significantly to maximizing uptime.

The fact that two teams verified the equipment independently of each other and shared results and comments added to the rigour of the approach, and this developed over time into an audit of each other's work.

PATIENT CENTRED

The equipment support plan optimized the equipment uptime. Very often minor problems with anaesthetic machines like small leaks in the patient circuit delay the starts of lists, and this can have a major impact on the scheduling of patient operations. The solution, which developed in-house expertise, ensured that at any time, the anaesthetists and anaesthetic nurse had expert technical assistance available to resolve often tricky issues quickly.

SELF-DIRECTED LEARNING

1. One factor that led to a successful outcome for this project was the fact that the hospital viewed the commercial company as a partner in delivering a complex technology solution for patients. Do you think that view of the medical equipment industry is prevalent in hospital clinical engineering departments? Justify your answer with four examples.
2. Can you detail how the experience of patients going for surgery might be effected by not having a process in place to rapidly resolve minor issues that might arise with an anaesthetic workstation at the start of a day's busy list?

CASE STUDY CS6.8: INCREASING VALUE BY CHANGING AN ESP TO BE PREDOMINANTLY OUTSOURCED TO MEDICAL EQUIPMENT SUPPLY COMPANIES

Section Links: Chapter 6, Section 6.2.3

ABSTRACT

Following the retirement of an expert, a Clinical Engineering Department was unable to continue to deliver value from an existing equipment support plan. The resulting solution required an increase in spending to deliver value.

Keywords: Surgical drills; Succession management; Increasing revenue costs

NARRATIVE

The Clinical Engineering Department at the centre of this case study had a long tradition of providing a preventive maintenance and repair service for dental and air-powered surgical drills. The service had been established by a senior clinical engineer who was also trained as a watchmaker and had a particular expertise in precision mechanics. As the number of drills owned by the hospital increased in line with an expansion in service, a second clinical engineer was trained to support the first, and the support plan extended to electrically power surgical drills. As well as the drills themselves, this pair maintained the driver units in the dental suites and theatres. The support for the drills, whilst shared between the two individuals, amounted to one full-time equivalent (FTE) post. Most of the work could be done in-house with a few drills each year having to be sent for manufacturer reconditioning. The service was highly regarded as these two individuals could quickly turn around minor problems, and the downtime associated with drills, as measured using a KPI, was acceptably low. The costs associated with this equipment support plan are presented in Table CS6.8A.

Following the retirement of the senior clinical engineer, another younger member of staff was assigned to replace him. Within 6 months, the delays associated with turning around the drills had increased and had given rise to two situations where surgery had to be rescheduled. This matter was raised through the hospital quality and risk committee and the MDC suggested a critical review of the drill support be undertaken. The review of the equipment support plan identified that the department as a collective did not have enough capability to support the equipment following the retirement of the acknowledged expert. Research into the cost of training up the existing staff led the head of the department to conclude that the cost was prohibitive, and the expertise, if developed in-house, would be hard to retain. The equipment support plan whilst once successful could no longer be viewed as adding value for two reasons. First, the benefits it previously brought were no longer able to be delivered, and second, the cost of the one full-time equivalent post was expensive to protect given the lack of ability to support the service.

Following a review of quotes from commercial companies, the department decided to outsource the scheduled and unscheduled components of the support plan to two companies in the form of an annual contract with each. The combined cost was £48,000 and this was a new expense for the hospital which had to be negotiated with the finance department. The risk to the scheduling of future surgery of not having an associated support capability for dental and surgical drills was not acceptable and the hospital sanctioned the increase in expenditure. The head of department also had to protect the existing one full-time equivalent post. Some of this post was assigned to support the drills' equipment support plan, for performance verification work and front-line support; however, most of the capacity was reassigned to support another equipment grouping's support plan. The costs associated with the new equipment support plan for the dental and surgical drills are presented in Table CS6.8B.

TABLE CS6.8A   Financial Summary of the Existing Equipment Support Plan

	CED Internal Costs		CED External Costs	
Equipment Support Actions	**Staff**	**Parts**	**Contract**	**Non-contract**
Performance verification				
Scheduled	€50,000			
Unscheduled		€8,500		€5,500
End-user support				
Subtotals	*€50,000*	*€8,500*		*€5,500*

Total cost	€64,000	
Cost of internal support	€58,500	91%
Cost of external support	€5,500	9%

List price of the equipment	€370,000
Cost-effectiveness index (CEI)	17%

TABLE CS6.8B   Financial Summary of the New Equipment Support Plan Which Relies on External Support

Equipment Support Actions	CED Internal Costs		CED External Costs	
	Staff	Parts	Contract	Non-contract
Performance verification				
Scheduled	€10,000		€48,000	
Unscheduled		€7,000		
End-user support				
Subtotals	*€10,000*	*€7,000*	*€48,000*	

Total cost	€65,000	
Cost of internal support	€17,000	26%
Cost of external support	€48,000	74%

List price of the equipment	€370,000
Cost-effectiveness index (CEI)	18%

ADDING VALUE

The service provided by the Clinical Engineering Department following the retirement of the expert was not delivering value. Whilst the costs remained the same, the ability of the two younger, less expert staff to deliver the benefits of a comprehensive in-house service was not the same as that of the retired expert.

To get the benefits delivered back to an acceptable level, the hospital had to increase expenditure and buy in supplier expertise. However, the final cost of outsourcing the service was backed off by the reduced requirement for the Clinical Engineering Department to assign resources to support the drills; even so, both cost and benefits increased.

Comparison of the two financial summary tables shows a significant change in the ratio of spend on internal to external support, going from 91% internal to 74% external, to deliver an adequate support service.

Whilst the clinical engineering view was that the new equipment support plan was slightly more expensive than the old, that view was not supported by the finance department who viewed this as an increase in expenditure of £48,000. The financial controller's position was that this could only be viewed as being a cost neutral if the clinical engineering staffing compliment for the department had been decreased. From the perspective of the financial controller, with no increase in benefits and an increase in costs, the value is decreased.

Benefits : Cost ∴ Value

However, the head of department was able to defend the change on the basis that it delivered increased value, that is, greater benefits in the form of a well-managed and timely support for the dental and surgical drills, and extra benefits by being able to assign the existing staff compliment to deliver an improvement in the equipment support plan for another group of equipment (in this case Dialysis).

Benefits : Cost ∴ Value

SELF-DIRECTED LEARNING

1. The financial controller and the head of department had different opinions on whether the solution delivered value. In your opinion which one was right? In your answer compare and contrast the value delivered from the drills' equipment support plan and the healthcare technology management programme as a whole.

2. If the result of putting the drills on contract resulted in longer than expected turnaround times, how would this affect the value delivered by the equipment support plan? If such delays did manifest, how could the hospital act to reduce them to acceptable levels?

3. Can you identify any other equipment group where greater value could be achieved by outsourcing the equipment support rather than it being undertaken by the Clinical Engineering Department?

CASE STUDY CS6.9: INCREASING VALUE BY CHANGING AN ESP TO FUND AND DEVELOP USER SUPPORT

Section Links: Chapter 6, Section 6.2.1; Chapter 7, Section 7.4.11

ABSTRACT

Following the introduction of new ventilation technology, the ICU and clinical engineering leads worked together to develop a new equipment support plan that included and prioritized the provision of user support for clinicians provided by the Clinical Engineering Department.

Keywords: Ventilators; User support; User training

NARRATIVE

All the ventilators for a 12-bed ICU were replaced as part of a planned replacement programme. The lead Intensivist was adamant that the extensive ventilation modes and functions available on the new ventilators would be exploited for patient benefit. Therefore, a comprehensive training programme for clinicians and nurses was planned for. The supplying company were contracted to perform the scheduled maintenance and routine performance verification with the contract costing €48,000 per annum. As part of this contract, the suppliers gave a commitment to provide a series of training seminars for all staff in the unit on the advanced features of the new ventilators, free of charge (FOC). The Clinical Engineering Department were already actively involved in user support for the ventilators in ICU. This role required half a full-time equivalent (FTE) post for front-line support, and this included a commitment to teaching, training and support for the simulation lab. The clinical engineers handled the front-line support which was predominantly user support and some technical repair which was associated with a small spend on spares (€4000). The costs for the equipment support plan for year 1 are tabulated in Table CS6.9A.

Towards the end of the first year, the Intensivist approached the head of clinical engineering asking if the clinical engineering support for teaching and training could be increased. The Intensivist was convinced that the efficacy of the unit as a whole had been enhanced by the new ventilation technology, particularly since it supported the introduction of novel weaning strategies not possible with the older ventilators. However, to make the training effective, the Intensivist wanted to increase the hours assigned by the Clinical Engineering Department to three activities: first, the management and running of the simulation lab; second, the time clinical engineering spent developing in-house training material for both the Anaesthetics

TABLE CS6.9A   Financial Summary of Ventilator Equipment Support Plan for Year 1

Equipment Support Actions	CED Internal Costs		CED External Costs	
	Staff	Parts	Contract	Non-contract
Performance verification			€48,000	
Scheduled				
Unscheduled	€25,000	€4,000		
End-user support	€5,000		FOC	
Subtotals	€30,000	€4,000	€48,000	
Total cost	€82,000			
Cost of internal support	€34,000	41%		
Cost of external support	€48,000	59%		
Number of items of equipment	20			
List price of each item	€40,000			
List price of the equipment	€800,000			
Cost-effectiveness index (CEI)	10%			

Department and the Postgraduate Diploma in ICU Nursing; and, third, their presence in the unit to provide one to one support at the bedside. Together the Intensivist and the head of clinical engineering made a proposal to increase the staffing compliment by a half a full-time equivalent post so that the clinical engineering support to ICU could be doubled.

The proposal was accepted in principle and the added value it would bring acknowledged; however, the financial controller challenged the proposers to find a way of funding the increase from within existing resources.

A new equipment support strategy was devised which gave prominence to the user support and training role. The annual service of the ventilators was easily within the competency of the Clinical Engineering Department, whose engineers had received training in how to do this work as part of the procurement process. So the responsibility for performance verification and scheduled maintenance was assumed by the Clinical Engineering Department and the contract with the supplier not renewed. This decrease in revenue spend as a result of dropping the contract was enough to get the financial controllers support for the expansion in the clinical engineering human resources. The full-time equivalent post (in reality a mix of individuals) was costed at €70,000 and an uplift in the parts budget was also required to purchase the necessary service kits annually.

There was no question over the quality of the work done by the supplying company, and their free of charge contributions to the teaching programmes were highly valued. However, the Intensivist and head of the Clinical Engineering Department felt that the new proposal delivered the same competent support and the added value of having more in-house resources to support teaching, training and ultimately extracting the benefit from the new technology. The figures are shown in Table CS6.9B.

ADDING VALUE

The real cost of the equipment support plan increased from €82,000 in year 1 to €87,000 in year 2. Expressing this as a percentage of the list price of the equipment the CEI went from 10% to 11%.

The most predominant feature of the costing comparison is a shift of resources going from external to internal suppliers of support. The diagram in Figure CS6.9A is useful in assessing whether the value increased. In year 1 the technical support was fully realized; however, the aspiration to support the teaching and training programme was only half

TABLE CS6.9B   Financial Summary of Ventilator Equipment Support Plan for Year 2

Equipment Support Actions	CED Internal Costs		CED External Costs	
	Staff	Parts	Contract	Non-contract
Performance verification			€0	
Scheduled				
Unscheduled	€70,000	€13,000		€4,000
End-user support			FOC	
Subtotals	€70,000	€13,000	€0	€4,000

Total cost	€87,000	
Cost of internal support	€83,000	95%
Cost of external support	€4,000	5%

Number of items of equipment	20
List price of each item	€40,000
List price of the equipment	€800,000
Cost-effectiveness index (CEI)	11%

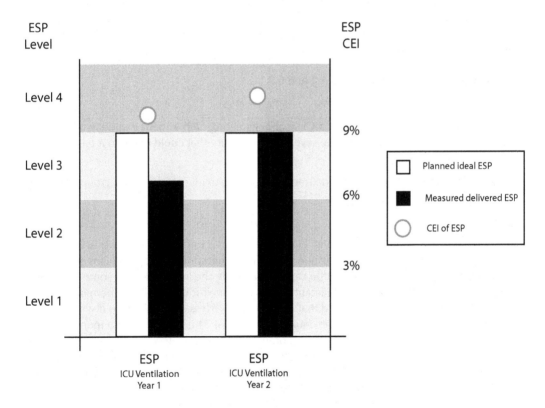

FIGURE CS6.9A   Comparison of the two ESPs for the support of the ICU ventilators.

realized. In year 2 whilst the CEI rose to 11%, the extra benefit of the user support was realized. Ultimately the Intensivist judged that the year 2 solution supported the ICU in improving its performance as a whole, and the role of the clinical engineering in supporting the delivery of high tech care was acknowledged.

This new ESP not only delivered comparable service support but also allowed a significant increase in the user support. Whilst costs increased, the benefits delivered increased further and the value was judged to have increased.

Benefits : Cost ∴ Value

SELF-DIRECTED LEARNING

1. In your opinion are there any new risks for the hospital in following the equipment support plan as set out for year 2? If so how could these risks be controlled and would the control measures increase the costs of the equipment support service?
2. Do you think the supplying company would continue to provide free of charge training support as part of the new equipment support plan?
3. How do you think this equipment support plan might evolve if in year 2 the training in the use of the new technology is effective and the needs for such a commitment to training decrease to half of that resourced for year 2 in year 3 and subsequent years?

CASE STUDY CS6.10: PRIORITIZING GROUPS OF EQUIPMENT FOR REVIEW OF THEIR ESPs

Section Links: Chapter 6, Sections 6.4.2 and 6.2.1

ABSTRACT

A scoring system was developed by a Clinical Engineering Department that allowed the prioritization of equipment groups by the severity of the injury that could occur to a patient if the equipment failed.

Keywords: Risk assessment; Patient-focused maintenance analysis; Equipment groups; Scoring systems

NARRATIVE

The head of the Clinical Engineering Department was concerned that, whilst the HTM Programme as a whole and its constituent individual equipment support plans were well structured and managed, there might be gaps. Specifically she asked whether some particular equipment, which might pose a significant risk to patients if it failed, should be more appropriately managed; she was concerned that it might receive less attention since it was not part of one of the high-cost/high-volume equipment groups. The senior management team in the department agreed, noting that the equipment support plans tended, as a priority, to be reviewed and developed for known high cost or life support items used in critical care areas. Whilst every year the HTM Programme improves and encompasses more equipment, there was the possibility that some patient risks were being overlooked.

The team undertook a scoring exercise to try and objectively identify equipment for review based on the potential impact of equipment failure on patients' health. (Note that this analysis was not to inform the assignment of resources, but to prioritize groups for review. The resources allocation would be part of that overall review.) Essentially this was a risk assessment process. However, the group made no attempt to estimate the probability of failure occurring, rather they focused on identifying risks where the consequence of the failure was high and could impact on patients' health.

Each equipment type (not each model) in the MEMS database was scored with either a one or zero under-five heading. Each heading was prompted by an identifiable risk to patients health should the equipment fail in use. Upon completion, the sum of the scoring for each type of equipment was calculated and any equipment with score of 1 or greater was reviewed.

The scoring headings were as follows:

A. Life support functionality:
   1 – if a failure of this device could result in the death;
   0 – if failure might impact on patient health but would not threaten the life of the patient.
B. Resuscitation functionality:
   1 – if a failure of this device could result in inability to resuscitate a failing patient;
   0 – if failure might impact on patient health but would not prevent resuscitation.
C. Is used to deliver a medication (gas, solid or liquid) or clinically used fluids:
   1 – if a failure of this device could result in medication or fluid delivery error;
   0 – if failure might delay treatment but would not result in medication or fluid delivery error.
D. Is used to perform a direct in vivo therapy on a patient's blood:
   1 – if Yes;
   0 – if No.
E. Delivers energies to the body during normal use that could result in immediate life-threatening injury if misapplied or the equipment output is out of specification:
   1 – if Yes;
   0 – if No.

Examples of the scores of some equipment (but not all) are presented in Table CS6.10A.

TABLE CS6.10A   Sample of Results of the Risk Scoring Exercise Undertaken for Various Types of Medical Equipment

Equipment Type	A Life Support	B Resuscitation	C Medication or Fluid Delivery	D Treatment Involving Patient's Blood	E Treatment Associated with Delivery of High Energy	Summed Score
Heart–lung machine	1		1	1		3
Intra-aortic balloon pump	1			1		2
Infusion/syringe pump	1		1	1		3
Rapid infuser		1	1	1		3
Defibrillator		1			1	2
Surgical tourniquet	1				1	2
Dialysis machine			1	1		2
Temporary external pacemaker	1				1	2
Surgical diathermy unit					1	1
RF lesion generator					1	1
Wall suction unit		1				1

In general, the analysis supported assumptions made in prioritizing groups of equipment for review. It identified equipment groups such as Infusion, Cardiac Perfusion, Defibrillation, Anaesthetics, Dialysis, Surgical Energy Devices and Ventilation as all being worthy of close inspection.

However, the analysis was useful in that it challenged perceptions about particular items of equipment which were not previously highlighted for review. For example, whilst the routine performance verification of the surgical diathermy units was in place, the analysis identified the need to consider the lesion generator used by the Pain Team in the same light, as it also delivers RF current. The temporary pacemakers scored the same as the ICU ventilators in terms of being a life support device, yet they were not part of any of the equipment support plans, and one for temporary pacemakers was immediately developed. The team was surprised that the surgical tourniquets scored the same as the defibrillators, and this prompted a review of the frequency with which the tourniquets were inspected.

The fact that any suction unit could play a pivotal role in a resuscitation event was known, but the exercise forced a rethink as to whether there was a need to develop a more proactive equipment support plan than that which was in place. The team identified that suction units on the resuscitation trolleys were more actively managed than wall suction regulators and that the difference was due to how the equipment was grouped for inspection, rather than its potential use in clinical practice.

The fact that none of the physiological monitoring or measurement equipment was scored in this exercise also gave rise to significant discussion, as that equipment group requires considerable support. The group decided to conduct a separate exercise to review the physiological measurement and monitoring equipment to assess the risks associated with its use, and which did not feature in this analysis.

The systematic approach taken led the team to consider not only the technology but also how it is used in practice. It forced the team to consider all uses of equipment in a clinical context where often the team fell back on grouping devices based on technology types. The ability to use the database ensured that all devices were given equal consideration.

The approach went a long way to refocusing the HTM Programme to be patient centred. The team identified that often priority was given to cost control or 'the squeaky wheel getting the grease' and the methodology allowed a review to be undertaken from a patient impact perspective, making the HTM Programme as a whole more patient centred.

ADDING VALUE

There was no significant ongoing cost to undertaking this exercise. Yet it identified and led to a number of quality improvement actions. Given that the analysis was focused on impact on patients, the exercise delivered added benefit in the form of better risk mitigation. Therefore, value was increased.

Benefits : Cost ∴ Value

SELF-DIRECTED LEARNING

The following is another version of the table presented earlier listing 20 other common medical equipment types (Table CS6.10B):

1. Score them using the same methodology set out in this case study.
2. Briefly write notes on either any difficulty you have in scoring particular equipment or results that surprised you.

TABLE CS6.10B   Additional Equipment to Be Risk Scored

Equipment Type	A Life Support	B Resuscitation	C Medication or Fluid Delivery	D Treatment Involving Patient's Blood	E Treatment Associated with Delivery of High Energy	Summed Score
Anaesthetic work station						
Video gastro-endoscopy system						
CT scanner						
Laryngoscope and blade set						
ECG recorder						
Surgical headlight						
Public area AED						
Glucometer in diabetes OPD						
Slit lamp in ED						
Multi-parameter monitor in ICU						
Hospital transport wheelchair						
Electronic thermometer						
Stethoscope on resus. trolley						
Fluid warmer						
Stand alone pulse oximeter						
CPAP unit in ward area						
Portable ultrasound unit in ED						
CO_2 surgical laser						
ECG stress test system in cardiology						
Biplane fluoroscopy system in cath lab						

3. In your opinion, are any of these items of equipment associated with a significant risk to the patient if they fail, even though they do not score highly using the methodology outlined here?

4. If so, explain the risk and describe how an equipment support plan could be developed to control that risk.

CASE STUDY CS6.11: PRIORITIZING GROUPS OF PHYSIOLOGICAL MONITORING AND MEASUREMENT EQUIPMENT FOR PERFORMANCE VERIFICATION

Section Links: Chapter 6, Sections 6.4.2 and 6.2.1

ABSTRACT

A scoring system was developed by a Clinical Engineering Department that supported the prioritization of physiological measurement and monitoring equipment for inclusion in a proactive performance verification programme.

Keywords: Risk assessment; Performance verification; Scoring systems

NARRATIVE

(For an introduction, see Case Study CS6.10)

Following a review of equipment that could cause injury to patients if it failed, the clinical engineering team were surprised to find that physiological measurement and monitoring equipment did not feature in the identified high-risk groups. Upon reflection the group identified that the thinking was that this type of equipment is diagnostic rather than therapeutic and so its failure is less likely to precipitate an immediate clinical crisis. In many cases, alternative equipment is available, or in some instances, the equipment is modular and failed parts can be quickly and easily swapped out. Also much of this equipment is electrical and electronic, and, due to standardization and improvement in design, this equipment tends to be reliable and fail safe. In fact, a review of the MEMS database showed that most of the problems associated with these systems were failures of accessories such as cables and transducers, also easily swapped out.

Nevertheless, the Clinical Engineering Department felt that the methodology used to identify serious injury risks was not sensitive enough to detect the important role the physiological monitoring and measurement played in the overall patient treatment. In many cases, this role included influencing the direction in which the care of the patient was progressed. In particular, the risk that a device appeared to be working correctly but was in fact producing a data error was identified as a potentially serious adverse event. So a new approach was taken to try and identify which physiological measurement and monitoring equipment was most in need of performance verification to ensure its accuracy. Consequently a second scoring system was developed to classify these types of equipment and test the assumption upon which the physiological measurement and monitoring equipment support plans were based.

The scoring system used had two components:

The first was used to score the impact that the equipment could have on influencing the direction of care. This was called score A and equipment was scored as 1 if it was used for data relay or secondary reporting (such as a central station in ICU). It was scored 2 if it was used for monitoring, that is to detect a change in status where the change is of more significance than the absolute value produced by the measurement. It was scored 3 where the equipment was used to make a critical measurement that was used in diagnosis, that is where the absolute value of the measurement could directly influence clinical decision-making.

The second score was called score B and this related to the clinical context within which the equipment was used. Where equipment was used in association with other devices and measurements, in an environment such as critical care where there are many staff trained in the use of the technology and results are nearly always interpreted in the light of other measurement, the equipment was scored 1. If the equipment was used in isolation, or in an area of the hospital where there are unlikely to be other measurements being made to contextualize the data from the equipment in question, it was scored 2.

The final score used to rank the equipment was the product of score A and score B. Examples of some of the equipment reviewed and their scores are presented in Table CS6.11A.

The results highlighted for the first time that identical equipment used in different clinical environments could prompt the need for different levels of performance verification. It suggested that it is not necessarily the more expensive and high-tech equipment, used in critical care environments, that needs the prioritization for performance verification.

The analysis also highlighted the importance of assuring the performance of the many physiological measurement systems used in speciality clinics such as cardiology, respiratory and neurology departments. However, in the hospital in question, each of these clinics was staffed by specialist-trained physiological measurement technicians who had expertise in using and verifying their equipment and who undertook a critical review of all measurements made.

TABLE CS6.11A   Sample of the Scores for Physiological Measurement and Monitoring Equipment

Equipment Type	Location	Score A 1 = Data Relay 2 = Monitoring 3 = Measurement	Score B 1 = Used in the Context of Other Measurements 2 = Used in Isolation of Other Measurements	Final Score
Vital signs monitor	General wards	3	2	6
Electronic thermometer	General wards	3	2	6
Handheld spirometer	Outpatient clinic	3	2	6
Audiometer	Outpatient clinic	3	2	6
Computerized spirometer	Respiratory measurement clinic	3	1	3
Ambulatory BP monitor	Home or Cardiology clinic	3	1	3
Multi-parameter monitor	ICU	2	1	2
Central station	ICU	1	1	1

For example, compare the score for the two spirometers: the relatively simple and inexpensive handheld one used by the triage nurse in the respiratory Outpatient Clinic and the more complex computerized spirometer used by the trained respiratory technician in the respiratory lab. The clinical engineering group judged that they could add more value by routinely checking the outpatient devices than they could by checking the computerized device which was under the supervision of a trained technician.

The existing equipment support plans tended to prioritize equipment in critical care areas. However, in discussion with clinicians they confirmed that in such environments no single measurement from one piece of equipment is likely to dictate the course of treatment; furthermore, there are many people involved in the care of the patient and they critically review the data, aware of the limitations of relying solely on technology. On the other hand, in ward areas, measurements of vital signs every 4 h were being made in isolation, very often by junior nursing staff with little experience. These measurements were being used to guide care and formed the basis of early warning scores for patients being cared for in general wards. Therefore, a monitor used to measure vital signs used in the ward was judged to require a higher level of performance verification than the multi-parameter monitor in ICU.

A central station in critical care scored 1 (Score A = 1 × Score B = 1). Whilst an electronic thermometer used in a ward setting to determine whether a patient had a fever scored 6 (Score A = 3 × Score B = 2).

As a result of the analysis, the Clinical Engineering Department reconfigured the equipment support plans for physiological measurement equipment to be grouped by area of use rather than by equipment type and prioritized performance verification of stand-alone measurement devices.

ADDING VALUE

There was no significant ongoing cost to undertaking this exercise. Yet it identified and led to a number of quality improvement actions. Given that the analysis was focused on impact on patients, the exercise delivered added benefit in the form of better risk mitigation. Therefore, value was increased.

↑   ↑

Benefits : Cost ∴ Value

SELF-DIRECTED LEARNING

1. This exercise highlighted the need to consider the clinical use of the equipment when developing the equipment support plans, rather than the device in isolation. Write notes on how you think such an approach would influence the development of equipment support plans for the following equipment:
   a. Ward-based CPAP ventilators;
   b. Home-based infusion pumps;
   c. Weighing scales used in Outpatient Departments.

2. Resources are finite and Clinical Engineering Departments can never achieve full support on every item of equipment, so risk assessment is necessary. If, following such a risk analysis, the head of Clinical Engineering Department proposed cancelling all performance verification of physiological monitoring in the critical care units for 1 year and instead proposed instigating a comprehensive performance verification programme for all the vital signs monitors and electronic thermometers used on the general wards, would you support or challenge that proposal? How would you argue for the position you took?

CASE STUDY CS6.12: MEASURING THE PERFORMANCE OF A CLINICAL ENGINEERING DEPARTMENT

Section Links: Chapter 6, Section 6.5

ABSTRACT

An established Clinical Engineering Department (CED) was working towards compliance with ISO 9001:2015. The project team required a set of performance indicators that would be communicated to stakeholders on a regular basis to show how the CED was performing both technically and organizationally.

Keywords: Key performance indicator; KPI; Assessment; Benchmarking; Analysis

NARRATIVE

A change in leadership within an established CED brought about a desire to achieve compliance with ISO 9001:2015, the ISO Quality Management Systems Standard. Part of this project required a set of key performance indicators (KPIs) to be agreed to monitor the effectiveness of the department, show improvement and highlight areas of concern.

In determining what should be measured and reported, the project team considered who the stakeholders were and to whom the CED reported. They came up with a list:

- The Medical Device Committee;
- Clinical users;
- Finance Department;
- Clinical engineers;
- Patients.

They then gave their attention to the technical tasks that were carried out by the CED:

- Responding to requests for repairs (unscheduled actions).
- Carrying out routine maintenance (scheduled actions and performance verification).
- Management of external contractors.

And finally, they looked at the softer side of the CED:

- Staff development.
- Staff training.
- Customer Survey results.

From their investigations they produced a list of functions and processes that could be measured. This list was considerably long and so they decided to look at their indicators in several ways in order to guide their decisions. The Audit Commission (2000) published a guidance on developing performance indicators including how they address the 'Three Es', of Economy, Efficiency and Effectiveness:

- Economy is the acquiring of appropriate material and human resources at the correct quality, at lowest cost. 'Are we paying more than we need to for this spare part?'
- Efficiency is the production of maximum output, using the supplied resource inputs. 'How much does it cost to maintain this equipment?'
- Effectiveness is the ability of a process to meet the required outcomes. 'Were our clinical users happy with the service they received?'

The performance indicators should be chosen to conform to specific characteristics, conveniently described by the abbreviation 'BARCUT V'.

Were the indicators chosen 'Balanced' or did they lean in one way whilst neglecting other activity areas of the CED? It was also important that they be 'Action' focused, and not created or measured just because it was technically possible; they needed to be designed so that actions could be taken on the basis of the data. Obviously the indicators had to be 'Relevant', especially to the Objectives of the HTM Programme and MDC. Then indicators also had to be 'Comparable'. Now there are two uses of indicators, internal and external. Internal indicators are used for detecting changes in the processes of the CED, and they do this well, as long as they are not changed very often. External indicators are used for comparing some element of the service to some other CED, and it is important that the comparison is valid. As the old saying goes 'Are you comparing apples with apples'? However chosen, they did still need to be 'Understandable'! As these indicators were to be published on corridor walls as well as to the MDC, they needed to be free of jargon, straightforward and clearly defined. They also needed to be easily producible so that they were 'Timely'. Finally, they needed to be 'Verifiable' for the purposes of audit.

These considerations led the CED project team to create a balanced scorecard of data from numerous sources focusing on four distinct areas of operation (Table CS6.12A).

The 'Service User' and 'Internal Management' sections of the scorecard were regarded as traditional methods of measuring performance. Introducing 'Continuous Improvement' was seen to be a softer indicator, especially considering the results of a satisfaction survey which would never simply be a pass/fail result. The 'Financial' section of the scorecard was created in response to a longer-term need by the MDC to manage the CED budget and eventually to compare costs with other similar organizations in a simple benchmarking comparison.

The CED felt that this reflected the main work areas and presented this to the MDC which accepted it as a starting position. In time, the balanced scorecard was supported by further, more specific indicators designed to measure performance in more detail as required to understand particular aspects of the CED's activities. In this way, data were available but did not have the time overhead when they were not required.

TABLE CS6.12A   Typical Balanced Scorecard of Key Performance Indicators for a Clinical Engineering Department

Service User	Internal Management
Response time from first contact to first action taken.	Scheduled maintenance completion. How much has been completed in a 12 month period.
Time to completion – from first contact to completion of job.	Quantity of repair jobs ongoing for over 3 months.
Continuous Improvement	**Financial**
Annual customer satisfaction survey.	Annual average cost of the CED per square meter of the organization served.
Time spent on CED technical training – hours per engineer per year.	Annual average cost of the CED per bed in the organization.

ADDING VALUE

Whilst there was a cost involved in the regular production of these figures, overall the benefit to the organization from the point of view of assurance provided was much greater.

Benefits : Cost ∴ Value

SUMMARY

By careful and logical analysis of a range of suggested indicators, the CED reduced these to a set that were meaningful, understandable and straightforward to produce. These were used to monitor and report on the function of the CED. Only minor changes were made after they had been set up, ensuring that the KPIs provided a means to assess the long term performance trends and not just a 'spot check' as is often the case. Monitoring of the KPIs by the CED leaderships ensured early warning of issues.

REFERENCE

Audit Commission. 2000. On target, the practice of performance indicators. London, UK: The Audit Commission. http://webarchive.nationalarchives.gov.uk/20150421134146/http://archive.audit-commission.gov.uk/auditcommission/subwebs/publications/studies/studyPDF/1398.pdf (accessed 2016-05-11).

SELF-DIRECTED LEARNING

1. Would you have selected the indicators suggested? Which ones would you suggest are added and why? Think about the stakeholder's needs and the reasons that they have for wanting information. If you were a patient, would you like to add targets to the indicators?
2. Is there a case for simply setting 'pass/fail' criteria and then having only exception reporting of problem areas to the MDC? What are the advantages and disadvantages of exception reporting?
3. If the indicators were to be used to compare data with another CED, where might errors in the data creep in? Think about scope of work, terminology and the varying needs of the clinical users in your answer.

CASE STUDY CS6.13: REVIEWING THE VALUE OF DIFFERENT ESPs WITHIN AN HTM PROGRAMME

Section Links: Chapter 6, Sections 6.6.3 and 6.5

ABSTRACT

At the end of 2012, the Clinical Engineering Department developed a chart for annually reviewing equipment support plans. In practice the CED focused their review efforts on the equipment groups it considered to have the highest clinical risk associated with failure. For each ESP, three pieces of information were plotted. The first two were qualitative, being the level of the ideal planned ESP and the assessment of the actual level of the ESP delivered in that year. The third was quantitative and was a calculation of the cost of providing the delivered ESP expressed using the cost-effectiveness index (CEI). The results were presented and discussed at a departmental forum. From this diagram, a quality improvement plan was developed which was to be implemented in 2013.

Keywords: Balancing an HTM Programme; Quality improvement plan; Optimizing equipment support plans

NARRATIVE

The ESP Value chart is presented in Figure CS6.13A, and a summary of the discussion relating to each equipment grouping analyzed follows.

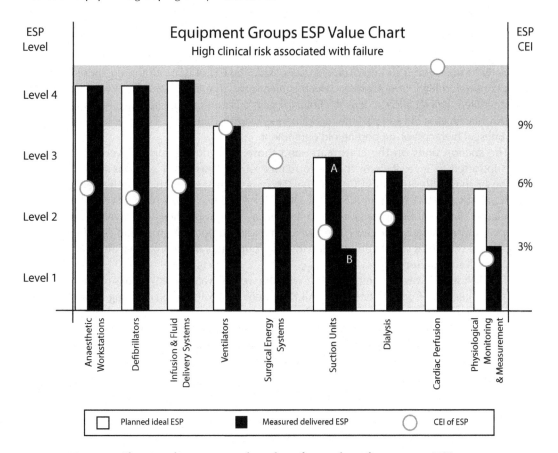

FIGURE CS6.13A   Chart used to compare the value of a number of concurrent ESPs.

The anaesthetic workstation ESP was planned to be delivered to level four support. External audit of the Clinical Engineering Department's work was undertaken by the equipment supplier and vice versa. Extensive user support was provided and all KPI targets were met. The plan was delivered in partnership with the equipment supplier's technical support department. This was achieved for a CEI of 6% which was considered appropriate given the level of user support provided.

The defibrillators ESP was planned to be delivered to level four. The performance of the programme and associated KPIs were reported to the hospital's Resuscitation Committee, and user support was provided through contribution to the Cardiac Life-Saving training programme. This plan was delivered in totality by the Clinical Engineering Department with a CEI of 5.2%.

The infusion ESP was to be delivered to level four. The performance of the programme and associated KPIs were reported to the hospital's Medication Safety Committee, and user support was provided through contribution to the Infusion and Medication Safety training programme. This plan was delivered in totality by the Clinical Engineering Department with a CEI of 6%.

The ventilation ESP was to be delivered to level three. The plan included extensive provision for user support and training. This plan was delivered in totality by the Clinical Engineering Department with a CEI of 9%. Some discussion took place as to whether the costs could be reduced further but no clear plan of action emerged. The department felt that given that this work was all being done in-house, there would be merit in looking for a way of having the work externally audited and this would progress this support plan to level four.

The surgical energy equipment (diathermies, harmonic scalpels, vessel sealing, etc.) had been identified in 2011 as an opportunity for improvement. The aim was that in 2012 this equipment group would be supported to level two. In 2012 the complete inventory had been through a performance verification, whereas in previous years only the surgical diathermy was proactively managed. This improvement was noted. The CEI of 7.5% for this equipment group was considered high for equipment which is electronic. The cause identified was the cost of unscheduled maintenance actions (technical repairs). The installed asset base of surgical diathermy machines is old and past the recommended replacement date. The high CEI was due to the unusual high cost of supporting old equipment.

The suction units on the crash carts and those used during surgery were planned to be supported to level 2, with the performance of the crash cart suctions also reported to the hospital's Resuscitation Committee. This was achieved, with the delivered support indicated by block A in Figure CS6.13A. However there was concern that the wall suction regulators were not being proactively managed as shown by their low level of delivered support represented by block B. The low CEI for this equipment group indicated that perhaps more resources should be assigned to this group and a more proactive approach taken to the wall suction regulators.

The dialysis ESP was planned for level 2 with the weekly water quality results reported to the hospital's Infection Control committee. All KPIs were met with the plan having an associated CEI of 4.5%.

The cardiac perfusion equipment (heart–lung machines, intra-aortic balloon pumps, etc.) were planned for level 2 support as the perfusionists were expert-trained users and they took care of reporting the QA of the machines within their own systems. Much of the equipment support plan for this specialist equipment is outsourced and all KPIs were met. In addition the Clinical Engineering Department helped interface this equipment to the anaesthetic record-keeping system as a project, reflected by a higher than planned level of user support. The CEI for this equipment gave cause for concern being 12% and it was agreed that the contract costs should be reviewed with the suppliers and opportunities for cost control explored with the suppliers, the perfusionists who are the users, and the CED.

The physiological measurement and monitoring equipment was planned for level 2 and this was not achieved. The sheer volume of equipment made it impossible to conduct the performance verification as planned. Also the CEI of this equipment was low implying that there were not enough resources being allocated to supporting this equipment.

OPPORTUNITIES FOR IMPROVEMENT

A quality improvement plan for 2013 was developed and summarized as follows:

- Look at a methodology for auditing the ESP for ventilators.
- Highlight with senior management the need to replace the ageing surgical diathermy equipment.
- Initiate a sweep through the hospital to examine and audit the condition of the wall suction regulators and following that, develop a position paper on how these devices should be included in the suction unit ESP.
- Review the Cardiac Perfusion service contract to look for opportunities to control the costs.
- Highlight to senior management the inability of the department to adequately perform performance verification of the physiological monitoring equipment that had proliferated throughout the hospital in the past 10 years – consider including this deficit on the hospital's Risk Register.

ADDING VALUE

There was no significant cost to undertaking this exercise. Yet it identified and led to a number of quality improvement actions which informed the better use of existing resources. It also identified areas where the CEI was high and where a more detailed investigation was merited to ensure that resources were being appropriately assigned and cost-effectiveness was being achieved. Therefore it increased the overall benefits delivered whilst also prompting cost control measures and in doing so increased value.

Benefits : Cost ∴ Value

SELF-DIRECTED LEARNING

1. Looking at the value chart for the equipment groups shown, can you identify any other opportunities for improvement?
2. Would you argue for or against reducing the level of user support provided to those using the anaesthetic workstations, infusion equipment and ventilators, in favour of introducing a proactive performance verification programme for the physiological monitoring and measurement equipment?
3. During the department discussion around the support of the general ward wall suction regulators, one experienced clinical engineer said, "We have run these wall suctions to failure for about 10 years and no serious incidents have ever occurred. If one fails, the nurses just take the one from the next bed and ring us. I don't see any reason to change our approach". Can you comment on the validity of their comments and consider what would you propose to do in relation to the management of wall suction regulators, if you were the head of department?

CHAPTER 7

The Extended Role of Clinical Engineers

Advancing Healthcare

CONTENTS

7.1 INTRODUCTION

The interlocking processes described in Chapters 5 and 6 make up a comprehensive healthcare technology management (HTM) system that efficiently manages the medical equipment in the healthcare organization. The objective of these processes is to optimize the delivery and support for medical equipment in healthcare. We saw in Chapter 1, Section 1.5.3 and in its Figure 1.4, that HTM has a dual remit, to support and advance patient care and to manage the equipment and technology used in healthcare. The equipment management remit is a positive, purposeful, continuing activity, predominantly a repetitive sequence of structured tasks that are known, can be foreseen and are planned. The management of these processes has an emphasis on constant improvement whether that is raising efficiency, reducing costs or improving quality. We have seen that these processes are informed, guided and controlled by regular review of key performance indicators (KPIs), and changes are made to the processes in the light of insights gained. For as long as the organization continues to use medical equipment, these processes will be required and will continue.

The clinical engineers who contribute to, manage and often lead these processes are in themselves an asset to the organization. Their expert guidance of the HTM processes adds value, and we will further discuss the development of clinical engineers in Chapter 8. Increasingly, senior management recognize that clinical engineers, having both systems engineering knowledge and in-depth understanding of the clinical environment, can contribute to aspects of the organization's management that go beyond routine equipment management and bring added value. Consequently, clinical engineers are often invited to take part in projects that could not be categorized as part of the routine HTM processes but usually fall into the broad supporting and advancing care remit.

Projects by their nature are different from processes. A project has a defined objective or deliverable, and the project ends when that outcome has been achieved. A project has a beginning, middle and end. At the end of the project, the team who were assembled to focus on its delivery will disband. The management of projects is focused on sequencing a series of tasks that deliver the outcome, rather than focused on the repetitive tasks characteristic of process management. Projects tend to be less standardized than processes and develop in response to meeting specific challenges with goals that include delivering the desired outcome on time, on budget and with the right quality. Projects are about doing something new or creating something new which can be an object, a report, an idea or a process. So projects tend to create change. Whilst processes are managed, projects need to be led and significant leadership is required to plan and execute a project. Where projects relate to the use of technology in healthcare, clinical engineers are often best placed to provide the project leadership. Project management tools such as PRINCE2® (PRojects IN Controlled Environments) distinguish between the executive leadership of a project and its day-to-day management. Clinical engineers, who are encouraged to learn project management methods and tools, may certainly be involved in the day-to-day management, but may also provide or share executive leadership with senior clinicians or senior healthcare management. This is one way in which clinical engineers can develop and advance healthcare.

In Chapter 1 we discussed how clinical engineers support and advance care through participation in research initiatives, often in collaboration with university-based colleagues. The temporary nature of such activity defines it as a project. In Chapter 5 we introduced perhaps the most common example of clinical engineers contributing to project management, namely their participation in projects focused on the acquisition of new equipment and systems. Whilst each acquisition is often considered the first phase of its associated equipment's life cycle, the acquisition phase for a particular procurement is by definition a project. The project is temporary; has the defined scope of selecting, buying and commissioning particular new equipment; has an agreed quality; must be to specification and fit for purpose and must be on budget and on time. The establishment of a temporary multidisciplinary project team for an acquisition project is a good example (Amoore et al. 2015).

Within a project team, people's roles and responsibilities are defined within the context of that project and may be quite different from those of their usual process roles, as has been noted in Chapter 5, Section 5.5.3. One of the advantages of working in a project structure is that, as part of the project initiation, the roles and responsibilities of individuals in the team assembled to deliver the project are defined. This allows for clinical engineers to play a part in the project for its duration that they may not play as part of their work in the ongoing HTM processes. For example, a clinical engineer might take a leadership role in the investigation of an incident and manage a multidisciplinary team made of professionals who would not normally report to a clinical engineer. This is an important point to understand. When entering a project, the usual reporting relationships can be suspended in favour of temporary ones that are tailored to the needs of the project. Again when the project is finished and indeed in working relationships outside the scope of the project, the usual reporting relationships apply.

Formal project management techniques provide a framework whereby issues such as the temporary change in reporting roles can be constructively negotiated. Control of projects is often achieved by breaking the project into tasks which are interrelated and then defining the inputs and outputs for each task along with defining the scope and resources required for each task. This is a classic systems approach (Chapter 2). By closely monitoring the progress of these tasks and their interrelationships, the project as a whole can be managed. Project management techniques also include defined methodologies for dealing with problems that might emerge during a project. For example, a specific critical task may require more resources to complete on time. The techniques provide a framework for identifying and managing risks to the project and defining reporting roles so that potential problems can be foreseen and dealt with effectively.

Whilst projects and processes have different characteristics, they do interrelate. As described earlier, an ongoing process such as the HTM process can prompt the initiation of a project, say an acquisition project or a decommissioning and disposal project. Also a change that might be required to improve an existing HTM process may need a project to be established to implement that change. So projects can and often do emerge from, and relate to, HTM processes. They often arise to deal with issues within an HTM Department.

In this chapter, we will explore some of the projects that clinical engineers typically participate in. These projects will be classified into two groups (Figure 7.1). The first will

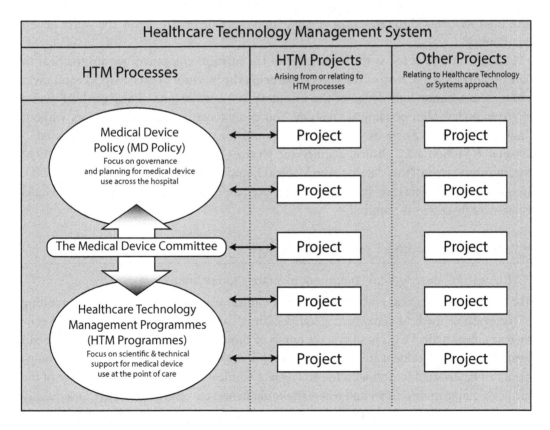

FIGURE 7.1 Diagram showing two types of projects to which clinical engineers contribute: those arising from HTM processes and those arising from advancing knowledge or the practice of delivering care.

be projects that arise from the equipment management processes described in preceding chapters. The second group of projects relate to those that are prompted by challenges that arise outside of the healthcare technology management processes, but which can usefully draw in the expertise of clinical engineers. So in describing the first grouping of projects, some overlap with the HTM described in preceding chapters is to be expected. However, the method of working in projects is different and merits an examination. The projects described in the second half of this chapter do not directly relate to the traditional equipment management role yet are part of a holistic healthcare technology management system that adds value to the healthcare organization.

The common thread running through all the activities described in this chapter remains broadly centred around the management and support of technology used in the provision of clinical care. However, we will see that there are many ways for clinical engineers to pro-actively support and advance care beyond the traditional equipment management activities they are often associated with. We will illustrate how clinical engineers' core training in engineering and systems thinking, together with their experience in interdisciplinary working, enable them to be effective as engineering experts, advocates for patient-centred care, managers, innovators and leaders. We believe that clinical engineers have a key role in knowledge management, often generating new knowledge and applying it to lead service improvements throughout the organization and potentially across the wider health community.

Fundamental to HTM is the knowledge of the medical equipment within the healthcare organization, the base of the knowledge being the inventory of its medical equipment assets. The accurate knowledge of the inventory is a cornerstone of any medical device strategy, policy, plan or tactical solution. You cannot manage medical devices without knowing what devices are used across the organization. Consequently, as we discussed in Chapter 4, Section 4.2.7, clinical engineering services have information systems that hold the inventory; we call this database the Medical Equipment Management System (MEMS). In many of the projects we discuss in this chapter, an accurate knowledge of the medical equipment inventory is required.

7.2 PROJECTS ARISING FROM OR RELATING TO HTM PROCESSES

7.2.1 Applying the Medical Equipment Inventory to Enhance Care

The continuing provision and development of clinical planning require an understanding of the available medical equipment resources. Thus, for example, the development of contingency plans to deal with sudden acute bursts of clinical activity following episodes of bad weather or a major accident are facilitated by knowledge of what life support and other equipment can be diverted to deal with the emergency. Clinical engineers have knowledge of the medical equipment inventory and where the equipment is currently deployed. Consequently clinical engineers should be included in the short-term project teams convened to respond to these acute changes in clinical demand. Case Study CS7.1 explores the response to acute increases in clinical demand in the Emergency Department (ED) and how clinical engineering managed to redeploy medical equipment to meet the increased demand.

When a clinical service is moved between premises to enable basic estates maintenance and upgrades to be carried out, a project team is often assembled to manage the movements. Clinical engineers support these projects through the deployment of medical equipment based on their knowledge of the available equipment.

Knowledge of what medical equipment is available will assist its possible redeployment to support changes to clinical services in response to changing clinical priorities, with clinical engineers joining or perhaps leading the project teams involved. Case Study CS2.1 discusses the work of the project team established to redesign palliative care services, jointly led by clinical engineering and palliative care, where knowledge of the equipment inventory was important in planning the deployment of resources. When the organization wishes to further develop a service, such as expand the number of critical care beds, accurate knowledge of the existing equipment provides one of the bases for planning what is required for the expansion. For example, Case Study CS1.1 describes the project convened to expand an endoscopy service where the clinical engineer's knowledge of the existing and planned increase in equipment and its resource requirements were vital.

7.2.2 The MEMS Database Supports Effective Medical Equipment Management

Medical equipment is a costly resource for healthcare organizations, for funding both its procurement and also for keeping it operational; the total cost of its ownership over its lifetime is complex and we have discussed this in Chapter 4. The medical equipment assets of a healthcare organization must be managed to ensure that the assets are effectively used to deliver healthcare. This requires knowledge of the assets available to the organization; the MEMS database provides this information. The asset information will be used to help ensure optimum deployment of the medical equipment, for replacement planning, and for responding to particular problems, with project teams convened to manage particular initiatives. We have already seen how knowledge of the assets can be used to manage acute increases in clinical demands (Case Study CS7.1).

An important aspect of the management of assets is to understand their utilization, with evidence suggesting that whilst some equipment is used to its full potential, the utilization of others is low. The MEMS provides the asset information for starting to understand equipment utilization. This will be supported by direct and indirect utilization data (e.g. number of imaging scans, number of dedicated consumables used or data from operating hour clocks). Many healthcare organizations have assembled project teams to address equipment utilization. For example, projects have been set up to study the use of the radio-frequency tracking of medical equipment to gather information on utilization and to plan improvements (Britton 2007). Other project teams have responded to the challenge of imbalances in equipment utilization by developing equipment libraries that facilitate sharing of medical equipment to improve utilization (Keay et al. 2015).

7.2.3 Planned Replacement Projects

Clinical engineers have important roles to play in the whole of medical equipment replacement projects (Figure 7.2). They will be involved in suggesting equipment for replacement based on life cycle replacement planning, helping to lobby for replacement funding.

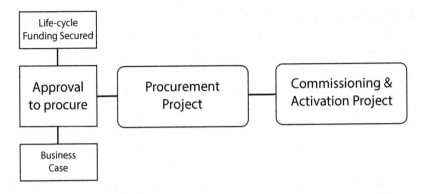

FIGURE 7.2   Planned replacement project.

They will help draft business cases, sometimes leading on their preparation. We will see in this section their important contribution to determining the equipment requirements and their guidance of the commissioning of procured equipment.

The Clinical Engineering Department (CED) will know the current distribution of equipment from the MEMS and from supporting it through its equipment support plans (ESP) (Chapter 6). Clinical engineers are therefore in a good position to make the business case to the MDC, or other appropriate body, that a planned and coordinated replacement project, to include standardization, should be put into place. Once approved, an appropriate clinical engineer may be in the best position to lead on the procurement project, drawing into a project team the variety of clinical users that can and need to contribute.

In many equipment replacement projects, particularly those that relate to widely used equipment types where there is no clear specialist clinical leadership, the clinical engineer may well be best placed to lead and coordinate the acquisition project. Examples would be the replacement of fleets of widely used equipment such as infusion devices, defibrillators or vital signs monitors. All these are general-purpose and essential equipment, used widely across a hospital by many different staff, but do not have a clear clinical 'champion' such as is the case with, for example, anaesthetic equipment or intensive care monitoring. Without coordination the risk to the organization and to patients is that, if the acquisition of such general-purpose equipment is left to individual wards or clinical departments, the results will be lack of standardization with a variety of equipment types in different areas, poor quality training and more expensive maintenance. A good example was the need to replace the thermometers used on the wards for routine temperature measurement, described in Case Study CS7.2.

The first task of the clinical engineer leading the team will be to select the team members. This is probably best done by deciding which professional and clinical groups need to be involved and then asking for appropriate nominees through their management team. It may be appropriate to suggest names of possible nominees. This ensures that all participants have the authority from their own profession and can also contribute as individually nominated experts. Almost everyone in the team will be operating outside of their normal role, reporting to the team leader for this project and reporting back through their managerial structure so that information is widely shared.

The exact composition of the team will vary depending on the objectives but, in general, appropriate clinical input is vital, whilst input from the Finance and Procurement departments will be essential for financial planning and to ensure the acquisition complies with the required procurement regulations including those pertaining to tendering.

At the start of the project, the clinical engineering leader will steer the team's thinking to clarify the required outcome, measurements or facilities, rather than focusing on what type of equipment to buy. Thus in the thermometer procurement example (Case Study CS7.2), the requirement was to accurately and reproducibly measure patient temperature rather than to decide from the start which particular type of thermometer to procure.

An important part of any equipment procurement project is clearly expressing the requirements through the equipment specification. We have discussed in Chapter 5 that an important role for the clinical engineer in the procurement process is to interpret and put into unambiguous terms the technical specifications required by the clinical user plus those required for the effective management of the equipment such as arrangements for maintenance and user and technical training.

Figure 7.2 summarizes the general equipment replacement process starting with gaining approval and funding for the replacement, progressing to the procurement project itself incorporating the specification, evaluation and selection and moving on to the commissioning. Figure 7.2 should be seen as a general illustration rather than a rigid structure. Thus the commissioning and activation project illustrated in Figure 7.2 may have two parts. One part is the need to plan the methodology of commissioning the perhaps large number of individual new devices and their subsequent ongoing maintenance. This work might be done by a different sub-project team perhaps entirely within clinical engineering, reporting to and coordinating with the procurement project team leader as well as to the internal clinical engineering leadership. The activation part, integrating the new equipment into the clinical workflow and training those who will use the new equipment (e.g. see Case Study CS2.1), might best require a separate sub-project team with clinical and clinical trainer members and with Procurement to ensure the logistics of consumable supplies. This sub-project team would also report to the overall project leader.

Key messages are therefore as follows:

- Clinical engineers may have a key role in project teams dealing with procurement issues, over and above their day-to-day roles.

- When in project roles, they will report to the project team leader in respect of that work but must be supported in it by their line managers.

- In appropriate circumstances, clinical engineers may lead such project teams and bring leadership, systems and technical skills to the team.

- Project teams need to be flexible in their membership and approach to their tasks.

7.2.3.1 Planning

Planning medical equipment procurements starts with developing its business case, appropriate in scale to the procurement size. For a small acquisition, say the purchase of an oxygen saturation monitor to replace one damaged beyond repair, many organizations require only a 'statement of need', a minimum business case, so that the person authorizing the expenditure knows what they are approving and why. Systems should be in place so that such 'mini' purchases are not subject to unnecessary and bureaucratic delays which can impact on patient care and cause unnecessary work for clinical staff.

Larger and more complex acquisitions may start with an outline business case that, once approved, will be followed up with a full business case. The full business case can be very complex especially if the healthcare organization has to seek external or government funds. Health economic and regional planning factors may come into play and the process can even become political.

An aspect of planning that links directly into the procurement phase, but may run alongside and inform the business case, is the identification of the most appropriate healthcare technology. In Chapter 5, Section 5.6.4, we provide a detailed bullet list of issues that must be considered as part of a procurement project. The detailed consideration of all these points will lead to the best and most effective decision. These considerations will help inform the formal tender specification.

Planning requires multidisciplinary endeavour to develop technical and functional specifications. This is predicated on having a clear understanding of the function required and the environment in which the equipment will be used. In some large acquisition projects, there may be a requirement to build new facilities. Where there are associated building works required or significant changes to the environment or services needed to support the new device or system, these should be planned from the start of the project. Other projects may involve the need for increases in staff or the initiation of a change management process relating to clinical work practice. All of these will need to be considered in the business case and actively planned and managed before, during and after the procurement and commissioning phases. Examples of procurement planning, where multidisciplinary teamwork was required, clinical practice required changing and the implications of procuring additional equipment had to be addressed, are discussed in Case Studies CS2.1, CS5.9 and CS7.2.

7.2.3.2 Tender Specification

Drafting of the formal tender specification will require expert multidisciplinary input. The more complex the project, the more comprehensive the input required to the specification which may have to include elements of the wider implications of the acquisition mentioned earlier. There is substantial risk of acquisition projects running into difficulties and not delivering optimally if the tender specification process is not well led and sufficiently inclusive.

Here again clinical engineers can make a significant contribution not least because of their familiarity with and understanding of technological developments. They will work with clinical, procurement and financial colleagues and should ensure that technical specifications are unambiguous and unbiased. The specifications should include adherence to the appropriate Standards, as discussed in Chapter 3, to ensure that the procured equipment complies with current requirements. Support issues such as warranty conditions, maintenance arrangements and user and technical training provision must be built into the tender and not left to be disadvantageously negotiated post-purchase.

The specification and associated tender documents may also require the inclusion of the criteria by which received tenders are evaluated. These are sometimes referred to as specification criteria, each with their own weighting factors: clinical functionality, governance, safety and patient considerations, technical considerations, support including after-sales support and whole life cost of ownership (Amoore et al. 2015). The criteria and their relative importance should be thought about and decided at this planning stage, with Case Study CS5.11 describing the use of tender criteria to guide decision-making.

Particular difficulties can arise with new build projects. Clinical users and the clinical engineers may have been properly consulted at the planning stage and may have produced detailed requirements, sometimes called 'room data sheets', but are then excluded from the site whilst construction continues. On handover of the site, they then find that requirements have been ignored or misinterpreted: electrical power points have been put in the wrong place or in insufficient numbers, patient observation windows between two rooms have been put in the wrong walls, etc. Effective and continuing liaison between the healthcare organization's project manager, the contractor and the ultimate users and their clinical engineering advisors is essential.

7.2.3.3 Procurement

In the wider replacement project, we have covered the important planning phase (Section 7.2.3.1) and the specification phase (Section 7.2.3.2); clinical engineers have vital roles to play in both. Each may require the formation of project teams to investigate the details in depth. The next phase involves issuing tenders to prospective suppliers, followed by the subsequent review and evaluation of the tender returns, often called the procurement phase.

The evaluation of the tenders is a crucial phase involving a multidisciplinary task in which clinical engineering must play a part. It is the clinical engineer who is best qualified to interpret each supplier's responses to the technical requirements. The clinical assessment must be discussed with the clinical users, remembering the strategic aims of the organization and of the particular departments where the equipment will be used. Closely aligned to this, the relevant patient and carer needs and general and specific considerations of patient safety should be considered. The technical characteristics of the equipment are important; does it meet the current Standards and what technical and functional aspects need to be assessed? Will this equipment be placed on the hospital information technology (IT) network and what impact does this have for the device, the network and other devices and systems already connected to that network? Equipment requires ongoing support, staff training, supply of accessories and consumables and maintenance; what arrangements are required and who will carry these out and who will be responsible for them over time? Financial aspects, acquisition and operating costs, require to be examined in the context of the resources available.

The multiple aspects involved require a structured, systematic, documented approach, with the various aspects including selection criteria clearly and simply detailed in the specification and procurement documentation (Amoore et al. 2015). This systematic approach should be adopted whatever the source of funding whether through capital or revenue funding or through donations. Problems can arise if any aspect is overlooked, potentially setting the scene for later ongoing clinical, technical and/or financial risks.

Judging of tenders is an activity that requires planning and careful execution (Case Study CS5.11). An important early stage is gathering detailed information about all the possible technology options. Some of the information will come in formal documents submitted by bidders describing their equipment and how it will meet the needs described in the tender documentation. But generally users will require to see and to use the actual equipment. This requires demonstrations of equipment to representative user groups or user trials. For large equipment, visits to other sites and communication with existing users may be helpful. Objectivity and transparency must be maintained. Undertaking such activities is time consuming and clinical engineers should be resourced and empowered to play a significant role as part of their formal responsibilities.

7.2.3.4 Commissioning

No healthcare technology should be put into clinical use without a formal commissioning process. This activity is part of both the 'Do' and the 'Check' phases in the quality management cycle (Chapter 2, Figure 2.10). The extent of commissioning will vary depending on the type of equipment and technology involved. All equipment will need verification against the purchase order and at least basic safety and functional checks. All, whatever the acquisition source, will need relevant data entered into the MEMS database. The steps involved are summarized in Figure 7.3, and except in the most uncomplicated cases, a short life project with a plan and a team should be set up to ensure that all the steps are carried out.

A straightforward situation (e.g. commissioning a new vital signs patient monitor to replace or add to identical ones already in use) will require little else than adding to the MEMS database and carrying out the required configuration and performing the safety and functional checks as detailed in the equipment support plan (Chapter 6, Section 6.2). The risk status of the equipment will have already been established, training of users will be in place (unless being deployed in a new clinical area), maintenance protocols and training will have been established and spares and test equipment will be available. Such a situation will not require a project approach and is likely to be a well-established process.

Whenever there is commissioning of new items of medical equipment, even if other items of the same type are already in use in the healthcare organization, it is prudent to check whether additional clinical training is required. Training of clinical user staff should be planned in advance. No equipment should be put into clinical use without adequate training of the staff that will use it. It may be necessary to delay the implementation of new healthcare technology until a predetermined percentage of relevant staff has been trained.

Where medical equipment is procured that is new to the healthcare organization or new to particular departments, a more detailed commissioning process is required. The first

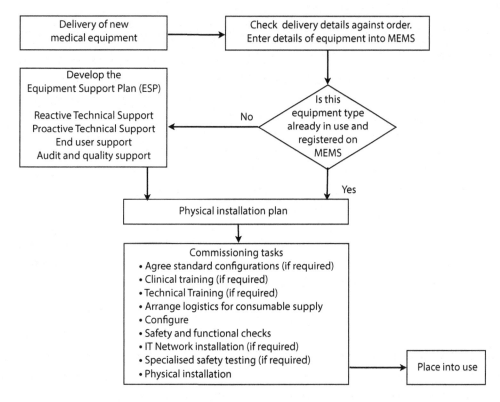

FIGURE 7.3 Flow chart for Commissioning new medical equipment. MEMS is the Medical Equipment Management System, ESP is the equipment support plan.

phase is to develop its support plan which is summarized in Figure 7.3 with the development of the technical and clinical support packages and the configuration plan (forming the basis of the ESP) and planning the physical installation. Establishing the risk status of the equipment will have to be considered and documented. Details of the ESP are covered in Chapter 6 and will not be repeated here; some aspects, including installation planning, will also have been covered in the tender specifications.

Configuration of medical equipment is a crucial part of the commissioning phase that has been described in Chapter 4, Section 4.4.1.3 and Case Study CS4.1. Where additional equipment similar to that already in use is being commissioned, it may be opportune to ask clinical staff if the configuration of the medical equipment is still optimal for its application.

The actual commissioning tasks are summarized in Figure 7.3. These start with assessing the need for clinical and technical training and, together with the appropriate department manager, discussing with the Procurement Department the logistical arrangements for the supply of any consumables required. Physical commissioning may include assembly of the equipment and programming the configuration. The functional and safety checks specified in the equipment support plans will be carried out.

Increasingly medical equipment is being integrated through IT networks into electronic healthcare recording systems, for example interfacing to specialized systems such as PACS systems or integrated clinical information systems (CIS). Linking of medical equipment to the IT networks should be carried out in discussion with the appropriate IT or eHealth staff in the healthcare organization. The integrity of any medical equipment to IT system links must be validated. When complex commissioning is required, a project plan should be drawn up with a nominated project manager, clear responsibilities allocated and clear sign-off criteria agreed (Case Study CS5.10).

Certain types of healthcare technology require complex and specialist testing and validation before being put into use. Radiation-generating equipment requires extensive radiation safety testing by specialist experts.

The physical installation of equipment must also be planned and executed in a formal way, to ensure not only the physical stability and security of the installation but also its ergonomic layout. With the goal of creating 'a healing environment', Thompson et al. (2012) have discussed and developed guidelines for intensive care design, noting that optimal ICU design can improve patient outcomes, reduce length of stay and reduce costs. The ergonomic placing of monitoring equipment in the ICU can have an impact on its effectiveness and on patient safety. We discuss in a Case Study (CS5.9) how the optimum architectural design of a CT imaging suite with twin CT scanners sharing a common control room and providing patient waiting, and examination areas can improve patient and staff flow and the effectiveness of the functional area. The placement of anaesthetic, monitoring and surgical equipment within an operating theatre can similarly improve the safety and effectiveness of patient care and reduce musculoskeletal stresses on clinical staff (Sheikhzadeh et al. 2009).

Clinical engineers should not shy away from getting involved in such details. The impact of the physical arrangements of the building and the position and layout of the equipment within the buildings must be considered at an early stage.

Once all the commissioning issues have been sorted out, healthcare technology enters an operational phase in it life cycle (see Figure 5.1). The Clinical Engineering Department's interaction with the technology will be through a cyclical series of processes which together form a tactical, that is 'day-to-day', approach to the operational management of healthcare technology. The details of this are dealt with in Chapter 6 where we look at specific HTM Programmes.

7.2.3.5 Discussion

Since the acquisition of healthcare technology should be evidence based, clinical engineers have a pivotal role in providing information, data and advice on many of the issues that need to be considered when making equipment procurement decisions. Clinical engineers can provide useful advice at the early planning stage, including the need for the equipment, the crucial first step. In providing this advice and decision-making, clinical engineers will apply and combine their knowledge both of the clinical requirements and of the technologies able to meet those requirements, supporting this with their knowledge and understanding of the status of existing equipment. The Clinical Engineering Department's links with senior management of the organization and the lead clinical staff for the different specialities will support the necessary dialogue to ensure effective equipment management decisions, whether these involve the transfer of existing equipment or procurement of additional or replacement equipment.

Furthermore, two aspects of clinical engineering training and experience can be brought to bear in this field. First, professional engineers are trained to take a systems approach to problem-solving and project management (see Chapter 2), to understand risk and risk management, to know about relevant regulations and safety standards (see Chapter 3), to be able to explain complex technical matters in a clear and understandable way and to work collaboratively in multidisciplinary teams, both formal and informal. Second, clinical engineers by training and experience are able to understand both the science involved in healthcare technology and the clinical applications and implications of the equipment involved. They are able to act as knowledge mediators between the technology and the clinical applications between the equipment and the clinician.

Finally, although much of Section 7.2.3 (Planned Replacement Projects) has focused on the contributions that clinical engineers will provide in these projects, often in leading roles, it is important to conclude by reminding ourselves that these are multidisciplinary projects. Clinical engineers need to ensure, particularly when leading these projects, that the views and opinions of the clinical staff are encouraged and listened to and help guide the process. We have noted that the project teams will include clinicians, preferably nominated by senior staff. The clinicians on project teams have the responsibility of helping to ensure that the clinical requirements of the equipment are included in the specifications and assessed during the evaluation phase. The healthcare organization may have a policy that patient representatives are included in these project teams. Patients can help guide equipment procurement planning, particularly where equipment is used in the community where the end users may be patients or lay carers. (Lay carers are not medically trained professionals, often family or friends.)

7.2.4 Managing Equipment Trials

Clinical engineers will often be involved in pre-purchase evaluation trials of medical equipment, often managing them (Case Studies CS7.2 and CS7.3). These are special cases of the local health technology assessment (HTA) processes that we will discuss in Section 7.2.5. These evaluation trials are not a general assessment of the efficacy and benefits of the technology, but specific practical evaluations of particular equipment by the clinical staff who would use it, in the environment where it is intended to be used, typically to compare different models of equipment prior to procurement. When carrying out local HTA, it is important to remember that HTA requires a multidisciplinary approach that considers and evaluates the social, economic, organizational and ethical issues of the health technology involved (WHO 2011).

Carrying out the evaluation within a healthcare organization requires careful management that covers safe introduction and use of the equipment, training of staff, logistics to ensure supply of consumables, infrastructure support including utilities and mounting and operating procedures. The evaluation of medical equipment in clinical trials may require ethics approval, and the ethics department in the healthcare organization should be consulted for advice. In some jurisdictions, specific regulations apply and these must be followed. This all requires a project to be established for the safe and effective management of trials. It is common to require the company supplying the equipment for evaluation to have indemnity insurance to cover problems that may be encountered during the trial.

Such projects are often led by clinical engineers. The project must first evaluate the equipment itself and then the clinical process within which it will be used. The purpose of the analysis is first to ensure that all relevant safety requirements are met and second to try and imaginatively foresee potential problems. As part of this evaluation, the clinical engineer will perform a technical and operational risk assessment before the device is put into use. They are likely to highlight risk issues that need to be controlled and identify where other allied health professionals, the infection control team for example, should be involved in risk assessment. The project should detail the exact methodology for the trial (Case Study CS7.3). This might include ensuring that adequate specialists from the supplying company are available during the trial to support and train the doctors and nurses who will be using the device under evaluation. Doctors may also request that the clinical engineer is present when the device is used both to facilitate the introduction of the device into the environment and clinical workflow and to assist with coordination of risk mitigation actions. Clinical engineering involvement may also be requested to evaluate the practical use of the technology in the clinical environment.

7.2.5 Local Healthcare Technology Assessment

Health technology assessment (HTA) systematically evaluates the benefits and costs of healthcare technologies (WHO 2011). It is a multidisciplinary process that examines the social, economic, organizational and ethical issues of healthcare technology, exploring the benefits and the life cycle costs delivered by new technologies. The assessment is a multidisciplinary process in which professionals from different backgrounds collaborate and come to a shared conclusion as to the benefits of new healthcare technology. HTA is typically undertaken at national levels.

At the local healthcare organization level, the results of national HTAs can inform equipment planning, but there is often a need for each organization to perform a level of local healthcare technology assessment before adopting a new technology. What might add value in one organization does not necessarily add value in another, and so it is important for each organization to critically review whether new medical equipment will deliver value for local stakeholders.

When novel acquisitions are being considered, it is usual for a healthcare organization to establish a project team to perform this local assessment and report it as a business case (Figure 7.4). The development of the business case will require both clinical and financial justification. The project to develop the business case is different from the project to procure the equipment and the skill sets of both might be quite different. The business case requires an independent review, assessing the merits of the technology and its local benefits and costs, concluding with a recommendation to either adopt or not adopt the technology. If the recommendation is to proceed to adopt the technology, then the business case will typically detail the conditions that adopting the technology requires, including clinical operational planning and financial planning. Clinical engineers often participate

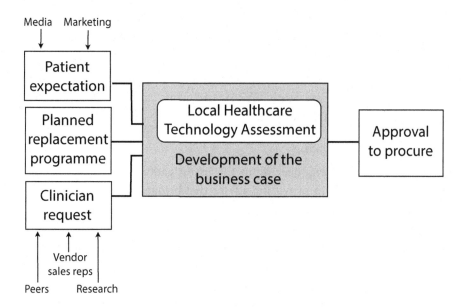

FIGURE 7.4   Local healthcare technology assessment.

in this activity (Case Study CS7.4) as they have an appreciation of both the clinical application of technology within the context of their own organization, and they are also aware of issues that affect the life cycle costs. When participating in such projects, the clinical engineer assumes the role of an independent expert and should not be influenced by their responsibilities as an equipment manager. The independent assessment by the project team is crucial as often the requests to senior management to adopt new technologies can be unduly influenced by marketing or public pressure. So both critical review and good planning to inform the procurement process are essential.

The introduction of new technologies requires a holistic assessment of benefits and risks, with clinical, technical and financial aspects considered. Health economics provides useful tools for these assessments, and cooperation between clinical engineers and health economists can clarify the net benefits of alternative technologies (Case Study CS7.5).

Clinical engineers may become involved in projects that seek to better understand the characteristics and limitations of medical equipment in clinical situations supporting directly and indirectly HTAs. Their contribution is particularly important because they understand the clinical context in which medical equipment is used. For example, whilst infusion devices are, from a technical point of view, simply pumps that deliver fluid from a container at a fixed rate, their involvement in many of the adverse events that involve medical devices suggests that the combination of the device, its accessories and the clinical situation requires greater scrutiny (Amoore and Adamson 2003). In a case study (Case Study CS7.8), we explore how the combination of an infusion pump and its accessories can cause the infusion system as a whole to miss-function.

Clinical engineers routinely use simulators of various sorts to test medical equipment. Simulators of the respiratory systems are used, for example, to check the function of ventilators, and ECG simulators are used to test ECG machines. Testing the operation of medical devices with test simulators can lead to improved understanding of the operational characteristics of the medical equipment if the clinical engineer adopts an enquiring mind. For example, questioning the measurement results when a non-invasive blood pressure (NIBP) monitor was tested with a simulator led to a greater understanding of the operation of these devices (Case Study CS7.6). This work can support health technology assessments, both locally and globally.

7.2.6 Adverse Event Investigations

The involvement of clinical engineering in the investigation of adverse events is a particular example of their role in providing advice that bridges the technical and the clinical, supporting the healthcare organization, its clinical staff and its patients (Case Studies CS2.2 and CS4.3). Clinical engineers also develop important links with regulatory agencies, manufacturers and national incident reporting centres that can be helpful in product enhancements following adverse events (Boutsikaris and Morabito 2014). The CED has a significant part to play in managing the aftermath of such incidents, investigating the root causes, advising on remedial action which might include technical measures or additional training, using the MEMS database to look for patterns. A particular emphasis should be on taking action to reduce the likelihood of recurrence.

Details of those adverse events involving medical equipment that can affect the safety and well-being of patients, visitors and other staff have been discussed elsewhere in this book (Chapters 2 and 4, Sections 2.4.6 and 4.4.2.4). The reader is referred to those sections for the details, including their various causes, emphasizing that the cause should not be simply attributed to device or user fault (Runciman et al. 1993; Reason 2000; Amoore and Ingram 2002; Jacobson and Murray 2007; Amoore 2014). Here we discuss assembling short life project teams convened in response to a specific adverse event, with the objective of identifying the causes and developing methods of preventing repetitions.

The clinical engineer will have an important role in these project teams, sometimes being asked to lead the team. After an incident has occurred, it is important to understand the underlying causes, and the clinical engineer's experience of the application of technology should guide the team away from simply attributing blame to equipment or user. Equipment or user faults may be involved, but the clinical engineer will want to understand the background, the details. More frequently the cause may result from complex interactions between device, user and patient within the context of the clinical environment and its supporting infrastructure. Case Study CS7.7 discusses the role of the clinical engineer in the project team investigating an adverse event.

Often the clinical engineer can take an objective approach to the investigations, trying to establish the facts. Even where user error was involved the clinical engineer should guard against the tendency to blame; rather, the objective should be to understand the causes of the user error, including the extent to which operating processes and procedures contributed. This can lead to development of methods to prevent recurrence, including seeking methods to mistake-proof the processes and procedures. (There is considerable literature on mistake-proofing the design of equipment and of processes; this is beyond the scope of this chapter but the interested reader could start their understanding of the process by reading Grout, 2007.)

The healthcare technology management process and its associated maintenance activities do sometimes contribute to adverse events. In these instances, the clinical engineer, as part of an investigation team, will need to be conscious of this possible cause and the priority of objectively assisting understanding the causes, even if those are maintenance failings, perhaps even where these might involve the clinical engineer's own department or actions. There will be occasions where the clinical engineer might wish to withdraw from such an investigation or part of an investigation if continuing involvement might be thought prejudicial to objective enquiry. Ethical considerations will dictate the involvement of the clinical engineer under such circumstances, with the clinical engineer guided by the objective of establishing the truth in the interests of patient safety.

Incident investigation project management is important, with the process often guided and governed by Standard Operating Procedures (SOPs) developed by the healthcare organization. The processes followed will depend on the assessment of the severity of the incident, with serious incidents involving death or major harm reported to senior management and Board. These SOPs will dictate the structure and method of operation of the project team, assigning leadership, often from the clinical leadership of the department where the incident occurred for incidents with less severe outcomes or from senior healthcare organization leadership for serious incidents. The relevant expertise to conduct the investigation should be assigned and this might well include clinical engineering (Case Study CS7.7).

The implications of the incident will guide the scope and depth of the incident investigation, but all incidents should be analyzed to uncover the underlying causes. This includes incidents in which no harm occurred, which are numerous; the lack of harm involved in them has led to them being referred to as 'free lessons'. They provide the opportunity to learn lessons without there having been any victims, anyone harmed by the incident.

There are a few important initial steps in responding to any adverse event. First, the care and safety of the person, often a patient, possibly adversely affected, must be the prime concern, with initial support given to them. Second, the evidence should be preserved, an aspect in which the clinical engineer can help take the lead, working with managers from the department where the incident occurred. Equipment involved in an incident should be quarantined, preserving where possible the setting of operating controls and leaving accessories and consumables intact. Data, records and patient notes should be included in the incident file. The initial details of the incident will be recorded in the organization's incident reporting system, linking in, as appropriate, with regional and national reporting systems. Third, care should be given to those clinical staff involved in the care of the patient when the incident occurred or who were involved or who witnessed the adverse event. These staff may be traumatized by the event, whether or not their actions or omissions contributed to it. These staff are sometimes referred to as 'second victims'; their feelings of hurt, guilt, inadequacy and failure will need to be handled with empathy and compassion. Staff within the Clinical Engineering Department may sometimes be 'second victims'. Clearly, where the cause of the incident was a deliberate or malevolent action by staff or visitors, then the legal system processes will take their course.

The CED should have procedures in place for handling adverse event investigations and the role of clinical engineers in them. The procedures must include appropriate provision for maintaining independence in circumstances in which the CED may have been responsible for maintaining the medical equipment involved.

Even when no incident has happened, clinical engineers have the responsibility for imaginatively foreseeing hazards or hazardous situations that may lead to risk. This may come from their good technical knowledge of medical equipment, from their knowledge of how equipment is used, from their involvement in user training or from insights gained from records in the MEMS of what can go wrong.

7.2.7 Medical Device Safety Alerts

Medical device governance includes the management of adverse events that occur within an organization and also includes managing the response to alerts received that may have been generated following incidents in other healthcare organizations. Formal systems for issuing these alerts have been developed by regulatory agencies, with alerts issued either by the manufacturer or by the regulatory agency.

As part of their post-market surveillance responsibilities (Chapter 9), manufacturers will issue alerts or warnings when they are aware of problems with their products. These take the form of 'Field Safety Corrective Action' (FSCA) or 'Field Safety Notices' (FSN) (GHTF 2006). These can be in response to quality control issues identified within the industry itself or from risks identified through the analysis of adverse events reported to the regulatory agencies.

Where regulatory agencies become aware of recurring incidents involving a particular device or types of devices, they may independently issue medical device safety alerts (the term used for these safety notices varies with time and between countries) to inform users of potential risks. Clinical engineers should make themselves aware of the regulatory alerting system in their jurisdiction. Safety alerts are also issued by non-governmental organizations such as the ECRI Institute (http://www.ecri.org).

Healthcare organizations need mechanisms for managing field corrective actions and safety alerts. Most will have some department at corporate level, possibly linked to its risk and quality management team, whose purpose is to proactively manage the risk associated with the use of medical devices. In the model we describe in Chapter 5, this function would sit within the Medical Device Committee (MDC), but it might also sit within a Risk Management or Health and Safety Committee. The dissemination of field corrective actions or safety alerts will be an ongoing process by one or other of these groups and must cover the whole healthcare organization. The system should also include a review of actions with a system of checking that those who receive these alerts do respond, taking the appropriate action. This review process is important because sadly repeat adverse events some involving patient deaths have occurred where subsequent investigations have revealed that the staff involved in the subsequent incidents had not been aware of the earlier incident or the safety alerts issued.

Whilst clinical engineers may hold responsibility for most medical equipment types, other groups in the healthcare organization may be affected: Facilities or Estates Management, Laboratory, Information Technology, Pharmacy, Radiology and Material Management. These groups will need to be informed of the alerts.

Occasionally the healthcare organization will need to initiate a project to respond to a safety alert, whether issued by the manufacturer or by a regulatory agency. The project's aim will be to identify whether the reported risk is relevant for its organization and, if so, to devise and implement a solution to control this risk. With their experience in systems approach and understanding of the clinical practice and environment, clinical engineers often take a leadership role in these projects. This is another example of how the project management structure allows the talents of individuals to be accessed and applied to developing solutions outside of their core remit. Whilst the Clinical Engineering Department may not be responsible for the ordering of endotracheal tubes, for example in the event of a safety notice being issued in relation to a potential risk associated with their use, the clinical engineer might well be tasked with leading the project group that assess the impact of this risk and mitigate it through taking action (Case Study CS7.8).

The triggers for the sort of medical device vigilance projects described earlier are not only external. The local safety and risk processes might just as easily trigger the formation of a medical device safety project. The HTM processes themselves gather huge amounts of data, and their analysis may also identify risks otherwise unseen. Sometimes these internal triggers or the analysis of an adverse event prompt the healthcare organization to formally report the risks to the regulators, working with them to understand the causes and negotiate with manufacturers and suppliers (Powell 2013; Boutsikaris and Morabito 2014) (Case Study CS7.8).

7.2.8 Projects to Improve HTM Processes

Clinical engineers need to develop a constructively critical approach to their HTM processes. They may do this as part of their quality management system's continual improvement process (Chapter 8, Section 8.4.2.1), often using techniques such as the Plan–Do–Check–Act (PDCA) method. We discussed continual improvement in Chapter 5, Section 5.10, and clearly it is incumbent on clinical engineers, particularly those who lead CEDs, to strive to improve the services that they offer (Case Study CS7.23). This will be done globally for all their activities, perhaps using key performance indicators and benchmarking to assist them monitor and measure progress. Quality improvement (QI) initiatives will also target specific aspects of their HTM services.

We described the equipment support plans (ESP) in Chapter 6 (Section 6.2), showing that each should be constructed as a quality cycle which is subject to regular review, assessing its cost-effectiveness and the benefits that it provides (Chapter 6, Section 6.2.3). The regular reviews of the ESPs help ensure that their associated costs bring real benefit, enhancing the value of this aspect of HTM. These assessments will help the CED leadership reviewing the cost benefits of the CED.

In their pursuit of continually improving the services offered, clinical engineers may embark on specific projects to look at defined areas of activity. Faced with increasing breakdown requests about certain items of equipment or of equipment from particular clinical areas might lead the CED leadership to instigate projects to investigate the reasons. The causes identified might be aged equipment that is no longer reliable and needs to replaced, or the need for additional training, perhaps because of new staff in an area. Equipment breakages might stem from poor handling (e.g. for flexible endoscopes) or from poor storage facilities (e.g. infusion devices 'dropped' into storage bins). We discuss in Case Study CS7.9 how a CED tackled the high costs associated with the replacement of accessories such as patient leads and probes used with medical equipment. The solutions included looking at system changes that reduced the risk of damage and particularly changing the culture and thoughts that clinicians had about these accessories, making them aware of their high costs.

There are two important lessons from Case Study CS7.9 that clinical engineers need to remember as they seek to improve their HTM processes. First, clinical engineers must develop and implement methods for measuring their processes. Measurements of costs will be appropriate in some cases, whilst performance indicator measures will support other improvement projects. Second, clinical engineers must learn to communicate and engage with their clinical colleagues. Measurement of compliance with meeting scheduled inspection targets revealed that the processes were failing, as we explore in Case Study CS7.10; solutions required recognition of the importance of sharing the planning with the clinicians who used the equipment for patient care and gaining their support for releasing the equipment for the technical checks. Recognizing the importance of measuring performance and cooperation with clinical colleagues will help clinical engineers improve their HTM processes. Case study CS7.23 stresses the leadership qualities required to have the vision to identify where improvements can be made and the courage to implement the required changes.

7.2.9 Support for Capital Build Projects with Medical Equipment Implications

The planning of modification to existing clinical facilities and the development of new facilities or indeed of new hospitals should incorporate consideration of the medical equipment that will be required to support the clinical services in those areas. Clinical engineers can and should be involved in the planning from an early stage. The expertise that the clinical engineer brings at this planning stage is twofold: an ability to look at plans and interpret how they will be when built, and the knowledge of the many different types of medical equipment that will be used in the buildings. Clinical engineers can also interact with and provide expertise to the hospital's Facilities department, and a good working relationship with them is essential.

Some medical technologies dictate specific build requirements for safe operation. If, for example, a new facility is to accommodate a medical laser, the build construction should be designed to ensure protection against inadvertent exposure to damaging intense laser light. Protections include door interlocks preventing entry to the room when the laser is firing and protection against reflections off walls. Warning lights should be placed on the entrance doors, linked to the laser system, advising that the laser is in use and that no one should enter. Clinical engineers will work with Facilities in ensuring that the build specification includes these functions and then test the systems once building work is complete. The system might also require three-phase supplies; clinical engineering may specify the position of the three-phase supply outlets for the Facilities department to install.

Areas where x-ray systems are used require radiation protection, often provided by lead-lined walls. Specialist advice from radiation protection experts is required when designing these rooms, with the advice dependent on the radiation intensity and the frequency of using the x-ray equipment. Similarly, specialist advice is required for the design and construction of MRI rooms and for their continuing operational use to protect against inadvertently bringing ferrous materials in proximity to the strong magnetic fields. The clinical engineer is not expected to have the expertise to be able to provide advice in all these areas but should be aware of the general principles and recommend that the appropriate specialists are consulted.

Particular Standards and requirement apply to the design and construction of the electrical supply systems in clinical areas such as operating theatres and critical care environments with the need for Isolated Power Supply (IPS) systems and Uninterruptible Power Supply (UPS) systems. IPS systems protect against abrupt power supply failure in the event of earth (ground) leakage currents that would otherwise result in fuse tripping. UPS systems ensure continuity of supply in the event of failure of the mains electrical supply. Whilst the Facilities department will have primary responsibility for the design and maintenance of these systems, advice from the clinical engineer will guide the process based on the requirements of the medical equipment in the rooms.

Similarly, other medical equipment may require specialized supplies, including anaesthetic machines and ventilators with their medical gases and dialysis machines that require ultrapure water. Some medical equipment may have air conditioning or cooling requirements. Involving the clinical engineer early in the planning stage can avoid plans or built facilities having to be modified at later times. Thus clinical engineers should be involved in the planning process for building developments where medical equipment will be installed or specialized protection is required such as radiation equipment or for medical equipment that requires fixed installations or specialized utility supplies.

In addition, clinical engineers should also be consulted for more general building development where medical equipment will be used, even where no specialized installations or protections are required. Aspects to which clinical engineers can contribute may involve usability or human factors and the physical interaction between different types of equipment. For example, they may look at plans and realize that the wall-mounted medical gas and suction points have been specified at a height that will result in them being struck by the bed head when the bed is moved, or they may realize that insufficient power points have been specified and these are planned to be all grouped together behind the bed position rather than at an appropriate height on either side.

Clinical engineers will also be needed to advise on how medical equipment will be mounted on walls or on ceiling-mounted pendants. The design of these mounts requires knowledge of the size and mass of the medical equipment and their power and utility requirements.

7.3 ADVANCING THE DELIVERY OF CARE: CONTRIBUTION BY CLINICAL ENGINEERS

Clinical engineers contribute to many projects outside of their defined HTM roles. They contribute as members of project teams and sometimes take leadership roles. The type of projects they contribute to and their role within these projects vary depending upon the experience and skills of the individual clinical engineer. Their involvement often arises from recognition by the organization's leadership of their knowledge and their insights into how medical equipment supports the delivery of care. They can act as facilitators of inter-disciplinary working groups, particularly in relation to projects that straddle the physical and life sciences. As engineers, their ability to analyze systems and to measure and inter-pret data makes them valuable contributors to research and development (R&D) or process improvement projects. They can provide insights and identify ways of analyzing problems and developing solutions that may not be obvious to clinicians. They may contribute at the macro level to projects to reconfigure services across healthcare organizations (Case Studies CS2.1, CS2.3 and CS2.4) or develop new systems. Based as they are at the point of care, clinical engineers can also contribute to quality improvement initiatives delivered within clinical care units. With their systems analysis and measurement experience, they can be valuable members of a project team focused on changing the way a clinical service is delivered. In this section, we will look at some of these project roles.

7.3.1 Innovating Care Processes and Quality Improvement

In 2001 the Institute of Medicine in the United States published the report *Crossing the Quality Chasm: A New Health System for the 21st Century* (IOM 2001). *Crossing the Quality Chasm* sets out six aims for healthcare and its improvement, namely for healthcare to be safe, effective, patient centred, timely, efficient and equitable. Healthcare organizations have, in general, lagged behind industry in implementing quality improvement initia-tives. However, that is changing with many organizations now having continuous quality improvement programmes that focus on innovating care delivery processes with the aim of improving safety, controlling costs and making the service more accessible and effec-tive for patients. The Institute for Healthcare Improvement is one of a number of organi-zations pioneering the improvement in healthcare quality (https://www.ihi.org). Quality in healthcare organizations is often evaluated by looking at structure, process and out-comes. It is well understood that one of the ways to identify deficiencies is the systematic collection, aggregation and analysis of several categories of data. Analysis of these data can then lead to root cause analysis (RCA). Once the cause of the deficiency is established, the risk of its recurrence can often be reduced by instigating a quality improvement proj-ect through the use of one of the well-established quality improvement processes such as Lean, Six Sigma or the Plan–Do–Check–Act (PDCA) cycles (Hughes 2008). Quality improvement projects are explored in Case Studies CS2.3 and CS2.4, whilst in Case Study CS7.23, we discuss how CEDs should be aware of opportunities for initiating improve-ments in care processes.

At all levels of the organization, staff should be encouraged to suggest and implement quality improvements. This is achieved through establishing a project and a team whose

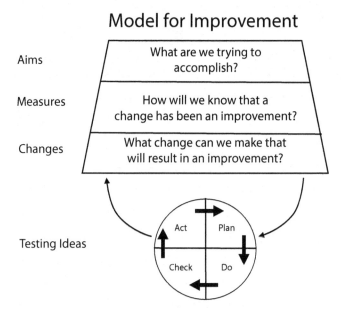

FIGURE 7.5 A Model for Improvement. (Adapted from Figure 1.1 in Langley, G.J. et al., *The Improvement Guide: A Practical Approach to Enhancing Organizational Performance*, 2nd edn., Jossy-Bass, San Francisco, CA, 2009. Copyright Clearance Centre, Inc. With permission.)

goal is to change the existing process for the better. A simple model for how a project team can do this is proposed by Langley et al. (2009) (Figure 7.5). First, state the aim: what is the goal, what is the quality improvement project trying to accomplish? Second, state the measurement to be used: how will an improvement be identified? Third, state the changes that can be made which will hopefully result in an improvement. Once this is done, the project goes into a testing phase and the Plan–Do–Check–Act cycle is used to make a change, test the effect of that change and measure the consequence. If the initiative is successful, it can be extended to other areas using the same methodology.

Macro-level projects which influence the environment and culture within which clinical care programmes are delivered will be run at senior management level. However, quality improvement initiatives can also be delivered at the micro level, within individual units or wards. In fact, it is at this level where the patient interacts with the provider of healthcare that quality, safety, reliability and efficiency are delivered and the patient experience of care is created. These front-line service delivery systems are called 'clinical microsystems' and are made up of small interdependent groups of people who work together regularly to provide care for specific groups of patients (Batalden 2015). Patient flows in Emergency Departments (ED) can be improved by systematically analyzing the clinical workload and developing and implementing improved methods of organizing the activity of the ED (Case Study CS7.11). Clinical engineers can apply their understanding of and application of systems analysis (Chapter 2) to advance care by analyzing and suggesting changes in clinical processes as described in Case Study CS2.4.

7.3.2 Research Projects

In Chapter 1 we defined research and development (R&D) as the term used to describe a group of activities that advance knowledge and deliver new technologies. Clinical engineers working within healthcare who participate in R&D activity do so as contributors to research projects. They will provide support for some aspect of the research project based on their core clinical engineering knowledge and skills. This may be by developing a prototype of a device, software or process. The need to ensure compliance with regulatory procedures requires that due diligence followed by good manufacturing practice is implemented (Case Study CS7.12). The role of the clinical engineer will vary from project to project, with the nature of project working allowing them to contribute in a manner distinct and beyond their HTM role (Case Studies CS7.13, CS7.14 and CS7.15). Project working supports clinical engineers working in collaboration with a variety of clinical and other colleagues in varied ways.

Some types of medical equipment in common use today arose from multidisciplinary research activity which included clinical engineers, for example the research project that led to the introduction of Patient-Controlled Analgesia (PCA) into routine clinical use. The team was headed by a consultant anaesthetist working with a research anaesthetist, a clinical engineer, a physicist/mathematician, a midwife and a clinical psychologist (Evans et al. 1976; McCarthy et al. 1976; McCarthy 1978, 1985).

Clinical engineers also initiate and lead research projects. Very often they identify a problem that needs to be solved and then form a project team to develop a solution. If the solution is to be tested, it will require a rigorous process of evaluation. Successful research will be published in peer-reviewed journals where the quality of the work is independently verified by peers before being published. Where commercialization of a development may be realized, the clinical engineer should consider patent protection prior to any publication, consulting with their research and development department for advice. It should be no surprise that often research projects led by clinical engineers have their genesis in a clinical problem and involve technology in the solution of that problem (Case Studies CS7.12 through CS7.15).

7.3.3 Innovation Projects

Innovation is the term used to describe the steps that bring a new technology or device from the research space into the marketplace where it can be applied for the good of all (Case Study CS7.15). It involves developing the solution discovered during the research activity so that it can be commercialized and bought to the market. It inevitably requires a commercial entity to be formed or identified who will work to develop the solution to be a viable product or service which can be sold. If a clinical engineer had the interest, motivation and skills to start a company to commercialize a discovery or new device, there is no reason why they should not do so. Alternatively, a clinical engineer might join an existing company specifically to work to bring a new technology to market. Doing so certainly is a worthy engineering pursuit and is clearly aligned with the definition of a clinical engineer using engineering and management skills to advance patient care. Any product developed would require being compliant with the relevant Standards and regulations as well as being commercially viable. However, even healthcare organization–based clinical engineers who are not employees of a commercializing company can play an important role in innovation. As part of the innovation process, a commercializing company will need to trial its new device and gather further data on its effectiveness. Clinical engineers may be involved in this prototype stage and then once again during the medical device clinical trial phase (Case Study CS7.15).

7.3.4 Clinical Informatics

Advances in computing, information technology and informatics continue to influence the practice of medicine with their application to healthcare services known as Clinical Informatics. Clinical informatics advances care by formatting, analyzing, evaluating and integrating information in such a way as to enhance individual and population health outcomes, to improve patient care and to strengthen the clinician–patient relationship (Gardner et al. 2009). Clinical informatics is interdisciplinary and operates at the intersection of clinical care, the health system and informatics and communications technology. Increasingly clinical engineers are taking a lead role in informatics projects. This is not surprising given their history of successfully supporting the introduction of new technologies to clinical practice. Since the 1960s clinical engineers and medical physicists have been enablers of the introduction of new and emerging technologies, be it CT scanners, minimally invasive surgery equipment or now informatics systems. Clinical engineers can and should assist and lead in aspects of clinical informatics projects since advancing patient care through the application of technology is central to their role.

Even though medical equipment is designed with digital interfacing to other equipment or systems in mind, the integration of a particular item of equipment into an information technology network, which is likely to include devices from a number of manufacturers, is not a trivial matter. To achieve true interoperability between equipment requires the creation of bespoke networks and systems, often requiring integration engines to facilitate secure data transfer between systems. A clinical information system (CIS) is the term used to describe clinical information technology systems that incorporate medical equipment and systems. These CIS are commonly used in theatres and critical care environments but are increasingly being used in lower acuity areas. These systems integrate data from medical equipment and other clinical systems like laboratory and radiology reporting systems. They also support computer entry of medication orders and records of medication delivery, records of procedures performed and medical and nursing notes. These CIS essentially computerize the complete patient record of care whilst the patient is in the unit supported by the system. Many healthcare organizations now aspire to having a complete network of clinical information systems or a master system that together provide a complete digital record of the patient care. Such systems are often referred to as Electronic Health Record systems or EHR.

The implementation of a CIS or EHR requires a change project whose aim is to improve the quality of care. Clinical engineers have a role in these projects. The fact that a healthcare organization has an IT department does not mean that solutions such as CIS or EHR can be delivered by that department alone. Rather, such projects should be viewed as clinical change projects supported by technology, in this case both information technology and medical equipment and systems. The projects require teamwork from those who have the knowledge and skills in a range of areas, with clinical engineering and IT departments sharing their knowledge and skills to achieve the solution (Case Studies CS2.3, CS5.10 and CS7.16).

Clinical engineers can contribute to CIS and EHR projects since they have a good understanding of information systems science, medical equipment, the systems being integrated and the clinical workflow supported by these systems (Case Study CS7.16). They are also aware of the safety issues associated with placing information technology infrastructure in the clinical environment. The clinical engineer's role extends well beyond managing the interface between the medical equipment and the information technology. Their skill in mapping processes, managing information technology assets and data, project management and experience in working in interdisciplinary teams makes them effective facilitators of the implementation of clinical information systems. Whilst good governance would suggest that such projects be formally led by and overseen by the clinician in charge of the unit, the clinical engineers often play a leadership role, bringing a synergy between the many partners involved (Case Studies CS5.10 and CS7.16).

What may not be as obvious to most observers is the impact that a CIS implementation project will have on the clinical workflow of the unit in which it is being implemented. Essentially, implementing a CIS transforms the unit into a paperless environment, with all processes being supported by the CIS software configuration rather than paper-based records. During the configuration process, the multidisciplinary teams must review existing practice and paper records and then imagine how the workflow can be improved and implemented on the software platform. This activity is of itself a quality improvement project which clinical engineers can facilitate and manage. In doing so, they assist clinicians to develop new ways of working that are supported by the information technology, rather than just computerizing existing paper-based systems. This configuration and implementation process is an opportunity for doctors and nurses to review and improve existing practice. Again, within the context of the project, its structures provide a framework for clinical engineers to play a new role, to use their analytical skills to review and redefine clinical workflows, as part of a multidisciplinary team. Once complete and the system is live and in use, the implementation project team will cease to exist and the system will be managed as a process.

A CIS creates a complete digital record of the care process for each patient, and this data is stored in a database sometimes called a data warehouse. The existence of this data creates an opportunity to use it as a measure of outcome during quality improvement projects (see Section 7.3.1). Quality improvement processes place an emphasis on measurement. These clinical record databases are a valuable resource allowing clinical outcomes to be measured and the care processes delivered to populations of patients analyzed. As the volume of information stored in CIS and EHR increases, it can be used to help clinicians and managers improve services. However, data need to be mined and analyzed to turn it into information which can inform decision-making. This creates the need for skilled individuals who understand the clinical data, can extract data from the database and analyze it. Clinical engineers can assist in mining these databases. By working on research or quality improvement project teams with doctors, nurses and managers, they can help measure the effectiveness of care. Where this yields evidence that suggests improvements can be achieved, a quality improvement exercise can be initiated and tested using the CIS (Case Study CS2.3).

7.3.5 Telemedicine

Healthcare organizations are constantly challenged to respond to the changing health and social needs of the populations they serve. In the Western world, with its well-reported shifting towards an ageing population, the organizations must adapt and develop new way of delivering care using nontraditional means. New ways of providing healthcare to remote regions across the world are required. Healthcare organizations are using clinical information systems to bridge the gap between regional and local hospitals to better manage scarce resources and utilize telecommunications to bring expertise to remote areas. They are also investigating how technology can bridge the gap between hospitals and patients in their homes, with a view to supporting elderly patients to live independently longer. Clinical engineers respond to these challenges not only through the development and implementation of assisted living devices but also through the development and implementation of specific medical IT systems. Telemedicine is the term used to describe projects which use information technologies and telecommunication networks to provide healthcare at a distance. Connected Health is the term used to describe projects that use readily available consumer technologies such as mobile/cell phones, Internet and web-enabled medical devices to deliver patient care and chronic disease management outside of the hospital to patients either in their own home or in primary care facilities. Whether the clinical informatics needs of the hospital lead it to implement clinical information systems within the organization or develop telemedicine or connected health projects in response to changing demand, there is a requirement for this activity to be carefully planned and executed (Case Study CS7.17). Implementing such systems is challenging and requires the establishment of a multidisciplinary team to undertake a change management programme that delivers a new working practice supported by the CIS. Nevertheless, the benefits delivered as a result of well-implemented clinical informatics projects are obvious both from the advancing care perspective and in improving the management of cost-effectiveness with the healthcare delivery system.

7.3.6 Major Incident and Business Continuity Planning

Healthcare organizations should have tried and tested business continuity plans for minimizing any disruption to services from major incidents. Two aspects should be considered: responding to major incidents and more general business continuity planning. Clinical engineers have an important contribution to both, able to apply their knowledge of the organization's medical equipment to move and deploy medical equipment where it is most needed during acute periods of demands.

7.3.6.1 Major Incident Planning

Major incidents may result in mass casualties arriving from earthquakes, serious traffic accidents, explosions and widespread fires, incidents at large events, terrorism or epidemics.

These can also cause massive increases in patients presenting at healthcare organizations, and it is sadly not uncommon for news bulletins to describe hundreds of injured filling local hospitals.

Hospital Boards have to carefully plan their resources to cope with 'normal' high level of demands, and their ability to carry resources for acute peak demands is limited. Consequently healthcare organizations have to develop plans that enable them to effectively cope with these surges in demand, whilst maintaining safe and effective care. Techniques included in plans will include deferring elective procedures, recruiting temporary staff and redeploying resources, including medical equipment. Case Study CS7.1 provides an example of clinical engineers responding to a relatively small, short-duration, increase in demand; major incidents will put greater pressure on resources and greater demands on clinical engineers to respond.

Major incident planning, as the phrase suggests, requires planning before the event and clinical engineers should be involved in the planning, providing advice on the medical equipment that could be redeployed to support the response to a major incident. Major incident planning also involves developing the command and control procedures for coordinating responses and the communication systems that will be required. Clinical engineers should be familiar with their organization's major incident plan including the command and control systems and the communication arrangements that support smooth and effective responses.

7.3.6.2 Business Continuity

Whilst major incident planning develops processes and procedures for responding to acute surges in demand for patient care from external incidents, business continuity planning addresses the resilience of the healthcare organization to continue to provide services when faced with unexpected or serious problems with some of its systems or resources. Examples of threats include failure of electrical supply to a hospital, disruption to hospital staffing perhaps caused by acute extreme weather, unexpected failure of fleets of vital equipment and safety alerts recommending withdrawal of particular models of equipment (Case Study CS5.7).

Business continuity plans will differ between clinical specialities. Renal dialysis services rely on the supply of ultrapure water, typically provided by on-site reverse osmosis water purification systems. The design of these systems should include service continuity in case of failure of part of the system. Critical care facilities and operating theatres are particularly dependent on continuing supply of electricity, with acute electrical supply failure raising real risks of loss of life. To protect against loss of supplied electricity those clinical areas particularly dependent on electricity to power their equipment are provided with uninterruptible power supplies (UPS), essentially banks of batteries with an inverter that will, without interruption, continue to supply electricity in the event of failure of the external supply and until the hospital's standby generator can take over the supply. UPS systems are typically designed to supply power for 30–60 min.

Business continuity plans should be regularly tested and reviewed. A practical, but sometimes overlooked, aspect of this is ensuring that UPS systems are checked. UPS systems left to themselves will fail, perhaps after 4 or more years as their batteries fail, particularly if the UPS system is housed out of the way in a confined room with risk of heating. Like any equipment, UPS systems require an equipment support plan.

7.4 CLINICAL ENGINEERING EXPERTS AND SPECIAL ADVISORS

Clinical engineers possess expert knowledge and skills in the application of medical equipment for healthcare. In this section, we explore examples of clinical engineers applying their expertise for advancing care and its safety. The topics covered are not comprehensive, and readers will be aware of other specialist contributions.

7.4.1 Application of Medical Equipment at the Point of Care

In-depth teaching of the physics and engineering principles that underpin the function of medical equipment and systems is generally not carried out in medical or nursing colleges. Yet everyday doctors and nurses use lasers, ultrasound, radio frequency electrical current, advanced sensors and measurement systems, devices which include digital signal processing and software algorithms to support the delivery of care. Associated with healthcare's increasing reliance on technology is a requirement for experts in engineering and science to support and advance the technology's clinical applications. Clinical engineers based at the point of care are rightly regarded as a valuable resource to those who use advanced technology to deliver care.

Clinical engineers may contribute to the diagnosis and treatment of patients by facilitating the application of new devices or novel methods at the point of care. Sometimes this takes the form of participation in a project, but it can just as often be a one-to-one involvement to deal with a specific event or simply a request for support. Clinical engineers may be asked to advise on the appropriate use of medical equipment used to care for specific patients with particular clinical conditions. This may take the form of advising on how best to use a device or on overcoming a limitation of a particular device. Often there is nothing wrong with the device itself; rather the challenge is the patient–device interaction and how to clinically manage a difficult patient condition with the device. Development of the solution may require a multidisciplinary team to convene and together develop a care plan. Clinical engineers' contributions to such a discussion can range from advice on the application of the devices used in the care of the patient, to the development of particular solutions for the patient in question. Clinical engineers can also help analyze or interpret measurements, particularly where measurement uncertainty, variability, suspected interference and other anomalies associated with the results are suspected.

In specialized areas such as renal dialysis, clinical engineers, perhaps called renal dialysis technologists, may be integrated with the clinical teams at the bedside, using their knowledge and skills to support the technology–patient interface (Case Study CS7.18). Direct support for the application of medical equipment is provided by clinical engineers in other areas including specialist clinics (respiratory, gastrointestinal) and may also be provided in critical care areas.

7.4.2 Managing Medical Equipment Alarms

The lack of coordination and management of clinical alarms on medical equipment has been recognized as a major clinical risk. The ECRI Institute has consistently over recent years listed hazards associated with medical equipment alarms as one of its top 10 medical equipment hazards (ECRI 2015). The Joint Commission in the United States called for all healthcare organizations in the United States to have medical equipment alarm policies and procedures in place by the start of 2016, with training and staff competence in the management of alarms (The Joint Commission 2013).

The alarm systems on medical equipment are designed to alert staff and caregivers of changes of patient conditions requiring attention, often urgent attention and of problems associated with the operation of medical equipment. Alarms can be grouped into three different types:

- Those that warn of changes in the clinical condition of a patient (e.g. heart rate too high);

- Those that remind carers to perform a task (e.g. change the bag on an infusion pump);

- Those that alert carers to a technical problem (e.g. invasive pressure transducer requires calibration or an ECG electrode has fallen off).

The proliferation of medical equipment with alarms requires careful management and control. On the one hand, poorly set alarms or alarms silenced can lead to critical changes in patient conditions not being detected, sometimes with fatal consequences (The Joint Commission 2013; ECRI 2015). On the other hand, triggering of alarms unnecessarily can lead to a cacophony of alarms and alarm fatigue (Mitka 2013). Alarm settings often lacked evidence for the limits which were not adjusted appropriately for particular patients and with staff knowledge and understanding of alarm management frequently lacking.

The Joint Commission reported that the factors contributing to adverse events associated with medical equipment alarms were: absent or inadequate alarm systems, inappropriate alarms settings (including not customized to the individual patient), alarm signals not audible and alarm signals inappropriately turned off. These were exacerbated by staffing issues: inadequate staff training on the medical equipment and its alarms, inadequate training on how to respond to alarms and inadequate staffing to respond to alarms. Technical and medical equipment design issues included the lack of integration between alarm settings between medical devices and failures of medical equipment.

We have discussed in Section 7.2.3.4 and Case Study CS4.1 the importance of carefully considering the configuration of medical equipment during its commissioning; this must include configuration of the alarms. By carefully designing the configuration conditions under which alarms will be triggered, there is evidence that clinical engineers working with their clinical colleagues can reduce the number of alarms by 89% (Whalen et al. 2014).

Technical problems can also contribute to adverse events with medical equipment. Clinical engineers should review their equipment support plans (Chapter 6, Section 6.2) to check that they include adequate testing of alarms and confirmation of alarm configurations. The alarm characteristics of medical equipment should be considered during the procurement of medical equipment, with attention to these in the specification and evaluation of medical equipment. Designers of medical equipment should be urged to improve alarm systems, paying attention to their usability and to incorporating where possible intelligent alarms that where appropriate, combine and integrate alarms – for example in multi-parameter physiological monitoring systems.

Staff training in the operation of medical equipment is frequently cited as a cause of adverse medical equipment events, with experience stressing the need for staff competence (Case Studies CS2.1 and CS4.2). The Joint Commission's Sentinel Event Alert in 2013 focused on staff training in the setting of alarms and their management. It also addressed the failure to customize the alarm settings of medical equipment that is 'attached' to a particular patient to the particular needs of the patient – an example of where patient-centred care is particularly relevant! Put another way, it is not sufficient simply to understand the technicalities of the medical equipment alarms but to apply the alarms appropriately for the particular patient. We discussed in Case Study CS2.1 the need to integrate technical and clinical applications training: with clinical alarms we have another example of integration between the technical and the clinical.

The management of medical equipment alarms requires multidisciplinary teamwork, with commitment and authorization from the healthcare organization's top management, administrative and clinical. The work of the Joint Commission shows that alarm management requires a clear understanding of the nature of medical equipment alarm systems, coordinated medical equipment alarm configurations, and staff who are trained and competent in managing alarms and who are supported by appropriate alarm management policies and procedures. Staff require training to set alarm levels appropriate for their patients with clear guidance on responding to alarms. This requires continuing effort and commitment and we describe, in Case Study CS7.19, an example of the convening of a project team to manage its alarms. Clinical engineers are urged to keep up to date on the topic, reading and taking account of the advice from professional organizations and regulatory agencies.

7.4.3 Risk Management

Managing risks associated with the application of medical equipment is a core part of the clinical engineers' HTM role. This is a vast field and many of its facets have been explored in detail elsewhere in this book. The management of risk associated with the maintenance and support for medical equipment underlines the development of the equipment support plans as discussed in Chapter 6 (Section 6.2). The whole framework of the HTM system as discussed in Chapter 5 and continued in Chapter 6 is designed to minimize the risks associated with medical equipment. The developments of Standards (Chapter 3), whilst also facilitating common requirements to support manufacturers developing medical equipment, are also designed to reduce the risks of medical equipment and to lay down essential requirements for safe operation. Various risk management strategies for HTM have been described, with Grimes (2015) recommending that clinical engineering departments identify the most common risks in their operation and manage them.

However, this section is thus not designed to discuss the HTM risk management itself but to recognize that some clinical engineers may be asked to assume greater responsibility for risk management that goes beyond the narrow risk management of the medical equipment itself. They may carry out this role within their own healthcare organization or may be involved at national level working with governmental and non-governmental organizations that seek to ensure the safe application of medical equipment.

We describe particular examples of clinical engineer's role in risk management in a few case studies. In Case Study CS5.12, we discuss the need for the clinical engineer to escalate identified risks to the Medical Device Committee. The clinical engineer, when faced with taking action on a safety alert that recommends removal of medical equipment from service, may have to weigh up the risks of continuing to use the equipment with those of depriving the patients of the availability of that type of equipment (Case Study CS5.7). Clinical engineers may need to adapt the methods for operating medical equipment and in doing so will need to assess the associated risks (Case Study CS7.20). Clinical engineers may need to assess and reduce the risks when disposing of large medical equipment, managing any residual risks, often seeking advice and support from others including their Health and Safety Department (Case Study CS4.4).

7.4.4 In-House Manufacture of Medical Equipment and the Electrical Safety Expert

In the course of implementing the HTM Programmes and in projects to support research and innovation, clinical engineers are often involved in developing new medical devices, many of which are electrically powered.

7.4.4.1 Knowledge and Understanding

Clinical engineers come from a variety of engineering backgrounds, but many are electrical or electronic engineers. This background gives them the basic knowledge and understanding to be able to develop expertise in electrical safety matters related to medical electrical equipment and patient safety. In order to develop this expertise, they need to have additional knowledge and understanding of the effects of electricity on the human body over a range of frequencies (Bruner and Leonard 1989; Wentworth 2009 – Chapter 2 of the report) and a knowledge of the formal Standards for medical electrical equipment, in particular the general Standard, IEC 60601-1 'Medical electrical equipment – Part 1: General requirements for basic safety and essential performance' (see Chapter 3). The version current in 2016 is Edition 3 (2006) plus amendment 1 (2012), but reference should always be made to the most up-to-date version. There is also some very useful explanatory material explaining the effects of electricity on the human body in the informative Annex A of IEC 60601-1 (under subclause 8.7.3 in the edition current in 2016) regarding the allowable levels of leakage current.

7.4.4.2 Medical Electrical Systems

The projects that a clinical engineer with expertise in medical equipment electrical safety is likely to get involved in can be quite varied. A common requirement is to assemble several items of equipment, which may include non-medical equipment (e.g. a printer or computer) onto a trolley and power them all from a single plugged-in mains cord. This sort of arrangement is called a 'medical electrical system' (MES), and there are a number of possible hazards associated with it. The IEC 60601-1 Standard recognizes this situation and deals with it in clause 16 and in the associated informative Annex A. It is also dealt with in Chapter 4 of Wentworth (2009). From these sources, a set of 'rules' can be derived. These are set out in Case Study CS7.21 that presents the challenge to the clinical engineer of assembling an MES.

The role of the expert clinical engineer in these MES projects is to seek to understand the clinical need for an MES and to explain to clinical users what is and is not appropriate, bearing in mind patient electrical safety. Sometimes the CED suggests to clinical users the need for bringing equipment together onto a single trolley so as to improve physical safety by reducing the number of trailing mains cords or eliminating unapproved multiple socket mains extension leads. The clinical engineer will agree and document a clinical user specification: what equipment the clinicians want to include, what sort of trolley they want to use, etc. The clinical engineer will then convert that user specification into an engineering requirement taking account of the electrical characteristics of the equipment involved. There may, for example, be a need for an isolating transformer to reduce leakage currents, something that the clinical user will not specify because it is outside of their knowledge, but is an 'implied' requirement. The engineering requirement will be transformed into an engineering specification and documented.

Frequently the CED is asked to facilitate such an arrangement and may even construct a custom-made trolley for the purpose. The clinical engineer expert in this area is likely to work with the CED staff doing the work, discussing methodologies and making suggestions, ensuring that adequate documentation is produced, arranging for and perhaps carrying out appropriate electrical testing and finally signing off the project.

7.4.4.3 Electrical Safety Procedures

Carrying out an electrical safety test is a fundamental technique that the clinical engineer must master. Most medical equipment electrical safety testing is relatively easy; once the clinical engineer has understood the test equipment and the medical equipment to be tested, the procedure involves little more than plugging the device to be tested into the tester and connecting, where applicable, any patient applied parts such as ECG leads. Automated test instruments will then automatically cycle through the test sequences, printing out the results and making them available for downloading to the MEMS if required. The automatic test sequence is typically paused or stopped if the test instrument finds a measurement outside the allowable limits.

However, attaching the parts of the medical equipment that touch the patient is not always straightforward, requiring ingenuity from the clinical engineer. We describe one such example in Case Study CS7.22.

7.4.4.4 In-House Manufacture or Modification of Medical Equipment

A more complex scenario than the development of a medical electrical system is the development or modification of medical equipment. An item of medical equipment may be required to be designed and built in-house, either because nothing suitable exists on the market to meet a specific clinical need or because the equipment is needed as part of a clinical research project. This and the modifications of equipment such that its basic purpose is changed are more complex situations.

The first thing to find out is whether such work is allowed by the relevant medical devices regulations in your jurisdiction without having to apply the full regulatory process to a one-off device. There is considerable variation and a changing situation round the world for in-house manufacture and use of medical devices. If the full regulatory processes are required, then a much bigger project is in prospect with potentially considerable administrative and technical expense which may not be justifiable for a non-commercial one-off device. Some jurisdictions may have in place a very much reduced level of regulation provided that the devices being made in-house are only going to be used within the same organization and are not going to be sold or transferred to other organizations. Within the CED, there needs to be someone who is knowledgeable about and understands the regulations that apply and can act as an expert advisor to the organization when such projects arise.

Not all device development takes place within the CED. There may be other departments making or modifying devices, for example within Rehabilitation Engineering or Physiotherapy departments. There may be research devices being made in associated university departments, including engineering departments. These can result in situations that are particularly difficult for the CED who often have not been involved in any of the design or development but at a very late stage are asked to assess the device for electrical safety before it is put into use on patients or volunteers. There have been cases where research projects have had to be put on hold because the device development has been done by expert and well-meaning researchers who have no expertise in medical device safety, including electrical safety. Often the researches are unaware of the risks. For example, a prototype device was developed for recording arterial blood pressure oscillations from an inflated cuff. The device was controlled by a standard PC. During inflation of the cuff, a software malfunction caused the software program to stop operating, but leaving the signal to the cuff inflation pressure pump at ON resulting in the cuff pressure increasing to well over 300 mmHg. Fortunately the technologist operating the equipment noticed the continuing cuff inflation and ripped it off. The designers of the research instrument were not aware of the safety Standard for blood pressure measurement devices that require an independent system to prevent overinflation of the cuff. This re-enforces the need for involving clinical engineering as early into a research project of this nature as possible.

Whatever the regulatory requirements and wherever carried out, a project to design and manufacture a medical device requires a well-planned process and a variety of skills. The classical, iterative, engineering design process should be followed with clear specifications, requirements and testing stages. Hazard identification and risk assessment is a vital part of the process and the relevant Standard, ISO 14971 'Medical devices: Application of risk management to medical devices', should be applied. A project of this nature is certain to be a team effort and so clear leadership is required and roles and responsibilities need to be clarified and assigned.

The clinical engineering expert who knows and understands the relevant medical devices regulations may well be assigned a wider responsibility within the organization to advise on device regulatory issues wherever they arise. In Case Study CS7.12, we explored the compliance with regulatory requirements that have been promulgated to ensure the safe and effective development of novel medical devices within healthcare organizations.

7.4.5 Advice for the Construction of New Buildings or Facilities

We have seen earlier in Section 7.2.4 that clinical engineers may get involved in project planning teams for the construction of new buildings or facilities or the reconfiguration of existing ones. In Section 7.2.4, we discussed this in the context of the clinical engineer supporting build contracts from the perspective of their HTM responsibilities. But clinical engineers may be asked to get involved in building developments from their wider understanding of healthcare processes. This may draw them away from their routine CED role, but the CED leadership must allow and support such activity because the expertise that the clinical engineer can bring will help lead to a better outcome.

Where new facilities are being built or existing ones upgraded clinical engineers can play a pivotal role in developing the design brief and acting as facilitators of conversations between, on the one hand, the architects and building engineers and, on the other hand, the clinical staff of the healthcare organization. Even where the architect and builder have experience in building medical facilities, there is a need to critically review the intended use of all medically used rooms and the facility as a whole. This is best done by a multidisciplinary team comprising representative of all groups who will use the space: doctors, nurses, allied health professionals, general support staff, patients and their carers. In this context, it often falls to the clinical engineer to act as the synergist between these groups and the contractors. The project will entail review of not only general layout and functionality issues but also the specification of services (electrical power outlets, lighting, medical gases, etc.) within the rooms.

It is beyond the competence of most healthcare professionals to adequately specify the engineering design of these new facilities, and so it may fall to the clinical engineer to represent the views of the clinicians and translate these into engineering specification for the contractors. In this regard clinical engineers act as internal independent consultants at the design and build phases and provide an ongoing advisory and quality assurance service to the healthcare organization. Ensuring that the design is integrated and practical and provides the best care solution is an often overlooked role that the clinical engineer is well equipped to provide. The clinical engineer may also be needed to ensure that patient flow movements are taken into account in the build designs. The clinical engineer can help bring together the technical and the clinical, ensuring that the design complies with human factors engineering design requirements.

The leadership role of clinical engineers may see them involved at senior levels when new facilities are being planned, providing general advice and guidance. Thus clinical engineers will be asked to join project teams developing new critical care areas, operating theatre complexes, endoscopy suites and renal dialysis facilities. On occasions, clinical engineers may be asked to lead these developments, assembling the appropriate project team, recognizing that each member brings specific expertise and contributions (Case Study CS5.9).

7.4.6 Wider Professional Involvement of Clinical Engineers

Sometimes clinical engineers can become involved outside of their own organization. Approval and support to do so are sometimes hard to gain, but there are definite benefits to the organization in allowing an appropriate amount of this activity.

We have mentioned in Chapter 3 that there are relevant professional associations in the fields of HTM and have detailed some of them in Chapter 3, Section 3.7.2. These are individual membership bodies, and many clinical engineers will be members of the relevant association in their own country. The work of these associations may influence government regulations on medical equipment and its management and clinical engineering professional matters such as formal schemes for the training of clinical engineers. Active involvement in their work brings benefit to the healthcare organization by ensuring that knowledge of developments in regulations and training schemes are brought back to the organization at an early stage and can be discussed and the implications considered and proposals fed back into the process.

A second area where clinical engineers can contribute significantly is in the development of Standards as discussed in Chapter 3. We described in Chapter 3, Appendix 3A, how formal Standards are developed. In our field, the Standards bodies, especially the international ones, IEC and ISO, tend to be dominated by manufacturers. It is rare to find clinical users involved in Standards' committees and even rarer to find a 'patient' representative (although obviously everyone taking part in the Standards-making process is a potential patient). Clinical engineers with their understanding of the technology and its clinical implications are very well placed to make a positive contribution to Standards development. This can either be through their national standards body (NSB) or by being willing and able to be nominated by their NSB to an international committee. It is worth noting that the present and previous chairs of the committee responsible for the IEC 60601-1 Standard are both clinical engineers.

Both these examples of external professional activities bring value to clinical engineers and to their employers by increasing the individual's knowledge in areas directly relevant to their HTM work, enabling them to better and more knowledgably perform their day-to-day duties. Within reason, such activity should be supported.

7.4.7 Medical IT Network Risk Management Expert

As more and more items of medical equipment are incorporated into IT networks, there is a corresponding need to carefully manage this activity. The placing of a piece of medical equipment onto an IT network can have unintended consequences for both medical equipment and the IT network. Healthcare organizations who integrate medical equipment into IT networks have a duty to do so in a managed and responsible way. This requires adopting a formal risk management process which clearly defines the roles, responsibilities and activities of all those involved. The purpose of the risk management process is to address safety, effectiveness and data and system security. Many healthcare organizations facing these challenges are looking to their clinical engineering departments for leadership in developing solutions. Clinical engineers have skills and experience in implementing technology in the clinical environment and have insights into how technology can support the healthcare system as a whole. The international Standard ISO/IEC 80001-1 specifically addresses this issue and is written to guide a healthcare organization in how it can best manage the process of incorporating medical equipment into IT networks. It defines a medical IT network as any IT network which includes a piece of medical equipment. The Standard is concerned with the management of such networks throughout the life cycle where there is no single medical equipment manufacturer assuming responsibility. The Standard does not outline technical solutions, but rather it proposes a particular approach to the management of the medical IT network as an asset. One of the roles it describes is that of a medical IT network Risk Manager; clinical engineers' familiarity with both the technology and its clinical application and their interdisciplinary working relationships allow them to take on this role (Case Study CS7.16). To be effective in this, as in many areas of their practice, clinical engineers must recognize the need for continuing development and education.

7.4.8 Non-Ionizing Radiation Protection Advisors

Some medical equipment emits ionizing radiation (e.g. x-ray equipment), whilst other equipment emits non-ionizing radiation (NIR), and patients and staff must be protected to ensure the safe use of this equipment and that the benefits of using the equipment outweigh the associated risks. Legislation requires that healthcare organizations control the exposure to ionizing radiation, and they may appoint staff to be (ionizing) Radiation Protection Advisors or contract with external advisors. Similarly, healthcare organizations must protect their patients and staff from exposure to NIR. Non-ionizing radiation at high doses has the potential to cause cell or organ damage – an obvious example is eye damage when exposed to high intensity light (direct sunlight or lasers). External advice may be sought or clinical engineers may be asked to become NIR protection advisors. This may be in association with the organization's health and safety committee.

Types of non-ionizing radiation include:

- optical energy, for example lasers and intense light sources;

- electromagnetic energy, for example MRI scanners and physiotherapy treatment devices;

- high-frequency radiation from electrosurgery machines;

- sound and ultrasound energy, for example diagnostic and therapeutic ultrasound equipment.

The clinical engineer who becomes an NIR safety officer or advisor for their organization must be competent in the role, including having the knowledge and understanding of any regulations that stipulate methods of protection and exposure limits for the different types of NIR. The basic principles for NIR safety are that any exposure to non-ionizing radiation must be justified by the patient benefit and that the intensity of the radiation should be minimized in relation to the optimum benefit. Protection of all, patients, staff and general public, who may be exposed to the radiation, is important, with protective equipment available, such as eye protection glasses. A general risk management principle is that any risks should be 'As Low As Reasonably Practicable' (ALARP) (HSE 2015), which leads to an important method of radiation protection in ensuring that radiation is kept to as low a dose as possible. Methods of reducing the NIR dose to the patient and operator include reducing the exposure time to the radiation, increasing the distance from the source and using shielding techniques.

Aspects that the clinical engineers should consider in exercising their roles as NIR protection advisors include:

- Identifying, analyzing and interpreting legislation and best practice guidance;

- Clarifying safe dose limits;

- Identifying the organization's devices that generate the radiation;

- Clarifying how the radiation can be monitored;

- Developing the appropriate equipment support plans (Chapter 6, Section 6.2) including their quality assurance programmes and clinical-use guidelines;

- Contributing to the development of the clinical pathways that utilize the radiation;

- Investigating control methods including environmental elements (room design and interlocks), equipment configuration, working practices and personal protective equipment;

- Reviewing management arrangements, safety guidelines and audits and incidents to capture current practice;

- Contacting professional bodies or peer groups for advice and support.

7.4.9 Consultant Clinical Engineer

The roles of consultant clinicians, or attending physicians as they are sometimes known within the United States and Canada, are well understood within the medical profession. They are the experts in a particular discipline who take the ultimate responsibility for the care and well-being of the patients in their discipline. They lead the practice of healthcare in that discipline, either for their organization or as independent practitioners.

Similarly, there are senior clinical engineers who practice at a consultant level: as experts in the discipline of clinical engineering, they have the knowledge, skills and experience that enable them to practice independently. They have the knowledge, skills, experience and ability to be consulted (hence the term *consultant*) in their area of expertise. They are the leaders in the field of clinical engineering, offering specialist advice, dealing with the more complex cases, teaching, directing research and innovation and managing and leading services. These functions could equivalently be described as taking ultimate responsibility for the integrity of the engineering and science that underpins the CED's range of services and the application of medical equipment in healthcare. By this we mean that they are not only responsible for their own practice but also have oversight of the integrity of service delivery across their clinical engineering workforce and indeed across the wider healthcare organization in terms of setting standards that relate to healthcare technologies.

Clinical engineers working or aspiring to this level will be recognized by the engineering profession in their country as professionals, with designations such as Professional Engineers (e.g. United States) and chartered engineers (e.g. United Kingdom). These designations indicate that the engineer has met the highest levels of professionalism for engineering in their country.

Within clinical engineering, this skill set, plus experience in healthcare technology management, enables engineers to work at this level not only to provide safe and effective services but also to generate new knowledge and drive change for patient benefit. In Chapter 2, Section 2.5.1, we presented Ferlie and Shortell's (2001) four-element model of healthcare (Figure 2.13). A distinguishing feature of the clinical engineer practising at consultant level is that they are active within all levels of the model:

- Contributing to direct patient care, contributing to the complex cases;

- Supporting the care team, optimizing and improving practice;

- Active at the organizational level – influencing policy and procedure, securing resources and managing risk;

- Providing input to the national knowledge base of their specialism, showing awareness of health strategy.

Practice at this level is characterized by a proactive approach: rather than waiting to be asked, the consultant will be proposing and leading improvements that benefit patients directly or indirectly by improving the procedures and processes operating within the healthcare organization, including within the CED. Case Study CS7.23 looks in more detail how the clinical engineer can work at this senior level.

In summary, we use the term *consultant clinical engineer* to indicate a clinical engineer who is an expert in the discipline of clinical engineering, who can provide the necessary leadership and who can act independently, leading and delivering services. As the word 'consultant' implies, this person has the knowledge, skills and expertise that enables them as a practitioner to be consulted when questions about healthcare technology management arise.

7.4.10 Other Technology Experts

Clinical engineers will develop general expertise and knowledge in order to practise their profession. Some may in addition develop specialized knowledge and skills in particular areas, for example applying medical equipment for specific areas of care, helping to manage clinical care in particular equipment-dependent specialities to optimize care (e.g. renal dialysis, rehabilitation engineering) and particular medical equipment types (e.g. anaesthetic, ophthalmic, endoscopy, patient monitoring, surgical). It is worthwhile for clinical engineers to recognize if they have developed any particular areas of expertise which they can apply for the benefit of patient care in their organization. They may, for example, have developed specialized expertise in understanding the characteristics and limitations of particular medical equipment. This may be on specific aspects of a device and its operational controls, or it may extend to a much more general knowledge of a particular device type. Routinely used medical equipment should not be overlooked. Technological developments have changed the technology behind routine vital signs monitors and the literature describes measurement problems and inaccuracies with automatic non-invasive blood pressure monitors and clinical thermometers (Amoore 2012; Vernon 2013). These are explored in more detail in Case Study CS7.24.

7.4.11 Continuing Training of Clinical Staff

Clinical engineers have important contributions to make to the continuing training of clinical staff (Case Study CS4.2). This will be carried out informally and formally. When clinical staff have problems in operating medical equipment or in understanding the properties and characteristics of medical equipment, they should be able to turn to their clinical engineering colleagues for help and advice.

Healthcare organizations do have formal continuous professional development schemes and perhaps even a 'training' department. These are often directed at nursing staff, but there is increasing recognition that physicians and other medical staff require continuous training and update of their competency certification to practise. Clinical engineers do have an important role to play in such formal training programmes. For example, the head of surgery, mindful of the risk of electrosurgery burns, may ask the clinical engineer to explain to the surgeons how electrosurgery equipment works, what are their risks and why and how to mitigate and manage the risks. The clinical engineer may provide regular lectures that cover the basic theory and demonstrate how the high-frequency current involved in electrosurgery can 'jump' gaps between a surgical tool and the patient.

It is not uncommon for clinical engineers to be asked to teach nurses and other professionals including anaesthesiologists the principles of safe and effective use of infusion devices. These formal training sessions may be accompanied by competency training certification which the clinical engineer may be asked to assist with.

7.5 REHABILITATION ENGINEERING

Clinical engineering encompasses a very wide area of expertises; within the profession in its wider context are those who specialize in particular areas of applying engineering to clinical healthcare. An example of a specialist area is rehabilitation engineering.

Clinical engineers working in rehabilitation engineering have patients referred to them with the aim of improving the functional mobility of persons with physical disabilities. They often work directly with disabled persons to assess their needs and design assistive devices to meet client needs in such areas as prosthetics, wheelchairs, seating, modified cars and electronic communication technology. They are actively involved in the assessment of clients, making measurements to assess their function often involving complex analysis, as well as the design and fabrication of adaptive equipment and assistive devices. Thus their practice of applying engineering principles to patients' needs can, and often does, bring them into direct clinical relationships with patients.

All the principles of managing healthcare technology that we articulate in this book apply to the various technologies, simple and complex, which are used in rehabilitation engineering. In addition, clinical engineers in rehabilitation engineering can contribute in the broader extended roles described in this chapter.

Engineers and scientists based in healthcare organizations providing rehabilitation engineering, clinical measurement and clinical informatics support may not use the term *clinical engineer* to describe themselves as each is a speciality in its own right. However, the boundaries between these activities and those of clinical engineering are blurred, and in practice many individuals work across these boundaries.

7.6 CONCLUSION

Clinical engineering's contribution to healthcare is far more than simply effectively and efficiently managing the Health Technology Management (HTM) processes and the individual medical equipment assets. These responsibilities are very important, with the safety and basic value provided by medical equipment reliant on its careful management. But clinical engineering is challenged to do more, to add value to the application of medical equipment for individual and population health. We see these twin roles of the clinical engineer clearly set out in Case Study CS7.23 which describes first how the consultant clinical engineer has a responsibility to review and develop the Clinical Engineering Department and its HTM responsibilities and second to look outward to proactively look for ways of enhancing the delivery of patient care within the healthcare organization and the evolving home care environment.

This chapter has outlined some of the ways that clinical engineers can advance healthcare in both of these twin remits. These are just some examples that illustrate the diversity and range of opportunities available to clinical engineers. The clinical engineer is encouraged to be proactive, to consult with clinical colleagues and understand the challenges facing them, in particular the challenges and opportunities relating to their medical equipment. Clinical engineers are challenged as to how they can improve the application of the medical equipment to enhance healthcare. It may be through innovative process and project work that goes beyond their everyday routine tasks, forging cooperative links with clinical colleagues. In doing so, the clinical engineers can satisfy both of their twin responsibilities, first to their own Clinical Engineering Department and its management of the healthcare technology and second to the wider healthcare organization which they serve.

REFERENCES

Amoore J. and E. Adamson. 2003. Infusion devices: Characteristics, limitations and risk management. *Nursing Standard*, 17(28): 45–52.

Amoore J.N. 2012. Oscillometric sphygmomanometers: A critical appraisal of current technology. *Blood Pressure Monitoring*, 17(2): 80–88.

Amoore J.N. 2014. A structured approach for investigating the causes of medical device adverse events. *Journal of Medical Engineering*, 2014: Article ID314138. http://www.hindawi.com/journals/jme/2014/314138/ (accessed 2016-05-04).

Amoore J.N., Hinrichs S. and P. Brooks Young. 2015. Medical equipment specification, evaluation and selection, Chapter 6. In *Quality in Clinical Engineering. Report 110*, ed. D. Clarkson, pp. 58–85. York, UK: Institute of Physics and Engineering in Medicine.

Amoore J.N. and P. Ingram. 2002. Learning from adverse incidents involving medical devices. *British Medical Journal*, 325(7358): 272–275.

Batalden P. 2015. *Transforming Microsystems in Healthcare*. Lebanon, NH: The Dartmouth Institute. https://clinicalmicrosystem.org (accessed 2016-05-04).

Boutsikaris S. and K. Morabito. 2014. Critical success factors for the clinical engineer when resolving medical device problems: A case review. *Journal of Clinical Engineering*, 39(4): 172–174.

Britton J. 2007. An investigation into the feasibility of locating portable medical devices using radio frequency identification devices and technology. *Journal of Medical Engineering and Technology*, 31(6): 450–458.

Bruner J.M.R. and P.F. Leonard. 1989. *Electricity, Safety and the Patient*. Chicago. IL: Year Book Medical Publishers.

ECRI. 2015. Top 10 health technology hazards for 2016. *Health Devices*, November 2015. https://www.ecri.org/2016hazards (accessed 2016-05-04).

Evans J.M., Rosen M., McCarthy J.P. and M.I.J. Hogg. 1976. Apparatus for patient-controlled administration of intravenous narcotics during labour. *The Lancet*, 1: 17–18.

Ferlie E.B. and S.M. Shortell. 2001. Improving the quality of health care in the United Kingdom and the United States: A framework for change. *Milbank Quarterly*, 79(2): 281–315.

Gardner R.M., Overhage J.M., Steen E.B. et al. 2009. Core content for the subspecialty of clinical informatics. *Journal of the American Medical Informatics Association*, 16(2): 153–157.

GHTF. 2006. Medical devices: Post market surveillance: Content of field safety notices. Study Group 2 of the Global Harmonization Task Force. GHTF/SG2/N57R8. http://www.imdrf.org/docs/ghtf/final/sg2/technical-docs/ghtf-sg2-n57r8-2006-guidance-field-safety-060627.pdf (accessed 2016-05-09).

Grimes S.L. 2015. Evolution of a risk-based approach to effective Healthcare Technology Management. *Horizons*, Spring: 34–42.

Grout J. 2007. Mistake-proofing the design of health care processes. AHRQ Publication No. 07-0020. Rockville, MD: Agency for Healthcare Research and Quality, U.S. Department of Health and Human Services.

HSE. 2015. ALARP "at a glance". London, UK: UK Health and Safety Executive (HSE). http://www.hse.gov.uk/risk/theory/alarpglance.htm (accessed 2016-05-04).

Hughes R.G. 2008. Tools and strategies for quality improvement and patient safety. In *Patient Safety and Quality: An Evidence-Based Handbook for Nurses*, ed. R.G. Hughes, Chapter 44. Rockville, MD: Agency for Healthcare Research and Quality, U.S. Department of Health and Human Services. http://archive.ahrq.gov/professionals/clinicians-providers/resources/nursing/resources/nurseshdbk/index.html (accessed 2016-05-04).

IOM. 2001. Crossing the quality chasm: A new health system for the 21st century. Committee on Quality of Health Care in America. Institute of Medicine. Washington, DC: National Academy Press.

Jacobson B. and A. Murray. 2007. *Medical Devices: Use and Safety*. Edinburgh, Scotland: Churchill Livingstone.

Keay S., McCarthy J.P. and B. Carey-Smith. 2015. Medical equipment libraries – Implementation, experience and user satisfaction. *Journal of Medical Engineering and Technology*, 39(6): 354–362.

Langley G.J., Moen R., Nolan K.M. et al. 2009. *The Improvement Guide: A Practical Approach to Enhancing Organizational Performance*, 2nd edn. San Francisco: Jossy-Bass. Copyright Clearance Centre, Inc.

McCarthy J.P. 1978. The logical control of intravenous drugs. MSc Thesis. University of Wales (Welsh National School of Medicine), Wales, UK.

McCarthy J.P. 1985. The Cardiff Palliator. In *Patient Controlled Analgesia*, eds. M. Harmer, M. Rosen and M.D. Vickers, pp. 87–91. Oxford, UK: Blackwell Scientific.

McCarthy J.P., Evans J.M., Hogg M.I.J. and M. Rosen. 1976. Patient controlled administration of intravenous analgesics. *Proceedings of the Conference on the Applications of Electronics in Medicine*. IERE Conference Proceedings No. 34, pp. 273–282, Southampton, UK, April 1976.

Mitka M. 2013. Joint Commission warns of alarm fatigue: Multitude of alarms from monitoring devices problematic. *The Journal of the American Medical Association*, 309(22):2315–2316.

Powell K. 2013. Hospital-based clinical engineers' contributions to the Food and Drug Administration's Medical Device Safety Network (MedSun) and to the Public Health. *Journal of Clinical Engineering*, 38(2): 72–74.

PRINCE2®. **PR**ojects **IN** **C**ontrolled Environments. https://www.prince2.com/uk/what-is-prince2#prince2-history (accessed 2016-05-04).

Reason J. 2000. Human error: Models and management. *British Medical Journal*, 320: 768–770.

Runciman W.B., Webb R.K., Lee R. and R. Holland. 1993. System failure: An analysis of 2000 incident reports. *Anaesthesia and Intensive Care*, 21: 684–695.

Sheikhzadeh A., Gore C., Zucerkman J.D. and M. Nordin. 2009. Perioperating nurses and technicians' perceptions of ergonomic risk factors in the surgical environment. *Applied Ergonomics*, 40: 833–830.

The Joint Commission. 2013. Sentinel Event Alert Issue 50: Medical device alarm safety in hospitals. https://www.jointcommission.org/sea_issue_50 (accessed 2016-05-04).

Thompson D., Hamilton D.K., Cadenhead C.D. et al. 2012. Guidelines for intensive care design. *Critical Care Medicine*, 40: 1586–1600.

Vernon G. 2013. Inaccuracy of forehead thermometers. *British Medical Journal*, 346: f1747. http://dx.doi.org/10.1136/bmj.f1747 (accessed 2016-05-04).

Wentworth S.D. (Ed.). 2009. IPEM Report 97: Guide to electrical safety testing of medical equipment: The why and the how. York, UK: Institute of Physics and Engineering in Medicine.

Whalen D.A., Covelle P.M., Piepenbrink J.C., Villanova K.L., Cuneo C.L. and E.H. Awtry. 2014. Novel approach to cardiac alarm management on telemetry units. *Journal of Cardiovascular Nursing*, 29(5):E13–E22.

WHO. 2011. Health technology assessment of medical devices. WHO Medical Device Technical Series. Geneva, Switzerland: World Health Organization. http://whqlibdoc.who.int/publications/2011/9789241501361_eng.pdf. (accessed 2016-05-04).

STANDARDS CITED

NOTE 1: Undated references to Standards are given. It is important to be aware always of the most up-to-date version. These can be found by searching the IEC or ISO website under the respective store tabs.

NOTE 2: Both the IEC and ISO are based in Geneva, Switzerland.

IEC 60601-1 Medical electrical equipment – Part 1: General requirements for basic safety and essential performance

ISO 14971 Medical devices. Application of risk management to medical devices

ISO/IEC 80001-1 Application of risk management for IT-networks incorporating medical devices – Part 1: Roles, responsibilities and activities

SELF-DIRECTED LEARNING

1. A physician wants to introduce a new clinical procedure requiring a novel health-care technology and asks for your help. What should you consider when preparing to respond? What questions would you ask the physician? Who would you also turn to that might be able to help?

2. Reflect on adverse events involving medical equipment that has occurred in your organization or that you have read about. How could you support the investigation process as part of the investigation project team? How can those involved in the incident be encouraged to identify and understand the underlying causes?

3. Advancing healthcare through project work: In this chapter, we suggest that the role of the clinical engineer is more than simply carrying out the healthcare technology management (HTM) processes and the associated medical equipment maintenance activities. Clinical engineers have visibility of the medical equipment and its application. Think of ways in which the application of the medical equipment in your healthcare organization can be improved. Taking just one example, discuss how you would develop a project to test this improvement idea and which clinical colleagues you would want to have on your project team.

4. Reflect on whether your healthcare organization benefits or would benefit from a clinical engineer operating at the consultant level. Identify areas where this level of expertise would advance patient care. Reflect on your ability to perform at this level and on any gaps in your skills and knowledge.

CASE STUDIES

CASE STUDY CS7.1: CLINICAL ENGINEERING PROVIDES ASSISTANCE DURING UNEXPECTED PEAKS IN CLINICAL DEMAND

Section Links: Chapter 7, Sections 7.2.1, 7.2.2 and 7.3.6

ABSTRACT

Unusually cold weather resulted in a high number of casualties arriving at the Emergency Department of the hospital. The normal work of the hospital was suspended in order to deal with this large influx; the Clinical Engineering Department (CED) was called upon to assist in the redeployment of medical equipment to priority areas.

Keywords: Business Continuity; Pressures; Redeployment of resources

NARRATIVE

During a particularly harsh weekend cold spell, the weather experienced was worse than fore-casted and, when compounded by the icy roads during Monday morning's rush hour traffic, resulted in a large influx of patients into the Emergency Department (ED) of the local hospital with injuries from slips and car accidents. Little notice of this change in the weather was given and so the usual activity of the hospital, with its outpatient, inpatient and operating lists, had been planned to go ahead as normal.

As attendance at the ED increased, senior management of the hospital met to decide whether to declare a major incident. Deciding against this, they activated their business continuity plans which included reopening a ward to cope with the post-operative recovery needed for the casualties. As part of this process, the CED was requested to assist in re-equipping the ward to a suitable standard with appropriate equipment. The CED was also asked by ED for additional equipment, including a defibrillator and two transport monitors. The ED equipment requirements, anticipated to be needed for only the Monday, were provided from the CED's emergency spares.

Using their knowledge of the medical equipment inventory and what would be required for such a ward, clinical engineers drew up a list of the equipment required:

- Bed x 20
- Vital signs monitor x 5
- Thermometers x 5
- Defibrillator x 1
- Hoist x 1

Infusion pumps would arrive with the patient from theatre and would initially be supplied from the equipment library.

Clinical engineers met with other department leads and found that beds were already in place in the closed ward, along with other furniture. The Hotel Services department sent cleaning staff to the ward to get it ready and make the beds. The Facilities department was tasked to make sure all lights and heating were operational.

Clinical engineers investigated the equipment required and found that by removing one vital signs monitor from each floor of the hospital, they could provide the required five without affecting patient care in those areas giving up a monitor. The thermometers were taken from a stock already held in the CED. These thermometers were held as replacements for faulty ones, provided by the company free of charge for this purpose – the decision was taken to worry about that issue later! A defibrillator was taken from the Training Department that was used for teaching but was fully maintained and fully functional. No spare patient lifting hoist was available, but as it was not expected to be needed frequently in this ward, it was decided that one could be borrowed from the adjacent ward if and when required. Within two hours the ward had been reopened and was receiving patients and in three hours was fully operational.

ADDING VALUE

The clinical engineers' sound knowledge of both the medical equipment inventory and the clinical equipment requirements for a typical ward enabled them to rapidly equip the ward that had to be reopened in an emergency and to supply the short-term equipment resources required by ED. This enabled the CED to add value to the service without incurring additional costs.

Benefits : Cost ∴ Value

PATIENT CENTRED

The patient was the focus of the work undertaken, with the medical equipment required for patient support understood and supplied to the ward in time for it receiving patients.

SUMMARY

A senior clinical engineer was able to join a multidisciplinary team tasked with setting up a temporary ward only using existing resources and to a tight timescale. Using only knowledge of likely clinical requirements and the current utilization and location of existing medical equipment, the clinical engineer was able to contribute to the overall establishment of the ward in a timely manner.

SELF-DIRECTED LEARNING

1. In the scenario of this case study, it was fortunate in that an empty ward was available for reoccupation. If such a space were not available, and a major incident declared, where else could a ward be established in your organization? What particular issues would need to be addressed or risk assessed from an equipment point of view? (Tip: consider infrastructure requirements.)
2. Sharing equipment between wards and departments in such a scenario seems logical. But how would you determine what was a safe level of equipment provision, and how does this differ in an emergency situation? Should it differ at all?
3. Would the requirement for equipment be any different if an incident had occurred with respiratory consequences for patients? What additional equipment may be required and how could this be obtained quickly?

CASE STUDY CS7.2: CLINICAL ENGINEER LEADING A PROCUREMENT PROJECT TEAM – REPLACING CLINICAL TEMPERATURE MEASURING DEVICES

Section Links: Chapter 7, Sections 7.2.3 and 7.2.4

ABSTRACT

The Clinical Engineering Department (CED) brought to the attention of the Medical Device Committee (MDC) a need to replace the clinical thermometers used on wards. Approval to proceed was obtained with a senior clinical engineer charged to lead a project to identify and procure a suitable standard clinical thermometer. A thermometer was selected and the methodology for its commissioning, distribution and ongoing management agreed. The outcome produced a cost saving.

Keywords: Procurement; Temperature measurement; Clinical thermometers; Tympanic membrane thermometers; Consumables

NARRATIVE

Whilst the hospital was mostly equipped with a particular model of tympanic thermometer, some wards used different technology devices. All types used a dedicated single-use disposable consumable for every measurement. All wards 'owned' their own thermometers; some had sufficient, some had spare but some did not have sufficient. Overall there was a shortage. The CED was responsible for maintaining the thermometers, keeping asset and maintenance records of all; it had dedicated calibration equipment for the predominant type.

The predominant device, a tympanic thermometer, became obsolete and the manufacturer offered an updated version. The CED was concerned that unless it took action, replacement of the existing model by its successor would happen in an ad hoc and uncoordinated way without taking the opportunity to evaluate possible alternatives and without the possibility of negotiating an advantageous financial agreement that included supply of the necessary consumables.

The senior clinical engineer proposed to the MDC that the opportunity be taken to assess the various technology options available as part of a procurement exercise aimed at standardizing the clinical thermometers, reducing cost and managing the devices across the organization through the CED equipment loan service. This was agreed and this clinical engineer was tasked with leading the project.

The clinical engineer's next task was to assemble a project team: nurses representing different clinical areas, the Procurement Department and CED. The first hurdle was to persuade the group not to have a preconceived idea of the type of technology they wanted but to start with a requirement specification: to measure patient temperature with an accuracy of ±0.1°C, in a reproducible manner, and with as little discomfort to patients as was consistent with the objective. The team agreed an estimated number of thermometry measurements made each year.

Various types of clinical thermometers are available (Davie and Amoore 2010). It was decided to evaluate tympanic and non-contact infrared thermometers, the latter having the advantage of not requiring a consumable for every measurement. A nearby hospital in a different healthcare organization that had recently purchased non-contact infrared thermometers would be asked for their experience.

The assessment concluded that tympanic thermometry devices were the preferred technology and this was agreed. Experience with the contactless devices had brought to attention a variety of operational problems including consistency of measurement, confirmed by the experience at the neighbouring hospital. One such device trialled was also not liked by patients.

The clinical engineer project leader then proposed a procurement model based on the supply of the consumables. Rather than seek tenders for the supply of temperature measuring devices and their disposable tip covers, the tender should be for the cost of a temperature measurement. The package to be supplied was to consist of:

- making available without charge a sufficient number of thermometers;
- making available a calibration device for CED to check any reported faulty thermometers;
- replacing without charge any thermometers found to be faulty;
- reimbursing the healthcare organization for the cost of batteries;
- making available additional thermometers as required;
- all to be paid for by an agreed price for an agreed number per annum of the single-use disposable, that is the cost of making a temperature measurement on a patient;
- the costing model to be reviewed annually to take account of actual number of disposable tips used.

Procurement was initially sceptical of this model but was persuaded that, though not previously used, the model was legitimate.

A tender document was issued for the annual purchase of 150,000 disposable tips. There was considerable interest from potential suppliers with several tenders submitted. The winning tender included a cost of $0.053 (5.3 cents) per disposable tip cover, compared with the previous cost of $0.09 (9.0 cents).

ADDING VALUE

The revenue saving is very clear:

Previous cost: 150,000 × $0.09 = $13,500 plus cost of thermometers plus CED support costs.

New cost: 150,000 × $0.053 = $7,950 with no cost for thermometers and reduced CED costs.

Follow-up annual review meetings led to further reduction in revenue costs.

The CED support costs were reduced because they no longer carry out repairs or adjustments on the devices. Devices that do not pass a simple calibration check are replaced free of charge. The CED holds spare thermometers to enable immediate replacement of faulty devices, with the stock of spares replenished by the supplier. The value is increased both by increasing benefits and by decreasing costs.

Benefits : Cost ∴ Value

SYSTEMS APPROACH

The process adopted by CED followed the systems approach of first clarifying and agreeing the objective (a robust method of accurate clinical thermometry) followed by carefully determining the specification, reviewing equipment that would meet the specification, carrying out user evaluations and assessing the evaluations leading to a preferred supplier.

PATIENT CENTRED

Patient considerations were included in the assessments. One non-contact infrared thermometer was rejected because the two red aiming beams intended to give a consistent indication of the distance from the patient's skin were upsetting for some patients who feared the beams would shine in their eyes.

SUMMARY

An experienced clinical engineer can play a leading role in identifying a need and then leading a procurement exercise. The clinical engineer identified an alternative model of managing thermometers, with an alternative means of funding with benefits to ensuring clinical availability of the devices and with financial benefits.

REFERENCES

Davie A. and J. Amoore. 2010. Best practice in the measurement of body temperature. *Nursing Standard*, 24(42): 42–49.

Gallimore D. 2004. Reviewing the effectiveness of tympanic thermometers. *Nursing Times*, 100: 32. http://www.nursingtimes.net/reviewing-the-effectiveness-of-tympanic-thermometers/204243.ful-larticle (accessed 2016-05-04).

SELF-DIRECTED LEARNING

1. Read the references earlier (Gallimore 2004; Davie and Amoore 2010), then search for more up-to-date papers on methods of measuring temperature including their accuracies.

2. Prepare a talk on clinical thermometry discussing the merits of different approaches and concluding with their evaluation.
3. Consider what other equipment types could be purchased using the procurement model based on usage of the equipment.

CASE STUDY CS7.3: LEADING A PROJECT TEAM EVALUATING MEDICAL EQUIPMENT

Section Links: Chapter 7, Section 7.2.4; Chapter 4, Section 4.3.2

ABSTRACT

The evaluation and selection of medical equipment for a healthcare organization is one of the Clinical Engineering Department's (CED) most important tasks. The process requires multidisciplinary insights from clinical users and clinical engineers and, particularly for home use devices, patients and carers. This requires assembling the evaluation and selection project team and leading the team through the process.

Keywords: Project team; Teamwork; Equipment evaluation

NARRATIVE

The chief surgeon asked the head of the CED to lead the replacement of surgical arthroscopy equipment, with the chief arthroscopy surgeon as clinical lead. The head of the CED convened a project team: two arthroscopy surgeons, two arthroscopy theatre nurses, the CED, infection control (for the hospital's central sterilization unit), Procurement and Finance. A patient representative was invited, but the patient representative council asked to be informed of problems and the outcome, but not to attend.

The team met and agreed the two objectives: to procure based on optimizing value and to procure three identical systems. The evaluation criteria were agreed, from which the specification was divided into separate sections: clinical effectiveness, human usability, training support, after-sales support, supply of consumables, technical and finance. A zero score in any section, for example lack of training support, would invalidate that tender submission. Given human usability's importance, it was agreed to ask the manual handling department to join the team and lead the ergonomic evaluation. Leads for each specification section were agreed, for example training and after-sales support were to be led, separately, by the two theatre nurses. Procurement and Finance agreed that the tender would ask for bids for both a consumables-based tender (3000–4000 operations per year, within price bandings of 200 per year) and a capital and consumable tender (c.f. Chapter 4, Section 4.3.2). Each tender would cost maintenance support, with options for partnership supplier and in-house technical support (Procurement to lead, supported by the CED and Finance).

Procurement with CED support undertook a scoping exercise revealing over five possible suppliers, too many for full clinical evaluations. A two-stage evaluation process was agreed: stage 1 would be a general call for responses from all suppliers of arthroscopy systems from which two products would be selected; stage 2 would be a full clinical evaluation.

The head of CED led the process, with regular meetings at which the leads of the different sections of the project reported on their progress. Timescales were set and a meeting held to finalize the tender specifications, with care taken to ensure that all opinions could be objectively and freely discussed, leading to a consensus on the tender wording and criteria weighting. The project team discussed the stage 1 tender returns, after which each section was evaluated and scored by its subgroup for presentation at the selection meeting. The suggestion that one submission be excluded because of an inability to provide user training was agreed. The individual scores and the sum of the weighted scores were discussed leading to two suppliers being short listed for clinical trials.

Clinical trials require the clinical use of the equipment in routine arthroscopy operations. The equipment was all approved for clinical use (FDA and European CE approval) and the hospital's ethics committee approved the clinical evaluation, noting that both suppliers had indemnity insurance to cover adverse events. Each supplier was allocated two weeks clinical evaluation, preceded by a two-day training. Each operation would be attended by the supplier's application specialist together with the project team's theatre nurses and clinical engineer. The manual handling lead attended half the operations for each supplier. After each operation, questionnaires were completed by the surgeon, the theatre nurse on duty and members of the project team attending the operation. The suppliers were not shown completed questionnaires, but both had been invited to help draft the questionnaire.

Following the clinical evaluation, the CED head collated the questionnaires whose contents had been kept confidential. Evaluations from the Infection Control, Procurement, Finance and CED were incorporated into a draft report presented to the full project team. Much debate and discussion ensued, with strong preference by one surgeon for the product with the lowest overall score. "But that is the product that we have used for the last decade and I refuse to change". The CED head had to reiterate the project's agreed remit, but feelings became so heated that a separate meeting with the surgical chief, the two arthroscopy surgeons and the CED head was called. This resulted in accepting that the product with the best overall rating be selected. The cost of the proposed system was similar to that of the previous system, but with enhanced benefits, particularly the human usability of the arthroscopes and the quality of the consumables.

Each prospective supplier received a summary of the findings. The supplier who had had the business for the past decade objected, threatening to ask their lawyers to formally object. Procurement called a meeting with the supplier to discuss; the CED head summarized the process and the supplier finally agreed to accept the result of the evaluation.

ADDING VALUE

The tender process was designed to ensure diligent and fair scrutiny of the products by staff with knowledge and experience in the equipment. Coordination of the project, to ensure that it met standards of openness was important. This ensured a result that added benefits to surgical arthroscopy without adding costs. The objectivity and knowledge of the CED regarding the medical equipment that is available and its applications can benefit equipment evaluations.

Benefits : Cost ∴ Value

SUMMARY

The chief surgeon commented six months after the evaluation: "The evaluation process was conducted systematically and appropriately, with the CED head ensuring carefully consideration of each aspect and that all involved were given the opportunity to discuss the products. Initial concerns about the preferred product were dispelled within a few months experience in routine clinical arthroscopy".

The CED head commented: "In accomplishing this procurement each of us took on roles separate from our normal line management responsibilities; this gave us the freedom and authority to robustly express our professional views in a forum of mutual respect and understanding".

SELF-DIRECTED LEARNING

1. Organize a meeting of your colleagues to discuss the previous selection process. How could the initial objection of the one surgeon have been avoided? What convinced the supplier who objected to the final selection that the process was fair and to drop their legal challenge?

2. Could the evaluation have been carried out without the time-consuming clinical trials? What would be the risks? What other methods could evaluate the ergonomics of the surgeons operating the arthroscopes?

3. Having a robust evaluation framework with clearly published criteria and weightings was crucial to getting this procurement completed. Case Study CS5.11 covers the evaluation and weighting criteria in more detail. Referring back to Case Study CS5.11, what criteria and with what weighting would you use when evaluating surgical endoscopy equipment? What reasons would you give to a potential supplier when queried about the criteria and weightings?

CASE STUDY CS7.4: INTRODUCING A NOVEL TECHNOLOGY

Section Links: Chapter 7, Section 7.2.5

ABSTRACT

A leading physician wished to procure tissue elastography equipment to replace liver biopsies for the diagnosis of hepatitis. The technique was novel for the healthcare organization which required that any proposed new clinical techniques be approved by its Hospital Technology Assessment Group to ensure that new developments are aligned with its strategic clinical planning. Following this approval, it could be submitted to the Capital Finance Committee for equipment funding allocation. The Clinical Engineering Department (CED) was asked to assist the physician with submissions to both groups. This study discusses the submission process that led to approval for the clinical application of the technique and equipment funding.

Keywords: Medical equipment; Assessment; Local health technology assessment; HTA; Novel equipment; Evaluation of need

NARRATIVE

Hepatitis affects millions of people worldwide, with growing evidence that early diagnosis can improve patient outcome (Ferraioli et al. 2014). The reference standard diagnostic method is liver histological analysis but this has disadvantages: the risks associated with invasive biopsies, observer analysis variability and costs of the procedure. Hepatitis is associated with liver fibrosis and increased tissue stiffness, and interest is increasing in determining this change in stiffness using shear wave elastography. The elasticity of the tissue affects the propagation of ultrasound waves through it, and hence, the technique can be used to estimate liver elasticity. Promising results had been reported in the medical literature.

The physician approached the CED asking how the equipment could be obtained. The clinical engineer explained the process and was asked by the physician to help with the submissions. Approval has to be obtained from the hospital's Hospital Technology Assessment Group (HTA-G) prior to introducing any new clinical technique or procedure. The submission requires information on the need for the procedure (patient population, patient benefit and alternative methods), the support required (clinic time and staff), resources required (medical equipment, funding for staff, funding for consumables) and impact on other departments including support departments. The clinical engineer suggested assembling a short-life project team to develop the submission proposal to the HTA-G, to which the physician agreed. The team (physician, lead nurse, clinical business manager, clinical engineer, Histology Manager, patient advocate, clerical support) met and after discussing the procedure agreed to submit the proposal, with the clinical engineer asked to lead the submission. The clinical engineer guided the various specialities through the documentation and liaised with Procurement and others as appropriate. The Finance checked the outline summary of the financial background to the proposal. The physician asked the clinical engineer to co-present the proposal to the HTA-G, chaired by the Chief Doctor and comprising medical and nursing leads. The submission was approved, with the Chief Doctor personally endorsing the proposal for submission to the Capital Finance Committee (CFC).

Submissions to CFC focus on the financing of the proposal, with requests for medical equipment requiring endorsement by the Medical Device Committee (MDC). The MDC approved, asking that the Histology Laboratory Manager be kept informed of progress. Submissions to CFC concentrate on the technical viability of the equipment, a survey of market availability of the equipment and financial details, including acquisition and operational costs. The clinical engineer assembled a new project team to develop the proposal to CFC; team members included those who submitted the proposal to the HTA-G plus Procurement, Facilities (accommodation and utility requirements) and the housekeeping manager; the Histology Manager asked for minutes and to be invited to attend only when required. The project team was charged with developing the submission and, if approval was granted, procuring, commissioning and installing the equipment.

CFC approved the submission that included a 5-year financial summary. Progress reports were required after 6 and 12 months. Procurement issued tenders for the equipment and the lead nurse formed a project team to plan the clinic's development.

Successful procurement and installation followed and the clinic started to see patients. Initial patient numbers were low, despite the physician discussing the benefits of the new technique to referring clinicians and community general practice doctors; they were initially cautious of its diagnostic accuracy. However, gradually patient groups heard about the non-invasive technique and, supported by a positive article in the medical journal, patient numbers increased.

Positive progress was reported after 12 months, with the increase in patient numbers pressurizing the physician to launch outreach services to community treatment centres and a community substance-abuse clinic. Two years later, the clinic had started a mobile outreach service three times a week, having had HTA-G and CFC approval for the development and associated financial approval.

SYSTEMS APPROACH

The systems approach enabled the clinical engineer to separate the two tasks, seeking clinical and financial approval on the one hand, commissioning and implementation on the other hand, allowing each project team to focus on the essential aspects that had to be addressed. In doing so they kept in focus the aim of enhancing diagnostic procedures for patient care.

ADDING VALUE

The clinical engineer added value by leading the physician through the approval submission processes. The total cost for the elastography assessment was 60% of that of the biopsy-histology alternative and was better tolerated by patients, particularly those requiring follow-up assessments. The knowledge and skills of clinical engineers can support the local assessment of medical technologies, enhancing the benefits of subsequent equipment procurements whilst reducing costs through robust tendering.

Benefits : Cost ∴ Value

SUMMARY

The clinical engineer, after understanding the requirements for the new proposed clinical procedure, guided the physician through the submission process. Approval was obtained for its clinical suitability, followed by funding approval. The clinical engineer convened project teams to develop the submissions and to guide the procurement and commissioning of the equipment.

The clinical engineer commented: "It was useful to separate the two stages with two distinct project teams, the one dealing with obtaining approval and the second with the commissioning process. Both tasks needed different skill sets, largely negotiating skills in the former, more practical implementation in the second. I was given the authority to lead both teams, but found that the project team environments empowered each team member to be innovative. We all felt positively engaged".

REFERENCE

Ferraioli G., Parekh P., Levitov A.B. and C. Filice. 2014. Shear wave elastography for evaluation of liver fibrosis. *Journal of Ultrasound in Medicine*, 33(2): 197–203.

SELF-DIRECTED LEARNING

1. What processes are required in your healthcare organization for obtaining approval for new clinical procedures that involve medical equipment? If none exist, what processes do you think should be required and why?
2. The clinical engineer leading this procurement separated it out into two distinct projects, with separate teams. Discuss this from a systems approach (c.f. Chapter 2) identifying the various elements involved in each project and how the relationship between elements affected the outcome. Link this to the SIMILAR process described in Chapter 2.
3. Using either a real or hypothetical example, draft a submission proposal for a new clinical service reliant on medical equipment, including in the submission the following headings: need for the development and alternative methods including current methods, patient benefits, resources required, departments whose support is needed, departments affected by the new service and financial analysis. Who would you assemble as your project team to develop the proposal?

CASE STUDY CS7.5: THE ROLE OF HEALTH ECONOMICS IN HEALTH TECHNOLOGY ASSESSMENT

Section Links: Chapter 7, Section 7.2.5

ABSTRACT

A basic tenet of health economics is to acknowledge the inevitable 'demand' for healthcare. Consequently, healthcare organizations are faced with deciding which treatments it is best to provide, recognizing that all cannot be afforded. Health economics is a method which can help decide which option achieves best value for money: 'How should I spend my finite budget to achieve the most health for the most people?' That is, it seeks the greatest good for the greatest number. The health economics of a new prostate surgical technique is explored.

Keywords: Health economics; Electrosurgery; Bipolar; Prostate; TURP

NARRATIVE

Benign prostate enlargement is a common medical problem in older men, leading to problems passing urine. When symptoms become persistent and bothersome, men are often offered surgical treatment under anaesthetic. The procedure, transurethral resection of the prostate (TURP), uses electrosurgery in which electrical currents cut away excess prostate tissue, widening the urethra to enable urine to pass freely. The well-established procedure uses a monopolar TURP (mTURP) system. The surgeon introduces a resectoscope through the urethra to direct electrical current from the active resectoscope electrode on the target prostate tissue. The current then disperses through the patient's body to a large conductive return electrode which completes the circuit. During mTURP, the bladder is continuously flushed to wash away blood and prostate tissue chips. The fluid must be non-conductive to avoid unintended electrical current burns in the bladder. A complication is excessive absorption of the irrigation fluid leading to a rare, but serious, condition of fluid overload known as transurethral resection (TUR) syndrome requiring critical care treatment.

A new bipolar TURP system (bTURP) has become available with both the active and return electrodes attached directly to its specialized resectoscope (Cleves et al. 2016). Because no electrical current disperses through the patient's body, normal saline (a conductive fluid) may be used for irrigation. Normal saline is not absorbed during surgery, so the TUR syndrome risk is eliminated. The urology surgeons requested funding from the Medical Device Committee (MDC) to purchase the specialized bTURP resectoscopes to enable them to switch to perform only bTURP procedures.

Good consistent evidence from randomized studies shows that both bTURP and mTURP are clinically equally effective, but bTURP has fewer complications; bTURP eliminates the risk of TUR syndrome and reduces the need for postoperative blood transfusion (Cleves et al. 2016). There is a suggestion from a single randomized study that bTURP may result in fewer hospital readmissions due to complications.

Using data from Cleves et al. (2016), a cost–benefit analysis was carried out assuming 150 TURP procedures per year.

EQUIPMENT COSTS

The equipment costs can be calculated from the capital, consumable and maintenance costs. Both the bTURP and mTURP procedures require the same electrosurgery generator, so this baseline equipment cost with the costs of its associated equipment support plan is the same for each procedure. bTURP requires the procurement of three specialized bipolar resectoscopes (total cost £25,200) with an expected 7-year working life, so their annualized cost is £3,600. The resectoscope supplier recommends an annual maintenance check at a cost of 5%, that is £180. The hospital already has the more robust monopolar resectoscopes, so we excluded their cost from the analysis as we wanted to explore the additional costs to the existing procedure. The total annual additional bTURP equipment cost is £3780 (Table CS7.5A).

TABLE CS7.5A   Summary of Annual Costs of mTURP and bTURP for 150 Procedures

	mTURP	bTURP
Costs of 3 bTURP resectoscopes	£0	£3,780
Consumable costs	£9,000	£24,000
Complications – critical care	£5,700	£0
Complications – blood transfusion	£2,970	£990
Subtotal	*£17,670*	*£28,770*
Estimated readmission costs	£66,720	£20,850
Total costs	£84,390	£49,620
Estimated annual savings of bTURP		£34,720

Each TURP procedure uses consumables (active electrodes). The cost per case for mTURP is £60 and for bTURP, £160. Table CS7.5A shows the annual costs for 150 procedures. All other operating theatre costs for the two procedures are equal.

CONSEQUENCES AND COMPLICATIONS

The complication data and associated costs are modified from Cleves et al. (2016). The cost–benefit analysis should consider the benefits and risks of the procedures. Both mTURP and bTURP are known to equally improve urinary symptoms. However, the risk of TUR syndrome is 2% for mTURP and zero for bTURP. TUR syndrome requires a 2-day admission to critical care costing £1900 per patient. A hospital performing only mTURP procedures would expect to treat three TUR syndrome cases at a total cost of £5,700.

The risk of needing blood transfusion in mTURP is 6% or 2% for bTURP. At £330 per transfusion, the costs would be £2970 or £990 for mTURP or bTURP, respectively.

Combining the costs of the complications with the equipment and consumable costs gives annual mTURP and bTURP costs of £17,670 and £28,770, respectively, for 150 procedures (Table CS7.5A).

The data on readmission rates for patients suffering post-surgical complications following TURP procedures are less robust, with only one randomized study available. It suggested readmission rates of, respectively, 16% and 5% following mTURP and bTURP procedures (Cleves et al. 2016). Each readmission costs the hospital £2,780, resulting in anticipated annual costs following mTURP and bTURP procedures of £66,720 and £20,850, respectively.

Table CS7.5A summarizes all the costs and the estimated annual saving of £34,770 achieved by switching to bTURP.

The estimated savings are uncertain because of the absence of robust readmission data. Sensitivity analysis can be used to assess the effects of data uncertainty. With only one uncertain data set, we can more simply determine bTURP's readmission rate for it to be more expensive. Assuming that the mTURP readmission rate is 16%, it is easy to show that bTURP would be more expensive if its readmission rate is greater than 12.5%.

ADDING VALUE

The health economics appraisal suggests that, compared to mTURP, bTURP costs less and provides better benefits to patients.

Benefits : Cost ∴ Value

SYSTEMS APPROACH

The clinicians requested funding for bTURP because they believed it offered improved outcomes for patients. They were very pleased to discover that overall bTURP would also cost less. However, the costs to some hospital departments would increase, with the Operating Theatre's revenue costs increasing because of the higher disposable bipolar electrodes costs. Cost-saving initiatives in hospitals can often run into problems that leave one department with higher costs, although the overall savings to the organization are greater. The Boards need to look holistically at the overall costs and reimburse departments facing higher costs from the overall savings.

SUMMARY

This example shows how health economics can assess costs and patient benefits to guide decision-making.

REFERENCE

Cleves A., Dimmock P., Hewitt N. and G. Carolan-Rees. 2016. The TURis system for transurethral resection of the prostate: A NICE medical technology guidance. *Applied Health Economics and Health Policy,* 14(3): 267–279.

SELF-DIRECTED LEARNING

1. Consider whether there are any other relevant factors that should be taken into account in the economic model.
2. What other clinical developments might benefit from health economics analysis? Whom would you need to work with to undertake the assessment?

CASE STUDY CS7.6: CLINICAL ENGINEER AND THE ENQUIRING MIND

Section Links: Chapter 7, Section 7.2.5

ABSTRACT

The twin remits of first managing healthcare technology and second supporting and advancing care requires that clinical engineers develop an enquiring mind, seeking to understand existing technological and clinical practices and asking how they can be improved. This may lead to collaborations with clinicians that advance the practice of healthcare. It may also lead the clinical engineer to contribute to a better understanding of how medical technologies operate.

Keywords: Enquiring; Questioning; Medical equipment; Understanding

NARRATIVE

The increasing move to using oscillometric non-invasive blood pressure (NIBP) equipment replacing stand-alone auscultatory sphygmomanometers and as components of multi-parameter monitors prompted the Clinical Engineering Department (CED) to procure an NIBP test simulator. However, when the simulator was set to a blood pressure (BP) of 120/80 mmHg and pulse rate (PR) of 60 bpm, the NIBP device being tested did not record a pressure of 120/80. Instead, it recorded systolic pressures that ranged from above 130 mmHg to below 100 mmHg when tested on the same repeated test pressure. This variability was found on other NIBP devices tested but did change with the make and model of the NIBP device. This prompted clinical engineers to doubt the simulator, but similar results were found with a different simulator.

Were the simulators faulty, were the NIBP devices faulty or was there a lack of understanding of the oscillometric technique and how the NIBP devices measure pressure? Calibration checks of the NIBP devices and measurements on subjects' arms suggested that they were functioning correctly. Research quickly revealed a lack of understanding of the technique.

This initiated a research programme starting with investigating the methodology of the oscillometric technique and developing links with clinical and scientific researchers in the field.

OSCILLOMETRIC NIBP TECHNIQUE

Oscillometric NIBP devices (and most automated sphygmomanometers use the technique) measure and analyze the small pressure pulses (referred to as oscillometric pulses) generated by the arterial flow underneath a pressurized cuff wrapped around a limb. The amplitude of these oscillometric pulses reaches a maximum when the pressure in the cuff equals the mean arterial pressure, that is at the cuff pressure which gives balanced pressures across the arterial wall. This is the basis for the oscillometric technique measuring the mean arterial pressure. The technique relies on proprietary algorithms for determining the systolic and diastolic pressures. These empirically derived algorithms analyze the changing amplitude and shapes of the oscillometric pulses in relation to the cuff pressure to determine the systolic and diastolic pressures (Amoore 2012).

Two consequences of the empirical indirect nature of the technique are important. First, different NIBP manufacturers use their own, often unique, empirical algorithms; hence oscillometric BP measurements may be device dependent. Second, the variability between consecutive measurements, even when presented with a constant pressure, varies between devices. The variability is a consequence of the sampling of the cuff pressure by the oscillometric pulses at the pulse rate. For example at a pulse rate of 60 bpm, the cuff pressure will change by 10 mmHg between sampling if the cuff pressure changes by 10 mmHg/s. Device manufacturers use sophisticated interpolation techniques to estimate the cuff pressures between the pulse-rate sampling of the cuff pressure and hence reduce the variability.

EXPERIMENTAL STUDIES

Systematic experimental studies were undertaken to better understand the operation of NIBP simulators, which were then at an early stage of development. For example, a series of experiments were designed to clarify the pressures (average and standard deviations) measured by different makes and models of NIBP devices when presented with the same repeated simulated waveform. This showed that the average and variability of the pressures recorded by different makes and models of NIBP devices varied when presented with simulated waveforms.

A key question requiring investigation was the ability of the simulators to test NIBP devices. An experiment was devised in which the BP was recorded from human upper arms using three different NIBP devices; the same three NIBP devices were then used to record simulated BPs. The three NIBP devices recorded different BPs, but the differences between devices were not the same for the simulated as for the human measurements. These experiments concluded that the NIBP simulator used could not validate (i.e. assess the systematic accuracy) of NIBP devices (Amoore and Scott 2000). Later experiments suggested that this was in part because of the artificial nature of the oscillometric waveform and that furthermore the oscillometric waveform shape varies between human subjects and may contribute to oscillometric–auscultatory BP measurement differences (Amoore et al. 2008); the oscillometric waveform is drawn by plotting the amplitude of the individual oscillometric pulses as a function of the cuff pressure. This led to work to develop a simulator that regenerated pressure pulses previously recorded from the cuffs wrapped around human subjects (Amoore et al. 2006).

IMPROVEMENTS IN THE OSCILLOMETRIC TECHNIQUE

Developers of NIBP devices have improved the accuracy of NIBP devices by improving their algorithms. The inherent variability associated with the sampling of the cuff pressure at the pulse rate has been reduced by sophisticated interpolation techniques. Leading physicians, supported by the relevant Standard (ISO 81060), have developed validation protocols which require agreement between the pressure measured by NIBP devices and 'gold standard' measurements of the BP (Ng 2013). The validation protocols and Standard have helped deliver improved accuracies of NIBP devices. Better understanding of the strengths and weaknesses of the technique has led to a better appreciation of its limitations. For example, there remains concern of the ability of oscillometric NIBP devices to measure accurately in patients with chronic renal disease and in certain maternity patients.

ROLE OF THE CLINICAL ENGINEER

It was important that the clinical engineer published the research results in refereed journals and presented at conferences. This shared knowledge and developed further research links, helping to improve the state of the art. It also helped generate research funding to make the work possible.

It is important that work of this nature is supported and recognized by heads of department. Managing medical equipment requires an understanding of how the equipment operates and its strengths and weaknesses. Clinical engineers can use their combined technical and clinical knowledge to better understand the application of medical equipment. Clinical engineers should show leadership in developing these deeper understandings.

Clinical engineers through their work on Standards bodies and through collaboration with professional groups of physicians can help understand the technology and its practical application, leading to improved NIBP devices and blood pressure measurement practices. Within their own organization, this work can help clinical engineers advise on the procurement of NIBP devices and their suitability for use with specific patient groups.

ADDING VALUE

Adopting an enquiring mind and pursing such investigations has a cost, but by doing so, clinical engineers can better understand the medical equipment they manage and hence advance benefits and increase value.

Benefits : Cost ∴ Value

SUMMARY

When a new NIBP test instrument was purchased the ambiguous results when testing NIBP devices prompted the clinical engineers to ask why? The questioning led to an improved understanding of the operation of NIBP devices that enabled the clinical engineers to better support the devices within their healthcare organization and to cooperate with other researchers to improve the understanding of the technology.

REFERENCES

Amoore J.N. 2012. Oscillometric sphygmomanometers: A critical appraisal of current technology. *Blood Pressure Monitoring*, 17(2): 80–88.

Amoore J.N. and D.H.T. Scott. 2000. Can simulators evaluate systematic differences between oscillometric NIBP monitors? *Blood Pressure Monitoring*, 5(2): 81–89.

Amoore J.N., Lemesre Y., Murray I.C. et al. 2008. Automatic blood pressure measurement: The oscillometric waveform shape is a potential contributor to the differences between oscillometric and auscultatory pressure measurements. *Journal of Hypertension*, 26(1): 35–43.

Amoore J.N., Vacher E., Murray I.C. et al. 2006. Can a simulator that regenerates physiological waveforms evaluate oscillometric non-invasive blood pressure devices? *Blood Pressure Monitoring*, 11(2): 63–67.

Ng K.-G. 2013. Clinical validation protocols for noninvasive blood pressure monitors and their recognition by regulatory authorities and professional organizations: Rationale and considerations for a single unified protocol or standard. *Blood Pressure Monitoring*, 18(5): 282–289.

STANDARDS CITED

NOTE: Undated references to Standards are given. It is important to always be aware of the most up-to-date version. These can be found by searching the ISO website.

ISO 81060   Non-invasive sphygmomanometers

SELF-DIRECTED LEARNING

1. Section 1.6 in Chapter 1 introduces the role of the clinical engineer in research and development, with Figures 1.6 and 1.7 showing the clinical engineers role extending beyond the healthcare organization to support innovation and the continuing developments of

technology post placing on the market. Can you summarize this case study in the light of these figures and structuring the summary around arrows A and B in Figure 1.6 and arrows E and H in Figure 1.7?

2. Is the enquiring mind of the clinical engineer limited to only equipment? How could this be evident in health technology management processes?

CASE STUDY CS7.7: THE ROLE OF CLINICAL ENGINEERS IN INVESTIGATING ADVERSE EVENTS

Section Links: Chapter 7, Section 7.2.6; Chapter 2, Section 2.4.2; Chapter 7, Section 7.4.11

ABSTRACT

Their knowledge and understanding of medical equipment makes the clinical engineer an important asset for investigating adverse events involving medical equipment. In contributing to these investigations, clinical engineers need to be mindful of the sensitivities around incidents and how best their knowledge and skills can be applied.

Keywords: Adverse events; Teamwork; Clinical engineer role; Patient safety

NARRATIVE

A patient suffered internal bleeding during endoscopy surgery for the removal of polyps from the colon. The procedure involves placing a wire loop around the polyps under endoscopic visualization. The wire loop, heated in a controlled manner, cuts through the polyps using heat, whilst simultaneously sealing, by coagulation, bleeding blood vessels. The cutting and coagulating are achieved by current, controlled from a surgical diathermy machine, which flows down a wire threaded through the endoscope's operating channel. The technique requires carefully balancing the cutting and coagulating currents to prevent excessive bleeding.

The excessive bleeding occurred when using a recently released new model surgical diathermy machine.

The investigation of this serious incident was led by the Assistant Director of Nursing (DoN) who convened a project team consisting of the surgical staff involved and the senior clinical engineer. At the initial briefing meeting, the known facts were outlined, following which each project team member was required to produce an interim report from their perspective. The Assistant DoN, supported by the Patient Liaison Officer, would brief the patient and family as the investigation progressed.

The clinical engineer was charged with investigating the details of the surgical diathermy and liaising with the national incident reporting agency and the supplier. As the Clinical Engineering Department (CED) had carried out the acceptance checks and configured the surgical diathermy machine, the representative of the national incident reporting agency was asked by the clinical engineer to oversee this aspect of the investigation.

The agency representative carried out a detailed analysis of the surgical diathermy machine. It had been quarantined after the incident, with the theatre charge nurse ensuring that no changes were made to any settings and that consumables were preserved in sealed bags. The senior clinical engineer contacted other healthcare organizations who had purchased this recently released surgical diathermy machine. The agency representative then convened a meeting of the supplier and the clinical engineer who had commissioned the equipment and who had drafted its equipment support plan.

Investigation revealed that the configured setting of the surgical diathermy machine had been agreed following a meeting between the supplier and the consultant surgeon, with the surgeon asking for a strong cutting current to overcome perceived lack of cutting power in the previous machine. As a result the balance of cutting to coagulating current was shifted towards cutting currents. Neighbouring healthcare organizations had used similar configurations. The investigation also revealed that this new machine had different operating controls – the foot pedal previously used to increase both cutting and coagulating current now increased only the cutting current. Training of the surgical team had been carried out by the supplier, who reported only a 60% training uptake; several surgeons did not attend, claiming knowledge of the technique. The surgeon who had used the machine during the procedure that led to the incident had not attended training. Interviews were held with the surgical team revealing that the surgeon had asked for the cutting power to be increased during the procedure.

Supported by the senior clinical engineer, the external representative produced a report summarizing the technical characteristics of the diathermy machine, its commissioning and configuration, the training provided and the conditions of operation at the time of the incident. This was presented to the Project Team which also heard reports from the surgical team and the Patient Liaison Officer.

The Project Team concluded that the incident was primarily caused by the lack of understanding of the characteristics of the new diathermy machine, poor uptake of training and poor configuration, with the strong recommendation that commissioning procedures be improved. The manufacturer added recommendations on device configurations to its documentation and developed improved training programmes, incorporating competency assessments, with reports on training uptake to be submitted via the head of the CED to the organization's management. The regional surgical network was advised that all the surgical team must attend training and be assessed for competence prior to using the equipment. CED amended its equipment support plan.

The role of the clinical engineer in this instance was to facilitate the investigation of the medical equipment and its commissioning by an independent investigator to ensure impartiality. The clinical engineer had also helped ensure that the equipment was appropriately quarantined after the incident.

After some adverse events, clinical engineers will take direct responsibility for investigating the medical equipment involved, provided that it will not involve investigation of the procedures and work practices carried out by the CED. Regular liaison between clinical engineering and those who manage adverse events should take place so that roles are understood and that the knowledge and skills of clinical engineering can support investigations.

ADDING VALUE

Clinical engineers, through their sound knowledge of medical equipment and its application, support adverse event investigations (Boutsikaris and Morabito 2014). Clinical engineers have the analytical skills and good communications with clinicians and manufactures that help identify the causes and can lead to safer products. By helping identify the causes and methods of preventing repetitions, they enhance value, adding benefit without adding costs.

Benefits : Cost ∴ Value

SUMMARY

Clinical engineers' knowledge and skills can benefit incident investigations. In exercising this role, the clinical engineer will be mindful of the need to call in independent investigators if the working practices of the CED are included in the investigation.

REFERENCE

Boutsikaris S. and K. Morabito. 2014. Critical success factors for the clinical engineer when resolving medical device problems: A case review. *Journal of Clinical Engineering*, 39(4): 172–174.

SELF-DIRECTED LEARNING

1. Prepare an outline report on an adverse event involving medical equipment, showing what aspects of the equipment and its history should be included in the investigation.
2. Why is it not always appropriate for clinical engineers to lead the investigation into medical equipment involved in adverse events?

CASE STUDY CS7.8: CLINICAL ENGINEERS AND REGULATORY AGENCIES WORKING TOGETHER TO INVESTIGATE ADVERSE EVENTS

Section Links: Chapter 7, Section 7.2.7; Chapter 9, Section 9.3.3

ABSTRACT

The investigation of adverse events involving medical devices can benefit from clinical engineers and the regulatory agencies working together. Clinical engineers understand their medical devices and the environment in which they are used; regulatory agencies know the national and international scene and have the influence to work with manufacturers and suppliers.

Keywords: Adverse events; Safety warnings; Safer healthcare; Clinical engineers; Regulatory agencies; Alarm; Infusion pump

NARRATIVE

Reports of occlusion alarms during syringe pump infusions were increasingly received by the Clinical Engineering Department. Deliveries of Patient-Controlled Analgesia (PCA) were being prevented, leaving post-operative patients in severe pain when their PCA infusions were interrupted by occlusion alarms. Deliveries of anaesthesia by Total Intravenous Anaesthesia (TIVA) pumps were disrupted by occlusion alarms during the course of the anaesthesia delivery preventing effective anaesthesia that relied on steady infusions at rates based on the pumps' pre-programmed models of the delivery of the anaesthetic agents.

Syringe pumps are designed with occlusion alarms to detect blockages in the infusion line between the syringe and the vein. Timely alarms are important to ensure that caregivers are promptly alerted to problems that are preventing medication delivery, encouraging low pressure settings for the alarms. However, too low a setting can lead to false nuisance alarms; the back pressure against the pump will rise, perhaps trigging an alarm, if a patient is raised, or the arm to which the infusion line is attached is raised relative to the syringe pump. When the alarm is triggered, the pump stops until the alarm has been attended to. Confounding the problem is that most pumps do not measure the pressure in the line; instead they estimate it from the force used to drive the syringe, perhaps by measuring the current required to drive the pump's motor. The back pressures against the motor are the friction (and stiction) between the syringe's plunger and barrel, the opening pressure of the anti-syphon valve, the pressure drop along the infusion line and cannula and the venous pressure (most infusions are delivered directly into veins) (Amoore et al. 1997; Amoore and Adamson 2003). From experience and studying the characteristics of pumps, infusion systems and patients, the back pressure alarm is typical set to an equivalent of about 500 mmHg. This is sufficient to overcome the venous pressure of up to a few mmHg, the pressure drop across the infusion line and cannula of about 100 to 200 mmHg (varies with viscosity of the infusion), the anti-syphon valve's opening pressure (100 mmHg) and the friction between plunger and barrel (100 mmHg), leaving an 'allowed' pressure reserve to accommodate the patient's arm being raised.

Infusions from both PCA and TIVA pumps had been delivered without problems for several years until suddenly clinical staff complained of increasing frequency of alarms. "What have you done to our pumps?" asked the pain control team increasingly desperate from their lack of ability to ensure patients' postoperative pain relief. The clinical engineers were hearing similar complaints from neighbouring hospitals and alerted the regulatory agencies. No changes had been made to the occlusion alarm setting and investigations were carried out. These revealed increased opening pressure of the anti-siphon valves on the infusion lines used in some hospitals and increased friction between the plunger and barrel of a popular brand of syringe. However, there was nothing to indicate on the part number of either consumable that the devices had changed.

The regulatory agencies questioned the two manufacturers involved. Neither was initially aware of anything that could cause a problem, though did concede some manufacturing changes. Further investigations did reveal that these manufacturing changes had, for the syringes, increased the friction, and for the infusion lines increased the anti-siphon valve's back pressure. But neither manufacturer had realized that these changes could affect the balances influencing the occlusion alarm settings in clinical practice. The problems were compounded in some hospitals that used both the particular brand of syringe and the particular brand of anti-siphon valve. A medical safety alert was issued for one of the products (MHRA 2014).

Resolution of the problem was achieved by collaboration between clinical engineers and regulatory agencies (Powell 2013; Boutsikaris and Morabito 2014). The clinical engineers understood the technical and clinical details of the medication delivery by syringe pumps; the regulatory agency had the national oversight enabling them to develop a wider perspective of the problem and the influence to discuss with the top management of the manufacturers involved. As a result, the manufacturers changed their product designs, reducing the plunger–barrel friction on the one hand and reducing the anti-siphon valve back-pressure on the other.

Clinical engineers must report problems to the regulatory agencies, who will otherwise be unaware of incidents and unable to use their powers to help deliver solutions.

ADDING VALUE

Clinical engineers, working with their colleagues in national regulatory agencies, can support the investigation of adverse events involving medical devices. Enabling clinical engineers to support this collaboration does require investment and cost of their time, but the benefits are greater than the costs, adding net value to the goal of safer healthcare.

Benefits : Cost ∴ Value

SUMMARY

The clinicians were relieved that they could resume use of their medical devices confident of trouble-free operation. The regulatory agency and the clinical engineers were pleased by their collaborative work and its result in solving the problems.

REFERENCES

Amoore J. and E. Adamson. 2003. Infusion devices: Characteristics, limitations and risk management. *Nursing Standard*, 17(28): 45–52.

Amoore J.N., Fraser S., Mather J., Rae A. and S. Storey. 1997. Occlusion pressure alarm setting on PCA pumps. *Anaesthesia*, 52(9): 917–918.

Boutsikaris S. and K. Morabito. 2014. Critical success factors for the clinical engineer when resolving medical device problems: A case review. *Journal of Clinical Engineering*, 39(4): 172–174.

MHRA. 2014. MDA/2014/003 Single-use syringes: Plastipak 50 mL Luer Lok single use manufactured by BD Medical. Medicines & Healthcare products Regulatory Agency, 17 January 2014. https://assets.digital.cabinet-office.gov.uk/media/5485ab2640f0b60241000203/con365967.pdf (accessed 2016-05-04).

Powell K. 2013. Hospital-based clinical engineers' contributions to the Food and Drug Administration's Medical Device Safety Network (MedSun) and to the Public Health. *Journal of Clinical Engineering*, 38(2): 72–74.

SELF-DIRECTED LEARNING

1. Choose an adverse event with a medical device that you are familiar with, either from your experiences or which you have read about. Could the risk of future recurrences be reduced by reporting it to the regulatory agencies? What benefits could your knowledge of the device and the clinical environment contribute when working with the regulatory agencies and what benefits can the regulatory agencies bring?

CASE STUDY CS7.9: CONTROLLING THE COSTS ASSOCIATED WITH THE MANAGEMENT OF MEDICAL EQUIPMENT WITH EXPENSIVE ACCESSORIES

Section Links: Chapter 7, Section 7.2.8

ABSTRACT

A Clinical Engineering Department (CED) concerned over the year-on-year rising costs of supplying medical equipment accessories undertook an exercise to analyze and control this expenditure.

Keywords: Medical equipment accessory; Cost control; Culture change; Collaboration

NARRATIVE

One of the roles of a CED was to supply high-cost accessories for medical equipment as required. The equipment included ECG machines, physiological monitors, heated humidifiers, vital signs monitors and electronic thermometers. The expensive accessories were ECG trunk and fly leads, pulse oximeter probes, main stream end tidal carbon dioxide transducers, blood pressure cuffs and thermistor probes. The CED's annual HTM Programme review highlighted the increasing cost of supplying these accessories. This led the CED to establish a quality improvement project to reduce this cost.

Analysis of the service record data in the Medical Equipment Management System revealed reasons for the supply of replacements:

- Accessory lost (34%)
- Accessory confirmed to have been damaged through misuse or poor storage (29%)
- Accessory failed due to normal ageing and wear and tear (37%)

This led the CED to conclude that the majority of the accessory replacements could be prevented. To better understand the causes of accessory loss or damage, the CED decided to start by working with users in two clinical units, the Emergency Department and the Intensive Care Unit.

INTERVENTION ONE: IMPROVING STORAGE AND TRAINING

The clinical engineers spent a week in each department looking at how the accessories were used and stored. A number of improvements were made to how accessories were stored: hooks were fitted beside patient monitors and baskets were fitted to mobile devices. As part of this exercise, informal in-service instruction was given to staff as to how to handle and store accessories. Posters were also put up in staff areas reminding all to take care of these accessories. After three months, there was no improvement or decrease in the demand for accessories from either unit.

There was one interesting outcome. In a conversation between a nurse and clinical engineer over replacement for a lost ECG trunk lead (which had an active signal acquisition module), it became clear that the nurse had no idea of the cost of the accessory (>$1000). Over the coming days, the clinical engineering staff informally surveyed clinical staff and found that none had a sense of the cost of any accessories, with most shocked to hear the purchase costs.

INTERVENTION TWO: RAISING AWARENESS OF COSTS

The CED put up, in the two units, posters that displayed the purchase cost of common accessories. When any accessory was supplied, it was put in a plastic bag with a large sticker on the front with its cost. The clinical engineers undertook further in-service training on care and use of accessories, deliberately emphasizing their costs after inviting users to estimate the purchase price of different accessories. The surprise the users experienced when they heard the costs led them to talk to each other about how much everything cost, raising awareness further.

In the three months after this intervention, the number of lost accessories reduced significantly and the number damaged in use fell by 40%.

INTERVENTION THREE: ROLL OUT THE APPROACH ACROSS THE WHOLE HOSPITAL

Following the success of the intervention in the two units, the CED rolled out the approach across the whole hospital. The roll out began with the CED setting up a stall at the annual clinical skills fair running a 'guess the cost of the accessory' quiz. (The hospital holds an annual two day 'Clinical Skills Education Fair' where all medical and nursing staff could attend the education centre and visit stalls set up by medical device suppliers and hospital trainers to give particular in-service training on devices and techniques.) The CED stall challenged staff to guess the cost of 10 medical equipment accessories, writing down their estimates. The stall was busy as staff joked and compared estimates. After staff had handed in their estimates, the clinical engineers revealed the costs, with most staff surprised and taken aback, echoing the response to intervention two in the pilot units. The CED stall at the clinical skills fair was followed up by a similar poster campaign to that used in the pilot sites to raise awareness of the cost of accessories, re-enforced by informal in-service training delivered in all departments.

In the year following the three interventions, the cost of supplying accessories fell for the first time in 3 years, to 58% of the previous year, a real saving of $48,000.

SYSTEMS APPROACH

The improvement started when the CED began measuring the supply of accessories. Once they measured it, they realized something was wrong and initiated interventions to control it. Continuing measurements provided feedback on whether the changes introduced were effective or not, identifying those which were effective.

CULTURE CHANGE

A culture had developed amongst clinical staff where there was less than optimal care taken of the accessories which were treated more like consumables. The culture had developed because staff were unaware of the cost of these items and clinical engineering supplied replacements without question. The causes were not lack of responsibility of staff, but a systems failing in which clinical staff were not made aware of the costs of accessories and in which CED distribute accessories without having controls in place. By doing things differently, communicating the costs and explaining that the accessories are actually components of medical equipment as opposed to consumables the practice across the hospital changed.

ADDING VALUE

The interventions reduced costs by raising awareness amongst staff of the costs of accessories and showing staff how to look after these valuable parts. Losses were reduced as were the damages to the accessories. This benefited care by reducing the interruption to care caused by failed or lost accessories. By engaging with clinical staff and training staff to be more careful of the medical equipment's accessories, clinical engineers can enhance the use of the medical equipment and reduce the costs incurred in replacing accessories.

$$\text{Benefits} : \text{Cost} \therefore \text{Value}$$

SUMMARY

At the start of this exercise, the CED took a technical view of the problem. Although they correctly identified the role of staff played in the contributing to the problem (lost or damaged accessories), they tried to solve it by improving the physical environment, the way accessories were used. In fact the real improvement came when they identified that this was a sociotechnical problem. What was missing in the wider system's soft elements was the knowledge of the cost of these items. Once this information was disseminated to those who use the equipment, the practice of staff changed due to the increased awareness that the impact of misuse or lost accessories had on hospital costs and the wider hospital system.

SELF-DIRECTED LEARNING

1. Can you identify any service or product provided by your CED that is not measured or controlled?
2. Can you identify and describe two Key Performance Indicators (KPIs) that you would recommend a CED use to ensure that they can measure and control the supply of medical equipment accessories?

CASE STUDY CS7.10: REORGANIZING THE PROCESS OF CARRYING OUT SCHEDULED INSPECTIONS

Section Links: Chapter 7, Section 7.2.8

ABSTRACT

Ensuring the safety and efficacy of medical equipment is an important part of the healthcare technology management (HTM) duties, but it is often difficult to achieve targets of testing all equipment on schedule. A Clinical Engineering Department (CED) addressed poor compliance with its targets by engaging with clinical staff in planning the process.

Keywords: Schedule inspection; Safety; Functional performance; Targets; Compliance; Team approach

NARRATIVE

Carrying out regular scheduled safety and functional performance checks on medical equipment is an important HTM activity. Manufacturers recommend the nature and the frequency of these checks in their maintenance manuals and regulatory agencies often call for these checks to be carried out, sometimes requiring compliance with manufacturer recommendations.

In practice CED often find it difficult to achieve compliance with these recommendations, failing to meet their targets. Reasons include the lack of resources and difficulty of accessing the medical equipment to carry out the work.

During its annual performance reviews, a CED was aware of failing to meet scheduled inspection performance targets and had tried various approaches including targeting equipment by type and targeting equipment by clinical area with varying degrees of success, but depressingly still remaining non-compliant. There was good compliance with some equipment deemed high risk such as anaesthetic machines, defibrillators and ventilators, but compliance for other equipment was difficult to achieve. The head of the CED convened a meeting of senior staff to discuss the problem and how to resolve. The CED served a major teaching hospital (about 800 beds) and a smaller children's hospital (about 150 beds), with a dedicated team serving the children's hospital. Compliance with scheduled inspection in the children's hospital was very good, with the assumption being that this was because of better resources and lower workload. However, as the senior staff discussed the problem, that assumption came under scrutiny. It was conceded that the environment of the children's hospital was more relaxed, with charity fundraising keeping the equipment replacement programme on track. However, the ratio of clinical engineering resources to equipment was no higher than in the larger hospital. "We know and work with our clinical colleagues in maintaining the equipment", commented the CED team leader of the children's hospital.

Could that be the key to improving the performance in the larger hospital where that collaboration was not embedded in the scheduled inspection programme but relied on CED determining the programme and how it would be carried out? The head of CED convened a small project team to investigate, asking two senior nursing managers with an interest in medical equipment to join the team. The nursing managers discussed with their colleagues revealing that the senior nurses in charge of clinical departments felt no involvement in a process in which clinical engineers appeared, seemingly at random to check equipment, often when the clinical departments had very busy patient loads. "Why don't you ask us before arriving to carry out whatever you clinical engineers need to do?" The response revealed that clinicians felt that clinical engineers were interrupting their clinical work by wanting to carry out technical task whose importance the clinicians did not appreciate or understand.

The project team suggested to the CED leadership that the lack of apparent willingness by CED to discuss and engage with the clinical departments was part of the problem. A proposal was accepted that included:

- Explaining to clinical leaders the need for the checks and what is involved;
- Meeting with clinical leaders at the early planning stage to discuss when it would be suitable for the CED to carry out the checks;
- Developing a programme that avoided checking in most areas during the busy winter months, with the exception of cardiac surgery whose planned elective surgery was reduced during the winter months;
- Increasing use of early mornings for checks in operating theatres;
- Regularly communication between CED and clinical departments, with CED reminding clinical departments a month before planned visits and providing lists of equipment to be checked;
- Colour-coded dots (different colour each year) placed on equipment to show what had been recently checked;
- After the scheduled checks had been carried out CED informing clinical departments what equipment had been checked and what still remained to be checked.

The changes were instituted, and a year later significant compliance improvements were achieved. The project team reviewed the results, consulting with senior clinicians who suggested modifications to the process of following up equipment not available for checking. These were put in place, with the determination to work on communication between the CED and clinical departments. The increased emphasis on achieving compliance with scheduled inspections impacted on carrying out urgent breakdown repairs, particularly when prioritization conflicts arose when notified about the availability of equipment in a critical care bedspace. Team leaders suggested that this was best resolved on a case-by-case basis considering the service demands at the time.

ADDING VALUE

Improving collaboration between CED and clinicians in planning and carrying out scheduled inspections of medical equipment achieved improved compliance. The change did not change CED costs, but did lead to improved benefits, both in meeting targets and also in cooperation with clinical departments.

$$\text{Benefits} : \text{Cost} \therefore \text{Value}$$

SUMMARY

A senior nurse manager commented: "Now we understand and feel able to support your attempts to access our equipment for carrying out checks. No longer do we feel you are intruding into our clinical work". Clinical engineers appreciated the cooperation with the clinicians, with those carrying out the checks reporting how clinicians, guided by the coloured dots, helped them access equipment, phoning them to advise when a clinical bay with its associated medical equipment was to be free so that the checks could be carried out.

SELF-DIRECTED LEARNING

1. Improved cooperation between clinical engineers and clinicians led to improvements in achieving scheduled inspection targets. How are the scheduled inspections managed in your hospital, how is the performance measured and how can the process be improved?

CASE STUDY CS7.11: IMPROVING AMBULATORY PATIENT FLOW IN EMERGENCY DEPARTMENTS

Section Links: Chapter 7, Section 7.3.1; Chapter 2, Section 2.6.2

Abstract: A multidisciplinary team was established to implement a new way of managing patients in ED to optimize access and reduce length of stay.

Keywords: ED; Quality improvement; Run charts; Multidisciplinary team; PDCA cycle

NARRATIVE

Emergency Department (ED) staff were frustrated that patients who could be dealt with quickly were filling up the waiting room, taking longer to be attended to, increasing waiting times and increasingly taking up more staff time. The delays led these patients to become agitated and demanding, causing the really sick patients to become distressed leading to a stressful and unsafe environment. A multidisciplinary team was established by the chief executive officer (CEO) as part of the hospital quality improvement (QI) programme with the aim of improving workflow through the ED by implementing a rapid access treatment unit (RATU) to cater for patients attending with minor illnesses or injuries. The objective was to provide a dedicated space and time in ED specifically for these patients to be dealt with effectively and efficiently; this in turn would allow the system to process the really sick patients in a timelier fashion and in a more appropriate environment.

A project charter was established incorporating an opportunity statement, project scope, expected benefits, outline composition of a project multidisciplinary team and key project measures. The charter aligned the project with the hospital's strategic objectives. The project plan involved multiple small tests of change using 'Plan–Do–Check–Act' (PDCA) cycles.

The team consisted of the lead ED consultant, the ED clinical nurse manager (CNM), a mental health social worker, the Risk Manager and the clinical engineer (CE). The ED consultant and CNM understood the clinical needs. The social worker investigated the human aspects including how the patients and staff dealt with and felt about the new service. The Risk Manager identified and minimized any risks. The CE used systems analysis expertise to look at the workflow and the processes involved and to provide data analysis giving both engineering and scientific support to the team. The CE's involvement in these activities recognized that quality control, process mapping, trending data and presenting the analysis in a meaningful manner are all part of the QI project methodology and core skills of clinical engineers. Both the Risk Manager and CE looked at the workflows utilizing root cause analysis techniques to analyze situations as they unfolded during the project.

The first task was to analyze the workflow prior to making any changes. Simple run charts were used to establish a baseline measure of the unit performance and to track progress as the small steps of change were introduced (Figure CS7.11A). Phase 1 focused on improving the management of patients with minor illness referred by General Practitioners. It aimed to reduce the waiting times for these patients on Monday afternoons by 50%, this being one of the busiest times for these types of patients.

The project aimed for certain 'Expected Benefits':

- Reduce waiting times for the patients with minor illness or injuries;
- Provide these patients with a more efficient and effective service;
- Reduce congestion in the Emergency Department.

Key project measures were established to check whether the RATU delivered these objectives. These measures included the waiting times and length of stay (LOS) of patients who are referred by their GP. Pre- and post-implementation satisfaction surveys were undertaken with both staff and patients to qualitatively assess the success of the changes made.

The RATU was established and its operation introduced in February, initially only on Monday afternoons as a test of change. This small change was tested plotted on the run chart (Figure CS7.11A). Run charts were used to compare the monthly mean waiting times and length of stay (LOS) of patients before and after the introduction of the RATU. Both quantitative measures fell in February giving an early indication that the intervention might be successful.

Qualitative measures also indicated early success. Both the patient and staff satisfaction surveys indicated that the RTAU was judged by all to be a more efficient and effective service. The reduction in the congestion of the Emergency Department on the Monday afternoons was commented on by many who responded to the surveys:

"RATU is good for staff in that patients can be streamed more easily. When the ED is very busy there can be a real sense of accomplishment in seeing the movement of people through the system. This contrasts with times when there is a sense of being stuck with a large volume of patients, many of whom are facing a long wait. The sense for staff of people moving through the system in an appropriate and speedy way seemed to be a factor in boosting staff morale." (Quote from one of the staff surveys)

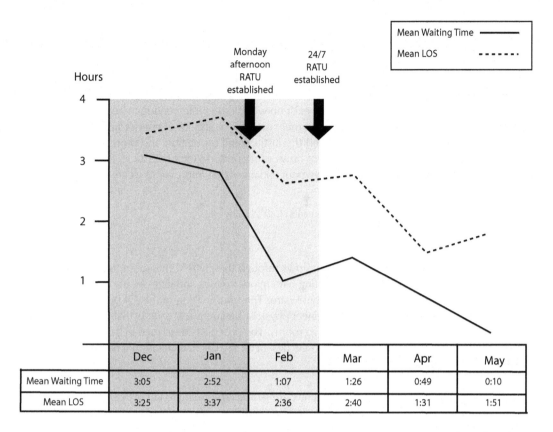

	Dec	Jan	Feb	Mar	Apr	May
Mean Waiting Time	3:05	2:52	1:07	1:26	0:49	0:10
Mean LOS	3:25	3:37	2:36	2:40	1:31	1:51

FIGURE CS7.11A   Run chart used to track progress in reducing waiting times and lengths of stay in the ED.

Once the measures of the changes introduced indicated early success, it was important to align this quality initiative with the hospital's Statement of Strategic Intent and the Hospital Service Plan. These incorporated safe and effective care, in line with the targets of the service delivery unit and with national standards. Consequently, a multidisciplinary presentation to the hospital's senior management team was delivered. This led to the decision to extend the RATU. The RATU opening times were extended to 24 h from March to May. The run charts were used to continue to measure the effect of this second phase of the project. The initiative continued to be effective, and consequently, it was decided to permanently establish the RATU.

The multidisciplinary team approach was central to the success of the project. "We had an excellent and diverse team drawn from several departments of the Hospital including the Clinical Governance Unit, the Radiology Department as well as staff from our own Emergency Department. The leadership and individual members of the team created an excellent atmosphere of teamwork. This culture helped the process of integrating the thoughts and knowledge of other professions in the hospital into the ED QI project thereby enriching the project to the extent that the solution produced was and is successful. It has stood the test of time: two years later it doubled in size, a measure of its success".

Whilst there was no new technology required or implemented as part of the solution, the clinical engineer's ability to understand the complex working dynamics of the ED and to collect and analyze data was valued by all. The clinical engineer played a lead role in presenting the project measures to the hospital's senior management team.

ADDING VALUE

The project was carried out by reassigning the roles and responsibilities of existing staff, so the quality improvement was achieved within existing resources. The interventions made during this QI project improved the quality of care. Access to physicians for those presenting with minor injuries was more timely. The experience of care for all patients and for the staff giving the care improved. At this time, no definitive data exist on whether the intervention increased or reduced costs. However, it helped the institution to meet national targets for ED performance particularly in regard to waiting times. This increased the institutional reputation and protected reimbursement. By systematically analyzing patient flows using performance measures, the clinical engineer was able to quantify quality improvement initiatives that improved the care of patients.

Benefits : Cost ∴ Value

SUMMARY

The deeper understanding of the patient flows through the ED and the consequent provision of a dedicated pathway for patients presenting with minor injuries and illnesses improved the effectiveness of the ED, enhancing the quality of care. The future vision of this QI project, driven by the multidisciplinary team with the engineer/physicist having a key role, is that RATU will be a permanent fixture in ED expanding to two rooms. The processes described in this case study are to be replicated in the establishment of a new design for the Clinical Decision Unit, part of the ED.

SELF-DIRECTED LEARNING

1. The run chart showed that the mean waiting time continuing to fall in the months after the RATU was established 24/7. Why do you think the waiting times continued to fall? The fall in the mean LOS reduced but not in proportion to the reductions seen in the waiting time. Why do you think that is?

CASE STUDY CS7.12: DEVELOPING A NOVEL MEDICAL DEVICE – PROCESSES TO SUPPORT GAINING REGULATORY APPROVAL

Section Links: Chapter 7, Sections 7.3.2 and 7.4.4

ABSTRACT

A need was identified to develop a new medical device to measure respiratory impedance during bronchoscopy procedures. The rationale to do so and the key steps to ensure regulatory compliance are presented.

Keywords: Medical device Standards; Medical device regulations; Novel medical device development; Technical documentation; Endobronchial lung volume reduction surgery

NARRATIVE

Clinical engineers will sometimes be asked to develop novel medical equipment to meet clinical requirements. The process of developing and making novel medical equipment requires compliance with ethical and regulatory standards. This case study will discuss these, using as an example the development of novel equipment for measuring lung function.

Endobronchial lung volume reduction (LVR) procedures are being developed to treat emphysema, but evaluations of their benefits have been inconclusive (Ernst and Anantham 2011).

Consequently, respiratory physicians wished to measure respiratory mechanics during LVR procedures to better assess their efficacy. The physicians wanted to measure respiratory system impedance, in particular the reactance which relates to the respiratory compliance, a sensitive indicator of the LVR procedure. The measurement is appealing to the clinician as it requires minimal subject cooperation and can be undertaken in a wide range of clinical settings, for example outside the pulmonary physiology laboratory.

The physicians approached the Clinical Engineering Department (CED) asking if it could develop equipment to measure the respiratory impedance in sedated patients by measuring the pressure and flow in the airway arising from an oscillatory airflow applied at the mouth (Michaelson et al. 1975; Oostveen et al. 2003).

The CED began by clarifying the clinical need. First, does the proposal have clinical merit? The physicians were asked to raise the proposal with the hospital's Clinical Review Board (CRB) that assesses the benefits against the costs and risks of proposed clinical developments. The CRB approved. Second, the CED wanted to investigate whether or not suitable equipment was commercially available. This required a clear functional specification of the proposed medical equipment. Systems that measure respiratory impedance are available commercially, but none could be identified that met two key criteria: a) measuring the impedance from sedated patients requiring minimal subject cooperation and b) impedance measurements at frequencies down to 1 Hz (Scott 1993).

Following CRB approval and confirmation that commercial medical equipment was not available, the CED and the respiratory physicians sought organizational approval to commence a research project. Approval was given and a project team was formed. The project commenced with exploratory simulation exercises of the low-frequency impedance measurements which proved invaluable in clarifying how improvements might be achieved. The organization's Research, Development and Innovation team were informed and research governance and ethical approval sought and obtained. The positive progress provided the evidence to successfully apply for financial support to develop a prototype device (IPEM 2014).

Developing medical equipment requires a robust approach to ensure compliance with medical device regulations, incorporating the associated quality control and risk assessment planning from the start of the project. The details of regulatory requirements vary between national jurisdictions and this case study will outline the UK requirements based on the EU Medical Devices Directive (MDD) (European Council 2007). However, whilst the general principles are applicable across jurisdictions, readers are advised to contact their local regulatory agency and seek advice from colleagues who have embarked on medical device developments.

Those developing novel medical devices must first determine whether the development is for in-house use only with no intention of 'commercialization', that is that there is no intention of placing on the market for use in other healthcare organizations either by sale or for free. If so, within the EU, the development can proceed without notifying the regulatory agencies; but note that the draft Medical Devices Regulation is likely to make notifying the regulators a requirement. Those developing the device should still follow good manufacturing practice, documenting their design criteria, product verification, adherence to Standards and risk assessment in the 'Technical Documentation', and manage the whole project through a clear robust quality management process.

In this case, clinical investigation of the concept was expected to require multicentre trials to prove the clinical effectiveness, but without general placing on the market. This required that the project be registered as an investigational medical device with the regulatory agency, in the United Kingdom, the Medicines and Healthcare products Regulatory Agency (MHRA 2014). If the clinical effectiveness is proven in the clinical trials, this registration would later help the process of applying for CE-mark registration to meet the requirements for placing the product on the market within the European Union in compliance with the contemporary European regulations regarding medical devices.

A 'Technical Documentation' file was started to document and control the management of the project. This records the start date of the project, often taken as the date when formal approval for commencement of the project was obtained. The Technical Documentation's structure is not prescribed but should be managed under formal document control and include the following:

- Clinical benefit:
  - o What patient and clinical benefits will the proposed device provide;
  - o Clinical need and justification;
  - o Clinical specification.
- Quality management system:
  - o Description of quality management system;
  - o Documentation control;
  - o Quality management strategy, audit and control;
  - o Commissioning procedures;
  - o Post-market surveillance planning.
- Risk management:
  - o Risk classification according to the regulations in your jurisdiction (in Europe, Annex IX of the MDD [European Council 2007]);
  - o Risk assessments, showing how risks are identified and minimized and clarifying any residual risks and how they are controlled;
  - o A risk/benefit analysis providing justification for any residual risks;
  - o Protection against malware and any software risks – if applicable;
  - o Data protection (Incorporate strategies to protect any personal data from the start of the design. For example, if the product records patient names or other details then the design process should seek to ensure that these data are protected if the product has to leave the healthcare organization. Protection could be by incorporating robust personal data deletion programmes or by storing the personal data on a data recording device separate from any operational software and that can be readily removed by the healthcare organization before the product leaves it.)
- Technical:
  - o Description, this section, like others, will develop as the technical design develops, with the file showing the technical development stages;
  - o Standards applied, showing how the device and its development is supported by and complies with the applicable Standards;
  - o Verification, detailing the technical verification through the various stages of the design development;
  - o Test reports.
- Ease of use:
  - o Human usability design criteria and assessments;
  - o Instructions for use;
  - o User training planning.
- Validation:
  - o Clinical trials, ethics approval, planning and control;
  - o Analysis of results and assessment;
  - o Verification;
  - o Reports.
- Product documentation, markings and labels.
- Declaration of conformity.

The Technical Documentation is an active file that both guides the development and documents it. It will develop as the project develops. Having its structure in place from the beginning can help ensure that the product complies with the necessary regulations. It is a good idea to discuss the structure and content of the Technical Documentation with colleagues who have been through the process.

It has not been the purpose of this study to detail the results of the development of the novel medical device itself but rather to focus on the essential steps to consider, the decision-making processes involved and the steps required to ensure regulatory compliance.

ADDING VALUE

Clinical engineers can add value to healthcare by aiding clinicians in identifying the right tool for the job, aided and abetted by their knowledge of clinical practice, available medical devices, the regulatory regime in their jurisdiction and the marketplace. If the need arises to develop a novel medical device, the CED can, subject to the right safeguards, knowledge, skills and funding being available, contribute to the design and development of innovative devices. Ensuring compliance with medical device development regulations will add costs but will help ensure that the device developed is safe and effective, bringing benefits and avoiding the higher costs that will be incurred if the regulatory requirements are left till the end of the project.

Benefits : Cost ∴ Value

SUMMARY

This case study describes the processes that should be followed to ensure that the development of novel medical devices is compliant with regulatory requirements designed to ensure that these developments lead to safe and effective devices. The 'Technical Documentation' file helps deliver and document a structured approach to the development. It must include a clear statement of the clinical problem, a clinical risk/benefit analysis, organizational support for the project and adherence to medical device design standards. It must show that development work is carried out within a formal quality management system. Research governance and ethical considerations must be addressed. The use of modelling at an early stage if possible, and a clear review of the literature, will ensure that subsequent project elements are built on sound foundations.

REFERENCES

Ernst A. and Anantham D. 2011. Bronchoscopic lung volume reduction. *Pulmonary Medicine*, 2011: Article ID610802. http://www.hindawi.com/journals/pm/2011/610802/ (accessed 2016-05-04).

European Council. 2007. Council directive 93/42/EEC (as amended) concerning medical devices. http://eur-lex.europa.eu/LexUriServ/LexUriServ.do?uri=CONSLEG:1993L0042:20071011:en:PDF (accessed 2016-05-04).

IPEM. 2014. Research and innovation awards. Institute of Physics and Engineering in Medicine. http://www.ipem.ac.uk/ProfessionalMatters/PrizesandAwards.aspx (accessed 2016-05-04).

MHRA. 2014. Guidance: In-house manufacture of medical devices. Medicines and Healthcare products Regulatory Agency. https://www.gov.uk/government/publications/in-house-manufacture-of-medical-devices/in-house-manufacture-of-medical-devices (accessed 2016-05-04).

Michaelson E.D., Grassman E.D. and W.R. Peters. 1975. Pulmonary mechanics by spectral analysis of forced random noise. *Journal of Clinical Investigation*, 56(5): 1210.

Oostveen E., MacLeod D., Lorino H. et al. 2003. The forced oscillation technique in clinical practice: Methodology, recommendations and future developments. *European Respiratory Journal*, 22: 1026–1041.

Scott R. S. 1993. Determination of respiratory impedance via the oscillatory airflow technique. PhD Thesis. University of Bath, Bath, UK.

SELF-DIRECTED LEARNING

1. Imagine you are embarking on the design of a novel medical device. How would you comply with the necessary regulatory, research governance and ethical requirements as outlined in this case study? What additional compliance processes would you need to fulfil to meet the requirements in your jurisdiction?
2. Describe the details of a quality management framework for medical device design and development, referencing appropriate Standards and guidelines. What requirements does the framework place on organizations and designers of medical devices?
3. How would you approach an enquiry by a clinician to address an unmet clinical need? How would you investigate the particular patient pathway, identifying the needs of clinicians and patients and the healthcare technologies involved? What alternative methodologies exist and contrast the pros and cons of existing practice with these alternatives?

CASE STUDY CS7.13: DESIGNING AND PROTOTYPING A DEVICE TO SUPPORT CLINICAL RESEARCH

Section Links: Chapter 7, Sections 7.3.2 and 7.4.4

ABSTRACT

A research study into chronic pain diagnosis required the development and realization of a bespoke medical device by the hospital's Clinical Engineering Department (CED).

Keywords: Support; Research studies; Brain mapping; fMRI; Medical device design; Chronic pain

NARRATIVE

MRI brain mapping studies are a type of functional MRI (fMRI) study where many areas of the brain are imaged at the same time to determine which anatomical sites of the brain are activated by particular stimuli, for example identifying those areas of the brain areas activated by the application of a painful stimulus. Brain mapping studies are optimized by synchronizing the acquisition of the fMRI images with the stimulus.

The Pain Medicine Department in a University Teaching Hospital wanted to conduct new research using fMRI brain mapping techniques into the perception of pain arising from a force being applied to the body. To optimize the brain mapping scans, the delivery of the force stimuli would have to be repeatable, randomized in time and synchronized with the fMRI scanner. The stimulus had to be shown to activate the recognized anatomical pain sites. No automatic method of delivering a force stimulus that can be electronically synchronized existed on the market or had been described in the literature. The Medical Director of the Pain Medicine Department approached the CED to see if they could design, develop and build such a novel device in their role as providers of scientific and technical support to clinical research.

A manually operated pneumatic apparatus that can apply a force to induce a pain stimulus had been described and used in unsynchronized MRI studies. It was based on a pneumatically operated linear actuator, which pushed down on the thumb nail. This Patient Applied Part was made of plastic with no ferrous or conducting parts. Adjusting the pressure of the air driving the linear actuator varied the intensity of the stimulus.

The CED built a pneumatically operated pain stimulus device based on this principle. The driving pressurized air was connected to the Patient Applied Part through a plastic tube, called the driving line which was passed through the waveguide from the control room into the scanning room. This was fed from a bespoke control box in the MRI control room. The control box was a novel electromechanical device that controlled the air pressure by using electronically controlled solenoid valves to switch the air pressure output to the applied part from one of three pressure regulators set at 1, 2 and 3 bar pressure. The control box enabled the different stimuli to be applied in a predetermined randomized sequence with a fixed stimulus duration and variable inter-stimulus duration. The switching was controlled by a microcontroller which was synchronized with the MRI scanner and into which the stimulus sequence was preprogrammed.

Development of novel medical equipment within healthcare organizations requires adherence to strict quality control standards (see Case Study CS7.12) including keeping a risk log during the design and construction phase. The risk management process must identify and minimize potential technical failures which could result in patient or user hazard, including the application of a prolonged stimulus. This risk log influenced several design decisions:

- The system was designed such that in the event of a loss of electrical power, it would fail safe; the driving line could not be pressurized in the event of power failure.
- The system included a mechanical over pressure relief valve. In the event that a regulator failed, the relief valve would open and protect the subject.
- Clear visual indications were designed to show the operator which pressure regulator was switched to the driving line at any given time so that the operator could monitor the progress of the stimulus sequence.
- A mechanical pressure gauge was connected to the driving line so that at all times, even in the event of a failure of the electronics, the operator in the control room could see exactly what if any stimulus was being applied to the subject.
- In the event of a microcontroller software failure, the operator could press a system stop and reboot switch on the front panel to halt the sequence and open the driving line to atmospheric pressure (no stimulus).
- In addition the operator could activate a mechanical manual override valve on the front panel to open the driving line to atmosphere.

The device was designed, constructed and tested over a period of 6 months, and, following verification, ethical approval was granted to test the device, connected to the MRI scanner, on human volunteers. This first part of the study showed activation of recognized brain pain sites in response to stimuli that varied with the applied stimulus strengths, confirming the operation of the system, with the random stimuli synchronized to the image acquisitions. The healthy volunteers described the pressure sensation as unpleasant.

Following the success of this trial on normal volunteers, ethical approval was granted to use the device to study a cohort of patients with chronic neuropathic pain, including investigating the effect of pharmacological therapeutic interventions on response to the pain stimuli. The device was permanently installed in the control room of the MRI scanner and responses determined from 40 subjects in an 18-month period, with the device never failing.

The voice of the lead clinical engineer:

"The design and construction of this device was informed by our experience as clinical engineers in maintaining medical equipment. We were skilled in the pneumatics as a result of all the work we do on anaesthetics and ventilators. We were aware of the safety issues in MRI as we work closely with our medical physics colleagues and also experience supporting the challenges of bringing ICU critical patients who need monitoring and inotrope support into the MRI magnetic and RF fields. Our department has a culture of trying to imaginatively foresee problems that might arise with equipment so the risk management process during design and build was second nature. Our head of department encourages participation in research. We are also encouraged to follow our interest in new technologies and it was one of my colleagues interest in programming microcontrollers that meant we had that expertise in the workshop to draw on."

Voice of the Medical Director of the Department of Pain Medicine:

"Chronic pain affects approximately 20% of the population. There is currently no diagnostic test available to help clinicians diagnose chronic pain objectively. This technique facilitated the clinical quantification of the different responses to pressure stimuli in volunteers and patients with chronic pain thereby resulting in a scan signature response associated with chronic pain. This technique opens the door to developing a scan which can provide a diagnostic test. Moreover it can also be used to assess response to treatment. This technique, if developed, would allow rapid outpatient treatment stratification and improve both speed of diagnosis and both efficiency and effectiveness of treatment delivery.

Clinical engineering developed the pressure stimulus device. It would not have been possible to carry out this work without their expertise. The clinical component of their expertise enabled them to understand clinical risk and to develop a system that had numerous safety features included. It also allowed an understanding of the problems of bringing equipment into the MRI scanner environment. That fact that there were no malfunctions or safety issues testifies to the success of the design. The reproducibility of the results between study cohorts demonstrated the robustness and reliability of the system. The understanding of both the clinical components and the engineering requirements to deliver the system was unique to clinical engineering."

SUMMARY

Whilst the research project was clinically focused, it could not be conducted without a bespoke apparatus being designed built and tested. The hospital's in-house clinical engineer's primary role was healthcare technology management, yet they were also available to support clinical research. Their ability to respond and meet the challenge presented by the clinical team was supported by the Head of Department who protected time for this endeavour. The experience gained from supporting medical equipment informed the design and build. An appreciation of risk management as applied to medical equipment guided it.

ADDING VALUE

The clinical engineers added value to the academic work of the teaching hospital by applying the skills and knowledge developed from their health technology management (HTM) role to the research project. In so doing, they demonstrated the complementary remits of HTM and advancing and supporting care and their ability to link the clinical with the technical for patient benefit. Cooperation between clinical engineers and clinicians improved the ability to diagnose pain bringing benefits to patients and clinicians that outweigh the added costs of the procedure.

Benefits : Cost ∴ Value

SELF-DIRECTED LEARNING

1. 'The CED described in this case study is the exception to the rule. Most departments would be busy with HTM and would not have the time or experience to be involved in this sort of research and innovation.' Would you agree with, defend or refute this statement? Give a five hundred word justification for your answer.

2. Multidisciplinary cooperation between clinical engineers and clinicians can create powerful synergies for advancing healthcare. How can this collaboration be encouraged and developed?

CASE STUDY CS7.14: DEVELOPMENT OF A PROTOTYPE SOLUTION FOR THE CLINICAL ASSESSMENT OF GAIT STABILITY

Section Links: Chapter 7, Sections 7.3.2 and 7.4.4

ABSTRACT

Clinicians can quickly identify deficiencies in clinical efficacy and clinical engineers are ideally placed to identify with those needs. Here a clinical need to better assess falls risk was progressed by an iterative application of scientific and engineering skills. A technology-based prototype solution was developed to quantify gait stability in older adults at risk of falls.

Keywords: Falls risk; Inertial sensors; Research solutions; Algorithm development; Validation; Gait analysis

NARRATIVE

The project started with a casual conversation between the clinical engineer and a leading consultant geriatrician and international falls expert. The problem, she explained, is that we don't have a good way of quantifying how stable a patient's gait is and therefore find it difficult to estimate their falls risk. We can observe the patient walking up and down the corridor or do the standard physiotherapy test (the timed up and go test) but really that's all we have. Second, the very best interventions aimed at preventing falls in older adults are only of limited effectiveness, and there have been no new initiatives to solve this problem in the last 30 years. Even a single fall can have a life-changing effect on an older adult. A hip or wrist fracture with resulting hospitalization, or a subsequent development of a fear of falling, can result in a downward spiral towards loss of independence and institutional living. More reliable ways of quantifying falls risk would benefit patients by identifying the need for therapeutic intervention at an early stage.

This presented the clinical engineer with a research challenge that stimulated further discussions with engineering colleagues and clinicians. A series of fortuitous events defined the way forward. First, a large geriatric research study was starting up in the research wing of the hospital. The clinical engineer requested to be assigned as the support engineer on the research programme and was granted a secondment by his Head of Department. He advised on and commissioned the medical equipment used in the study whilst simultaneously learning the language and tools of the comprehensive geriatric assessment. Second, the research programme was part funded by Intel® as an industrial partner whose digital health group had been developing an open-source prototype body-worn inertial sensor (http://www.shimmersensing.com). As part of the development cycle, the clinical engineer worked closely with Intel® engineers to identify clinical applications for the sensors. Technology for falls risk assessment was agreed as a research theme in the programme.

The clinical engineer was assigned to research the role of inertial sensors as an instrument for quantifying gait stability and falls risk.

SCIENTIFIC APPROACH

A review of the literature identified that high stride-to-stride variability was emerging as a promising predictor of falls risk. Can variability in stride time be accurately measured with an inertial sensor? The first development was to apply and validate software to capture the sensor signals during gait using the Bluetooth protocol. The reliability and limitations of real-time data streaming using Bluetooth were tested in a long corridor in the clinical environment. The raw accelerometer and gyroscope signals available from sensors require signal processing and the development of algorithms to transform them into meaningful outcome measures. MATLAB®-based algorithms were scripted to extract stride times and stride time variability from features identified in the signals. More advanced algorithms exploring the structure of variability in a high-dimensional state space reconstruction were also developed and tested. Through a series of gait laboratory experiments, comparing with motion camera technology as gold standard, it was shown that:

- Stride-to-stride variability could be measured accurately using an inertial sensor;
- The measure reflected instability in walking as evidenced when the balance of young adults was perturbed whilst walking on a treadmill.

Returning to the clinical environment, the prototype sensor and algorithm were trialled successfully in a pilot study on older adults. The sensor was shown to be well tolerated in older adults, easily attached to the lower limbs and unobtrusive during gait trials. Based on the pilot data, the clinical engineer developed a gait protocol for the measurement of stride time variability and led a full study comprising 400 adults. Using logistic regression statistical analysis, the clinical engineer showed that stride time variability correlated positively with conventional measures of gait and balance and other measures of health-related decline previously determined from the geriatric assessment.

An important outcome from the research study was that the clinical team recognized the value of quantifying stability in gait and supported the establishment of a clinical gait analysis laboratory in a newly built 'Centre of Excellence for Successful Ageing'. This initiative allows further development of algorithms and validation of the sensors in older adults at risk of falls. Continued clinical engineering involvement in this facility will enable expansion of the Clinical Engineering Department activity and further research themes to emerge.

ADDING VALUE

In this case, the clinician knew that a novel technology approach was needed; however, she was not aware of emerging technologies or how to apply them for patient benefit. The clinical engineer with core knowledge of the principles of biomechanics and motion, software and signal processing, and acquired knowledge of the health issues and limitations of older adults could identify and develop prototype solutions. Whilst in the long term it is hoped that the research will lead to reduced incidence of falls and hence lower costs, in the early stages, the collaboration does incur costs, but these are outweighed by the benefits. The collaboration between clinician and clinical engineer improved the knowledge about the risk of patient falls, improving the ability of the clinicians to care for their patients.

Benefits : Cost ∴ Value

PATIENT CENTRED

Primary research by the clinical engineer led to the establishment of a clinical gait analysis facility for the assessment of patients at risk of falls. The availability of quantitative gait analysis in a clinical environment enhances the clinical services available to patients.

SUMMARY

Clinical engineer's voice:

"This case study highlights a number of factors that are essential for innovative clinical engineering. Firstly, the willingness to step forward out of the comfort zone as a clinical engineer and engage with consultant colleagues at a research level is fundamental to identifying important clinical needs.

Secondly, the ability to be able to interface between clinical need, R&D, industrial partnership and innovation is a unique skill that should be developed as part of the professional development of clinical engineers.

Keeping up to date with emerging technologies and evaluating opportunities to apply them as innovative solutions is an important aspect of the role of the clinical engineer who is uniquely placed to visualize their practical application in the clinical environment. Being adaptable and decisive in the face of opportunity, along with a supportive Clinical Engineering Department, are all essential factors to enable high-level clinical engineering R&D."

SELF-DIRECTED LEARNING

1. This case study illustrates the importance of collaboration between disciplines. This collaboration is illustrated in Figure 1.5, Chapter 1. Discuss this case study in the light of Figure 1.5, in particular the importance of the Life and Physical Sciences working together to advance care.
2. Figure 1.5 illustrates that clinical engineering is just one of a number of related disciplines, each of which is a strand of the wider discipline of biomedical engineering. Identify the different range of skills and knowledge applied during this research study and how many of these map onto other strands of biomedical engineering listed in Figure 1.5.

CASE STUDY CS7.15: COMMERCIALIZING THE OUTPUT OF RESEARCH INTO A NEW PHYSIOLOGICAL MEASUREMENT TECHNIQUE

Section Links: Chapter 7, Sections 7.3.2 and 7.3.3

ABSTRACT

When clinical engineers work collaboratively with clinicians, research developments can follow. These innovative partnerships may lead to the commercialization of novel medical equipment or techniques, benefiting wider healthcare.

Keywords: Research; Innovation; Physiological measurement

NARRATIVE

This case study describes the experiences of a clinical engineer working in support of a busy Gastrointestinal Endoscopy day surgery service. His role included equipment management, but importantly also support for research into new methods of delivering care using technology.

He was based in the day surgery unit and maintained the endoscopes and associated imaging systems, gastro-manometery, pH monitoring, lasers and ionizing imaging systems. In supporting many research activities, he regularly assisted with the application of new technologies during clinical procedures. The clinical engineer had had experience in the maintenance of many clinical technologies before taking up this particular role and had just completed a master's degree in bioengineering that included a significant research component.

Supporting the Advancement of Care

Whilst working in the endoscopy unit, one of the clinicians informed the clinical engineer of a novel surgical technique for suturing valves in the digestive system that was performed endoscopically. The clinical engineer sought information on this technique and invited the company developing this technique to the unit to discuss its application in clinical practice. As part of this visit, the technique was demonstrated to the clinicians and clinical engineer. Given the Day Surgery's commitment to research and assessing emerging technologies, as well as the local skills of the clinical engineers, the unit became a development site for the device. This formal agreement between the hospital and the industry was led and project managed by the clinical engineer involved. The clinicians and clinical engineer worked together with technical support from the designers to research the technique's effectiveness and published the first one-year follow-up on patients treated using this technique (Mahmood et al. 2003).

Original Research

Involvement in this research required the clinical engineer to identify how measurement of the competence of sphincters in the digestive system was performed. He identified that at the time there was no reliable method to make these measures. This prompted him to try and develop a method of assessing sphincter competency endoscopically. Initial studies were carried out in the local hospital to see if new parameters could be determined. Through literature searches and participation in various research conferences, the clinical engineer became aware of a group in Europe who were using a technique for measuring the cross-sectional area of lumens in the gastrointestinal tract. Following a number of collaborations with this group, the clinical engineer took leave of absence from his hospital job and, funded by a research grant, went to work with the university group to complete the research. Whilst working in the university, he was able to consolidate his knowledge whilst still working with leading-edge clinicians and researchers to develop the principles of the novel physiological measurement.

The research showed that it is possible to inflate a cylindrical shaped non-conducting bag in a valvular region in the digestive system and use the voltage change across electrode pairs on the outside of the bag to estimate the cross-sectional area of the narrowest opening and at the same time measure pressure in the bag. Developing the concept to include multiple electrodes along the bag allowed the competence along the length of the sphincter to be assessed. The technique was tested in an animal model and published (McMahon et al. 2004, 2009).

Innovation

Having developed a new clinical tool, the research team group were motivated to bring the idea out of the research environment and into clinical practice. To do so, they had to engage with industry. They established a license agreement to develop the concept into a product with a small medical device company. Over the next two years, whilst the product was in development, the clinical engineer returned to his hospital role and continued to research the application of the discovery. During this period, he also acted as a consultant to the company now commercializing the idea. This technology was subsequently developed into a commercial medical device known as EndoFLIP® (Kwiatek et al. 2010; Crospon 2016).

The voice of the clinical engineer:

"As clinical engineers we are not only in a good position to carry out applied research with our colleagues, be it clinicians or other health care professionals, but we are in a great situation to demonstrate how this research can have impact. When we talk about impact we must consider innovation. Where research is focused on academic expertise, advanced education and knowledge, innovation is focused on ideation, design, need and real world application. For research to have impact in medicine it most often needs to be linked with innovation. In fact if applied clinical research is to help more than a small handful of patients it must lead to innovation. Broad minded and creative clinical engineers are in an ideal position to link good research, particularly research that levers off technology, with innovations for the benefit of patients. This concept and its development going forward is, in my opinion the key to improving care and optimising the use of technology in medicine in the future. What was most rewarding for me was that as the product developed I was able to continue researching applications of the technology. A number of grants allow for clinical engineering support for this research to continue, and two students, one an engineer and one a speech therapist have completed PhDs on further applications of the technique."

ADDING VALUE

This hospital-based clinical engineer created added value by bringing a new technique to the market. The costs associated with all this work were significant, and the ultimate test of the value of the invention will be whether it is adopted and used in clinical practice.

However, the unique contribution of the clinical engineer in identifying the need and developing a solution through both research and innovation delivered benefits for future patients and so added value.

Benefits : Cost ∴ Value

SUMMARY

This case study illustrates that through participation in the multidisciplinary clinical team, the clinical engineer was able to bring technical know-how, not just on current and available technologies but also on those in development and, in fact, leading others generating the ideas for new technologies.

The clinical engineer was involved in providing consultancy work to the company developing the commercial device. This was a great opportunity to get insights into the commercialization of a concept, the costs involved and the type of hurdles to be overcome.

It highlights the importance of clinical engineers being involved with academia. This creates a network of interaction with other engineers, scientists and other clinicians outside of those worked with directly. It also provides an avenue and support for research activities such as clinical studies that might be linked to the technology, it gives access to technology transfer offices where specialists can advise on idea protection, and it also provides a link to grant funding.

REFERENCES

Crospon. 2016. http://www.crospon.com (accessed 2016-05-04).
Kwiatek M.A., Pandolfino J.E., Hirano I. et al. 2010. Esophagogastric junction distensibility assessed with an endoscopic functional luminal imaging probe (EndoFLIP). *Gastrointestinal Endoscope*, 72(2): 272–278.
Mahmood Z., McMahon B.P., Arfin Q. et al. 2003. Endocinch therapy for gastro-oesophageal reflux disease: A one year prospective follow up. *Gut*, 52(1): 34–39.

McMahon B.P., Frøkjaer J.B., Drewes A.M. et al. 2004. A new measurement of oesophago-gastric junction competence. *Neurogastroenterol Motil*, 16(5): 543–546.

McMahon B.P., Jobe B.A., Pandolfino J.E. et al. 2009. Do we really understand the role of the oesophago-gastric junction in disease? *World Journal of Gastroenterology*, 15(2): 144–145.

SELF-DIRECTED LEARNING

1. What are the barriers to clinical engineers collaborating with industry in developing innovative medical equipment? How can the barriers be minimized and what protections need to be put in place to safeguard the interests of the clinical engineer, the healthcare organization and industry?

CASE STUDY CS7.16: ESTABLISHING A RISK MANAGEMENT FRAMEWORK FOR AN INTENSIVE CARE UNIT'S MEDICAL IT NETWORK

Section Links: Chapter 7, Sections 7.3.4 and 7.4.7

ABSTRACT

As part of the upgrade of a clinical information system, the need for a named medical IT network Risk Manager was identified. The clinical engineer, who was part of the multidisciplinary team supporting this system, took up the role of Risk Manager.

Keywords: Medical IT network; Risk manager; Clinical engineer; Clinical information system

NARRATIVE

Clinical information systems (CIS) are computer-based systems that collect, store, process and present the clinical information required to deliver patient care. A CIS in an intensive care unit (ICU) often interfaces to bedside electromedical equipment as well as other clinical systems such as Laboratory and the Radiology Information Systems. This case study relates to the management of a CIS in a 40-bed ICU. The CIS is under the governance of the ICU's Medical Director, who is supported by a multidisciplinary team (MDT) consisting of doctors, nurses, pharmacists, laboratory scientists, information technology professionals and clinical engineers. The only full-time MDT members are two nurses who are system administrators; they are custodians of the configuration and provide ongoing training and user support.

In advance of a planned CIS upgrade, a risk management exercise was undertaken by the MDT to mitigate possible associated risks. The first step was a holistic analysis of the CIS to identify and describe all its constituent parts which revealed that it had both hard and soft elements. The hard elements included medical equipment, IT networks, bedside medical grade PCs, servers and ICT integration engines and software and database applications. The soft elements included the MDT and the processes established to manage and regularly update the system configuration. The CIS interfaces with other technical systems owned and operated by staff in other hospital departments; the soft elements included staff who manage these other interfaced systems. The external companies who support the CIS and would perform the upgrade were also identified as soft elements. The MDT characterized the CIS as a complex sociotechnical system consisting of people, processes and technology that together deliver an information management process which directly supports patient care.

The CIS incorporates a medical IT network defined by the IEC 80001-1 Standard as an IT network that incorporates at least one medical device. The Standard recommends that a hospital managing a medical IT network should appoint a medical IT network Risk Manager to be responsible for the risk management process used to maintain the safety and effectiveness of the medical IT network.

TABLE CS7.16A   Example of a Risk Assessment Log

Risk Description	Risk Category	Initial Risk		Initial Risk Rating	Control Measures	Person Responsible for Action
		Likelihood	Impact			
Risk of incorrect configuration – data entry.	Safety efficiency	Possible (3)	Moderate (3)	9	Minimum two individuals assigned to process. Personnel assigned to task have established experience in tasks to be undertaken.	Configuration team
Device interface – risk of incompatibility with devices or server.	Efficiency	Unlikely figure (2)	Minor (2)	4	Testing of device interface prior to go live (including software revision review).	Clinical engineering

The MDT agreed the need for a medical IT network Risk Manager reporting to the ICU Director. Who should take on the role? The ICT department members of the MDT had solid IT and network knowledge but lacked experience of supporting equipment at the point of care and the clinical application of the CIS. The two nurses, who acted as system administrators, had excellent knowledge of the clinical operation of the system but were inexperienced in the technical aspects of interfacing and network management. The clinical engineer was the only MDT member who had both the required clinical and technical knowledge and the expertise in the risk management of technology placed in clinical environments. The MDT, supported by the ICU Director, nominated the clinical engineer as Network Risk Manager.

The clinical engineer began by convening a meeting of all those staff who manage systems interfaced with the CIS, such as the lab system. The purpose was to share information and foster an understanding of the possible unforeseen consequences that change in one system might have on another. From this developed a shared vision of how the upgrade would be planned and implemented to ensure that safety, effectiveness and security were considered and managed through the project.

The upgrade was broken into work streams: server replacement, interface mapping and validation, network configuration, application upgrade and training. Risks identified within each of these streams were shared at the combined MDT and mitigation actions identified and implemented.

As part of the upgrade project, an assessment of the associated risk was led by the clinical engineer in their capacity as medical IT network Risk Manager. In doing so, the clinical engineer worked in close partnership with the system administrators. The risk identification process took input from the MDT. Risks were identified by considering the individual system elements in turn (Table CS7.16A).

ADDING VALUE

Engineers are acutely aware of the need to mitigate against adverse events that can arise from the use of technology. By acting as medical IT network Risk Managers, they can add value by minimizing the risks associated with the use of ICT technology in care. As the clinical engineer described it:

"As clinical engineers the equipment we are responsible for impacts directly on the care provided to patients. This means that we are intimately familiar with assessing the risk associated with technology and its use for patient care. We are constantly considering risk as part of our daily assessments. Given the complexity of any medical IT network the task of medical IT network Risk Manager is daunting. While clinical engineers are skilled and well placed to fulfil such a role, they must be supported to do so by Top Management and an engaged and empowered MDT that can consider all facets of potential risks."

There is a cost associated with the involvement of the clinical engineer in being the medical IT network Risk Manager, but the benefits outweigh the cost.

Benefits : Cost ∴ Value

SUMMARY

The risk management of a medical IT network is an information-rich task that requires in-depth and up-to-date knowledge of the medical IT network, an appreciation of the changes in the usage of technology and a practiced familiarity with the application of the risk assessment process to technology in the clinical environment. Clinical engineers have these skills which, combined with their experience of working in multidisciplinary teams, makes them potential candidates to fulfil the role of medical IT network Risk Manager.

REFERENCE

IEC 80001-1. Application of risk management for IT-networks incorporating medical devices – Part 1: Roles, responsibilities and activities. Geneva, Switzerland: International Electrotechnical Commission.

SELF-DIRECTED LEARNING

1. In this case study, the clinical engineer assumed the role of the medical IT network Risk Manager. This role might also be undertaken by others from different professions. Compare and contrast how the role of clinical engineers would be different if the medical IT network Risk Manager was (a) an IT Engineer and (b) a clinical nurse manager.
2. Teamwork is at the heart of managing the complex sociotechnical systems. Consider other sociotechnical systems in healthcare and describe them, discussing the team members involved and what each discipline contributes to its safe and effective operation.

CASE STUDY CS7.17: SUPPORTING A CONNECTED HEALTH INITIATIVE AIMED AT IMPROVING PATIENT ACCESS TO CLINICAL SERVICES

Section Links: Chapter 7, Sections 7.3.2 and 7.3.5

ABSTRACT

A multidisciplinary team developed and tested a new way of delivering speech and language therapy using a connected health approach

Keywords: Connected health; iPad; Speech therapy for children; Supporting research; Systems approach

NARRATIVE

Children with cleft palate or similar impairments often have speech errors that can have an impact on a child's intelligibility and quality of life. The standard of care aims for normal or near-normal speech by school entry. This requires timely access to speech therapy services. However, there

is a well-acknowledged lack of speech therapy services for preschool children in many jurisdictions. Studies suggest that parents can be trained to successfully deliver therapy at home to
improve the articulation of children with non-structurally related speech impairment.

Two Cleft Specialist speech and language therapists established a 'Parent Led Articulation
Therapy programme' (PLAT) (Sell et al. 2009). Studies on speech therapy interventions using
video conferencing have shown that this can lead to positive speech outcomes equivalent to
that of face-to-face therapy. The Cleft Specialists invited a clinical engineer to join them to set
up and to evaluate a connected health solution, which is an integral part of support for the PLAT
programme. The vision is to deliver a high-quality service that uses both family and therapist
time more efficiently than the conventional face-to-face speech therapy service delivery model.
Parents wanted the system to be user friendly and portable within the home. The speech and language therapists required high-quality audio reproduction and appropriate video quality to allow
them to assess and monitor the child's speech and the child's response to therapeutic prompts
during the session.

Initially the clinical engineer's role in the project was to assist with the analysis and selection of the technology for the connected health dimension of the project. The first steps were
to describe the connected health solution as a system, identifying all its elements and how the
elements would come together to meet the requirements of the children, parents and speech
and language therapists. This stage included modelling the system, including its hardware and
software, using flow diagrams. The user interface was included in the modelling of the system;
this enabled the operation of the system in its clinical context to be analyzed.

The clinical engineer proposed Apple iPads as the technology platform. This decision was
based on their ease of use, portability and the fact that the PLAT programme could be aug-
mented by the use of Apps available for the iPad platform. Two popular applications Skype
and FaceTime which are easy to use and support two-way audio/video phone calls were
evaluated. Specifications were compared, followed up by evaluating the quality of the video
and audio when used over domestic Internet service provider networks. Independently, two
therapists (one involved in the study and one not involved) judged FaceTime software as
the easier to use. The audio quality from the FaceTime app was independently judged to be
superior by the clinical engineer and the two Cleft Specialist speech and language therapists.
All three also rated video quality on FaceTime as superior, in particular the synchronization
between video and audio which is of great importance when assessing speech impairments
remotely. Consequently, FaceTime was chosen as the software.

The quality of the audio from the iPad's internal speakers was judged to be insufficient to
allow for accurate assessment of speech by the therapists. The clinical engineer suggested using
professional broadcast monitors and headphones instead, and this greatly increased the quality
of the system from the speech and language therapists' point of view.

User tests were first conducted with volunteers acting as parents in a number of locations
remote from the therapists. This identified that the positioning of the iPad in the child's home
had a significant impact on the audio and video quality. The position relationships of the iPad
to the child and parent, to sources of ambient noise and to local lighting were all found to be
important. In response, the clinical engineer developed a User Guide for parents outlining how
to optimize the positioning of the iPad and the child in relation to sources of light and noise.
This user guide also showed parents how to connect the iPad on a WiFi network and how to
initiate or receive a FaceTime session with the therapist.

A second test was undertaken with four parents and their children. These were recruited
from the cohort attending cleft lip and palate clinics and who were undertaking training in
PLAT as part of the feasibility study. The trial results were favourable, and parents reported the
quality of the connected health component of the sessions as 'adequate-to-good' most of the
time. Overall the system was evaluated as easy to use by both parents and one of the therapists.

The solution as a whole was judged to be adequate for remotely monitoring PLAT sessions by the speech and language therapists.

The success of the feasibility study led to the establishment of a larger two-centre randomized controlled trial of this PLAT connected health solution. This controlled trial is ongoing at the time of this book going to press.

ADDING VALUE

The clinical engineer's involvement in this research supported the development of a new way of delivering speech therapy that solves access problems of parents and children. The clinical engineer foresaw potential problems and developed both technical and training solutions in advance of the pilot programme. In doing so, this optimized the benefits that could be realized from the connected health solution. This work was undertaken part time, contributing to the clinical engineer's role in supporting and advancing care in the hospital, so no extra costs were associated with his involvement.

Benefits : Cost ∴ Value

The research speech and language therapists commented:

"During a previous collaboration with the same engineer, knowledge had been gained about lighting, positioning and location of cameras and microphones for high-quality audio/video recordings, in order that therapists can see and hear the subtle and complex aspects of speech production in structurally related disorders (Sell et al. 2009). In the present study, the engineer advised on how the hardware works and the best ways to optimize audio and video quality for interaction with the child and his parent/carer. He advised on how to position the iPads in order for the therapist to view tongue movements for speech, to best interact with the child and the parent/carer and also how to set up the iPad for optimal FaceTime interactions.

As the project evolved, some issues arose around the use of the iPads. Speech and language therapists do not have technical training and usually have insufficient experience of using technology in therapy. The collaboration and support of the engineer has ensured optimal use of the connected health solution and the use of technology as a supplementary tool in home-based therapy."

SUMMARY

The clinical engineer joined the team with a view of supporting a research study aimed at advancing how care is delivered. He took a systems approach and whilst he rightly identified the need to evaluate different hardware and software solutions, he went further and analyzed how the soft elements of the system (the people) used the hard elements (technology) so that the solution as a whole was optimized. Central to the approach was to work in an interdisciplinary fashion with both the therapists and the parent volunteers.

REFERENCE

Sell D., John A., Harding-Bell A., Sweeney T., Hegarty F. and J. Freeman. 2009. Cleft Audit Protocol for Speech (CAPS-A): A comprehensive training package for speech analysis. *International Journal of Language & Communication Disorders*, 44(4): 529–548.

SELF-DIRECTED LEARNING

1. The work undertaken by the clinical engineer in this project followed a systems engineering approach. Can you identify which parts of the narrative related to each step of the INCOSE 'SIMILAR' guide for such projects (Chapter 2, Section 2.2)?

CASE STUDY CS7.18: CLINICAL ENGINEERING SUPPORT AT THE POINT OF CARE – HOME DIALYSIS

Section Links: Chapter 7, Section 7.4.1

ABSTRACT

Clinical engineering provides support to the Dialysis Unit in a major tertiary level teaching hospital. This includes support for patients using home dialysis. The support is primarily technical but also includes training patients in the use of their home equipment. This requires a good understanding of the clinical aspects of renal function and renal dialysis.

Keywords: Renal dialysis; Home healthcare; Equipment management

NARRATIVE

The Dialysis Technical Services (DTS) team within the Clinical Engineering Department provides support to the Renal Dialysis Unit in a tertiary level teaching hospital. This Unit serves a fairly large, mixed urban and rural, geographic area with a population of about a million people. The Unit makes use of hospital-based dialysis equipment, equipment in a satellite unit and home dialysis machines for those patients for whom an installation at home has been deemed to be suitable and appropriate.

Support for home dialysis presents additional challenges:

- There is direct contact and interaction with patients in their homes.
- The equipment is spread over a wide geographical area.
- The equipment is life supporting, though not instantaneously so.
- The equipment is complex, requires electrical and plumbing installation and scheduled preventive maintenance.

Direct personal contact with the same patients over extended periods of time is not something that most clinical engineers experience. Therefore, some additional training and guidance for the DTS staff is important. Issues such as confidentiality, tact, personal manner and dealing with distressing circumstances need to be considered.

The DTS team had been involved with the clinical staff in the selection of the equipment and have had all the necessary technical training to enable them to service the equipment. Before the final clinical decision is made to provide a home dialysis facility for a particular patient, a senior DTS clinical engineer will liaise with the home dialysis coordinator and carry out a site survey to establish whether the home is physically suitable for a dialysis installation. Such aspects as size and layout, electricity supply and incoming water pressure must be assessed. If an installation is to go ahead, DTS will draw up a detailed plan, taking the patient's views into account and liaise with the specialist installation contractor. The work of the contractor, once authorized, will be monitored by DTS and finally signed off by the team leader. The DTS team carry out the installation and commissioning of the medical equipment.

Through in-house seminars and external courses, DTS staff have a good understanding of the clinical aspects of renal dialysis and have worked with the dialysis nurses to develop and help deliver training courses for patients who are to dialyze at home. This includes ensuring that the patient has a clear understanding of the normal functioning of the machine and the meaning of, and response to, any alarm signals that might come up. The training is first done in the hospital unit and then reinforced in the patient's home just prior to the first home dialysis. This first home dialysis is scheduled so that rapid technical help is available should it be necessary, and in all cases a renal nurse will be in attendance.

Scheduled services of the equipment are managed through the MEMS database. The relevant DTS staff have available estate cars stocked with the necessary spares and test equipment, one of which is a 4 × 4 to cope with emergency rural call-outs in bad winter weather. A Key Performance Indicator (KPI) based around the completion of scheduled services within an agreed time window is kept under review.

Home patients have a contact number for the DTS team, one for normal hours and one for out of hours. All calls are first discussed with the patient and an appropriate response for non-scheduled service agreed. Because the exact timing of a dialysis session is not immediately life-threatening for a patient, the response to calls can be prioritized and planned to efficiently use resources, but always with the agreement of the patient. These responses are also monitored through the use of suitable KPIs.

ADDING VALUE

Home dialysis is not suitable for all patients. It requires considerable input from the patient and their carer but in many cases enables a more normal life to be continued. The health economics of hospital versus satellite versus home dialysis is under constant review, but where home dialysis is provided for some patients, well-structured clinical engineering input brings benefit to those patients at some additional cost. Clinical engineering support for home dialysis patients adds value.

Benefits : Cost ∴ Value

SYSTEMS APPROACH

The service described developed over a number of years, but the process involving all stakeholders with clear goals and well-planned steps along the way is a good example of a systems approach.

PATIENT CENTRED

Patients have to be fully engaged if home dialysis is to be successful. A good working relationship between them and the clinical engineering staff is vital for efficient and mutually acceptable performance.

CULTURE AND ETHICS

Close working relationships with patients brings with it a need for a good understanding of ethical and confidentiality issues and an appropriate personal approach.

SUMMARY

Providing technical and support services to patients using complex medical equipment at home has additional challenges, not least the close working relationship with the patients, which has to be managed.

SELF-DIRECTED LEARNING

1. Devise some appropriate KPIs to monitor performance in regard to handling and responding to calls for non-scheduled service.
2. What are the implications of the wide geographical area noted previously?
3. What is meant by 'satellite dialysis'?

CASE STUDY CS7.19: PROACTIVE MANAGEMENT OF MEDICAL EQUIPMENT ALARM SIGNALS

Section Links: Chapter 7, Section 7.4.2

ABSTRACT

Medical equipment alarm signals alert caregivers to patient and equipment conditions requiring attention. However, the plethora of alarms experienced by users can lead to alarm fatigue on the one hand and the risk of missing vital alarms on the other hand. This can have serious healthcare repercussions. Urgent calls for careful systematic alarm management were made by a hospital Board.

Keywords: Medical equipment alarm signals; Teamwork; Alarm systems; Visual and auditory alarms; Medical device configuration

NARRATIVE

A non-executive Board member asked the Chief Executive Officer (CEO) to report on the hospital's procedures for managing medical equipment alarms, noting a serious failure of an arrhythmia alarm in the coronary care unit (CCU) and the regulatory agency's call for alarm management improvements. The CEO, backed by the Medical and Nurse Directors, asked the Medical Device Committee (MDC) to form a multidisciplinary project team to investigate.

The MDC recognized the problem's enormous scope, welcoming the opportunity to tackle it methodically with the Board's authority. Alarm fatigue was a real problem with too many alarms triggered and complaints from patients and visitors of staff frequently ignoring alarms. The MDC debated the objective, concluding that medical equipment should be configured with evidence-based default alarm settings, operated by staff who are competent and knowledgeable in clinical alarms, supported by policies and procedures appropriate to the different specialities using medical equipment. Some important groundwork was in place, particularly the medical equipment configuration programme initiated by the Clinical Engineering Department (CED). This programme was standardizing and recording the configurations and software status of all the organization's medical equipment.

A core project team was formed, co-led by the Clinical Engineering Chief and a senior anaesthetist, with three senior nurses (critical care, theatres, general wards), a clinical engineer, a Risk Manager and a patient representative. The project team reviewed the recommendations of The Joint Commission (2013) and the literature and agreed the need for:

- Heightened awareness of the importance of medical equipment alarm signals;
- Alarms that are set to alert, inform and guide action;
- Control of alarm default settings, authorized with records kept;
- Training clinical staff in managing alarms;
- Training in selecting the appropriate alarm settings for a particular patient's condition;
- Authorization for changing alarm settings and for silencing alarms;
- Clinical engineers including alarm setting configuration and checks in their equipment support plans;
- Regular review and audit, including lessons to be learnt from adverse events in which deficiencies in alarm management were identified;
- Listening to and surveying the opinions of clinicians, patients and visitors.

The diversity of the medical equipment suggested no single implementation plan but an over-arching policy supplemented by speciality-dependent policies and procedures. The core team would co-opt speciality groups to investigate and propose solutions for their own areas – for example maternity and neonatal care. A systems engineering approach was adopted with a proposed solution based around the elements of agreed hospital-wide policies and procedures linked to and supporting speciality specific policies and procedures. The process would start by gathering information from clinicians, patients and visitors, looking at them in relation to patient care from the patient's bedside to develop hospital-wide and department-specific solutions (Figure CS7.19A).

Figure CS7.19A shows relationships between hospital-wide and department-specific processes, each helping to inform the other. Priorities had to be set: general principles that would inform hospital-wide policies and procedures should be agreed, their development guided by work in priority departments. The hospital's chief doctors and others were canvassed for identifying the priority areas for alarm management. CCU was an obvious candidate with the patient's representative pointing to the distressing effect on patients and visitors of the 'continual sound of alarms' in neonatology. The nurses asked that a group of general medical wards be included, with particular concern about alarms in private rooms and from telemetry monitors. The anaesthetists offered to investigate surgical critical care areas and theatres.

The processes of developing each department's recommendations were agreed with one or two members of the core project team working with a team from each department

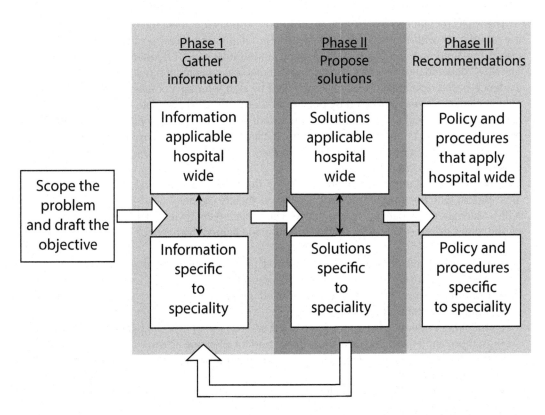

FIGURE CS7.19A Processes for developing the hospital-wide management of medical equipment alarm systems. Phases 1 and 2 iteratively refined the information required to develop solutions.

nominated by and working under the authority of its chief physician. The general approach was agreed:

- Identify all medical equipment with alarms, linking to the configuration standardization initiative and using the inventory database known as the Medical Equipment Management System.
- Canvass clinicians, patients and visitors for ideas on how to manage alarms.
- Taking note of human factors that influence setting and responding to alarms, recognizing that they may vary between disciplines.
- Analyze and use clinical and technical evidence to inform alarm defaults.
- Agreed settings to be signed off by the appropriate chief doctor.
- Develop training programmes.
- Audit to check continuing progress and clinicians' knowledge and understanding.
- Report regularly to the Board via the MDC.

The project team found areas of good practice, but areas where little attention was paid to medical equipment alarms and where the default approach was to silence the alarms, with sometimes only a cursory check as to what triggered the alarms. Of particular concern was the almost tacit acceptance in certain clinical areas that alarms, mainly nuisance alarms, were a fact of life to be tolerated. The team drew on good practice reported from the literature and in nationwide meetings on alarm management that showed that appropriate alarm settings and correct medical equipment operation could dramatically reduce the alarm frequencies. The alarm management initiative was universally welcomed by staff as was the recognition of the need for training. Recommendations were also made about assessing alarms usability when evaluating medical equipment prior to procurement.

Particular attention was paid to the views of patients and visitors on the alarm. Patients and visitors spoke of high levels of alarm noise in critical care areas, preventing rest, often making sleep impossible and hindering recovery. It was disconcerting to see that staff appeared to ignore many of the alarms, but reassuring to see them rushing to attend to other alarms. 'Why, when some alarms do not require any action, are the alarms allowed to happen?' said one patient in a critical care ward; 'I can't get any rest because of the continual noise'. A parent spending long periods of time with his son in a neonatal intensive care unit spoke of the disheartening effect on already anxious nerves of the continual buzz of alarms. 'My baby I know is not well, but all these alarms, seemingly continuous and many seemingly ignored, makes me more worried that my baby is not doing well'.

Patients' comments were found to be particularly useful for raising awareness, both at CEO level and in clinical areas, of the importance of appropriate management of medical equipment alarms.

ADDING VALUE

Initially the effort involved in setting up the processes for managing the alarms added costs, with dedicated staff time required to investigate and develop solutions. However, the benefits outweighed the costs which were anticipated to decrease as the policies and procedures were agreed across the organization and embedded in routine practice. The process added value to alarm management.

Benefits : Cost ∴ Value

SUMMARY

Supported by the hospital's top management, clinical engineers worked cooperatively with clinicians to develop policies and procedures for managing the configuration of medical equipment alarms and the response to alarms. The MDC was able to report to the Board on the improving position, with the Board calling for 6-monthly reports on medical equipment alarm managements, problems and successes. The MDC reported on a system that was based on awareness of the importance of alarms, competent staff, evidence-based alarm settings, control and authorization.

REFERENCE

The Joint Commission. 2013. Sentinel Event Alert Issue 50: Medical device alarm safety in hospitals. https://www.jointcommission.org/sea_issue_50 (accessed 2016-05-04).

SELF-DIRECTED LEARNING

1. Investigate the medical equipment alarms in a particular area of your organization. Review the settings and the documentation, including authorizations for the settings. Are the settings still appropriate? Canvas the views of clinicians. Visit the areas and observe the practical experiences of alarms. Review any adverse events associated with alarms. Document and reflect your findings suggesting, if appropriate, improvements that could be made.
2. Review the policies and procedures for determining the default alarm configurations for new medical equipment. Are the policies and procedures complied with? Suggest improvements if appropriate.
3. Visit a ward or department and put yourself in the place of a patient or visitor. What can you hear and see relating to alarms? How do patients, visitors and clinicians seem to be reacting to them?

CASE STUDY CS7.20: DEVELOPING AN AID FOR THE DELIVERY OF PHOTODYNAMIC THERAPY AND ASSESSING THE RISKS

Section Links: Chapter 7, Sections 7.4.3, 7.3.3 and 7.4.4

ABSTRACT

The delivery of photodynamic therapy (PDT) used in the treatment of cancerous cells involved clinical engineers in the design of a surgical aid to reduce the physical demands on clinical staff. An analysis of the risks involved was undertaken.

Keywords: Design; Surgical aid; Risk assessment; Clinical use

NARRATIVE

Photodynamic therapy (PDT) involves the administration of light-sensitive drugs which are later activated by exposure to a particular wavelength of light to initiate therapeutic activity. PDT has been shown to be effective in treating cancerous tumours (Dougherty et al. 1998). The light-sensitive drug is administered to the patient after which the patient has to remain for several days in almost absolute darkness to avoid triggering the drug as it is transported through the body by the blood circulation. Laser light is then shone into the cancerous tumour, activating the drug in that area. After the activation, the patient must then again avoid exposure to light till the remaining drug in the body is metabolized.

Clinical engineers were responsible for setting up the laser generator and its associated fibre optic cables in a darkened room designated for this purpose. The treatment required a calculation to determine the optimum light power output and the time over which it would activate the drug. The light had to be delivered a set distance from the treatment site in order to deliver the correct dose and care was required to ensure that no unaffected tissue was exposed. The surgeon calculated the required dose from knowledge of the cancerous tumour, and the clinical engineer was asked to time the procedure.

The clinical engineer observed that it was difficult for surgeons to keep their hands still during the required long exposure period, typically several minutes. The surgeon's hand holding the laser fibre to direct the laser light was observed to move and the surgeon had to stop, allowing his arm and hand to recover, before proceeding. It was also noted that the patient, who was only given a local anaesthetic, sometimes moved due to inherent discomfort. The net consequence was exposure of non-malignant tissue and the intended malignant tissue not receiving the prescribed calculated light exposure.

Clinical engineers investigated the problem and developed a stand with an articulated arm that could gently hold the fibre and keep it in place during the exposure. This was positioned and monitored by the surgeon, but it reduced the stress on the surgeon's arm and hand, minimizing the resultant sway and misdirection of the laser beam. The surgeon had to continue to monitor the position carefully as the patient could also move, but freed from having to hold the laser fibre, the surgeon could concentrate better on the patient, keeping the laser light directed to the malignant tissue.

Clinical engineering and the surgeon both undertook a risk assessment of this adaptation and found that risks had changed when using the stand. It had:

- Reduced risk of shake by surgeon leading to improved accuracy of laser delivery and faster delivery due to removing the need for breaks;
- Increased risk of miss-delivery if the surgeon does not pay full attention to the treatment site and does not compensate for any patient movement.

These risks were recorded and discussed with the Medical Device Committee who accepted the risk assessment.

ADDING VALUE

Using experience gained from other areas, the clinical engineer was able to create a tool that gave immediate benefit at a very low cost. The risks were assessed and found to be acceptable, improving the outcome for the patient and supporting the surgeon's administration of the therapy.

Benefits : Cost ∴ Value

SYSTEMS APPROACH

Using the knowledge that treatment time, power output, distance and treatment area were so important in the delivery of a successful treatment, clinical engineers devised a mechanical aid that minimized variability.

PATIENT CENTRED

The patient's comfort was improved through decreased treatment time, and the improvements in accuracy enabled the patient to get the best treatment possible.

SUMMARY

A clinical situation that was noticed by the clinical engineers was able to be improved by the creation of a mechanical aid to assist laser delivery. By working together, the clinical and technical staff reduced patient treatment time and improved the risk score of the procedure.

REFERENCE

Dougherty T.J., Comer C.J., Henderson B.W. et al. (1988). Photodynamic therapy. *Journal of the National Cancer Institute*, 90(12): 889–905.

SELF-DIRECTED LEARNING

1. How would you feel about raising a problem that you have observed with a senior member of clinical staff? What steps would you take and why?
2. What do you think are the advantages and disadvantages of involving different professions in writing a risk assessment?
3. Considering the close working of clinical engineering and medical staff in the case study, can you think of any other areas in a hospital, perhaps in a specific department, where close working such as this is done, or should be?

CASE STUDY CS7.21: THE 'RULES' FOR PUTTING TOGETHER A MEDICAL ELECTRICAL SYSTEM (MES)

Section Links: Chapter 7, Section 7.4.4

ABSTRACT

The Clinical Engineering Department (CED) was asked for advice on converting a consultation/examination room into an endoscopy room. They proposed assembling onto an endoscopy trolley all the necessary equipment, medical and non-medical, all powered from a multiple socket outlet. They were authorized to design and construct the appropriate 'medical electrical system' (MES; see the IEC 60601-1 Standard, subclause 3.64). Their understanding of the wider implications of managing endoscopy equipment led them to propose a further room conversion so that cleaning facilities were close to this new clinical endoscopy service. The combination of technical knowledge of how to put an MES together, outlined in a set of 'rules' in the Annex to this case study, and an ability to analyze the endoscope usage allowed CED to support the development of a new effective service.

Keywords: Endoscopes; Medical electrical systems; Patient environment; Leakage current; Touch current; Single-fault conditions; Multiple socket outlet

NARRATIVE

In the hospital's Ear Nose and Throat (ENT) department, outpatients who needed endoscopic examination were admitted to the Day Surgery unit. However, increasing demand was putting strain on the service and delays for patients.

The ENT's clinical director suggested to the hospital's Medical Director and the Director of Planning that an ENT endoscopy facility should be established in one of the examination rooms in the ENT outpatients department (OPD). This would ease access for patients, improve patient flow management and allow for better scheduling in Day Surgery.

The examination room was available but had not been designed as an invasive patient investigation area. It had very few mains power sockets and, whilst its size was acceptable, the endoscopy equipment would be close to the patient, within the 'patient environment' requiring equipment to comply with the strict medical equipment electrical safety Standards. The CED was asked to investigate whether the room could be configured to support the proposed ENT endoscopy outpatient service.

The ENT endoscopy equipment consisted of light source, video processor, video monitor, video printer, PC and suction unit, all requiring mains electrical power. All but the suction unit were required to be functionally interconnected, and all were required to be in the same room and close to the patient.

The CED investigated and advised that the equipment could all be mounted on a suitable trolley, powered from a single mains socket via a multiple socket outlet also mounted on the trolley. They were authorized to go ahead with this project. They used their knowledge of electrical safety Standards and of 'rules' derived from them (set out in the Annex to this case study) to design, construct and test a 'medical electrical system' (MES) built onto a trolley suitable and safe for use in the 'patient environment'. The system was successfully built and documented for future maintenance and perhaps development. The equipment support plan was developed.

Their knowledge of not only the technology but also its application highlighted a secondary problem. Reviewing the proposed outpatients service from a systems perspective, they identified the need for a local ENT endoscope cleaning facility. Without this cleaning facility, unnecessary delays would be introduced, caused by the time required to transport ENT endoscopes back to the Day Surgery Unit for cleaning. This would undermine the hoped for improvement in patient flow. Working with Infection Control, the clinical engineers suggested converting a nearby unused pharmacy preparation room into a scope cleaning and disinfection room. This would enable achievement of the desired increase in patient throughput with the purchase of only two more nasoendoscopes.

ADDING VALUE

CED's input resulted in a successful development whilst controlling and minimizing the increased costs. Whilst the costs increased, the benefits increased proportionately more, increasing value to the patients; benefits were less stressful patient flow through the system and reduced waiting time for diagnosis.

CEDs are often asked or may suggest assembling several items of equipment into an MES. Doing so will add value by improving safety or usability or both but must be done in such a way as to be compliant with electrical safety Standards.

Benefits : Cost ∴ Value

SYSTEMS APPROACH

Clinical engineering took a systems approach to developing a solution, focusing on the objective of improving patient flow. They used their electrical safety knowledge to assemble the endoscopy equipment safely onto a single trolley, appropriately designed to be used in the patient room. They also used their knowledge of endoscopy workflow to suggest and plan the development of an endoscopy cleaning and disinfection room to ensure that endoscopy processing did not obstruct the improved patient throughput.

PATIENT CENTRED

The CED's objective focused on improving patient care. They achieved a solution that was safe for patients, whilst considering ease of use for staff, with consequent patient benefits, and that secured the improvement in patient flow resulting in reduced delays to diagnosis.

SUMMARY

The ENT surgeon commented in a later report to the Medical Director: "CED have provided us with a system that is easy to use and enables us to keep up with the increased patient demand. They also managed to persuade Finance to fund two more nasoendoscopes".

ANNEX: 'RULES' FOR COMBINING ELECTRICAL EQUIPMENT INTO A MEDICAL ELECTRICAL SYSTEM (MES)

This annex deals with some of the technical factors to be considered when combining several items of medical electrical equipment with one or more items of non-medical equipment (e.g. computer or printer), all powered from a plugged-in multiple socket mains connection device defined in the IEC 60601-1 Standard as a 'multiple socket outlet' or MSO.

The 'patient environment' concept is important to understand: only equipment whose normal and single-fault Touch Currents are within the allowed medical equipment limits can be situated within the patient environment. The IEC 60601-1 Standard defines the patient environment as 'any volume in which intentional or unintentional contact can occur between a patient and parts of the medical electrical equipment or medical electrical system, or between a patient and other persons touching parts of the medical electrical equipment or medical electrical system'.

Therefore, six rules for developing the MES are suggested:

Rule 1: Any complete MES must have Touch Current from any part not exceeding 100 μA in normal conditions and not exceeding 500 μA in single-fault conditions (e.g. with the protective earth connection broken).

Issues to Consider

1. If the protective earth conductor is broken, normal earth leakage current (ELC) becomes single-fault Touch Current from parts of equipment that would normally be protectively earthed (see Figure A.2 in IEC 62353:2014).
2. When multiple items of equipment are plugged into an MSO, the ELC from all the equipment adds together in the protective earth conductor of the MSO power cord. The combined ELC possibly exceeds the 500 μA limit for plugged-in equipment with accessible protectively earthed parts. Therefore, the single-fault Touch Current would be exceeded. This can occur even if all the equipment is medical electrical equipment. If even one MES item is non-medical, the combined ELC is very likely to exceed 500 μA; non-medical equipment has an allowable ELC of up to 3.5 mA.

Solution: Steps must be taken to reduce the ELC of the MES. The simplest and most effective design is to supply some or all of the equipment via a separating or an isolation transformer. A suitable isolating transformer was used in this case.

Rule 2: It should not be possible for other equipment to be easily plugged into the MSO.

Issues to Consider

1. Plugging additional or different equipment into the MSO changes the configuration of the MES, altering its safe design specification.
   Subclause 16.9.2.1a) of IEC 60601-1 says that the sockets of the MSO must either be accessible only with the use of a tool (e.g. see IEC 60601-1 Annex I Figure I.1) or must be of a type not referenced in IEC/TR 60083.

Solution: MSOs made up of appliance outlets of the C13/C14 type to IEC 60320-3 are acceptable.

Rule 3: The protective earth resistance to the mains plug earth (ground) pin from any exposed protectively earthed part of the MES (including the trolley) must not exceed 0.2 Ω when new. (0.3 Ω is allowed in service; see IEC62354:2014 2nd Edition).

Issues to Consider

1. The protective earth pathway to the mains plug's earth pin from each earthed equipment included in the MES includes the resistance of its mains cord plus the resistance of the system mains cord from the MSO.

Solution: It may be necessary to use short power cords within the MES and use a system power cord of greater cross-sectional area than is necessary for current-carrying purposes in order to reduce its resistance.

Rule 4: Never directly electrically connect a patient to non-medical equipment.

Issues to Consider

1. If a patient were connected to non-medical equipment, patient leakage current and the separation of the patient from live voltages would not meet the safety requirements set out in IEC 60601-1, increasing the risk to the patient.

Rule 5: Any electrical connection to a patient must be to medical electrical equipment with a Type BF or Type CF Applied Part.

Issues to Consider

1. The 2nd Edition of IEC 60601-1 allowed an electrical connection to a patient to be made via a Type B (earth referenced) Applied Part, but the 3rd Edition stipulates that such connections must be F-Type (Floating).
2. Older medical electrical equipment may still be in use with Type B electrical patient connections, but it is good practice not to use such equipment in an MES.

Rule 6: Document carefully the design of the MES including the details of its constituent equipment. Document the electrical safety test results of each individual item, including any non-medical equipment, and of the MES as a whole.

Issues to Consider

1. Putting together, an MES is an example of in-house manufacture of medical equipment even when all the individual equipment is commercially sourced.
2. Documentation, including test results, is important for being able to demonstrate that the design complies with the requirements of the various Standards.
3. The documentation may be a regulatory requirement in your jurisdiction.

STANDARDS CITED

NOTE: Undated references are given. Always refer to the most up-to-date version.

IEC 60320-3. Appliance couplers for household and similar general purposes – Part 3: Standard sheets and gauges

IEC 60601-1. Medical electrical equipment – Part 1: General requirements for basic safety and essential performance.

IEC 62353. Medical electrical equipment – Recurrent test and test after repair of medical electrical equipment.

IEC/TR 60083. Plugs and socket outlets for domestic and similar general use standardized in member countries of IEC.

SELF-DIRECTED LEARNING

1. Consider the definition of 'the patient environment'. Would you consider a computer mounted in a fixed position in the corner of an operating room to be within the patient environment? Whether Yes or No, document your reasoning.
2. Consider and explain how powering equipment from a separating or an isolation transformer reduces the ELC.
3. Look up IEC 60601-1 Annex A subclause 16.9.2.1d) and be clear about the difference between a separating and an isolation transformer.
4. Look up IEC 60601-1 and be clear about the meaning of 'earth leakage current', 'touch current', 'patient leakage current', 'protective earth', 'Type B', 'Type BF' and 'Type CF'.

CASE STUDY CS7.22: ENSURING THE ELECTRICAL SAFETY OF MEDICAL ELECTRICAL EQUIPMENT SYSTEMS BASED ON RELEVANT STANDARDS

Section Links: Chapter 7, Section 7.4.4; Chapter 3, Section 3.5.4; Chapter 6, Section 6.2

ABSTRACT

This case study discusses the requirements for testing a new item of surgical equipment to ensure the electrical safety of its applied part.

Keywords: Electrical; Safety; Tests; Maintenance; Standards

NARRATIVE

A phacoemulsifier is a specialized item of medical equipment designed to remove a lens from the eye by using ultrasonic aspiration. The machine was inspected upon arrival and found to have many patient applied parts, footswitches and fluid accessories and is connected to mains electricity. In many ways, it is the clinical engineers' perfect storm of concern: mains electricity, fluids and intimate patient contact.

Part of its acceptance testing will be the creation of an equipment support plan (ESP) (Chapter 6, Section 6.2) that will define its required testing regime. It is likely that the Class of the equipment and Type of its applied parts and the nature of its operational use will require it to have a high level of maintenance, including electrical safety checks. The manufacturer recommended annual maintenance services, to include an electrical safety test.

The equipment was declared by the manufacturer to be made to the requirements of the IEC 60601-1 General Standard and the IEC 80601-2-58 Particular Standard (see Chapter 3, Section 3.5.4 for an explanation of the General Standard and Particular Standards). These are type test product standards and do not include any specific requirements for acceptance testing.

The IEC 62353:2014 Standard (2nd Edition Medical electrical equipment – Recurrent test and test after repair of medical electrical equipment) provides guidance on the tests to be carried out. This 2nd edition explicitly includes in its subclause 5.3.4.1 and Annex A.3 the individual measurements of earth, touch and patient leakage currents, based on the test configurations in the IEC 60601-1 Standard, as an acceptable method of electrical safety testing. Another useful guidance document is published by the Institute of Physics and Engineering in Medicine as Report 97 (Wentworth 2009).

It is common for acceptance test procedures to include recording the electrical safety results, storing them as the baseline measurements of protective earth resistance, insulation resistance, earth leakage current, touch current and, where applicable, patient applied part leakage currents.

Measurements of the applied part leakage currents in this case presented the first hurdle as the patient applied parts, in common with other surgical equipment, are sterile. One cannot carry out a patient connection test without a patient connection! This was resolved by obtaining a set of patient applied parts from theatre and keeping them in the Clinical Engineering Department (CED) for use as required.

The next hurdle involves the fluids. How and where do these flow and what do they come into contact with? The best advice when safety testing medical equipment is to set it up as if it were just about to be used on the patient. Only by doing this will the test be relevant. Testing the phacoemulsifier requires priming the sets with water or saline and attaching the aspiration probe. Connections for testing the applied part were made to the metal probe through which the fluid flows. However, in this as for certain other medical devices, direct electrical connections to the applied part may not be immediately available; achieving them requires creativity from the clinical engineer. For example, it is possible to measure the touch and patient leakage currents at a pulse oximeter probe by wrapping aluminium foil around it.

Finally, the accessories were investigated – in this case a foot pedal electrically connected through a cable. Details of the cable connections were not included in the supplier's documentation, and it was not clear whether they included an earth/ground, perhaps to the metal casing of the foot pedal. It is important to feel able to contact manufacturers or suppliers for specific advice when unsure; clinical engineers are not expected to be experts in all aspects of all the equipment they maintain.

However, ensuring the electrical safety of medical equipment goes much further than an acceptance test. Regular testing is required by many regulatory agencies. Good notes made at the time of the acceptance testing can be of real benefit later in the machine's life cycle as tips and information can be passed from engineer to engineer to enable efficient testing in the future. Records kept can be referred back to in case of queries surrounding future test results.

IEC 62353 requires that test results are recorded (subclause 6.1). In particular, acceptance tests are very useful because they give a set of 'as-new' reference results against which to compare subsequent results. Test result values that have increased significantly from the reference values should trigger further investigation. Test result values that are significantly less should raise questions about the test set up, for example significantly lower earth leakage current values would suggest an alternative earth pathway, perhaps via a network connection.

It may be sufficient to simply record a pass on a routine test if results are within say 10% of the reference values, but test results after repair should be recorded. If the test is borderline or failing, then a more in-depth look should be made. Some electrical safety testing equipment is now able to connect with electronic record-keeping, so results can be stored with the minimum of fuss and managed by software.

Electrical safety testing is carried out to minimize the risk of electric current injury. The danger from what may appear as apparently low-risk devices was highlighted recently (EFA 2015) when a doctor and patient suffered severe injury from a damaged plug-in power supply unit supplying an auroscope. Incidents like these re-enforce the need to manage the risks associated with medical electrical equipment.

ADDING VALUE

There is a cost to carrying out electrical safety tests, but these can be combined in most cases with the equipment's routine planned maintenance. Regulatory agencies are keen to seek assurance that this testing regime is carried out, providing reassurance that patients, staff and visitors are all safe from electrical hazards.

Benefits : Cost ∴ Value

SUMMARY

We have seen how complex medical equipment is acceptance tested from an electrical safety point of view. Difficulties in testing in the field can be overcome with creativity and understanding. Regular testing is a bread and butter activity for CEDs.

REFERENCES

EFA. 2015. Estates and facilities alert EFA/2015/015. Issued 16 November 2015. UK Department of Health. https://www.cas.dh.gov.uk/ViewandAcknowledgment/ViewAlert.aspx?AlertID=102395 (accessed 2016-05-04).

Wentworth S.D. (Ed.). 2009. IPEM Report 97: Guide to electrical safety testing of medical equipment: The why and the how. York, UK: Institute of Physics and Engineering in Medicine.

STANDARDS CITED

NOTE: Undated references are given.

Always refer to the most up-to-date version. These can be found by searching the IEC or ISO website under the respective store tabs.

IEC 60601-1 Medical electrical equipment – Part 1: General requirements for basic safety and essential performance

IEC 62353 Medical electrical equipment – Recurrent test and test after repair of medical electrical equipment

IEC 80601-2-58 Medical electrical equipment – Part 2-58: Particular requirements for the basic safety and essential performance of lens removal devices and vitrectomy devices for ophthalmic surgery

SELF-DIRECTED LEARNING

1. Taking two examples of medical equipment, a phacoemulsifier and an auroscope, both mains connected, how do you think their testing regimes may differ? Think about handling, portability and training for the users of the equipment. Does this make a difference?

2. How can the clinical engineer cope with the possibility of damaged equipment causing harm? What other colleagues may be able to assist in preventing damaged equipment from being used clinically?

CASE STUDY CS7.23: THE ROLE OF THE CONSULTANT CLINICAL ENGINEER IN SERVICE REFORM

Section Links: Chapter 7, Sections 7.2.8, 7.3.1, 7.4.9 and 7.6

ABSTRACT

Consultant clinical engineers can have an important role in ensuring that their service continuously evolves to meet changing organizational needs. The organization's strategic direction in turn is usually dictated by demands for the provision of healthcare, economic factors and external health policy considerations. This case study shows a consultant clinical engineer proactively developing the healthcare technology management (HTM) service so that it is optimized to support the organization and directly to support patient care.

Keywords: Business planning; Service redesign; Service optimization; Workforce planning

NARRATIVE

The descriptor 'consultant' in the term *consultant clinical engineers* describes those clinical engineers who are leaders in their field and experts in the discipline of clinical engineering and who have the knowledge, skills and experience to be able to work independently and to be consulted for advice about the practice of clinical engineering. Their knowledge and skills enable consultant clinical engineers to take responsibility for the delivery of clinical engineering services, for leading, managing and directing the science and technology that underpins the application of healthcare technologies. Consultant clinical engineers are leaders who proactively lead and manage medical equipment to support and advance care, delivering positive changes and improvements rather than passively managing the medical equipment. The word consultant is used to describe senior clinical engineers practising as leaders and who can lead significant organizational change by influencing what technologies to adopt for the healthcare organization and also the range of services offered by the Clinical Engineering Department (CED).

This case study focuses on a consultant clinical engineer led HTM service in a 500-bed district general hospital, managing over 15,000 medical devices valued at over £60 million. The hospital faced financial difficulties and adverse criticism from regulators, particularly relating to aspects of its patient care. Additionally the national regulatory authorities had published a 5-year vision stressing increased care in the community. In response, the hospital's Board embarked on an improvement strategy in which all departments were challenged to improve clinical services whilst reducing costs. How could the HTM service respond to and support the improvements?

The CED service provided medical device life cycle management and a range of professional advisory services, such as handling safety alerts, advising on equipment replacement and medical device adverse event investigations. The consultant clinical engineer leading the service proactively reviewed the existing services in the context of the challenges faced, analyzing the resources available, in particular the workforce knowledge and skills. The hospital's economic, clinical and future needs were analyzed and contextualized to clinical engineering. This capacity and capability review was undertaken with a patient perspective in mind – how could clinical engineering benefit patients through the optimal application of healthcare technologies? This review resulted in a future vision for clinical engineering serving the organization:

- Continue striving for excellence in the life cycle management and risk minimization of medical equipment.
- Continue to evolve the CED to offer comprehensive HTM services, incorporating a range of clinical support services.
- Investigate healthcare technology solutions to minimize hospital admissions and speed up patient discharges.
- Further develop services to support home healthcare.

Essentially the vision was for a clinical engineering service with a continuing quality improvement ethos and a twin remit: first, recognizing and strengthening its core HTM activity, and second, being outward looking to the needs of the healthcare organization and using its knowledge and skills to support and advance clinical services (c.f. Figure 1.8, Chapter 1). The vision for the later part of this twin remit was to deliver more clinical support services and become active in the management of personalized technologies for patients. Providing patient-focused services was to be at the core of both service remits.

In terms of personalized technologies, the vision was to provide rehabilitation engineering services to which previously the local population had little access; this aligned with the wider hospital's strategic planning to improve rehabilitation services. Further community outreach came from the request from physicians to take over the issue of nebulizers to patients in the community from a failing private company. This would further develop the home care services already provided by the CED. Locally, within the hospital, the CED saw a real opportunity of meaningfully supporting hospital-wide clinical services by leading, in cooperation with the resuscitation team, an overhaul of the management of its resuscitation equipment.

The CED workforce review identified that many of the envisaged changes could be accommodated by extending the roles of existing technical staff, with appropriate training support, together with developing the role of support workers. However, it highlighted that rehabilitation engineering expertise was needed. A workforce proposal was developed and support obtained from a national training programme initiative to take on a young clinical engineer who had successfully applied for a senior leadership training programme. This training programme was a national initiative to enhance rehabilitation service capabilities; it complemented the department's vision of providing rehabilitation engineering services to the local community.

These proposals grew out of discussions with hospital executives and senior clinicians and from organization-wide quality improvement planning meetings. Clinical engineering was proactive in engaging and contributing to forums and taking stock of how, where and when they could contribute to organizational improvement. A timeline was drawn up showing what initiatives were achievable in the short term within available resources and those with a longer delivery time frame. From this planning, the department embarked on a journey to evolve its service provision, with priority targets as follows:

- Improved partnership working with hospital users in medical equipment management;
- Procurement of new adult resuscitation trolleys with modern cardiac defibrillators incorporating the latest biphasic waveforms and support for cardiopulmonary resuscitation in line with cardiac resuscitation guidelines;
- Managing the stocking of the resuscitation trolleys in cooperation with Pharmacy;
- Managing the local community care nebulizer service;
- Developing a local rehabilitation service, with engagement events with patients, health professionals, local and national clinical engineering experts to better understand the needs and to scope potential opportunities to improve local rehabilitation services for patients.

The proactive involvements described in this case study illustrate how the consultant practitioner can lead the development of a holistic clinical engineering service. The core HTM services must be led and managed effectively, looking to where improvements can be made, recognizing the importance of an ethos of continuous quality improvement. But engaging with and satisfying the remit of supporting and advancing care (the other branch of the twin remit) requires that the leader of the CED has the vision, expertise and courage to develop the clinical engineering service to satisfy the wider needs of its healthcare organization and indeed beyond to the wider healthcare environment.

What attributes are required for a clinical engineer to lead, initiate and see through to implementation improvements in these twin remits of healthcare technology management? Leadership, certainly, will be required, leadership backed by expertise in the field of clinical engineering, with vision, understanding, innovation, scholarship, communication and persuasion, determination and the courage to drive through improvements, encouraging staff and stakeholders to implement the changes. These clinical engineers are visionaries who can develop new methods of working and have the courage and strength to implement them. As communicators, these leading clinical engineers will be able to listen to and understand the needs of patients, their fellow healthcare professionals and the healthcare organization and develop innovative practices to support them.

ADDING VALUE

Clinical engineers who are experts in clinical engineering have the ability of analyzing existing services in the context of healthcare organization's needs and aims. This enables these leading practitioners to take responsibility for developing and implementing service improvements benefiting the healthcare organization and the patients. The services of these expert practitioners, sometimes described as condultants, come at a price, but the benefits delivered outweigh the costs.

Benefits : Cost ∴ Value

PATIENT CENTRED

The expert leading clinical engineer has a comprehensive appreciation of how to lead, develop and manage clinical engineering services for the benefits of patients.

SUMMARY

An example of the role of a leading expert clinical engineer (described as a consultant) in proactively redesigning services for patient benefit has been outlined. Such advanced level practice is aimed at adding value to the healthcare endeavour through the optimal use of healthcare technologies for patient benefit.

SELF-DIRECTED LEARNING

1. What are the key components that make up the role of the advance senior clinical engineer that we have described as a consultant? Discuss the benefits this role can bring to a service, perhaps contrasting this with the roles of existing senior clinical engineers.
2. Develop a workforce plan for the next five years for your service, considering opportunities for new roles. Develop some examples of new or additional roles and develop new job descriptions to accompany them.

CASE STUDY CS7.24: MEASUREMENT ERRORS WITH VITAL SIGNS MONITORS

Section Links: Chapter 7, Section 7.4.10

ABSTRACT

Measurements of blood pressure, temperature and oxygen saturation are routinely made in all healthcare sectors, with medical equipment providing these functions available on the high street for home use. However, the measurements are increasingly indirect and accurate measurements require an understanding of the technology, the associated accessories and how the technology should be used.

Keywords: Medical equipment; Measurement errors; Clinical engineer role

NARRATIVE

A patient attended a community general practice surgery for a routine health check as part of a health improvement initiative. The check revealed a high blood pressure (BP); this was confirmed on multiple readings and the patient was prescribed medication for BP control. Follow-up monthly visits to the practice's nurse-led clinic found normal BP, but on the next visit to the physician the BP was again found to be high. The physician suspected possible white coat hypertension (anxiety induced high BP in presence of a physician) and referred the patient to the nurse clinic for confirmatory measurements – which found the BP normal.

That evening the physician reflected on the measurement and on other patients for whom he had recorded high BPs, several of which had led to anti-hypertension medication. Exploring the literature on measurement accuracy, articles on false high measurements attributed to incorrect cuff size jumped out at him. Back in his surgery he observed that he only had a small adult BP cuff, selected because most of his patients were frail and elderly; his nurses informed him they had a range of cuff sizes, selecting the appropriate cuff for the patient's arm circumference. He recalled that the patient whom he had diagnosed as hypertensive was obese with a large upper arm. BP measurements with too small a cuff, he had learnt from the literature, can lead to faulty hypertension diagnoses.

The physician worked regular sessions in the local hospital and on his next visit checked the availability of cuff sizes – not all departments had a range of cuff sizes, with several nurses unaware of the importance of cuff size. Raising it at the next physician meeting, he found general concern with other physicians asking about clinical thermometry, pointing to an article in a medical journal that discussed 'normal' measurements in a patient with an obvious fever (Vernon 2013). 'We rely on medical diagnostic technology with its ostensibly authoritative numeric display' echoed around the room. 'Perhaps those clinical engineers who introduce these technologies could help' and the physician offered to speak with them.

The clinical engineer explained that indirect technologies were the basis for most current vital signs measurement devices, pointing to a letter to a medical journal (Amoore et al. 2013). The problem is compounded by the apparent precision and authority of the digital displays on these devices; this can blind clinicians to the measurement, the scope for errors and the relationship between what is being measured and the measurement result. Typically these technologies incorporate proprietary algorithms that process the measured signals to deliver the result. The effect is that models of vital signs monitors from different suppliers may yield different values for the vital signs of BP, temperature and oxygen saturation. Furthermore, the measurement algorithm in certain devices can be altered in the device's configuration menus, with many unaware of the configuration options and the implications of changing the algorithm. For example, a thermometer can be configured to record either an oral, ear or direct equivalent temperature, with resultant temperature measurement differences of 0.2°C or more. Anecdotally it was reported that in one hospital two neighbouring wards, both serving the same medical speciality, had their thermometers configured in different ways.

The physician asked the clinical engineer's help in raising the need to ensure accurate vital signs measurements at the next Board meeting. Board members were concerned that measurements taken for granted could easily be erroneous and asked the Medical Device Committee (MDC) to investigate, adding this to their MDC Action Plan (Chapter 5, Section 5.4.1).

The MDC's Project Team reported back with a range of suggested improvements: careful selection of medical devices that evaluated accuracy and the effects of human usability on accuracy, consistent evidence-based configuration decisions, clinical staff training and maintenance support that looks holistically at the vital signs measurement equipment and their accessories. The Board welcomed the report, noting that it was directed at vital signs measurement devices, but asked if, once these suggestions had been implemented, the MDC could assess what lessons could be learnt from this for the management of other medical equipment.

ADDING VALUE

Clinical engineers, through their knowledge of routinely used medical technologies, can improve diagnostic accuracy, enhancing the quality of patient care at no additional cost. Clinical engineers should use this knowledge and insight when helping select medical equipment and when helping train clinical staff. By standardizing the medical equipment in a healthcare organization, and their configuration, clinical engineers can further contribute to improved diagnosis. The structured processes set up to manage medical equipment through the MDC helped reach a solution that was applied across the whole hospital.

The physician asked the clinical engineer if similar support could be provided to physicians' private practices, recognizing that this would incur costs. It was agreed that detailed discussions and planning would be required through the Physicians' County Group.

Benefits : Cost ∴ Value

SUMMARY

Diagnostic errors are often overlooked particularly those associated with routine vital signs measurements. It is important that the characteristics and limitations of medical equipment and their accessories are fully understood by clinical engineers who can use this knowledge to help ensure accurate measurements.

REFERENCES

Amoore J.N., Davie A. and D.H.T. Scott. 2013. Routine vital signs: What are we measuring? (letter) *British Medical Journal,* 2013;346:f1747.

Vernon G. 2013. Inaccuracy of forehead thermometers. *British Medical Journal,* 346: f1747. http://dx.doi.org/10.1136/bmj.f1747 (accessed 2016-05-04).

SELF-DIRECTED LEARNING

1. Survey the blood pressure cuffs available in general wards in your healthcare organization. Are a range of cuff sizes available? Do clinicians understand the importance of using the correct cuff size?

2. Does your healthcare organization have a standardization policy for its basic vital signs measurement equipment? If yes, how can standardization be maintained? If not, how can standardization be achieved?

3. Prepare a talk for your clinical engineering colleagues on vital signs monitoring, exploring the technology used in your healthcare organization with explanations on how the measurements were made and what could cause erroneous measurements. From this suggest methods for improving the accuracy of these measurements.

CHAPTER 8

The Clinical Engineering Department

Achieving the Vision

CONTENTS

8.1 INTRODUCTION

Central to this book is that medical equipment is essential for the delivery of healthcare. Consequently, all the equipment in a healthcare organization must be managed effectively to enhance the ability of the healthcare organization to care for its patients. Professional institutions such as the Institute of Asset Management (https://theiam.org) argue that assets, and in healthcare the medical equipment are assets, should be managed in such a way as to achieve maximum benefits for their organization. We have argued that the management of the medical equipment assets, termed *healthcare technology management* (HTM), requires a structured systems approach, and Chapters 2 and 4 through 6 have discussed the mechanisms for achieving the goal of effectively managing the medical equipment assets for the benefit of patients, basing the approach on the asset management standard ISO 55000. But we have also shown that the clinical engineers who manage the medical equipment have a broader role to use their knowledge and expertise to advance care.

These activities all require an effective and efficient Clinical Engineering Department (CED). There is no one correct operational blueprint for clinical engineering departments, with structures varying both within and between countries. Often, the structure is rightly influenced by the organization's overall governance structure and the place of the CED within it. The structure may be influenced by the department's history but should be reviewed regularly to ensure it remains fit for purpose.

In this chapter, we discuss the CED, recognizing that it requires operational and leadership structures that support its important and far-reaching roles, namely its twin remits of managing the medical technology and advancing care. We will also look at the internal structures of a CED recognizing that they may well be dictated by circumstances outside of the control of the CED leadership. Within the CED, there are a variety of ways of organizing the staff and a variety of mix of staff that it may be beneficial to employ. We will look at all these especially in the context of the leadership of the department.

8.2 THE MISSION, VISION AND VALUES OF THE CLINICAL ENGINEERING DEPARTMENT

Many organizations produce mission, vision and value statements that seek to clarify their objectives (mission), what they want their organizations to look like (vision) and the principles guiding the conduct of their operations (values). They are carefully constructed descriptions that seek to inspire their own staff and give confidence to their customers. Consequently, CEDs should develop these statements co-operatively with their staff, helping to ensure that the staff share the sentiments and aspirations of the statements, recognizing the common mission, striving to achieve the vision and at all times carrying out their duties in fulfilment of the values.

Thinking about drafting and refining these three statements is a very good starting point for the development of a quality management system (QMS) for the department. The Standards that support QMS have been discussed in Chapter 3, and we will explore this further in Section 8.4.2.1.

8.2.1 The Mission Statement

All CEDs should be able to clearly state their purpose. A mission statement is a short statement that clearly articulates what the CED aims to do and for whom. Developing its mission statement requires the department to reflect on and focus clearly on its central role. In CEDs where the activities are diverse, this can be challenging.

The mission statement should consider the department's outputs, clients and values. The outputs could be summarized as 'activities which support and advance care through the application of technology'. The clients are the patients, their families and carers and the clinical and corporate staff of the healthcare organization. The values might include a statement that all the work of the department would be done to the highest standard possible with compassion and respect for all, be ethical and based on evidence in the clinical, engineering and scientific fields. A mission statement is meant to be read by all not just the members of the department and so, when writing a mission statement, try to do so in a way that ensures those outside the department can understand it and get a sense of the department's priorities and what it aims to deliver.

8.2.2 The Vision Statement

A department's vision is the long-term objective it ideally would like to see if its work is successful. A vision should motivate and enable members of the department to see how their effort contributes to the overarching department's purpose. Vision statements might include:

> "The department will assist our clinical colleagues to deliver safe and effective care to patients by supporting them through the safe and effective application of technology in medicine. The department aims to be a centre of excellence in healthcare technology management and to share its experience and knowledge with the wider clinical engineering community."

8.2.3 The Value Statement

A department's values are its guiding principles. They should reflect the beliefs that really matter to the department and its members and that inform how things are done. Statements that might be considered would include: 'The department will value engineering and scientific excellence and apply it to the management of the technology for which it is responsible in an ethical and sustainable way'.

It might also state that work will be based on current evidence-based engineering principles including a focus on the environment and sustainability. The work of the department will be prioritized to meet the needs of patients and their families and carers above all. The department recognizes and values the development of technology and will support staff to engage in research, innovation and professional development. The department recognizes that its work is interdisciplinary and encourages and supports individuals contributing to projects that support and advance care by focusing on outputs that are beyond their Healthcare Technology Management role.

8.3 THE CLINICAL ENGINEERING DEPARTMENT'S POSITION WITHIN HEALTHCARE ORGANIZATIONS

The position of the CED within a healthcare organization should be designed to help it best deliver its services, supporting it in meeting its mission, vision and values as discussed earlier. The design's starting point should recognize the importance of medical devices for healthcare delivery and the CED's role in their management by applying structured processes that add value to their use for the care of patients, supporting the clinical teams using them. This will recognize that CED serves not just one clinical department or speciality but the organization as a whole. These principles should guide decisions on the CED's position within the organization.

However, it is recognized that CEDs have evolved over time within healthcare organizations, often from small groups of dedicated individuals, and the position of CEDs within the organizations reflects these historical developments. Hence, the CED's position varies between and within countries. In some jurisdictions, including the United Kingdom and Ireland, the professions of medical physics and of clinical engineering are closely linked, and within the United Kingdom, a single professional body, the Institute of Physics and Engineering in Medicine (http://www.ipem.ac.uk) supports both. Thus, in many tertiary care UK and Irish healthcare organizations (see Chapter 1, Section 1.2), CEDs are major sections of medical physics and clinical engineering services. Such departments are usually regarded as clinical support departments alongside others such as pharmacy, radiology and laboratory services, and this does seem to have an effect on how clinical engineering is perceived by top management. There are many examples of CEDs that are very positively perceived by top management who value their contributions to the effective running of the healthcare organization. Whilst the structural position of the CED has an impact, it is often also the personal contribution of the CED leadership and membership that transforms the reputation of the department from being simply a repair and maintenance team to a department that supports and advances care as well as managing the medical equipment. This is why, from Chapter 1 onwards, we stress the twin remit of clinical engineers in HTM, giving examples of both remits through the project work described in Chapter 7.

However, in some healthcare organizations, the medical equipment management services have developed from within Facilities departments. From there many have developed from being simply maintenance teams into clinical engineering departments, perhaps called by different names, providing the range of patient-focused support discussed in this book.

In the United States, the pattern is not dissimilar but without the strong links to Medical Physics. CEDs in large tertiary centres, often linked to universities, are well engaged and influential in their organizations. The American College of Clinical Engineering (ACCE; http://accenet.org/Pages/Default.aspx) provides a well-established and active professional society.

In some healthcare organizations, CED staff are not employees of the healthcare organization but are provided on contract by commercial organizations that specialize in this work.

The Association for the Advancement of Medical Instrumentation (AAMI; http://www.aami.org) is a not for profit multi-membership organization whose mission is 'the development, management, and use of safe and effective healthcare technology'. AAMI runs courses, conferences, continuing education programmes and collaborative initiatives and also provides resources, including peer-reviewed journals, technical documents, books, videos, etc. An AAMI-sponsored forum in 2011 debated a suitable consensus name for the work and concluded that Healthcare Technology Management was the most appropriate. The full report is available (AAMI 2011).

Besides its physical reporting structure, it is important that the CED becomes integrated into the healthcare organization's operations. The CED must have a strong ethos that encourages collaboration between its staff and the clinical and managerial staff of the healthcare organization. Clinical engineers can support and advance care by communicating and collaborating with their clinical colleagues (Morschauser 2014), and we have shown in this book, particularly in Chapter 7, examples of this collaboration in practice.

Whatever its name or its place in the structure of the healthcare organization, the principles we discuss in this book are applicable. The level of influence that a CED has is primarily influenced by the quality of support that it provides. Whether the CED is an internal department of the healthcare organization or is a contract-provided service, there is a need for the expectations of the service to be clearly stated and its performance to be measured and monitored. We have discussed the role of the Medical Device Committee (MDC) in ensuring that this happens in Chapter 5. However, the willingness of the CED leadership to engage, to be proactive and politely assertive in the areas in which clinical engineers have demonstrable competence, and to work collaboratively with other professionals will all contribute to its profile and ability to support patient care.

8.4 THE STRUCTURE OF A CLINICAL ENGINEERING DEPARTMENT

The CED's structure should help all members of the department understand and fulfil their role. It should make transparent the contribution of each of the department's component sections, how the sections relate to each other and to the whole to deliver the department's objective.

It is not unusual for CEDs to grow from small groups of motivated individuals where the mission and organizational structure was implicit. However, over time this shared understanding can be eroded as a result of staff changes, changes in scale or the need to realign with the overall hospital organizational structure. So it is important to revisit the department's organizational structure and design periodically, perhaps routinely every five years and in response to major changes in the healthcare organization that it serves. Whilst department structure and design are interrelated, they are different, and clinical engineering leaders should understand this when undertaking department review. In Section 8.4.1, we will discuss the organizational structure, following that in Section 8.4.2 with a discussion on the organizational design.

8.4.1 Illustrative Organizational Structures of the CED

The organizational structure is the framework from which the department delivers its services. It is focused on the teams, the people in them and how they work together. It should determine who performs various tasks and how these tasks and the people performing them relate to one another. The organizational structure lists the roles and responsibilities of the different job positions in the department and also details the reporting arrangements, identifying who can make decisions and at what level. The structure should be a real-world account of what is actually in place and not an idealized or hoped-for solution.

The Head of the CED will need to ensure that the department's structure is appropriate for providing the services it offers. The internal structures will depend on the size of the department. A small department might have one single technical team, whilst a large department serving a large healthcare organization, perhaps with multiple hospital sites, might have several teams. The CED might also have different teams concentrating on specific functional equipment types (e.g. renal dialysis equipment) or on specific clinical specialities (e.g. maternity and neonatal services).

8.4.1.1 Centralized and Decentralized Structures

Centralized organizational structures focus management authority and decision-making in a hierarchical fashion. Often, they have a single leader or management team at the top of the structure and authority and information flow from this top level of management to working groups or individuals. Decentralized organizational structures have a number of groups, working in parallel each managing itself as a stand-alone working team.

Small departments often operate in a centralized manner because of small staff numbers and consequently with little need for complex management structures. In centralized organizational structures, the head of department may be the only manager in the department and will make decisions, communicating them as instructions for the working teams to implement. The approach can work where good communications and a co-operative spirit are encouraged, with the department's small size fostering an ethos of shared endeavour to a common aim.

However, the greater size of larger departments makes it more difficult to achieve this spirit of shared endeavour in a centralized organizational structure. Central decision-making by the department head can often feel remote at the point of contact between the department and the delivery of its services. The decision-making authority of working teams, composed of motivated and skilled professionals, is reduced to how to implement the instructions. This model does not work well for organizations which are diverse and staffed with individuals who practise autonomously. For that reason, many CEDs adopt a decentralized structure.

In departments with a decentralized structure, all members and grades of staff are empowered to make strategic decisions for themselves and their teams, and then the information and consequences of the decisions are reported up to the head of department. This approach of allowing and enabling decision-making to happen at the interface between the department and its clients, the patients and clinical staff means that decisions are better informed by the real needs of those who use the service.

Decentralized structures run the risk of lack of coherent practices with different parts of the department working to different objectives and in different ways. Strong supportive encouraging leadership is important with tools such as regular local and senior leadership team meetings designed to develop the department's co-operative working in accordance with its shared mission, vision and values (Section 8.2). A quality management system and registration to Standards such as ISO 9001 should be considered as one way of ensuring control and consistency across decentralized services.

Healthcare systems are complex, and so a CED which has to be able to respond to requests for different types of services may develop complex organizational structures. Where the complexity of its organizational structure risks the efficient operation and delivery of the department's services, it is important that the structure be reviewed and perhaps revised to more simply reflect its operational requirements. Case Study CS8.1 describes the review of a department that became too complex as it responded to changes in its healthcare organization.

The organizational structure may sometimes appear to be neither centralized nor decentralised, but rather the management structure seems to adapt and change depending on the task. Such a flexible adaptive approach has the advantage that members or teams have the freedom and ability to creatively respond to unpredictable challenges. This approach can be powerful when dealing with an environment that is changing rapidly. However, a flexible and adaptive approach does not remove the need to work within quality management systems, protocols and equipment support plans (ESPs) as has been discussed in Chapter 6.

Where a department is spread out, often in different buildings, and staff are based at the point of care, the decentralized structure with its inherent delegated authorities and responsibilities has much merit. These departments tend towards more adaptive, flexible approaches due to the need to be able to respond to unpredictable demands and changing environments and to support highly educated and motivated staff to self-organize to deliver excellence at the point of care. Where departments organize themselves in a way that allows individual members of staff to have autonomy, then the stated departmental vision, value and mission statements become particularly important. Where individuals are self-directed, the values are often the primary influence for the direction that work takes.

8.4.1.2 Organizational Chart

It is not the intention of this book to be prescriptive about the structure of the CED but rather to clarify general principles. Figure 8.1 summarizes the principles of its management chart, identifying in general the reporting systems that will help define the position of the CED within the healthcare organization.

Figure 8.1 concentrates on the reporting arrangements for the CED to the healthcare organization that it supports. In Section 8.4.1.3 we will discuss in more detail the external reporting arrangements of the CED, and how the informal links with, for example the Medical Device Committee, support the strategic HTM Programme outlined in Figures 3.3 and 3.4 in Chapter 3.

Within the CED itself, Figure 8.1 shows leadership from the Head of Department and the department's leadership team to its functional groups: one or more service teams delivering the HTM and advancing care services supported by an administrative team and guided by a quality and audit function. The service teams in a decentralized structure may have considerable delegated authority and responsibility and might operate in different physical buildings, particularly in a large healthcare organization spread across different sites. However, even in a medium-sized hospital, the service teams might work from

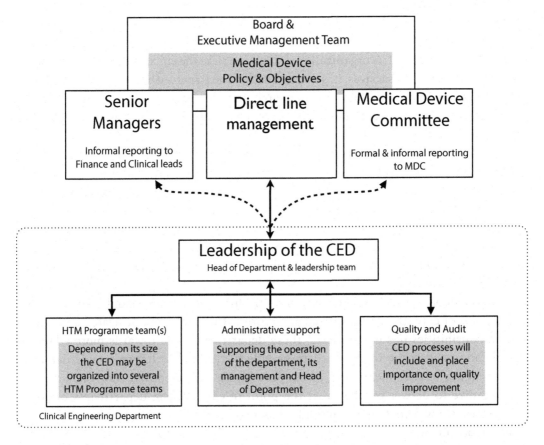

FIGURE 8.1 Outline of the organizational structure of a Clinical Engineering Department (CED) showing the structure of the department itself and its formal and informal reporting links.

different workshops, for example, small teams dedicated to service support for groups such as maternity and children services, critical care and theatres and renal dialysis. The physical proximity of the service teams to the clinical services can support the advancing care role of the clinical engineers, with the CED's senior leadership important to keeping the department as a whole behind its common mission, vision and values.

The operation of the CED will require administrative and clerical support, helping ensure its efficient and effective operation and freeing its leadership from administrative duties. Depending on the department's size, a call handling or helpdesk service might help efficient operation. Its administrative support will ensure links with support departments within the healthcare organization such as Human Resources (also known as Personnel) and Payroll services.

The CED structure adopted in one healthcare organization may be very different from that in another, with differences between jurisdictions and even within jurisdictions. The CED leadership should recognize that the object of the structure is to optimally provide services and, where good evidence suggests changes, should be open to re-organization. Case Study CS8.1 analyzes the structure of a CED, inviting the reader through its self-study exercises to reflect on its strengths and weaknesses.

8.4.1.3 Reporting Arrangements

The key to any department is its reporting arrangements, those within the department itself and those of the department within its healthcare organization. The department needs a clear formal line management reporting structure which embeds the CED within its healthcare organization. Through this the CED will raise departmental operational issues with the healthcare organization's management.

Often, the challenge in leadership of a CED is the different reporting arrangements with line management responsibility through a formal link and with perhaps less formal but nonetheless important reporting arrangements with other groups. Figure 8.1 illustrates this with two examples of informal reporting links. We have discussed in Chapter 4, Section 4.5.3 the important role of the Medical Device Committee (MDC) (recognizing that this may have different forms and titles in different healthcare organizations). In Chapter 3, we described in Figures 3.3 and 3.4 the reporting arrangements involving the Medical Device Committee and the healthcare organization's Board that are central to the delivery of the HTM strategy. We recognize this link in Figure 8.1 showing the informal link between the CED and the Medical Device Committee. The Head of the CED will discuss governance of the medical equipment with the MDC and such discussions may be formal when agreeing priorities or reporting on KPIs. The Head of the CED may have financial responsibilities for medical equipment procurement and maintenance for which it might report directly to the Director or Deputy Director of the Finance Department. The Head of the CED will also have informal reporting arrangements with senior medical and nursing leadership and with members of the healthcare organization's executive team. The Head of the CED will need to recognize the boundaries of these informal reporting arrangements, respecting the formal line management reporting structure. In Chapter 5, Section 5.5.3, matrix management has been proposed as a method of supporting different reporting arrangements, but the Head of the CED will need to work through the arrangements with their line manager (Figure 8.1).

8.4.2 The Organizational Design of the CED

The organizational design of a CED should detail the functions it delivers and outline the plan which will support the activity. Here, the focus is on describing the goal of the activity and then detailing the plan of how that goal will be achieved. It is less focused on the individual relationships and more focused on how the department or teams within it deliver their services, the outputs for the patients, clinicians and hospital. A good leader of a CED will analyze all the tasks to be performed and the goals of the department and then develop groupings of people and resources to best and most efficiently achieve those outputs.

How a CED is designed depends on its environment. Some department design is predicated on special teams with particular knowledge and skill in support technologies (dialysis, anaesthetics, disinfection and sterilization). Others might design a department to support particular clinical delivery microsystems (Cardiology, ICU, maternity services, and Emergency Department). Still others might design around the need to support a number of different physical locations. In reality most organizational designs evolve or adapt in response to challenges and the changing needs of the organization. A degree of pragmatism is necessary. It is very rare that there is the freedom or opportunity to design an ideal situation. Regardless, the goal of organizational design is to set into action the people and resources that make up the organizational structure to deliver the mission. Neither the organizational structure nor design should be considered fixed; either or both should change over time to ensure the department remains effective and relevant, able efficiently to meet its objectives. Such flexibility will enable the CED to respond to changing demands for clinical engineering services. These can arise from, for example, the development of additional clinical services within the healthcare organization. Alternatively, the CED might be asked to provide clinical engineering services for other autonomous healthcare organizations (Case Study CS8.2).

8.4.2.1 Quality Management System

We have emphasized in this book the importance of continuous quality improvement. The organizational design of the CED should incorporate quality control systems including a quality management system.

The CED department will have an internal operational function that runs its quality scheme and measures the operational performance. The quality scheme may or may not be subject to external audit, but a formal quality management system provides many essential tools for the effective running of the CED including processes for obtaining feedback from those for whom the department provides services. Key Performance Indicators (KPIs) will be measured and reported with processes in place through the Head of CED and team leaders to rectify problems identified and improve services.

A department may wish to instigate a formal quality management system, based on, for example, ISO 9001:2015. This gives the service the ability to demonstrate that it has been independently audited, which brings benefits both internally and externally. This Standard was discussed in Chapter 3, Section 3.5.3.

8.4.3 Some Thoughts on Physical Structures

The CED staff need physical structures from which to deliver services including laboratory and workshop facilities, offices and training and meeting areas. As with the position of the department in the organization's managerial hierarchy, so its physical place depends on historical developments. Opportunities sometimes present themselves to improve the location of the CED; so some discussion of the advantages and disadvantages of different arrangements is worthwhile. These can be summarized in three general models for in-house support plus a fourth service delivery arrangement:

1. The department is largely based in single premises within a major hospital having all or most of the facilities needed grouped together.

2. The department has a main base in the hospital, largely providing the office administrative and meeting/training facilities with some or many clinically focused teams based very close to their specialist clinical areas, for example, a team supporting the neonatal unit located close to that unit.

3. The department is largely housed in a single premises not located on a major hospital site, but having all or most of the facilities needed grouped together.

4. A small core CED, largely administrative but clinical engineer led, which manages external service contracts for all technical support and keeps all the necessary technical and financial records.

In model 3, the clinical engineering staff are often direct employees of the healthcare organization but could be employed by an external clinical engineering services company.

There are of course variations on these general models. For example, the CED may be responsible for two or more hospitals geographically separated, either within a large urban area or in a rural area that are perhaps up to 60 km (40 miles) apart so the CED may have two or more bases each generally conforming to model 1. Alternatively, the CED within one healthcare organization may offer its services to another healthcare organization, with the terms of reference of its service provision governed by a service level agreement (Case Study CS8.2). This would require those developing the service level agreement to consider the physical facilities required to deliver the services.

It is worth considering the pros and cons of each model as set out in Table 8.1.

8.5 STAFF ROLES AND RESPONSIBILITIES

We have seen in this book that the life cycle support of medical equipment required to deliver value is complex and interdisciplinary. Clinical engineers play a lead role in providing this support and are integral to its success. Appropriately qualified, trained and motivated staff are vital to an effective HTM service.

Within engineering as a discipline there are specialist strands of knowledge and practice such as electrical, civil and mechanical. Within each of these there are different types of engineering practitioners each of whom is an expert in a different aspect of the application

TABLE 8.1   Four General Organizational Models for a Clinical Engineering Department

Model Description	Advantages	Disadvantages
1. Department largely based on a single premise within a major hospital having all or most of the facilities needed grouped together.	• Economies of scale • Flexibility of use of workforce • Good peer support • Shared test equipment • 'Supporting patient care' role can be managed • Rapid response from on-site staff	• Often located in the basement or the Facilities yard, well away from clinical areas • May be difficult if serving multiple sites
2. Department has a main base in the hospital, largely providing the office, administrative and meeting/training facilities with some or many clinically focused teams based very close to their specialist clinical areas.	• Very close to clinical areas • Clinical engineering staff well integrated with clinical staff • Emphasizes the 'supporting patient care' role • Rapid response from on-site staff	• Small teams vulnerable to staff shortages • Technical staff become very specialized • More difficult peer support • Duplication of test equipment • Often crowded facilities
3. Department largely housed in a single premises not on a major hospital site or remote from the main building having all or most of the facilities needed grouped together.	• Economies of scale • Flexibility of use of workforce • Good peer support • Shared test equipment • Effective when supporting multiple sites and home healthcare	• Remote from hour-by-hour contact with clinical staff • 'Supporting patient care' role more difficult but not impossible • May be difficult to provide a speedy response to issues • May require a 'duty clinical engineer' on major sites
4. A small core CED, largely administrative but clinical engineer led; manages external service contracts and keeps all technical and financial records.	• Small head count of directly employed staff	• Essential that there is a clinical engineering input independent of the contractors • Can be a complex relationship • Do the service providers have the best interests of the organization at heart? • 'Supporting patient care' role may be absent • Slow response compared to having clinical engineers on-site

of engineering to solve problems. Within engineering it is well established that different and complementary types of practice experts are required if the profession as a whole is to think of solutions for problems and make them a practical reality. The International Engineering Alliance (IEA 2013) has recognized this and described the need for engineering technicians, technologists and professional engineers (PE). The levels of educational attainment and the range of competencies for each of three general engineering roles have been agreed between the engineering institutions in 14 countries in three 'accords' negotiated through the International Engineering Alliance known as the Washington Accord

(1989) for professional engineers, the Sidney Accord (2001) for engineering technologists and the Dublin Accord (2002) for engineering technicians. Further general details are available at http://www.ieagreements.org/ and a detailed paper is also available (IEA 2013).

Similarly, within clinical engineering there are engineering technicians, technologists and professional engineers.

8.5.1 Clinical Engineering Technicians

Clinical engineering technicians are the experts who deliver the technical solutions on the ground. They are the people to go to when there is a technical problem to solve. By complimenting their engineering education with practical experience, they are highly effective at delivering solutions. In practice clinical engineering technicians maintain medical equipment, help prototype new equipment, perform highly complex quality control and support research and development through supporting the conducting of experiments and collecting data. They also take responsibility for managing the delivery of complex schedules of work. Most will have completed at least a two year higher education engineering course and will have developed technical skills in electrical, electronic or mechanical engineering. This will have been supplemented by external or internal general training in basic physiology and specialist training on the equipment which they will be looking after. Some will have the educational qualifications necessary to become a clinical engineering technologist and may develop their careers in that direction by gaining competence through further learning and experience. In the United States they are usually called Biomedical Engineering Technicians (BMETs) or Biomeds. Job titles tend to vary between jurisdictions.

8.5.2 Clinical Engineering Technologists

Clinical engineering technologists also focus on the applied and practical applications of clinical engineering. Engineering technologists study engineering in a broader context than their professional engineering colleagues (see Section 8.5.3) and with a greater focus on practical application. They play a key role in transforming a conceptual solution into a realizable design that can be put into action. They are often employed in engineering management roles or as team leaders. They are the experts to whom other technical members of staff refer for some specialist types of equipment. They will have the experience and clinical knowledge to be confident in advising clinical staff about equipment-related problems. Their work is likely to be less routine and more 'problem-solving' both in technical areas and in managerial areas. Such engineers are well qualified to a graduate level or equivalent and have competence and experience gained over a number of years. They are likely to be signed up to a registration scheme, either specifically for clinical engineering technologists such as the Register of Clinical Technologists in the United Kingdom (http://therct.org.uk/) or with a general engineering institution who have a registration level for engineers meeting the Sydney Accord protocols (IEA 2013).

In the United States a programme of certification that is, in general terms, equivalent to the technologist level is run and accredited by AAMI (2016).

8.5.3 Clinical Engineers

Clinical engineers will have come through an engineering programme where the focus is on theory and advancing knowledge. Such programmes have a higher requirement on mathematics, learning engineering theory in depth and understanding how to access and contribute to new engineering knowledge through research. It is now common for engineers to have a master's as well as a bachelor's degree. Clinical engineers are experts in developing new solutions at the conceptual level. They typically are concerned with how to apply new discoveries to advance care. In the United States clinical engineers can receive accredited certification through the American College of Clinical Engineers, with a similar process available in Canada. Engineers in the United States can be accredited as Professional Engineers (PE) with a similar UK processes leading to the Chartered Engineer (CEng) designation.

In this book we have used the term *clinical engineer* to include all professional engineering and technical staff involved in clinical engineering, because all are working in the clinical engineering profession. In larger CEDs, particularly those that are also involved in collaborative R&D, the Head of the CED is likely to be a registered professional engineer as described earlier. They may also have a small number of similarly qualified and registered clinical engineers on their staff, probably in specialist roles such as service planning with other clinical disciplines, service development and complex incident investigation or in the provision of direct clinical services.

Some of these may practise at a consultant level; as experts in the discipline of clinical engineering they have the knowledge, skills and experience that enable them to practise independently. They have the knowledge, skills, experience and ability to be consulted (hence the term *consultant*) in their area of expertise. They are the leaders in the field of clinical engineering, offering specialist advice, dealing with the more complex cases, teaching, directing research and innovation and managing and leading services. This has been discussed in more detail in Chapter 7, Section 7.4.9.

However, most of the clinical engineering leadership team are likely to be well-qualified engineering technologists, probably registered at an appropriate level. Indeed, in smaller departments, all of the leadership roles, including the Head of the CED, may be fulfilled by this category of engineer.

8.5.4 Clinical Engineering: A Continuum

If you visit any established CED, you would find all three of these types of engineering people working together to meet the department's mission. However, the collaborative nature of engineer working might make it difficult to identify at first glance who is fulfilling the engineering, technologist and technician roles. That is not unusual. In reality, most individuals are developing their own knowledge and experience through their practice, and work interactively with colleagues so that conceptual solutions are imagined, realized and implemented using a team approach that allows for the edges of each type of engineering activity to be blurred. Also, remember that there is more to being an effective engineer than an engineering qualification. Other personal skills such as ability to communicate, leadership, systems thinking, analytical or creative ability also play a part and, in a well-functioning team, individuals with these particular skill sets should be able to apply them.

Sections 8.5.1 through 8.5.3 have concentrated on the roles of technicians, technologists and engineers, based on the Washington Accord (IEA 2013). It is important to draw attention to the essential work that is carried out by support workers assisting these professionals in their roles, helping with routine tasks.

So to summarize, clinical engineering teams are made up of a mix of individuals each with a particular engineering expertise and other attributes, who should be led and managed to apply their collective talents to the realization of the benefits of the medical equipment for patient care. Therefore, in this book when we say clinical engineer, we mean all individuals who contribute to clinical engineering activity. We will continue to do that but the reader must remember that the name has a particular meaning in certain countries and is used differently in different parts of the world.

In particular, in the United States, the name clinical engineer is used almost exclusively for those professional engineers working in healthcare technology management. The American College of Clinical Engineering (ACCE) has a definition and explanation of clinical engineers and their role and functions on the College's website (http://accenet. org/about/Pages/ClinicalEngineer.aspx). Despite the potential breadth of the definition, the description focuses mostly on the 'engineering' aspects and appears not to recognize that clinical engineers may also be directly involved in the application of engineering skills to clinical problems outside HTM (the supporting patient care role), research and development or working directly with patients. We explored some of these advanced roles in Chapter 7.

In the United Kingdom, similarly qualified clinical engineers are involved in exactly this HTM work, but in addition some may be involved directly with patients in, for example, making physiological measurements or designing, prescribing and fitting engineering solutions such as special seating or communications devices for patients that have physical disability problems. This is more in line with the explanation of 'Biomedical Engineering' given by Bronzino in the Introduction and Preface of the second edition of his *Biomedical Engineering Handbook*.

> "Biomedical engineers apply electrical, mechanical, chemical, optical, and other engineering principles to understand, modify, or control biological (i.e. human and animal) systems, as well as design and manufacture products that can monitor physiological functions and assist in the diagnosis and treatment of patients. When biomedical engineers work within a hospital or clinic, they are more properly called clinical engineers."

> BRONZINO (2000)

Because of this potential direct involvement with patients, clinical engineers in the United Kingdom who are professionally qualified at the master's level and have undergone defined training are included in the general category of Clinical Scientists all of whom have to be

registered with the Health and Care Professions Council (HCPC) (http://www.hcpc-uk.
co.uk/). Standards are emerging that illustrate the various practice proficiencies required
of clinical engineers. For example, the UK Academy of Healthcare Science has published
standards of proficiency under the banner of Good Scientific Practice for improving quality and protecting patients (AHCS 2014). Though written to guide healthcare scientists in
the United Kingdom (where clinical engineers are included in the healthcare science profession), these standards are universally applicable and include those of being able to communicate effectively, which we will discuss later in this chapter in Section 8.8, and being
able to practise safely and effectively.

The United States' ACCE and the United Kingdom's HCPC requirements for registration, the levels of knowledge, understanding, competence and application generally match
up with those set out in the Washington Accord requirements (IEA 2013).

8.5.5 Summary: The Staff of the CED

Appropriately qualified, trained and motivated staff are required for an effective HTM
service. As a generalization, CEDs need four types of staff. In leadership positions, they
need postgraduate qualified engineers with formal professional engineering registration
and graduate educated engineering technologists with relevant registration. Then there is
a need for engineering technicians who may have graduate equivalent qualifications combined with technical craft qualifications. Finally, there is a need for support staff, some
with administrative and clerical skills and, depending on the size of the department, some
with IT, business and commercial skills. Much also depends on leadership which we will
deal with in the next section.

8.6 LEADERSHIP IN THE CLINICAL ENGINEERING DEPARTMENT

The responsibility and authority for developing, instigating, reviewing and improving the
Healthcare Technology Management (HTM) Programme and its equipment support plans
(ESPs) described in Chapter 6 lies with the leadership of the CED. As we have seen earlier,
CEDs come in many shapes and sizes, dependent upon local needs and circumstances and
historical developments. However, in all departments the Head of Department, the leader,
needs to develop a departmental structure and operational processes and ethos that enable
it to fulfil its mission.

The leader should have a vision for the department, guiding the staff on the journey to
achieve the objectives. The astute leader will interpret the overall objectives of the healthcare organization, the Medical Device Policy and the Medical Device Committee into a
local vision for the CED, from which can be developed its goals and targets which all its
staff can relate to and strive to achieve. The leadership of the CED will include both the
head of department and, in larger departments, the leaders of its teams. Some large departments may have a Technical Services Manager overseeing the routine equipment management processes, freeing the head of department to engage in the strategic issues. So what
skills do the leadership of the CED need in order to develop the strategic vision and manage the day-to-day activities?

8.6.1 Setting the Strategic Objectives of the CED

Unless the department has a clear strategic mission, vision and value, its direction will be unclear and vague. It is the responsibility of the leadership to set the operational agenda for the department, providing a clear direction and purpose of the department and its staff. We have discussed in Section 8.2 the importance of the mission, vision and value statements of the department. The Head has a primary responsibility for setting the strategic objective of the CED that will be based on and reflects these underlying statements that provide the ethos for the CED and hence the principles that underpin its operation.

Thus, the leadership of the CED must have vision; the vision must be based on a solid understanding of the contribution of medical equipment to healthcare delivery and that equipment can be used to enhance healthcare. The leadership must understand the delivery of healthcare and the structures within the healthcare organization. Clinical engineering bridges the clinical and the technical, and the CED leadership must support and develop the structures that form that bridge. With a strong technical and scientific foundation, the leadership must have a good grasp of the clinical requirements and continue to develop expertise in both. The leadership should also understand the Standards, regulations and guidelines applicable to the service (Chapter 3).

8.6.2 Communication Skills

Communication plays a major part in the leadership's 'toolkit'. Internal communications within the CED are important to maintain a sense of involvement and purpose. This may be achieved through informal team meetings, newsletters or more formal department meetings where topics such as performance, improvements, operational pressures and developments in the wider organization are presented and discussed. In whatever form communication takes place, it is important to allow for everyone to have an opportunity to take part and be valued. In a larger department, this may have to be done by cascading information and discussion down from a more formal 'senior leadership team meeting' to individual team meetings.

The CED leadership will also need to develop communication links with patients and to ensure that the patient focus is kept fresh within the CED. The leadership might liaise with patient advocates who will provide a patient viewpoint on actual activities and experiences, perhaps through them including patients themselves in giving feedback on their healthcare experiences, particularly where the experiences relate to medical equipment. At other times the leadership may arrange for patient 'stories' to be read-out and discussed at meetings. There will be times that clinical engineers will communicate directly with patients, and the leadership must ensure that the staff are trained to do so. Direct communication with patients is particularly applicable where home healthcare is being supported, for example, in providing technical support to home dialysis patients (Case Study CS7.18) or delivering special custom seating to wheelchair-bound clients (Case Study CS5.2).

Above all else a culture of continual improvement needs to be established within the department. From the routine questioning of self and others to questioning whether activities could be improved in any way comes a leaner and more agile department better suited and capable of delivering the objectives and ultimately enhanced care to the patient.

8.6.3 Organizational Ability

The leadership of the CED has the responsibility for organizing the deployment of its resources and staff to efficiently achieve the department's objectives. Rather like the conductor of an orchestra who knows and understands the characteristics and musicality of each instrument (and those who play the instruments), the CED leadership should know the teams and orchestrate them to perform the tasks.

In organizing the workflow, the leadership should be able to be supported by the department's structure which should have been designed to efficiently carry out the tasks. We can turn again to our orchestral analogy, with the various parts of the orchestra arranged to complement the fullness of its music, the drums at the rear and the string instruments to the front. The leadership will organize the workload to make best use of the competencies and characteristics of the staff. This may require dividing the workload between staff groups, making best use of their skills.

The leadership will also need to plan and prioritize the workload, arranging the activities over a period of time (perhaps a year), ensuring that staff time and resources are allocated to meet the required scheduled and unscheduled tasks and the various facets of HTM. This planning should allow for the unexpected and include contingency arrangements for staff leave and vacancies and for major incidents affecting the healthcare organization (e.g. major environmental disasters or road or air accidents).

The leadership will see that there is a wide range of medical equipment in use and within that range there are a depth of complexities and consequently different levels of competencies required to undertake the work. For example, when arranging for the maintenance of a particular medical device, the leader might reflect on the skills required for different tasks. Whilst a fully qualified and experienced clinical engineer will be able to maintain service and perhaps train clinical staff on this equipment, it might be beneficial to arrange for the more junior members of the department to do the routine maintenance whilst leaving the more complex repairs for more experienced staff. The leadership of the CED needs to sensitively manage the work so that the greatest efficiencies can be made without jeopardizing personal development or succession planning.

8.6.4 Developing Individuals

The professionalism of the CED as a whole is not just a product of the technical knowledge and skills of the individuals within it. It also includes the softer skills such as teamwork, communication, tact and diplomacy. A 'well-rounded', capable clinical engineer is one who not only attains such a broad range of skills but also seeks to maintain and improve them over the years.

The leadership of the CED should make use of every opportunity to develop the staff members in a multitude of ways to build up professionalism and so provide a richer service and ultimately influence patient care. This may be through formal training schemes, perhaps accredited and managed nationally or through less formal local schemes. Ongoing learning for all staff is usually called continuing professional development (CPD), and there are various systems available to assist and guide the individual in identifying and recording the learning within an activity. The responsibility of this learning is the individual's, and many professional institutions offer and require participation in such CPD schemes to maintain membership or registration. The membership of professional institutions and/or registration of clinical engineers within the department lend a weight of authority and value to the organization and should be strongly encouraged. The development and encouragement of clinical engineers should be a key component of the day-to-day operation of the department, but this need not be considered as onerous.

8.7 THE TRAINING ROLE OF CLINICAL ENGINEERS

8.7.1 Developing the Workforce

As we discussed in the previous section, senior members of the CED have a responsibility for developing and training their more junior colleagues. Indeed, on a more general level, every member of the CED can learn from other members, and most Heads of Department will acknowledge the learning they gained from members of their team, whatever their organizational status in the department.

Departments will need to develop more formal training systems for their staff. This will include general induction training for new staff members which will typically complement organization-wide induction training. Besides the induction training, the Head of Department will need to set up CPD programmes that develop the members of staff in clinical engineering practice. For example, consider engineers coming into the profession with engineering qualifications, for example, at either B level or M level. These new entrants will not be, nor will they consider themselves to be, fully trained unless they have followed a recognized clinical engineering training scheme. They will need additional education and training to become *clinical* engineers, with, as we have said already, knowledge and understanding of technology and of the clinical implications of that technology. Those who have done specific clinical or biomedical engineering degrees may have a small head start, but conversely, their engineering knowledge may be a bit less. So all will require further training aimed at increasing their knowledge and understanding.

The exact nature and arrangements for such continuing training and possible regulation will depend on the jurisdiction in which you are working and on the type of employment role you have. We have described in more detail in Section 8.5 that, through the International Engineering Alliance (http://www.ieagreements.org/), there are three internationally recognized grades of engineers, professional engineers, engineering technologists and engineering technicians.

For the professional engineer grade, there are formal in-service examined graduate training schemes established, usually on a voluntary basis, through professional bodies such as the ACCE in the United States (http://accenet.org/CECertification/Pages/Default.aspx) or through government ministries of health with a statutory implication. In the United Kingdom, for example, you cannot by law call yourself a 'Clinical Scientist (clinical engineer)' unless you are registered on an independent but government recognized register of health and care professionals (http://www.hcpc-uk.co.uk/).

Acceptance on to the register requires compliance with knowledge and skills competences stipulated by the registration system.

For engineering technologist roles, the situation is less formalized. Voluntary registration schemes for technologists have been developed in both the United States and the United Kingdom. Employers do not usually require that their technologists are registered, but clinical engineers working as technologists are encouraged to consider the benefits of registration, enhancing their professional standing.

The situation with those employed in engineering technician roles is even less formalized, and most training will be done in service and will focus on training for specific tasks, building on the basic skills and knowledge of the technician. The roles and training may be suitable for apprentices who may also be given the opportunity to develop their educational qualifications and their skills and thus progress in the profession.

Most training will be carried out in service by more senior clinical engineers. In a well-established department, the training will be planned, assessed and recorded, and it will be reviewed and updated as necessary. It may take the form of internally delivered training courses on, for example, electrical safety and testing, or it may be one-to-one instruction on the servicing of particular types of equipment. Where appropriate, staff might be asked to undertake specific training at colleges of higher education or universities. Clearly, the clinical engineers who operate as expert advisors (Chapter 7) have a responsibility to pass on that expertise; this might be by co-opting junior colleagues to work with them thus spreading knowledge by association, or it might be by more formal seminar presentations.

The CED should be a learning organization, teaching, developing and enhancing the knowledge and skills of its entire staff.

8.7.2 Contributing to Academic Programmes

Clinical engineers working in a healthcare organization may also get involved in teaching at university level at B or M levels or equivalent qualifications or supervising doctoral students. This may be through contributing individual lectures within their own expertise on clinical or biomedical engineering or medical physics courses, or by working with colleagues to deliver a complete module on such courses. In some cases, honorary university contracts or joint appointments are made available.

The value of such work is in engaging with intelligent students, promoting the clinical engineering profession and inspiring them to see clinical engineering work as valuable, ethical and interesting. Some master's level students are often more mature, looking to increase their qualifications or make a career change and the opportunity to influence and guide them is both personally rewarding and of longer-term benefit to clinical engineering.

CEDs should consider providing work placements to students, but these need to be carefully designed and supervised. Work placement or research projects may arise directly from the clinical engineering work and can be a valuable way of answering questions or investigating problems that might otherwise have to be deferred within the normal workload of the department. These may be at all levels including PhD, and there are examples of very useful developments of practical use to service delivery coming out of such projects (Rogers 2009).

A more mundane but real benefit of being involved in university teaching is that the library facilities of the university are thereby made available and sources of information such as access to peer-reviewed journals and Standards is much easier.

8.8 COMMUNICATION

Previously in this chapter communication has been discussed as one of the leadership skills required. However, good communication is the responsibility of everyone in the CED, and there are almost as many ways to communicate as there are people with whom to communicate. The clinical engineer has a central communication role in HTM because the clinical engineer understands both the technology and the clinical context in which it is used. They are able to act as an interpreter between the clinical and technical worlds. Figure 8.2 shows how the clinical engineer is centrally positioned between the clinical, technical and patients' viewpoints.

Communication with clinical users, patients and technical colleagues takes place in many ways. Clinicians may want to discuss how to effectively use medical equipment with clinical engineers. Often this will be informal, chance meetings in corridors or wards, but will also be formal. In these circumstances clinical engineers will often need to apply skills in communicating technical topics in a way that is readily understood by non-technical staff. Clinical users will often wish to report a medical device for unscheduled maintenance, and this can be done by using the phone, to a dedicated telephone number and help desk, or perhaps through an automated web-based intranet link to the Medical Equipment Management System. People tend to prefer dealing with others directly, so the telephone remains a popular choice, with answering machines and web reporting left to the hours when the help desk is not staffed.

Clinical engineers may have direct and indirect contact with patients, on some occasions being alone with patients (as when called to support equipment in the home environment) or supporting clinical staff managing patients to whom equipment is connected. The clinical engineer will be conscious of the concerns and anxieties of the patients and

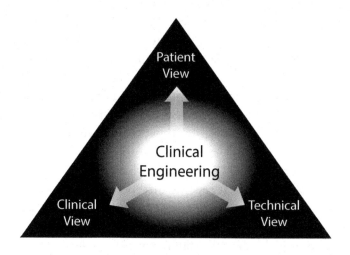

FIGURE 8.2 The clinical engineer is centrally positioned to assist communication.

their carers, be conscious of their own limitations in terms of clinical knowledge of the patients and their health and disease status and be conscious of the scope and limitations of their professional responsibilities. The clinical engineer can put the patient or relatives at ease, manage the situation and present the department, organization and profession in a good light.

Technical communication is common between the CED and external service vendors. When resolving and clarifying issues and querying part numbers, orders and outstanding repairs, primarily by telephone and e-mail, it is important to build up good working relationships with these vendors. From time to time incidents may need to be reported to national vigilance bodies, and this might involve a web-based submission supported by documentation.

Supported by Figure 8.2, this section began with the central role that clinical engineers have in communication about the use of technology in medicine; the clinical engineer, who is the custodian of the medical equipment on behalf of the patient, guiding the clinicians, linking the clinical and the technical for the sake of the patient. Communication will take many forms, but, whatever its nature, it must be delivered in a manner that is understandable to the recipient, be they clinician or patient.

8.9 CONCLUSION

Clinical engineers have important opportunities to serve healthcare with a twin remit that includes managing the medical equipment and supporting and advancing healthcare (Chapter 1). To realize these opportunities and effectively provide the twin remit requires that clinical engineers operate from a firm base. In this chapter we have discussed the position of the Clinical Engineering Department within the healthcare organization and how the position is important for it to achieve these twin objectives. The CED must have appropriate communication and reporting links with the healthcare organization, balancing line management reporting with wider links that enable it more effectively to serve the delivery of healthcare.

The CED must be guided by a strategic vision, supported by mission and value statements, which set the ethos for its operation. These will guide the staff working in the department to operate professionally in the interests of the patients and the healthcare organization. To support the staff, the CED should have the appropriate structure that empowers the individual members to deliver effectively. The department should be based in appropriate premises, recognizing that different models of provision may work best for different circumstances. The structure should provide an environment that encourages a learning organization that continuously seeks to improve using quality management methodologies.

Staff are the key resources of the CED without which it cannot meet its objectives. The staff need supportive leadership that sets the direction and role of the department, and the staff need to be encouraged and supported to develop and enhance their skills, both technical and clinical. The CED leadership must include workforce planning to ensure that the department retains and develops the staff required to undertake its tasks, achieving its mission and realizing its vision.

REFERENCES

AAMI. 2011. Report of future forum to identify a unified name and vision. Arlington, VA: Association for the Advancement of Medical Instrumentation. http://s3.amazonaws.com/rdcms-aami/files/production/public/FileDownloads/HTM/Final_Future%20_Forum.pdf (accessed 2016-05-12).

AAMI. 2016. About certification. Arlington, VA: AAMI Credentials Institute. http://www.aami.org/professionaldevelopment/content.aspx?itemnumber=1134&navItemNumber=577 (accessed 2016-05-12).

AHCS. 2014. Improving quality, protecting patients. Standards of proficiency for Healthcare Science Practitioners. Leicestershire, England: Academy of Healthcare Science. https://www.ahcs.ac.uk/2014/07/academy-publishes-standards-for-practitioner-register (accessed 2016-05-12).

Bronzino J.D. (Ed). 2000. *The Biomedical Engineering Handbook*, 2nd edn., Vol. 1. Boca Raton, FL: CRC Press.

IEA. 2013. Graduate attributes and professional competencies, Version 3. Wellington, New Zealand: International Engineering Alliance. http://www.ieagreements.org/GradProfiles.cfm (accessed 2016-05-12).

Morschauser M. 2014. Improving patient safety through collaboration between clinical staff and engineering staff in hospitals. *Journal of Clinical Engineering*, 39(3): 129–131.

Rogers P.D. 2009. Modelling the crashworthiness of specialist wheelchairs. PhD thesis. University of Glamorgan, Wales, UK.

SELF-DIRECTED LEARNING

1. What is your personal vision for clinical engineering in healthcare? Consider your personal ambitions and your understanding of the challenges facing healthcare in your own organization. How can you merge your personal ambitions and aspirations with the challenges facing healthcare to support improved application of medical equipment for population and individual health care?

2. How can medical equipment be managed to enhance its value? In this book we present the challenge of managing medical equipment through a structured methodology (based on ISO 55000) that enhances the value of the application of the equipment for patient care. How should this challenge be met considering first the key principles that should underline the Healthcare Technology Management (HTM) processes? In what practical ways can HTM enhance the value of the application of the medical equipment? You could base your discussion on one or all of the elements of the medical equipment journey in healthcare (e.g. its acquisition, operational or disposal phases).

3. Consider the structure and organization of your Clinical Engineering Department, or of a department that you have heard about. Constructively, review the department, looking at how well it serves its healthcare organization. What component parts (perhaps called sections) does the CED require to effectively provide services and how should the structure ensure effective interrelationships between these component parts? What are the important internal relationships between the component parts? From this analysis suggest improvements that can be made. How would you test the suggested improvements before attempting to implement them?

CASE STUDIES

CASE STUDY CS8.1: ANALYSIS OF A CLINICAL ENGINEERING DEPARTMENT USING SYSTEMS ENGINEERING METHODOLOGIES

Section Links: Chapter 8, Section 8.4.1; Chapter 2, Section 2.6

ABSTRACT

The merging of three hospitals into one healthcare organization created the opportunity for reassessing their Clinical Engineering Departments (CED), now to be reformed as one department. The process started with a review of the ethos and objectives of the new CED followed by assessing how to structure it to achieve those objectives and to create it from the former departments.

Keywords: Clinical Engineering Department; systems engineering; structure; objective; service review

NARRATIVE

The merging of three neighbouring hospitals, one large and two small, created a new CED from three distinct departments each with different approaches and operational methods, with the head of the larger hospital's CED asked to lead the new department. The challenge was to create a united team from groups wary of each other, concerned about the loss of their individual identities. They were also conscious of the requirement to support the new merged organization aspiring to be the best care provider in the region, providing high-quality tertiary and community care.

The CED head called the team leaders from the three former hospitals together: what should we aim for, what should be our mission and our vision and values (Chapter 8, Section 8.2)? Consensus was soon reached: to deliver award-winning clinical engineering services for the benefit of patients attending any of the hospitals and receiving community care. This was followed by asking how, in practice, could this be achieved. Think of the CED as a machine, the head told the team leaders; what components, what elements does the machine need for it to accomplish its tasks? This required careful analysis of the tasks that were really important.

First, the CED must manage the medical equipment, taking a life cycle approach, ensuring it is safe, functional and available where and when needed. The best of the different approaches from the former hospitals should be adopted, linking in with the structural changes of the merged organization. The larger hospital was concentrating on maternity services and critical, neurosurgical and cardiothoracic care, leaving one hospital to concentrate on orthopaedics and day surgery and the other on cancer services and chronic geriatric care. Community care would be led from the hospital providing day surgery. These discussions led to agreement on the formation of local teams in a decentralized structure (Chapter 8, Section 8.4) with considerable local autonomy and responsibility, united by the common mission and values. This agreement helped defuse the unease from the loss of the former independence and suggested the formation of workshop teams based primarily on hospital site (geography) with elements of equipment specialization (critical care and theatres, maternity and neonatal, general ward, community care, renal dialysis). These workshop teams would form the core components through which the CED would deliver its Healthcare Technology Management (HTM).

These ideas were discussed with all the staff in the now large CED; lively discussions were held, with the decentralized structure welcomed as an opportunity to still develop local loyalty, whilst recognizing the common working and objectives. Where appropriate, rotation of staff between teams would provide additional experience and potential promotion and succession planning opportunities.

The CED senior leadership team met with the Chief Executive Officer (CEO), Director of Finance, Medical Director and the Nurse Director to discuss how the CED could best serve the new organization. The CED leadership explained how they wished to optimize the benefits available from the medical equipment assets. This required not just that the equipment be kept safe and functional, but that the finances available for the medical equipment (procurement and operational) be carefully managed to enhance value. They asked that the organization's top management trust the CED to deliver the medical equipment services, analogous to the delivery of pharmaceutical services by the Pharmacy department. Two of the former hospitals had required the clinical engineers to obtain higher level authorization for even low-cost external parts requisitions; the CED asked for an overall budget allocation based on equipment replacement value which would also cover external service contracts. The CED would work with the Finance Directorate to manage this budget with auditable accounts using their Medical Equipment Management System (MEMS) database. The CED also asked for the management of the medical equipment procurement budget, allocating the budget between the clinical services in association with the heads of the clinical services. The CED showed how they planned to organize the allocation of the budget for the procurement of medical equipment.

After the successful meeting with the organization's top management, the CED leadership discussed the other components that they required to deliver the services. Central services included the management of the MEMS database, including analysis and reporting from it. This would be provided by an internal team that included quality control as the CEO had stressed the need for external quality audit. Each workshop team would provide informal medical equipment training, but it was decided to have a small training team that would also structure the training for the CED's own staff. A central team would also be set up to support the management of the medical equipment procurement budget, including specifications, evaluations and selections, and a two-person team formed to manage external contracts. Central administration support would also be needed.

All the CED's components (workshop teams, quality control, training, contract management and medical equipment budget) would be encouraged to be innovative. In particular, they would be urged to think creatively about what services could be developed to enhance the deployment of the medical equipment to advance healthcare. The renal dialysis team had several ideas, in particular how they could work with clinical colleagues to enhance home dialysis.

The CED leadership presented a report to the CEO summarizing the proposed CED with line management accountability to the Director of Support Services. The report also requested the development of an HTM structure that included a Medical Device Policy and a Medical Device Committee (Chapters 4 and 5). The CEO asked that the Head of CED present this to the Board for discussion and approval.

ADDING VALUE

The CED leadership showed to the Board how, through a structured approach, it could enhance the benefits of the medical equipment assets and reduce their operational costs.

$$\text{Benefits} : \text{Cost} \therefore \text{Value}$$

SYSTEMS APPROACH

The analysis began by first clarifying and agreeing the CED's objectives followed by examining its component parts (its workshop teams and central support sections) and how they worked together to deliver the objectives looking at processes, inputs and outputs.

SUMMARY

Teamwork and a willingness to share ideas were key to developing a unified leadership that co-operated to deliver a shared mission and vision underpinned by agreed values.

SELF-DIRECTED LEARNING

1. The redevelopment of the CED started with agreeing to its objective, where it wanted to be and the values that would underpin its service delivery. What are the benefits and what are the risks of spending time discussing and debating the mission, the vision and the values of a CED? How can the risks be minimized and managed?
2. How could using the CATWOE (Chapter 2, Section 2.2) approach facilitate bringing the different clinical engineering teams together? The CATWOE approach emphasizes Transformation that focuses on the Customers, recognizing the Actors and the Owners involved, whilst taking a World view that is subject to External constraints and demands. Discuss, in this situation, the different aspects of C, A, T, W, O and E.
3. Merging differing CEDs can bring both benefits and disadvantages. Can you identify both from the points of view of typical stakeholders such as a clinical engineer, head of department and clinical professional?

CASE STUDY CS8.2: HOW ONE ORGANIZATION IS DELIVERING A SERVICE TO ANOTHER THROUGH A SERVICE CONTRACT

Section Links: Chapter 8, Sections 8.4.2 and 8.4.3

ABSTRACT

Two neighbouring healthcare organizations, one a large hospital with an established Clinical Engineering Department (CED), the other, a smaller community facility without a CED, worked together to provide a cross-site service and increase value in both the organizations.

Keywords: Organization; contract; service; external

NARRATIVE

The established CED in the large hospital was approached by the Facilities Director of a neighbouring smaller healthcare organization to provide a clinical engineering service both to assist their compliance with regulatory requirements and to improve service efficiencies and patient safety. The smaller organization, spread across several community sites, had no detailed inventory of its medical equipment showing their locations across the sites, limited record keeping processes and no strategic medical equipment management plan.

The first stage was to clarify and define the scope of the services to be offered. Determining the scope required an equipment inventory which the CED led, drawing up the inventory detailing the medical equipment and their locations, recording the details in the CED's Medical Equipment Management System (MEMS). Preliminary assessments of the equipment's condition were recorded using two descriptors, age (new, midterm, old) and condition (good, average, poor, very poor). The inventory compilation identified a challenge of the proposed service: supporting relatively few items across multiple locations.

It was agreed that the service's scope was to maintain the medical equipment on the agreed inventory through a mixture of in-house and sub-contracted contracts. Minor repairs of a low cost were included, but any significant repairs would be charged separately and reported on a monthly basis. The accounting system would be developed to ensure clarity and separation of the costs associated with the work done by the CED for each healthcare organization. Therefore, a logical separation of parts, materials, sub-contracts and labour costs associated with the two different services was implemented. However, it was agreed that the costs of minor parts would not be explicitly charged but included in a general overhead fee.

The scope also required that added value components be identified, with agreement reached that the CED would manage external safety warnings and assist in the investigation of adverse events. The CED argued that their expertise could be important for the process of specifying and selecting medical equipment and this was agreed.

Once the scope of the services to be provided had been agreed, it was necessary to clarify the governance and quality control arrangements that would govern the contract. These would be based on agreed Key Performance Indicators (KPIs) (Chapter 6). The KPIs were discussed between both organizations: what was important for the 'customer' to know, and what is practical to measure and realistic to deliver. The agreed set contained a mixture of indicators:

- Scheduled maintenance achieved – 95% of the equipment to have its scheduled maintenance carried out on time;
- Response to ad hoc repairs – response within 4 hours (working hours) for critical equipment and within 48 hours for non-critical equipment;
- Safety warnings attended – all safety warnings attended to in the time limit set by the safety warning;
- Financial reporting – quarterly financial reports.

A financial schedule was drawn up which would clearly show the costs incurred for each location with a consolidated invoice submitted each quarter, along with individual invoices for ad hoc repairs. Regular contract meetings would be held to report on the delivery of the services at which amendments to the services could also be discussed.

These agreements provided the basis for drafting a legal contract making clear the responsibilities of both organizations.

The CED required resources to provide the services. A specific team of clinical engineers was established, and these engineers received specific training for certain areas, such as mental health facilities and prisons. The new service required the provision of new facilities including workshop vans with complete kits of tools and test equipment and a specific workshop with easy access to the parked vans. This was established on the periphery of the large hospital site where some unused buildings were available. The vans were leased for the duration of the contract so that the start-up costs could be kept as low as possible.

A separate budgeting account was created to list income and expenditure associated with the contract, showing costs incurred for parts ordered, sub-contracted service contracts, transport and direct labour costs. Costs could be monitored for the whole service and also at the level of individual service tasks.

Delivery of a new service from an existing service has the advantage of being able to share overhead resources, including the head of the existing CED, HR, utility costs and premises overheads. These were identified and charged in the hourly rate set for a clinical engineer.

Accepting new clinical engineering service costs entails an acceptance of risk transfer. The income from the management of 'customer' equipment is not just a matter of cost plus profit. Within the contract hourly rate, an allocation was set aside to compensate for the increased level of risk that the CED and its parent organization now carry. This is not a trivial aspect of the costing as equipment management carries a significant responsibility. Both organizations need to acknowledge this, recognizing the overarching responsibilities that the CED's healthcare organization now accepts, and the additional risk; a suitable financial levy needs to be set in order to compensate for this.

As the delivery of the contract progressed and equipment was received, removed and replaced, changes were processed through a series of variation documents agreed at regular contract meetings. The ownership of the medical equipment still resided with the smaller organization, and they managed the replacement of equipment internally with expert advice from the CED. Initially, evaluation and trial of equipment were not considered part of the contract, and this aspect remained with clinical colleagues.

Communication was key to the successful running of the contract and having a named and authorized contact at each organization to channel communication through avoided confusion and accountability issues.

ADDING VALUE

The financial cost to the customer organization increased from the starting position where very limited medical equipment maintenance was provided. Sharing services with a larger hospital improved efficiencies and benefits included improved availability of safe functioning medical equipment. In addition, the experience and knowledge of the clinical engineers have assisted in other areas, such as procurement of medical equipment where previously there was no expertise. The benefits of good Health Technology Management with governance and compliance come at a cost but value still increases overall.

Benefits : Cost ∴ Value

SYSTEMS APPROACH

The development of a clinical engineering service from scratch is a time-consuming and costly business. This case study has shown that the knowledge and skills possessed by one organization can be utilized quickly by another, with systems that work for one only requiring marginal adjustment for the other.

PATIENT CENTRED

The safety, suitability and availability of medical equipment for patients are a pre-requisite for safe patient care. The use of skills and knowledge possessed by one organization in another organization, in the way that has been described, enables a safer patient environment.

SUMMARY

The opportunity to manage the medical equipment of another organization needs to be viewed with care, and all costs and other variables need to be identified to enable such a project to run smoothly and not to the detriment of one party. A successfully run contract needs to include continual monitoring and regular communication throughout the term.

SELF-DIRECTED LEARNING

1. Can you think of any added value that a CED brings to such a contract rather than a series of external service contracts managed by administrators?
2. What particular issues can you identify when clinical engineers are working off-site in remote areas? How could you manage this?

CHAPTER 9

Medical Device Governance

CONTENTS

9.1 INTRODUCTION

'Medical device governance' covers the processes, practices, Standards and regulations that together help ensure the safe and effective application of medical devices for patient care. This term requires some explanation. Two terms, medical device and medical equipment, are used when discussing healthcare technology. They are defined in Chapter 1, Section 1.1.1. 'Medical device' is used here rather than 'medical equipment' as it includes the single-use consumable items (e.g. syringes and infusion lines – see Case Study CS7.8) that are used with medical equipment. Medical device governance covers this wider remit. The word 'governance' covers the rules and practices shared by industry, regulators, hospitals and funders that support the application of technology in healthcare to meet the objectives of all the stakeholders including citizens as funders and users of the service.

9.2 THE EXTENDED LIFE CYCLE FROM DESIGN THROUGH OPERATIONAL USE TO END OF LIFE

Medical device governance is broader than the processes that govern healthcare technology within healthcare organizations. It extends to the processes operated within the design and manufacturing industry and to the environmentally appropriate processes used to dispose of devices at the end of their life. In Chapter 4, we discussed the life cycle of medical equipment from the perspective of its operational life within healthcare organizations.

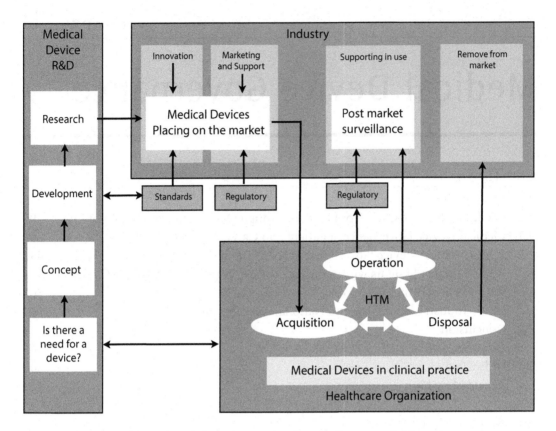

FIGURE 9.1 Manufacturer's and clinical operational view of equipment life cycles.

Discussion of all the processes that support 'medical device governance' requires that we consider the extended life cycle including the pre-operational use phases within industry. This extended life cycle can be thought of as comprising the life cycles from the perspective of the manufacturing industry and from healthcare organizations (Figure 9.1). The safe and effective application of medical devices requires diligence over this extended life cycle. It is perhaps helpful to briefly clarify these two perspectives before we look in detail at the wide range of processes that ensure medical device governance.

The industry life cycle begins with clarifying if there is a clinical need for the device and whether it will be commercially viable. This identification of need will have later parallels within individual healthcare organizations as they clarify whether a particular device will bring value to the organization by enhancing the care that it can deliver for its patients. This will involve discussions on the concept of the proposed device, typically with staff in healthcare organizations. We discussed in Chapter 1, Section 1.6 the important role that clinical engineers can have in these developments and the necessary research to clarify the validity of the concepts. The manufacturer will commence a sequence of concept and prototype designs, testing and improving the proposed device. During this process, the design team will refer to Standards (see Chapter 3) and seek approval from regulatory authorities (e.g. the U.S. 510k process or in the European Union, CE marking to show compliance with the Medical Devices Directive – see Chapter 3 for details) (Sorenson and Drummond 2014).

Successful product developments will lead to commercially available medical devices that will be marketed for use in healthcare, including selling and support. An important aspect of the medical device governance from the perspective of industry is its 'Post-Market Surveillance' (Daniel et al. 2015; Vockley 2015). This puts the responsibility on industry and those who place medical devices on the market to continually monitor the application of the products whilst in operational use, investigating any problems that might occur and, where necessary, informing regulatory authorities and issuing safety warnings.

At some point, the manufacturer may choose to withdraw support for the medical device, providing notification advice for its End of Life (EOL). The manufacturer will typically give a few years' notice of the EOL of its product. This will be based on one or more of several reasons: development of new products, the device's functional obsolescence, lack of ability to support and lack of continuation of supply of some of its component parts. We discussed the equipment replacement planning in more detail in Chapter 5, Section 5.6.3 and in Case Study CS5.8 with manufacturer's declaration of end-of-support being one of the criteria used in replacement planning.

The clinical operational life cycle of medical equipment has been discussed in detail in this book from several perspectives. Figure 9.1 summarizes the healthcare organization life cycle as composed of three elements, the acquisition, the operational use and the disposal of the equipment.

The extended life cycle is a complicated narrative with interactions between industry and healthcare organizations at all phases. We have seen in Chapter 2 how the systems engineering approach enables us to conveniently analyze complex systems such as this, exploring the processes at various depths appropriate to the analysis.

9.3 MEDICAL DEVICE GOVERNANCE, A PATIENT AND CARER PERSPECTIVE

With this background, we will explore the various processes that, combined, seek to ensure that the medical devices prescribed and deployed for the care of a patient are appropriate, safe and effective. These processes will extend over the full industry and healthcare organization life cycles. We will discuss these processes as concentric rings from the perspective of patient and carer, with the inner ring consisting of these processes more visible and immediate to the patient and carer. The middle ring describes processes in healthcare organizations whilst the outer ring summarizes the industry processes. The three rings summarize the processes, each of which is labelled with an alphabetical code and brief title. The text details these processes, dividing them into subsections arranged by their concentric circle.

9.3.1 Medical Device Governance at the Point of Care (Inner Circle)

At the point of care, the patient and carer will have several expectations of the medical devices they encounter. From a general perspective, they will expect devices that are clean, functional, safe, effective, easy to use, not frightening but reassuring and preferably aesthetically pleasing. However, when considering medical device governance, we need to look more deeply and in more detail.

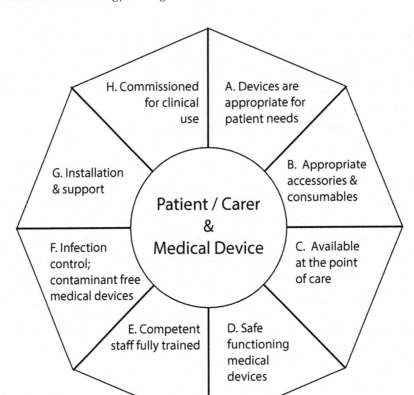

FIGURE 9.2 The inner ring of the A–Z of medical device governance; processes at the point of care. The elements A to H are described in the text.

First, the medical devices provided for their care should be appropriate for their needs ('A'; i.e. process A in Figure 9.2). This is achieved by those prescribing the devices being knowledgeable about the patient's needs, knowledgeable about the devices that are available and hence able to prescribe those best suited to the patient. There should be a natural match between the function of the devices and the patient's needs. The appropriate equipment ('A') and its accessories and consumables ('B') should be available at the point of clinical care ('C'); logistical arrangements for the supply of consumables to the place of care must be in place to ensure this availability, particularly important for home healthcare, but also required for hospital care. Availability at the point of care (time and place) is important, with estimates suggesting that perhaps as many as one in six of adverse events involving medical equipment are caused by the failure to ensure this availability (Amoore 2014). The logistic arrangements for consumable supply should have been planned prior to procurement and reviewed regularly.

Medical equipment and devices should be safe and functional ('D'). The Healthcare Technology Management (HTM) Programmes provide this assurance, with processes and procedures such as the Equipment Support Plans (ESPs) (Chapter 6) checking the functional integrity and safety of the equipment. Those who operate medical devices must be able to do so competently ('E'), knowing the characteristics and limitations of the devices. It is not uncommon for problems with medical devices to be ascribed to lack of operator competence. Competency-assessed training is important, but we will see later that the selection ('L' – see Figure 9.3) and design processes ('W' – see Figure 9.4) need to ensure that the medical devices meet Standards of human usability such as IEC 62366-1. Healthcare organizations, supported by medical device suppliers, need to have available competency-based training programmes. These training programmes will be required when the medical devices are first commissioned, but continuing training programmes, with training records, must be provided throughout the operational life of the devices to ensure that competence is maintained and that new staff are trained (Case Study CS4.2). Clinical engineers will often have an important role to play in training clinical staff.

Infection control is a major risk factor in healthcare, with the sick more vulnerable to infection than the healthy. The Clinical Engineering Department will work with the Infection Control team and their clinical colleagues to ensure that the medical equipment and its accessories are kept clean and free from contamination ('F'). Whilst infection control considerations have led to increasing use of disposable accessories, from surgical instruments to routine transducers for invasive blood pressure measurements, the need for cleaning of reusable devices remains. Manufacturers typically include cleaning instructions in their operating manuals. Medical devices should not normally be returned to clinical engineering workshops soiled; procedures to ensure safe decontamination of devices that have been subject to possible internal contamination should be in place. External cleaning between episodes of patient care is also important, and clarity should be reached over who is responsible for this.

Patients will expect to see equipment that is safely mounted and installed ('G'). This includes its physical ergonomic layout within the environment of care; poor ergonomics have led to adverse events, with clinical staff not able to see and operate devices properly. The mounting and positioning must be robust, such that devices do not fall from physical mounts. The introduction of medical devices into the clinical environment, often called the 'commissioning', must have been properly planned and executed ('H'). The commissioning, discussed in more detail in Chapter 7, Section 7.2.3.4, will include configuring the devices to ensure that their settings and features (including the alarms) are appropriate for the particular environment of care (Case Study CS4.1).

9.3.2 Support Systems within the Healthcare Organization (Middle Circle)

Behind the scenes from the immediate patient/medical equipment interface, the healthcare organization will have procedures in place to ensure governance of all their medical device assets (Figure 9.3). The patients and the Board of the healthcare organization will expect the medical devices to be competently managed in accordance with agreed polices. The Medical Device Policy and Strategic HTM Plan ('I') are useful tools for exercising this governance. These are discussed in more detail in Chapters 4 and 5 and may be supported by tools such as responsibility matrices showing who is responsible for what equipment. We have described an overseeing committee ('J') that we call the Medical Device Committee (MDC) (Chapters 4 and 5). The MDC is delegated by the Board and responsible to the Board for ensuring good governance of the medical equipment in compliance with its Medical Device Policy (Case Study CS5.3).

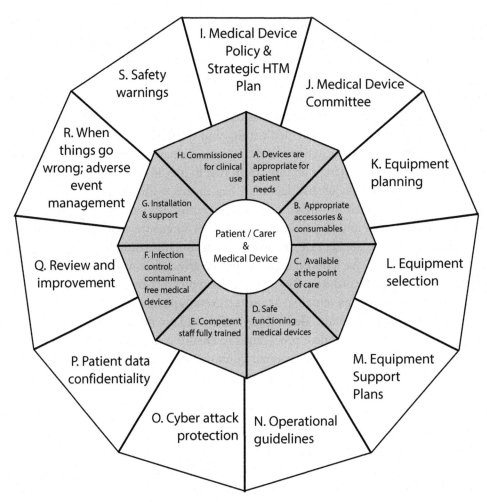

FIGURE 9.3 The middle ring of the A–Z of medical device governance; support systems within the healthcare organization. The elements I–S are described in the text. The inner circle of processes at the point of care is also shown.

The A–Z of medical device governance processes starts with ensuring that the medical devices supplied for use at the point of care are appropriate for each individual patient's needs ('A' – Figure 9.2). This requires careful medical device planning ('K'), compliant with the changing strategic aims of the organization and the state of the art of medical technology that recognizes its expected useful lifespans. This in turn is reliant on prudent financial planning for its procurement and replacement, including developing good understanding by the organization's Finance Department of the whole life cost of medical equipment (Chapter 4, Section 4.2). This may be achieved by developing and implementing, in co-operation with Finance, a rolling replacement programme that predicts future budgetary demands, but that has the flexibility to accommodate changing clinical priorities (Case Study CS5.6).

Acquiring the appropriate medical devices is not achieved without effort and planning. We have discussed in Chapter 5 the importance of including users and patients in the evaluation and selection of medical devices. The selection process ('L') should be rigorous to ensure the acquisition of the most suitable devices for the clinical care that is provided. Unless suitable medical devices are acquired, it may not be possible to prescribe the appropriate devices that match the needs of the patient, process 'A'. The acquisition processes, as we have seen in Chapter 5, Section 5.6.4, start with the identification of need, in turn informing the detailed specification from which evaluation criteria are developed to guide the selection (Case Study CS5.11).

The governance of the medical devices in operational use requires technical equipment support plans (Chapter 6) ('M') and clinical operational guidelines ('N'). The ESPs are typically developed and managed by the Clinical Engineering Department. The clinical operational guidelines and procedures will include care plans and clinical care checklists; examples including routine checks of fluid delivery from infusion pumps and operational integrity checklists of anaesthetic and surgical equipment prior to commencing surgical procedures.

In Chapter 6, Section 6.2.2, we described the ESP as a holistic support package that encompasses both technical and clinical support, suggesting that the development of the ESP should involve co-operation with clinical staff. This will enable linkages to be established between the ESPs and the clinical operational guidelines, each supporting the other. Thus, for example, the MRI ESP (Case Study CS6.1) includes as explicit elements, the clinical support and patient support components. This collaborative approach between the clinical and technical care for the equipment can also benefit user training on medical equipment, where the training packages can be developed to incorporate both clinical and technical elements, with clinicians and clinical engineers encouraged to attend both. The collaboration can in turn help drive improvements in care pathways (Case Studies CS2.4 and CS2.5).

Information flow increases the effectiveness of healthcare, and consequently, medical equipment is increasingly connected and networked through Information Technology (IT) networks. However, this is not without risk, opening the medical equipment systems to vulnerability to attack from malicious software (ECRI 2010, 2014; Wirth 2011; Maron 2013; Wu and Eagles 2016). The FDA and other regulatory agencies have issued warnings on the vulnerability of medical equipment to cyberattack. The ISO 80001-1 Standard has been developed to guide healthcare organizations in the risk management of networked medical equipment (Eagles 2008). Clinical Engineering departments will need to work with their IT and eHealth (WHO 2016) colleagues and with suppliers to ensure that their medical equipment is protected against malicious attack ('O'). The FDA has emphasized the shared responsibility of healthcare organizations and medical equipment suppliers for ensuring medical equipment protection (FDA 2009). Healthcare organizations are cautioned against independently adding antivirus protection fixes to medical equipment as these may adversely affect the operation of the medical equipment; team work between healthcare organizations and suppliers is required.

Co-operation between clinical engineering, IT/eHealth and Information Governance colleagues is required to ensure that medical devices do not become sources for breaching patient data confidentiality ('P'). Medical devices may acquire and store patient data including patient identifiers such as name, date of birth and hospital number. Healthcare organizations have a duty of care for the personal data that they acquire, including the data acquired by medical devices. Conventional processes for managing the security of patient data on personal computers and conventional IT systems are not always appropriate for medical devices. A particular risk is the short-term use of medical devices within a particular healthcare organization. For example, medical devices may be loaned for a particular clinical purpose or for evaluation; during this period of clinical use, the device may acquire personal data from patients and also staff. The healthcare organization needs to ensure that the data are erased from the medical device prior to the device's removal from the healthcare organization (Case Study CS4.5). There are similar concerns about patient data on medical devices that require to be sent away for external repair. It is typically not possible to safely remove and destroy the hard disk or other internal storage system from the medical device prior to its return to the supplier without adversely affecting the operation of the medical device. This is because the same internal storage medium may hold the equipment's operating system and the patient data. Healthcare organizations should work with suppliers to agree processes for ensuring the safe removal of patient data from medical equipment prior to agreeing to evaluate or procure medical equipment. Clinical engineers have a responsibility to work with their IT/eHealth and Information Governance colleagues to preserve the confidentiality of patient and staff information.

In the background, underpinning and supporting the organization's medical equipment support systems will be reviews and audits of how well the equipment is supporting patient care and the need for optimizing, refreshing and replacement ('Q'). We discussed in Chapter 5, Section 5.4.1 the continual improvement of HTM processes through the Plan-Do-Check-Act cycles.

9.3.3 When Things Go Wrong

An important part of the 'behind the scenes' procedures for ensuring good governance of the medical equipment will be the procedures put in place to manage problems or incidents, some of which might cause harm to patients, visitors and/or staff ('R'). We have seen in Chapter 5 that the policies for the management of adverse events involving medical equipment are included in the Medical Device Policy (Chapter 5, Section 5.3.2.12). The topic is so important that it has been discussed elsewhere in this book (Chapters 2, 4 and 7). The nature and causes of these adverse events will not be discussed here but the reader is referred to the literature (Amoore and Ingram 2002, Jacobson and Murray 2007, Amoore 2014). Medical equipment is involved in a small but recognizable number of adverse events. Because of the nature of the medical equipment and its use, the implications of these adverse events can be severe. Clinical engineers have an important role in reporting incidents, in the project teams formed to investigate incidents (Case Study CS7.7) and in holistically looking at the causes of incidents (Case Study CS2.2) to develop methods of preventing recurrence, including reporting to and working with regulatory authorities (Powell 2013; Boutsikaris and Morabito 2014; Morschauser 2014).

Safety Warnings are issued by governmental organizations, regulatory authorities and suppliers advising healthcare organizations and users of problems with medical devices or their operational use ('S'). Healthcare organizations must have procedures in place to respond to these safety notices, taking the appropriate actions. The need to respond to these safety warnings is included in the Medical Device Policy (Chapter 5, Section 5.3.2.13).

9.3.4 Industry, Standards and National Agencies (Outer Circle)

Beyond the healthcare organization, within industry and national regulatory agencies, processes are in place to ensure the integrity of medical devices (outer circle). Governance starts at the beginning of the medical device development cycle (Figure 9.1), from the very start, when the idea for a new or revised medical device is being conceived. At the very start, industry should ask itself what benefit the proposed product will provide for patient care; will it improve on existing methods of healthcare delivery ('T')?

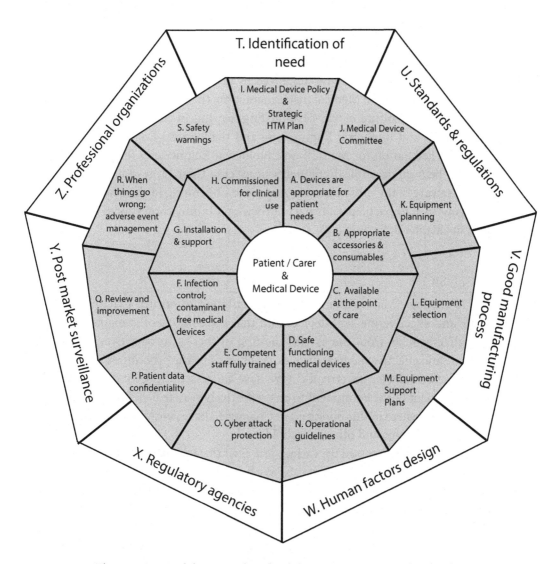

FIGURE 9.4 The outer ring of the A–Z of medical device governance: role of industry, Standards bodies and National Agencies. The elements T–Z are described in the text. The inner and middle circles (Figures 9.2 and 9.3) are also shown.

Standards play a very important role in ensuring the safety and efficacy of medical equipment ('U') and we have seen in Chapter 3 the various types of Standards and how, combined, they work together to improve medical equipment. In addition, there are Standards that guide good manufacturing practice ('V'). The design and manufacturing team will need to be aware of the human usability of the equipment, taking into account the intended environment of use and the anticipated skills and competence of end users ('W'), recognizing that some end users will be patients and their carers without professional clinical knowledge and skills.

Supporting the equipment manufacturers are national regulatory agencies ('X') such as the FDA in the United States and medical device regulations such as those prescribed in the European Union (Chapter 3). Post-market surveillance ('Y') keeps an eye on how equipment actually performs in practice (Wright and Datlof 2010; Sorenson and Drummond 2014; Daniel et al. 2015; Vockley 2015), with action taken to address safety and efficacy problems that may arise.

Individual clinical engineers will be supported by professional organizations ('Z') such as the American Association of Medical Instrumentation, the American College of Clinical Engineering, the UK Institute of Physics and Engineering in Medicine, the Australian Society for Medical and Biological Engineering and the Clinical Engineering section of the International Federation for Medical and Biological Engineering (http://www.ifmbe. org). These can help guide clinical engineers practicing their profession. Clinical engineers should be aware of the professional organizations operating in their jurisdiction and should be encouraged to participate as part of their continuing professional development as discussed in Chapter 8, Section 8.7.

9.4 CONCLUSION

Healthcare Technology Management (HTM) is designed to ensure that patients benefit from safe and effective medical devices and that the equipment is managed to cost-effectively realize benefits from the assets. The first of these two objectives, namely the provision of safe and effective medical devices, is achieved by the procedures and processes that make up 'medical device governance'. (We have addressed the second of these objectives, the management of the equipment to cost-effectively realize benefits from the assets in Chapter 4 and elsewhere in this book and in Chapter 10 have summarized in more detail how HTM benefits patients and other stakeholders.)

In this chapter, we have discussed the various processes that together enable medical devices to be used safely and effectively for patient care. We have explored this from the perspective of the patient whose care is supported by the medical devices. At the point of care, patients will want to be supported by appropriate medical devices operated competently by trained staff, confident that the devices are properly maintained and kept clean with procedures in place if problems develop. The processes in the inner circle of Figure 9.2 are designed to ensure this.

But these processes need background support. The appropriate medical devices will not be available unless the healthcare organization has equipment planning and selection strategies implemented. The healthcare organization should manage its assets strategically and we have emphasized from Chapter 1 the need for a structured approach, suggesting that this be based around the ISO 55000 group of asset management Standards. A structured approach based around a Medical Device Policy, owned by the healthcare organization's Board, is advocated (Chapter 5) and supported by a Medical Device Committee. Holistic plans for supporting all the medical equipment, which we have termed the equipment support plans, ensure that the equipment is kept safe and functional, and with clinicians trained in their operation. Procedures must be in place to manage risks and to respond when problems with medical devices occur.

Good practices within healthcare organizations themselves would be compromised and undermined if the industry supplying the medical devices does not ensure its own good governance. Industry is supported by Standards and regulations, with increasing advice on the principles of human factors design to ensure that medical devices are intuitively easy to use, with the risk of incorrect use minimized – the so-called mistake-proof design concept. Manufacturers and suppliers have a continuing responsibility for their products, with post-market surveillance processes in place to support them in exercising this responsibility. Standards are also maintained and supported by professional organizations that help those working in the field ensure their continuing professional development.

Thus, those receiving support and healthcare from medical devices are sustained by processes close to the point of care, at the infrastructure level of their healthcare organization and by the wider medical device industry and the standards and professionalism that supports it. Safe and effective medical devices require and are provided by supportive rings of processes and procedures.

REFERENCES

Amoore J.N. 2014. A structured approach for investigating the causes of medical device adverse events. *Journal of Medical Engineering*, vol. 2014, Article ID 314138, 13 pages, 2014. doi: 10.1155/2014/314138 (accessed 2016-08-11).

Amoore J.N. and P. Ingram. 2002. Learning from adverse incidents involving medical devices. *British Medical Journal*, 325(7358): 272–275.

Boutsikaris S. and K. Morabito. 2014. Critical success factors for the clinical engineer when resolving medical device problems: A case review. *Journal of Clinical Engineering*, 39(4): 172–174.

Daniel G., Colvin H., Khaterzai S., McClellan M. and P. Aurora. 2015. Strengthening patient care: Building an effective national medical device surveillance system. The Brookings Institution, Washington, DC. http://www.fda.gov/downloads/aboutFDA/CentersOfMedical ProductsandTobacco/CDRH/CDRHReports/UCM435112.pdf (accessed 2016-05-12).

Eagles S. 2008. An introduction to IEC 80001: Aiming for patient safety in the networked healthcare environment. *IT Horizons*, 4: 15–19. Published by AAMI, http://my.aami.org/store/detail.aspx?id=IT2008 (accessed 2016-05-21).

ECRI. 2010. Anti-malware software and medical devices. *Health Devices*, 39: 360–365.

ECRI. 2014. Top 10 health technology hazards for 2015. https://www.ecri.org/Press/Pages/ECRI-Institute-Announces-Top-10-Health-Technology-Hazards-for-2015.aspx (accessed 2016-05-12).

FDA. 2009. Cybersecurity for networked medical devices is a shared responsibility: FDA safety reminder, November 2009. http://www.fda.gov/MedicalDevices/Safety/AlertsandNotices/ucm189111.htm (accessed 2016-05-12).

Jacobson B. and A. Murray. 2007. *Medical Devices: Use and Safety*. Edinburgh, UK: Churchill Livingstone.

Maron D.F. 2013. A new cyber concern: Hack attacks on medical devices. *Scientific American*, June 25, 2013.

Morschauser M. 2014. Improving patient safety through collaboration between clinical staff and engineering staff in hospitals. *Journal of Clinical Engineering*, 39(3): 129–131.

Powell K. 2013. Hospital-based clinical engineers' contributions to the Food and Drug Administration's Medical Device Safety Network (MedSun) and to the Public Health. *Journal of Clinical Engineering*, 38(2): 72–74.

Sorenson C. and M. Drummond. 2014. Improving medical device regulation: The United States and Europe in perspective. *Milbank Quarterly*, 92(1): 114–150.

Vockley M. 2015. Keeping an eye on medical devices postmarket surveillance enters new age. *Biomedical Instrumentation & Technology*, 49(6): 378–392.

WHO. 2016. Health topics—eHealth. http://www.who.int/topics/ehealth/en/ (accessed 2016-09-06).

Wirth A. 2011. Cybercrimes pose growing threat to medical devices. *Biomedical Instrumentation & Technology*, 45(1): 26–34. doi: 10.2345/0899-8205-45.1.26 (accessed 2016-05-12).

Wright E. and S. Datlof. 2010. Adverse event reporting in the EU and the USA: Similarities and differences. *Journal of Medical Device Regulation*, 7(3): 14–22.

Wu F. and S. Eagles. 2016. Cybersecurity for medical device manufacturers: Ensuring safety and functionality. *Biomedical Instrumentation & Technology*, 50: 23–34.

STANDARDS CITED

Note 1: Undated references to Standards are given. It is important to always be aware of the most up-to-date version. These can be found by searching the IEC or the ISO website – see Chapter 3.

IEC 62366-1   *Medical devices, application of usability engineering to medical devices*

IEC 80001-1   *Application of risk management for IT-networks incorporating medical devices – roles, responsibilities and activities*

ISO 55000   *Asset management – overview, principles and terminology*

ISO 55001   *Asset management – management systems – requirements*

ISO 55002   *Asset management – management systems – guidelines for the application of ISO 55001*

SELF-DIRECTED LEARNING

1. Prepare a talk for your colleagues on medical device governance showing how, in your healthcare organization, the HTM processes in place help ensure safe and effective application of medical equipment. Then critically evaluate a few of those processes, discussing how they actually work in practice. Analyse them, perhaps using systems engineering methods or using techniques such as the Plan-Do-Check-Act cycle to suggest and test improvements to the processes.

CHAPTER 10

Healthcare Technology Management Adding Value

CONTENTS

10.1 INTRODUCTION

An underlying theme of this book is that the merit of Healthcare Technology Management (HTM) should be judged on the value that the medical equipment and devices provide as well as what the management programme delivers for the stakeholders. The stakeholders will include the patients who are the ultimate beneficiaries of healthcare. The clinicians who use medical equipment to deliver healthcare are also stakeholders as are those charged with the management of healthcare organizations and those who fund these organizations and systems. The stakeholders require that the value of the medical equipment assets is enhanced, thereby actively supporting and advancing care, bringing benefit to both patients and the healthcare organization. Amongst the stakeholders that clinical engineers benefit through effective HTM, it is worth including the medical device industry. Whilst the industry is strictly speaking not a stakeholder served by the Clinical Engineering Department (CED), the role played by clinical engineers within healthcare organizations does benefit the industry, and the medical device industry benefits the delivery of healthcare. Benefitting stakeholders requires HTM that is not simply passive, but an active process that aims to realize the enhanced value that is available from these assets.

In concluding this book, we summarize our thesis that effective HTM can bring this benefit to the stakeholders by actively managing the medical equipment to both support and advance care. We support this conclusion here with a brief overview of the contents of this book, enabling this chapter to be read on its own.

10.2 OVERVIEW OF THIS BOOK

In Chapter 1, we presented a vision of clinical engineers advancing care through a structured approach to managing medical equipment and devices designed to add value for patient care. 'Value' is defined as the benefits derived from the medical equipment and devices for the organization in relation to the costs incurred. The structured approach is based around the Asset Management Standard, ISO 55000, which puts at its core the objective of managing the assets in such a way as to add value. There has been much debate about how best to organize healthcare systems to ensure that they are affordable and effective. These debates are not new and are unlikely to fade. It has been suggested that healthcare organizations can benefit from a structured 'systems' methodology, and in Chapter 2, we discuss the systems engineering approach with its focus on identifying the purposes and the objectives of healthcare systems and processes. Standards and regulations provide the foundation of good practice in HTM, and we discuss these in Chapter 3.

Chapter 1 introduced the concept of Value, the ratio of the benefits obtained from an asset divided by its associated costs, and we return to this theme of adding value through effective HTM throughout this book.

Chapters 1 through 3 set the general background scene from which we discuss practical concepts and approaches to HTM. We moved on in the next part of this book, starting in Chapter 4, by reviewing the management of medical equipment across its whole life cycle, from which we discuss the real whole life financial costs associated with medical equipment; this shows that it is not simply the acquisition costs that must be considered when evaluating medical equipment for procurement, but that those selecting medical equipment must investigate the costs of maintaining, using and managing the equipment. From there we proceeded, in Chapters 5 and 6, to explore in more detail the processes of managing the medical equipment, what is commonly called Healthcare Technology Management (HTM). HTM aims to increase the value of the medical equipment assets for the healthcare organization, linking the HTM objectives and operations to the strategic aims of the organization. We showed how this can be achieved with a structured approach in which the organization's Board and Chief Executive Officer (CEO) exercise governance of the medical equipment through their Medical Device (MD) Policy and an interdisciplinary group we call the Medical Device Committee (MDC). We recognize that the concept of a MDC may be new to many readers but suggest that it is similar to the governance structures put in place for managing pharmaceuticals in many healthcare organizations (e.g. through a drugs and therapeutics committee). We then, in Chapter 6, looked in more detail at the principles that underlie the practical support for medical equipment through Equipment Support Plans (ESPs).

We have emphasized from the beginning that clinical engineers have a twin remit: first, to manage the technology, and second, through their knowledge of the technology and its clinical implications and their knowledge of systems engineering, to support and advance patient care more broadly. We explore these twin remits further in Chapter 7, showing how they are often achieved by convening specific projects that enable appropriate, often multidisciplinary, groups of staff to concentrate on developing solutions to particular projects, and we have illustrated some of these scenarios with case studies.

The clinical engineers who provide HTM services require resources, and in Chapter 8 we described the principles behind the organizational structures of CEDs.

Finally, in Chapter 9, we summarized the processes that together provide medical device governance, which are the processes that ensure safe and effective application of medical devices for patient care.

10.3 HTM: BENEFITTING STAKEHOLDERS

10.3.1 HTM Benefitting Patients and Carers

The patients and their carers are the ultimate raison d'être and beneficiaries of healthcare. HTM must be directed and focused to maximize the benefits that the medical devices and equipment assets offer patients and their carers, recognizing the responsibility of achieving this objective ethically and sensitively, ensuring person-centred HTM.

Whilst patients may not be aware of the details of the HTM processes that are required, they rightly expect that appropriate management systems should be in place. Just as those entering an aircraft can confidently expect the aeronautical engineers to have done a complete and professional job, patients have a similar expectation, albeit implicit, regarding HTM. So the silent and often invisible work of clinical engineers in delivering holistic HTM is to the benefit of patients.

Patient- and carer-centred healthcare starts with the recognition that patients and their carers are participants rather than simply recipients of care. The practice of HTM should recognize and support this important approach to healthcare. Thus, the acquisition of the medical equipment should begin by asking what benefits the equipment will provide for patients and their carers.

Where medical equipment is to be used in the home environment, it is perhaps clearer that patient and carer considerations should be included in the specification and evaluation criteria. Patient and carer considerations also need to be included when planning medical equipment for use within hospitals. It is widely accepted that good hospital architectural design contributes to an environment that promotes a sense of well-being for staff and patients. Well-designed rooms, the use of natural light, good interior design and the use of art can all contribute to an environment that makes the lived experience of being in hospital better for all. Clinical engineers should work with architects and clinical staff to sensitively design medical rooms and the deployment of medical equipment within the rooms, considering not only the intended clinical use of the equipment but also how it impinges on the experience of the patient during their stay.

Just as good architectural design makes for a better healing environment so also can good medical equipment design. Medical equipment can, by its very nature, have an adverse psychological effect on patients and their families. Those unfamiliar with hospital environments find the sight of family and friends connected to medical equipment disconcerting and sometimes frightening, particularly in equipment intensive areas such as Critical Care. Good ergonomics can improve the aesthetics of the physical connections between patient and medical devices such as patient monitors and infusion devices, in addition to minimizing the limitations on patient movements of cables and sensor wires. Good configuration of alarm settings can go a long way to reducing the inevitable stress and disturbance for patients associated with alarm errors. Manufacturers are increasingly aware of the need to address patient concerns in their design criteria. This requires them to work within the constraints of the science underpinning the medical equipment whilst considering patient aspects. A good example is the efforts that manufacturers of MRI scanners have made to minimize the noise and claustrophobic nature of the MRI experience, even including this in their advertising literature.

Explanations to patients and visitors of the purposes of medical equipment can help. One of the authors has a vivid memory of being called, out of hours, to replace a faulty multi-parameter patient monitor in a critical care ward. When the monitor failed, patient monitoring had been immediately transferred to a transport monitor. The clinical engineer arrived to find the wife of the patient, already distraught by her husband's illness, further worried that the smaller transport monitor was not providing as much care as the larger monitor. She frantically urged him to get operational the 'better' monitor so that it could look after her husband as quickly as possible. As clinical engineers support medical equipment, they must keep in mind the needs and feelings of patients and their families and friends.

Patients and their carers may, to varying extents, operate or help to operate their equipment; this is not just applicable for medical equipment used in the home environment. The descriptor 'patient-controlled analgesia' (PCA) pump implies patient control of a medical device that delivers (pumps) controlled amounts of medication under patient control. These PCA devices are often deployed after surgical operations to relieve pain, with the patient taking direct control of administering the pain relief through a handset or other controller. Lay carers may also support care in hospitals. This is not uncommon in neonatal and paediatric wards where parents may be encouraged to contribute to their child's care, perhaps including helping to apply or operate some medical equipment. Proper training and support are required for these carers to ensure that they can safely and effectively help manage the equipment.

Similarly, maintenance arrangements should consider the needs of patients and carers, not just in the home environment but also within the hospital. It may include helping patients and carers to manage problems with equipment and how to report them. We advocate the development of holistic equipment support plans that recognize that clinical engineers, clinicians, patients and carers have a contribution to make to the maintenance of medical equipment.

Including patients and their carers in the evaluation of medical equipment and as members of medical equipment committees such as the MDC can help educate clinical engineers and others of the needs of patients and carers and how the HTM process can be directed to their benefit.

Including patients and carers as partners in their healthcare does not of course imply that the professionals abrogate their responsibility of care for the patients. The professionals have the knowledge, skills and access to resources to direct and provide care, and this needs to continue to be acknowledged and respected.

10.3.2 HTM Benefitting Clinical Staff

The clinical staff should benefit from HTM that ensures the availability at the point of care of the most appropriate medical equipment and enhances the value of that equipment through ensuring that the maximum benefit is derived from it. There has been a long-standing perception that clinicians are not adequately involved in the evaluation and selection of the equipment they use, and too often that lack of perceived involvement is the reality. Critical care nursing staff being present whilst patient monitoring equipment is being evaluated in their ward but not being asked to comment nor being involved in the selection process sadly does happen and does not bode well for a successful outcome. Surgeons being told that this is the surgical endoscopy system they will use without being able to evaluate the usability characteristics of different systems can lead to problems later.

Clinicians from all relevant professions (not just doctors) should be members of procurement project teams, starting with their input to the identification of need and the specification and evaluation criteria. The procurement project should adopt a multidisciplinary team approach, with clinicians included in the evaluation teams, recognizing that different groups of clinicians may want to be involved in different ways. Their participation can highlight and prevent medical equipment with poor ergonomic design being procured. The active participation of clinicians in the procurement process can help ensure that the equipment provides the optimum benefits for their delivery of healthcare. Clinicians' advice on the physical mounting of medical equipment in the clinical area can also benefit their ability to work ergonomically with the equipment, avoiding repetitive strain injury and ultimately benefitting both staff and patients. Some equipment may need to be portable with associated ergonomic requirements: the displays and controls of a patient monitor to be used for patient transport trolleys must be able to be seen and readily accessed during patient transfers. Medical equipment must be designed with operator controls logically arranged to help ensure safe operation. Active participation by clinicians during the evaluation process can identify poor operator controls and operator controls that can lead to hazardous situations. For example, a nurse may recognize during the evaluation of a blood pressure monitor that it is easy to inadvertently enter a maintenance mode that inflates the cuff to a high pressure to test for airway leaks. If this is done whilst a cuff is placed around a limb to measure the blood pressure, the high cuff pressures generated can lead to severe pain or bruising to the patient. This is a good reason to veto that particular device; good practice would ensure that constructive critical comments are given to the supplier. Clinicians' familiarity with the practical operation of medical equipment is an important experiential knowledge base for medical equipment evaluation.

HTM should incorporate effective training for end users in the operation of their medical equipment, and the training must include an understanding of the characteristics and limitations of the technology. For example, it is not sufficient for clinicians to know how to operate a pulse oximeter, but they should also understand how the device measures the arterial oxygen saturation so that they can be aware of contra indications such as conditions where its measurements may be unreliable. They should also understand any feedback information from the device that suggests problems. Pulse oximeters, to continue that example, may display the pulse waveform or an indication of pulse strength that may confirm valid monitoring or indicate a problem. Effective clinician training, which incorporates technical training on the medical equipment and also its clinical application, enhances the safe and effective deployment of the medical equipment. It thereby minimizes those reported faults where investigation reveals no faults and also minimizes adverse incidents associated with the use of medical devices and equipment. Clinical engineers can play a very effective part in clinical user training.

Clearly, clinicians will benefit from medical equipment that is properly maintained and thus safe and effective in use. Clinicians also benefit when medical equipment is configured to match, as closely as possible, their clinical practices. Configuration of medical equipment is an important part of its commissioning, ensuring that the functional characteristics of the equipment are best suited to the circumstances of clinical use. Configuration will include adapting the characteristics (layout and colours) of the visual display, enhancing the prominence of clinical signs important to the users. It will also tailor functional attributes such as the permissible flow rates of infusion pumps. Careful configuration of the alarm settings on medical equipment can better ensure that alarms that are triggered are real and require intervention, minimizing the false alarms that can disturb healthcare processes. Standard configurations for all equipment of the same type used across the healthcare organization add significantly to patient safety and effective user training. Clinical engineers are well placed to facilitate agreement on such organization-wide configurations.

10.3.3 HTM Benefitting the Healthcare Organization

Much of the day-to-day HTM work is directed to individual clinical departments to provide them with the maximum benefit from their medical equipment. This will be shown in clinical engineers responding to calls about faulty equipment, equipment operation, training and requests for additional or replacement equipment. This is valid, and through its support for individual departments, the HTM benefits the organization as a whole.

The healthcare organization also benefits directly from HTM processes that competently and effectively manage and control all phases of all its medical equipment over their full life cycle. These processes must be designed to support the strategic aims of the organization in an inclusive manner, recognizing the objectives of senior clinical and executive staff. The executive and non-executive members of the Board will require assurance that the medical equipment is effectively utilized for healthcare delivery. Structured HTM processes provide frameworks that reassure the Board and senior management of the quality of the management of the medical equipment for which they are responsible. These structured processes provide methods that enable the Board to scrutinize the management of the medical equipment and to obtain clarification when questions arise as to their management.

Leaders of CEDs should foster links with the senior management staff, understanding their priorities whilst at the same time informing senior management of those aspects important for medical equipment, its management and its application. Clinical engineers should not be afraid to be proactive in recommending changes that could benefit patients or add value to the organization. The CED is often well placed to comment on organizational-wide initiatives that could benefit the care process. This mutual sharing of priorities, information and concerns, supported by groups such as the Medical Device Committee, can foster the necessary collaboration and understanding that optimizes the HTM for the benefit of the whole organization.

10.3.4 HTM Benefitting the Funding Providers

The objective of HTM is to enhance the *value* of the medical equipment assets, that is, maximizing the *benefit* to be derived from the equipment in relation to the *costs*, an objective whose realization can be achieved by adhering to good asset management principles such as those outlined in ISO 55000. Clearly, funding providers will be keen to see reduced costs, but they are likely to be keener to see the funding that they provide effectively used, generating enhanced *benefits*. Optimizing the balance of *benefits* and *costs* does not imply opting for the lowest cost.

The benefits delivered from medical equipment vary depending upon the equipment and the clinical context in which it is used. However, where benefits delivered are maximized, one can reasonably expect to see an improvement in quality of care or the experience of care, or an increase in the quantity of care that can be delivered to the population served by the institution. Thus, for example, a better measurement might improve diagnostic accuracy. A better design of a piece of equipment might reduce the risk of an adverse event. A better equipment support plan might mean less downtime and the organization better able to meet the demands for healthcare.

Funding providers want to see tangible benefits realized from the assets. The HTM should not simply passively manage the medical equipment, but manage it in such a way as to advance and support patient care. This will often require that the culture and processes of care change to take advantage of improved functionality of medical equipment. An example from the application of technology in everyday life might help illustrate this. Wired telephone equipment required users to phone from designated telephone areas and phone booths. Consider the implications of retaining that culture and practice when using mobile phones, not exploiting the 'mobile' functionality. Applications of modern medical equipment without considering how its functionalities can improve the processes of care is rather like confining the use of mobile phones to designated phone points. Clinical engineers are well placed to encourage clinicians and hospital managers to enhance the benefits of patient care that technological advancements offer.

This emphasis on *value* will apply to all phases of the medical equipment life cycle. Procuring the lowest cost medical equipment does not necessarily lead to the greatest benefit; indeed, some procurement regulations do not allow selection to be based on cost alone. In some jurisdictions, procurement guidelines promote the selection of equipment based on the most economically advantageous tender, taking value and life cycle costs into account not just the purchase price. Medical equipment must be managed recognizing its whole life costs which incorporate the acquisition, operational and disposal costs, assessing them against the whole life benefits. It is this whole life benefit that the HTM processes are designed to optimize. Different types of medical equipment may require different equipment support plans, but the different processes and plans should be proactively managed to ensure a holistic HTM Programme that optimally supports the equipment to serve the delivery of healthcare for the benefit of patients.

10.3.5 HTM Benefitting the Medical Device Industry and Regulatory Authorities

Clinical engineers employed within healthcare organizations work closely with representatives of the medical device industry throughout the whole life cycle of medical equipment from acquisition, through its operational use and its end-of-life disposal. At the often competitive acquisition phase, the medical equipment industry benefits from clear advice and help from clinical engineers as the industry seeks to offer its most suitable products to meet the healthcare organization's needs for equipment to support patient care. This will extend to the support that the clinical engineers provide industry in the organization of selection evaluation trials and final discussions on the equipment to be ordered and the commissioning of the selected equipment. The clinical engineer will need to be objective and business like in their dealings with the medical device industry particularly during this competitive phase and also throughout the life cycle of the equipment within the healthcare organization.

The industry also benefits from clinical engineers through learning about the practical use of medical equipment in delivering care and more particularly in finding solutions to problems. The medical device industry has a continuing duty of care for the equipment it supplies, summarized in the phrase 'post-market surveillance' (PMS). Clinical engineers support industry carrying out this duty of care. The support can be particularly useful where medical devices are involved in adverse events. The clinical engineers will need to be particularly objective, recognizing that their responsibilities are first to their healthcare organization and its patients and staff, then to regulatory authorities and only then to the medical device industry. The regulatory authorities gain much from working with clinical engineers investigating adverse events involving medical devices.

Clinical engineers, as we have seen in Chapter 1, also benefit the medical device industry as it seeks to develop new products. The industry often looks to clinical engineers for advice and support in introducing these new products. Through their work in Standards organizations (Chapter 3), clinical engineers also advance the state of the art of medical equipment and its effectiveness and safety.

10.4 DISCUSSION

This concluding chapter rounds off the call in Chapter 1 for clinical engineers to manage the medical equipment in such a way as to add value for:

- The patients, for whom the medical equipment supports and delivers healthcare;

- The clinicians who use the medical equipment as tools for providing healthcare;

- The healthcare organization that provides the care;

- Those who fund the healthcare organization.

Underpinning these added benefits and ensuring the safe and effective application of medical equipment are several important themes. The discipline of engineering aims to solve real-world problems, to make the world a better place for all. Engineers are skilled not only in generating new abstract ideas to solve problems but also in translating them into reality. They can do this in a structured and cost-effective way for the benefit of all. These aims and skills apply particularly to clinical engineers, bringing the benefits of technology to healthcare, bridging the technical and clinical knowledge spaces through their understanding and skills in both areas. Clinical engineers apply these skills within the healthcare system to improve the treatment and care of those most in need in society. It is a challenging and rewarding endeavour.

All too often those not familiar with the practice of clinical engineering judge it on the basis of what they see. The physical presence of most CEDs would lead the casual observer to believe the work was predominantly a maintenance one. Whilst technical maintenance is an important part of the work of clinical engineers, it does not define the role completely. In this book, we have set out how medical equipment should be managed in a wider context, one that focuses on delivering real benefits for patients, their carers and families. Further, we stress that the process of managing the medical equipment should achieve and enhance the value of the equipment for the healthcare organization. We align the process of managing medical equipment (HTM) with well-established asset management practices that are used by engineers in other industries and set out in the ISO 55000 suite of Standards.

We promote managing the medical equipment assets in a holistic way, one that is, informed by evidence of what is appropriate for today's application of the equipment, not just practice as it was in the past. The word 'application' in the previous sentence is important. The focus is not simply on the equipment, particularly not on the equipment for its own sake. Rather, it recognizes the important role that the application of medical equipment has in supporting healthcare; the challenge to clinical engineers is to use their knowledge and skills to support and advance healthcare through the effective application of medical equipment. Thus, we discussed the twin roles of the clinical engineer: first to manage the medical equipment and second to support and advance healthcare.

The supporting and advancing care role of clinical engineers stem from them being creative problem-solvers. This opens up a myriad of scientific and technical support opportunities that, whilst related to technology, extend way beyond the traditional asset management function. We have touched on some of these opportunities in Chapter 7 and elsewhere throughout this book. However, there are many more, and as technology advances and changes, new roles and opportunities to contribute will emerge. Clinical engineers have always played a role in advancing care by ensuring the appropriate adoption and use of technology in healthcare. Their challenge is to continue to extend the boundaries.

The interdependence of these two activities, asset management and creative problem solving, is important. Both inform and reinforce each other, and it is hard to imagine how a CED can stay effective without participating, to some degree, in both.

Clinical engineering is an interdisciplinary pursuit drawing as it does on both the physical and the life sciences. It is a place where the enquiring mind will find much to occupy it. There remain and will always be challenges to overcome, processes to improve, technology to deploy more effectively and patients to care for. So engineers of all types who have the opportunity to enter and work in this field will find it a stimulating and rewarding sector of engineering within which to practise.

Clinical engineers face both an exciting challenge and a wonderful opportunity: to enhance healthcare through the structured management of its medical device assets that add value for the benefit of patients. Healthcare faces enormous challenges, confronted on the one hand by the demands from rising populations with increasing healthcare needs, whilst on the other the limited resources in a world facing the need for sustainable growth. The challenge and opportunity are to manage and apply medical devices to meet the healthcare demands within the tight resource constraints. At the same time, the emphasis on person-centred care also requires that healthcare meet the specific and unique needs of individuals.

Advances in technology have transformed human life; this is not new, but accelerating developments in recent decades, particularly in communication technologies, have opened up undreamt of opportunities. Technology has also transformed medicine, with medical equipment opening up new opportunities for delivering care, making what was once unthinkable now routine. But healthcare technologies are challenged to do more, to make possible what seems impossible, namely, providing populations with good quality affordable care whilst ensuring that specific individual needs are catered for. And this includes, where appropriate, moving care out of hospital institutions and back into the community. 'Back into the community' because in the past centuries much of the care took place within the community. The smart phone systems of today have put control of communication technology, and more, in the hands of the individual, transforming communication from early telephones that were wired in place and required those using them to contact the operator to be connected. The challenge for medical technology is to make similar transformations, placing care in the hands of the patients, their family and friends.

In healthcare, clinical engineers act as advocates for technology. However, first and foremost as with all health professionals, clinical engineers are advocates for patient safety. Engineers recognize, more than most professions, the essential attributes that ensure safe and effective medical equipment. Clinical engineers also recognize that effective application of medical equipment is not achieved by simply providing the equipment. Rather, the benefits achievable by the introduction of medical equipment often require culture changes in the healthcare processes. Hence, clinical engineers will act as change agents supporting the application of proven medical equipment. In doing so, they increase the benefit delivered.

Many disciplines within a healthcare organization contribute to the holistic Healthcare Technology Management system and its delivery, but only clinical engineers contribute to all its elements. Their remit covers the development of the overall strategic policy for managing the medical equipment including the structured elements of the Medical Device Policy, the Strategic HTM Plan and the Medical Device Committee (Chapter 5). Complementing their strategic role is their practical remit of implementing the day-to-day management of the medical equipment through detailed ESPs. Clinical engineers manage the whole life cycle of medical equipment within the healthcare organization, from acquisition through operational use to end-of-life disposal. Furthermore, clinical engineers and the CED will develop links with the senior management and clinical leadership of their healthcare organization; these links help promote and ensure effective management of the medical equipment over its life cycle.

The healthcare system that we all want is one that continuously evolves to improve outcomes for patients; this means that as new technologies develop, they will be adopted to support new ways of delivering care. Clinical engineers will support the adoption of the new technologies and the development of new clinical care methods. At the same time, the collaborative and interdisciplinary roles of clinical engineers will also evolve and develop as they lead, manage and support the application of medical equipment and devices to advance healthcare.

10.5 A CALL TO ARMS

The five authors of this book have, between them, over 175 years of experience of clinical engineering, largely in hospitals and also in academia and industry. We have experience of working at all grades of clinical engineering in the national health services in the UK and Ireland as technicians, technologists and professional clinical engineers. We have participated in the wide range of HTM activities: carrying out scheduled inspection; diagnosing and repairing faulty equipment; supporting and training clinicians; specifying, evaluating and selecting medical equipment; managing adverse events; being active in research and development; teaching and managing CEDs and exercising responsibility for medical equipment budgets. Since the 1970s, we have seen significant change in the technology used in medical equipment. Wonderful advances in diagnosis and therapy made real by developments and innovations in material science, engineering and bioengineering. As new technologies found their way into medical equipment, the clinical engineering teams we were part of have had to adapt and change. Change was necessary both to understand the engineering and science underpinning the equipment and to appreciate how it can best be applied in clinical practice. Paradoxically, the one constant we have seen as working clinical engineers over the past 30–40+ years has been change.

As we come to the end of this book, we would ask readers to be open to and embrace change. It is likely that at least some clinical engineering practices, carefully developed and implemented many years ago, could be improved by adjusting the approach to recognize how technology, clinical practice and healthcare delivery within society have all changed. We invite all clinical engineers to examine their practice and challenge the assumptions upon which it is based. Throughout this book, we have emphasized the importance of operating within a quality system where practice is regularly reviewed to ensure it is fit for purpose. If following such a review, no change is warranted, all well and good. However, where change is indicated we encourage clinical engineers at all grades to lead on implementing the necessary change and, using their core skills as engineers, to do so for the good of the patients and people we serve. It may not always be easy to do; however, we should all try and be advocates for good evidence-based practice.

We encourage clinical engineers to extend themselves to support technology and systems which, at the time of writing, are considered new challenges: supporting Medical IT networks, supporting the increasing deployment of medical equipment into the community, analyzing and improving care processes in the emerging role for clinical engineers and researching, developing and innovating medical equipment for the advancement of bioengineering. Through case studies we have illustrated how clinical engineers can contribute to these activities. This requires an openness and ability to work in new ways; these should be fostered and encouraged by leaders of clinical engineering services.

We are mindful that, to be competent and effective in responding to new challenges, clinical engineers need to have a solid grounding in engineering, in system science and in clinical engineering practices. Continued education and training is required, and we encourage clinical engineers to engage in continuous professional development (even when they retire). It is important to contribute to the education and training of the next generation of engineers.

The HTM activity associated with the technical maintenance and user support of large fleets of equipment is important and necessary. It should be conducted in a rigorous manner and be evidenced based. Whilst a significant driver of this activity is cost control, we have demonstrated that this activity also increases the quality and availability of care delivered by the application of technology. It not only assures positive outcomes but also proactively reduces risk of negative ones.

In Chapter 1, we wrote of how both the technical maintenance and user support and also the advancing care role were the two pillars upon which a holistic patient-centred HTM Programme is delivered. The metaphor of the twin pillars of an arch with patient-centred care as the keystone has served us well in structuring the argument for both pillars in this book. Whilst that conceptualization is useful, it is incomplete. In reality, these two activities are so intertwined as to be blurred in practice . A modern HTM approach would embrace aspects of both, and the knowledge and experience gained in either are intimately connected with and relevant to the practice of the other. What is important is that a holistic approach is taken, based on good engineering practice and keeping the needs of the stakeholders to the fore.

The key attribute of engineers and by extension clinical engineers is that we seek to make the world a better place by solving problems, or perhaps better stated as developing solutions to challenges. Clinical engineering is an exciting and rewarding pursuit for an engineer: you are constantly challenged to adapt and support change. Therefore, everyday provides an opportunity for learning and exploration. Whilst often the focus of the activity is related to equipment, the application of that same equipment and how it contributes to the care of people is never far away. Good clinical engineering supports good care and is a rewarding and valuable activity.

Glossary

Term	Synonyms	Definition	Further Reference
Asset		A thing or entity that has potential or actual *value* to an organization.	ISO 55000:2014 §3.2.1
Asset management		A system employed to manage the whole life cycle of (usually physical) objects having a value to the organization.	ISO 55000:2014 §3.3.1 'coordinated activity of an organization to realize value from assets'
Asset management objectives (AMO)		Asset management objectives are set by the *Healthcare organization* to achieve specific measurable results. These also inform the *Strategic HTM plan* and are contained within the *Medical device policy* (Asset management policy).	Strategic HTM Plan = Strategic Asset Management Plan in ISO 55000:2014 §3.3.2 Medical Device Policy = Asset Management Policy in ISO 55000:2014 §3.3.1 + §3.1.18
Asset management plan		*HTM Programmes* which are designed by technical service departments (e.g. the Clinical Engineering Department) and approved by the *MDC*.	ISO 55000:2014 §3.3.3 'Documented information that specifies the activities, resources and timescales required for an individual asset, or a grouping of assets, to achieve the organization's asset management objectives'.
Asset management policy		The *Medical device policy*	ISO 55001:2014 §3.1.18 'Intentions and direction of an organization as formally expressed by its top management'
Biomedical engineer		Usually used for *Clinical engineers* employed in research and development.	
Biomedical engineering		The application of engineering to the medical field.	See Bronzino J.D. (Ed.), 2000, *The Biomedical Engineering Handbook*, 2nd edn., Vol. 1, Boca Raton, FL: CRC Press

(Continued)

Term	Synonyms	Definition	Further Reference
Biomedical equipment technician (BMET)	Biomed	Used in North America as an equivalent to Clinical engineering technologist/technician. See Chapter 8, Section 8.5.	
Black box testing		A method for testing a software system that does not require detailed knowledge of the structure of the code but concentrates on the functional performance of its inputs and outputs. This may involve selecting a set of valid and invalid input data and operational interactions and checking for valid output responses. Also known as functional testing. c.f. *Grey Box Testing*	
Clinical engineer		A qualified person who understands the technical aspects of *medical devices* and equipment and their clinical implications and is able to manage and advise on their use within a *healthcare organization.*	See also Bauld T.J., 1991, The definition of a clinical engineer, *J. Clin. Eng.*, 16:403–405
Clinical engineering		A profession responsible for applying engineering to the management of the life cycle of *medical devices* and equipment within a *healthcare organization* and their interface with and support for the clinical *user.*	
Clinical governance		Assurance that the needs of the patient are being met in the best way and that improvements are sought and acted upon.	See also the Health and Safety Executive, Ireland (www.hse.ie/eng/about/Who/qualityandpatient-safety/Clinical_Governance/)
Clinician		A qualified medical professional: doctor, physician, nurse, *licensed/allied health professional.*	
Connected health		A model for *healthcare* delivery that uses technology to put the correct information in the correct hands at the correct time. It facilitates the provision of *healthcare* remotely.	

(Continued)

Term	Synonyms	Definition	Further Reference
eHealth		'The use of information and communication technologies (ICT) for health. Examples include treating patients, conducting research, educating the health workforce, tracking diseases and monitoring public health' World Health Organization. 2016. Health topics – eHealth. http://www.who.int/topics/ehealth/en/ (accessed 2016-09-06)	See also *Connected health* and *Telemedicine*
End user		A patient or carer who uses *medical devices* themselves, as opposed to a clinical *user*, or *operator*	
End-user support		A term used to describe technical or application support given to patients and carers who use *medical devices* and equipment themselves. The term is also used by suppliers and manufacturers to describe the after-sales service they offer to *healthcare organizations*.	
Equipment management		A system employed to manage the whole life cycle of physical objects having a value to the organization.	
Equipment support plan (ESP)		A document which sets out the range and scope of maintenance and holistic support actions to be delivered mainly by the Clinical Engineering Department for a specific device or group of devices. Also includes details of the information sources upon which the plan is based and the means by which the plan's effectiveness can be assessed.	
Grey box testing		A method for testing a software system which involves some knowledge of the structure of the software code and internal details of the program. This enables a more thorough debugging of the software. c.f. *Black Box Testing*	
Healthcare		The process of diagnosis, alleviation and/or treatment by a *clinician*.	

Term	Synonyms	Definition	Further Reference
Healthcare organization		A regulated organization licensed to provide diagnosis, alleviation and treatment of medical conditions.	
Healthcare technology		The application of organized knowledge and skills in the form of devices, medicines, vaccines, procedures and systems developed to solve a health problem and improve the quality of life.	WHO: World Health Assembly resolution WHA60. 29 May 2007
Healthcare technology management programme	HTM Programme	A collection of processes delivered by a competent provider and designed to deliver appropriate scientific and technical support with the aim of ensuring *healthcare technology* remains safe and effective in clinical practice.	Healthcare Technology Management Programme = Asset Management Plan in ISO 55000:2014 §3.3.3
Health system		Commonly a national structure managed by a government or other organizations for the purposes of *healthcare* delivery.	
Home care	Community care	A model of *healthcare* delivery that seeks to address the needs of the patient within a home setting, usually supported by other carers.	
Hospital		A physical location where *secondary* or *tertiary care* is delivered.	
Human factors		'Human factors refer to environmental, organizational and job factors, and human and individual characteristics which influence behaviour at work in a way which can affect health and safety. A simple way to view human factors is to think about three aspects: the job, the individual and the organization and how they impact people's health and safety-related behaviour'. (Health and Safety Executive, United Kingdom)	
Human usability		See *usability*	
IEC		International Electrotechnical Commission Deals with all formal *Standards* in the electrical fields.	
Innovation		The process of translating an idea, research discovery or invention into a physical product, service or practice that creates value and which may be placed on the market.	

(*Continued*)

Term	Synonyms	Definition	Further Reference
Information and communications technology (ICT)		Similar to *IT* but with a focus on communication technologies such as Internet, wireless, mobile data.	
Information technology (IT)		Commonly used to described hardware and software used on and with, or communicating with, computers of any type.	
ISO		International Standards Organization Deals with all formal *Standards* other than those in the electrical field.	
Jurisdiction		The legal authority in which the *healthcare organization* is operating.	
Key performance indicator (KPI)		Measurable data produced by the *healthcare technology management programme* with the intention of monitoring activity, showing improvement and giving assurance that the system is working as intended.	
Licensed health professional	Allied health professional (may include different groupings of professionals in different jurisdictions)	'An individual who has successfully completed a prescribed program of study in a variety of health fields and who has obtained a license or certificate indicating his or her competence to practice in that field'.	Jonas W.B., 2005, *Mosby's Dictionary of Complementary and Alternative Medicine*, St Louis, MO: Elsevier Mosby
Medical device		An article, instrument, apparatus or machine that is used in the prevention, diagnosis or treatment of illness or disease, or for detecting, measuring, restoring, correcting or modifying the structure or function of the body for some health purpose. Typically, the purpose of a medical device is not achieved by pharmacological, immunological or metabolic means.	WHO Introduction to medical equipment inventory management. *WHO Medical device technical series.* 2011; page 3 WHO (http://www.who. int/medical_devices/ definitions/en/)
Medical device committee	MDC	A committee of the Board of management of the *healthcare organization* which has strategic responsibility for the management of the organization's *medical devices*.	
Medical device committee action plan	MDC Action Plan	A dynamic document that sets out how the *MDC* is overseeing the active asset management of the organization's *medical device* and *equipment* assets and keeping track of progress.	No direct equivalent in ISO 55000 but can be regarded as a dynamic *asset management plan*

(*Continued*)

Term	Synonyms	Definition	Further Reference
Medical device policy	MD Policy	Contains the principles about *HTM*. It sets the organization's top level requirements for *medical devices* by aligning them with the organization's strategic aims. The *MDC* drafts it, but the Board owns and approves it. The MDC reviews it and makes recommendations to the Board.	Medical Device Policy = Asset Management Policy in ISO 55000:2014 §3.3.1 + §3.1.18
Medical engineer	Clinical engineer	Used interchangeably in some organizations with *Clinical engineer*.	
Medical equipment		Typically a reusable *medical device*, usually powered. (WHO: Medical devices requiring calibration, maintenance, repair, user training and decommissioning – activities usually managed by clinical engineers. Medical equipment is used for the specific purposes of diagnosis and treatment of disease or rehabilitation following disease or injury; it can be used either alone or in combination with any accessory, consumable or other pieces of medical equipment. Medical equipment excludes implantable, disposable or single-use medical devices.)	WHO Introduction to medical equipment inventory management. *WHO Medical device technical series.* 2011; page 3 WHO (http://www.who.int/medical_devices/definitions/en/)
Medical equipment management system (MEMS)	Database	A computerized relational database used for the management of *medical equipment* within the organization. Typically contains an inventory and details of equipment plus work records, documentation, staff details. Contains report functions and maintenance scheduling functions.	
Medical IT network		'An *IT* network that incorporates at least one medical device'. Should be isolated from other networks through security measures.	ISO 80001-1:2011 §2.16
Medical specialist		Doctors or physicians that have undertaken further training in a specialized field of medicine.	
Operator		A person who has received appropriate instruction in the use of a *medical device* or *equipment* sufficient to enable them to use it in a certain situation. The Operator can be a *clinician* or a patient, or can be a *clinical engineer* maintaining the equipment.	Operator person handling equipment IEC 60601-1:2013 §3.73

(Continued)

Term	Synonyms	Definition	Further Reference
Operator manual	Instructions for use (IFU); Operators guide; Directions for use (DFU)	Documentation provided to the *user* so that they are able to use the device correctly and safely. Contains explanations of controls and indicators, alarm settings, basic troubleshooting and decontamination methods to be used.	
Paramedic		A *licensed healthcare professional* trained specifically in emergency medicine but is not a qualified doctor/physician.	
Performance verification	Performance check; Testing	Scheduled actions which are put in place to assure that equipment which appears to be working is performing to specification. Does not include the replacement of parts.	See Chapter 6, Section 6.2.2a
Planned preventative maintenance (PPM)	Scheduled Actions; Planned Maintenance, Preventative Maintenance (PM)	The process of carrying out actions designed to reduce the risk of failure of equipment in service. May include the regular replacement of parts. Often *performance verification* is carried out at the same time.	See Chapter 6, Section 6.2.2b
Primary care		Usually the first point of contact for non-urgent medical attention. Commonly situated in the community.	
Quality cycle		An iterative Plan–Do–Check–Act approach to get closer to a solution.	
Replacement asset cost		The current financial cost of replacing an asset.	
Research and development (R&D)		An activity that seeks to innovate, develop and introduce new processes or techniques.	
Residual value		The value of *medical equipment* when the asset has reached the end of its usefulness. The value may not be financial.	
Risk management		The processes by which hazards are identified in a particular setting and then a reduction in the likelihood of potential harm is sought and identified.	See also ISO 14971:2007 §2.2 'Risk management uses policies, procedures, and practices to systematically analyze, evaluate, control, and monitor risk'

(Continued)

Term	Synonyms	Definition	Further Reference
Scheduled maintenance	Scheduled actions; Planned preventive maintenance (PPM); Routine service; Preventive maintenance (PM)	See *planned preventive maintenance.*	
Secondary care		Services provided by *medical specialists* to whom patients have been referred, or where emergency medical services are located.	
Service manual	Service guide; Test procedures	Documentation providing technical information sufficient to verify that the device is performing correctly (as a minimum). Usually contains schematics, parts lists, test procedures, etc.	
Stakeholders		'Person or organization that can affect, be affected by, or perceive themselves to be affected by a decision or activity' Those persons or organizations that have an interest or concern in the system, object, purpose, etc., being described.	ISO 55000:2014 §3.1.22
Standard		An agreed criterion to which parties adhere for the purposes of commonality or A formal document issued by ISO or IEC for the purpose of standardization in a particular field.	
Strategic asset management plan (SAMP)		In this book, we use the term *Strategic healthcare technology management plan (Strategic HTM plan)* for the ISO 55000 SAMP.	ISO 55000:2014 §3.3.2 'documented information that specifies how organizational objectives are to be converted into asset management objectives, the approach for developing asset management plans, and the role of the asset management system in supporting achievement of the asset management objectives'

(Continued)

Term	Synonyms	Definition	Further Reference
Strategic healthcare technology management plan Strategic HTM plan		This is the top level plan devised by the *MDC* as to how, at a strategic level, *HTM* will be managed in line with the *Medical Devices Policy*. It authorizes different groups to develop *HTM Programmes* to be responsible for types of equipment. It is formulated, owned and reviewed by the *MDC*.	ISO 556000:14 §3.3.2 Strategic Asset Management Plan
Technical training		Education for the *clinical engineer* to enable him/her to competently undertake work on *medical equipment*. This may be to a certain level of complexity such as first-line fault identification, regular servicing or repair.	
Telemedicine	Telehealth; Remote medicine	The technology used to remotely monitor, diagnose or treat a patient usually through an Internet connection where distance is an obstacle or for mutual benefit.	
Tertiary care	Specialist care	Highly specialized care for those referred by primary or secondary care, for example, Oncology centres.	
Theatre	Operating theatre; Operating room (OR)	A clean area where surgery is carried out. Different terms with the same meaning are used in different jurisdictions.	
Total cost of ownership	Cost of ownership; Whole life cost	See *whole life cost*.	
Unscheduled maintenance	Repair	The process of bringing an item of *medical equipment* back within manufacturers' specification following a breakdown.	
User		A person who has received training to a level sufficient to be able to use equipment as intended in his/her situation. See also *end user*.	
Usability		The ease of use and learnability of a human-made object. Considers whether the use of the device is counter-intuitive and could therefore lead to use errors.	
Value		The importance or worth of something. Not necessarily financial.	
Whole life cost	Total cost; Total cost of ownership	The sum of all costs, direct and indirect, in regard to medical equipment; these result from its specification, acquisition, lifetime running costs and include its eventual disposal cost.	

CROSSWALK

HTM Terminology		ISO 55000 Terminology
Asset Management Objectives	↔	Asset Management Objectives
Medical Device Policy	↔	Asset Management Policy
Strategic HTM Plan	↔	Strategic Asset Management Plan (SAMP)
MDC Action Plan	↔	No direct equivalent
HTM Programme	↔	Asset Management Plan(s)
Equipment support plan	↔	Asset Management Plan(s)

Index

Printed in the United States
by Baker & Taylor Publisher Services